SYMPOSIA OF THE
SOCIETY FOR EXPERIMENTAL BIOLOGY

NUMBER XXXII

SYMPOSIA OF THE SOCIETY
FOR EXPERIMENTAL BIOLOGY

The Journal of Experimental Botany
is published by the Oxford University Press
for the Society for Experimental Biology

SYMPOSIA OF THE
SOCIETY FOR EXPERIMENTAL BIOLOGY

NUMBER XXXII

CELL–CELL
RECOGNITION

Published for the Society for Experimental Biology

CAMBRIDGE UNIVERSITY PRESS

CAMBRIDGE

LONDON · NEW YORK · MELBOURNE

Published by the Syndics of the Cambridge University Press

The Pitt Building, Trumpington Street, Cambridge CB2 1RP

Bentley House, 200 Euston Road, London NW1 2DB

32 East 57th Street, New York, NY 10022, USA

296 Beaconsfield Parade, Middle Park, Melbourne 3206, Australia

First published 1978

Printed in Great Britain at the
University Press, Cambridge

Library of Congress Cataloguing in Publication Data

Main entry under title:

Cell–cell recognition

(Symposia of the Society for Experimental Biology; no. 32)

Papers from a symposium sponsored by the Society for Experimental Biology,
held at Oxford, 31 August–2 September 1977

Includes index

1. Cellular recognition – Congresses

2. Cell interaction – Congresses   I. Society for Experimental Biology (Gt Brit.)

1I. Series: Society for Experimental Biology (Gt Brit.) Symposia; no. 32

QH 302. S 622 no. 32 [QH 604.2] 574'.08s [574'.8'76] 77–28646

ISBN 0 521 22020 3

CONTENTS

[v]

FOREWORD

When the committee that planned this symposium (Jack Hannay, John Dale, Lewis Wolpert, Dai Rees, Chris Duncan and myself) met at Imperial College, London, we agreed that it would be timely to hold a meeting on cell–cell recognition, not merely because of the growing interest in this field but also because (*i*) a number of disparate and rival hypotheses have developed which need parade before the public in a single site, and (*ii*) because we seem to be seeing possible ideas that may unify the field peeping out from the accumulating pile of experimental results.

The three main hypotheses of cell–cell recognition are: first, the specific adhesion theory which is discussed by Burger, Wiese, Crandall and Pierce in this volume; second, the concept of differential adhesion championed in this volume by its original protagonist Malcolm Steinberg with support from Garrod and his colleagues; and third, the relatively young concept of interaction modulation proposed by myself and discussed to some extent by Maria de Sousa.

If we ask the question 'What is the fundamental nature of the cell–cell recognition event?', it will be seen that the answer is particularly the concern of the articles by Katz and Heslop-Harrison, and that the genetical aspects of the question are treated by Peter Newell's paper and by my own.

Some important areas of biology concerned with recognition by and of cells have had to be omitted because of the precept from the Society of Experimental Biology, who sponsor these symposia, that we invite no more than eighteen speakers. For this reason such topics as platelet adherence, fungal parasitism, first-set graft rejection in invertebrates, fertilisation in animals and many aspects of immunology have been left out. My apologies to those who would have liked to contribute from these areas and to those who would like to read about them. We judged that work in these areas though of importance did not reveal the present-day problems of the field as clearly as the areas we selected. The areas of omission are however alluded to by the writers in this volume and to some extent by the contributors to the poster session whose abstracts complete this volume.

We felt that it could be salutary to examine two areas in which there is some doubt about the extent of recognition. The first of these is neuronal organisation, where the extent of specific cell–cell recognition is under argument, the contributions from Pierce and Horder representing very different ideas about this. The second is the question of whether second-set graft rejection occurs in invertebrates, which is reviewed by Dales.

One almost novel theme introduced in this symposium is the possibility that graft rejection occurs in plants and Yeoman has contributed on this subject.

We deliberately chose to play down the work on classical antibody producing systems because the volume of such work is very considerable and almost over-exposed to the scientific public in many symposium volumes. Instead we have looked at the phenomena of cellular immunology which involve recognition and this involves lymphocyte interactions both amongst their various castes but also with other cells. The papers by Ford, de Sousa, Katz and myself particularly address themselves to this area.

If we move so that we view biology from the traditional viewpoints of zoology, botany and microbiology, we can see that the symposium parcels up in another mode.

(1) Plant recognition systems, considered by Crandall, Wiese, Heslop-Harrison and Yeoman.

(2) Slime mould recognition, treated by Newell and Garrod.

(3) Recognition in metazoan invertebrates, discussed by Burger and Dales.

(4) Morphogenetic recognition in vertebrate systems, reported on by Steinberg, Hoover, Pierce, Horder and myself.

(5) Recognition in lymphocytic and related systems, described by Ford, Katz, de Sousa, Greaves and myself.

(6) Adhesive recognition in vertebrate cells, explained by Rees and Hoover.

I should like to express particular thanks to Maria de Sousa, who, though resident in New York, was much involved in the planning of the symposium, to Peter Newell who was a most careful local secretary charged with all the important arrangements of accommodation as well as being a contributor, and to Jack Hannay who was a most stalwart support to the meeting in many ways. I should also like to thank Mrs Margaret Smith in Glasgow University who carried out most of the secretarial work to get the symposium volume born.

The symposium was held at Oxford from 31 August to 2 September 1977: the rainy weather ensured that our minds did not notice the other charms of Oxford so that we were perhaps unusually scientifically objective.

ADAM CURTIS
Editor of the thirty-second
symposium of the Society for
Experimental Biology

CELL–CELL RECOGNITION: MOLECULAR ASPECTS. RECOGNITION AND ITS RELATION TO MORPHOGENETIC PROCESSES IN GENERAL

By MAX M. BURGER, W. BURKART,
G. WEINBAUM* AND J. JUMBLATT

Department of Biochemistry, Biocenter, University of Basle,
Klingelbergstrasse 70, CH 4056 Basle, Switzerland

INTRODUCTION

Exactly seventy years ago H. V. Wilson (1907) summarized his work at the Marine Biological Laboratory in Woods Hole in an article entitled 'On some phenomena of coalescence and regeneration in sponges'. It was an article that laid down the seeds for a new research area and gave rise, directly or indirectly, to most of our recent progress in the vastly developing field of cell–cell recognition in embryonal development.

When Wilson cut a marine sponge into pieces, squeezed it through bolting silk and let the resulting cell suspension settle to a glass dish, he observed that the cells attached and eventually reconstituted a small functional sponge. When cells from the red sponge *Microciona prolifera* were mixed with those from the yellow sponge *Cliona celata*, they sorted out and formed separate sponges, recognizable by their pigment. The techniques were refined over the next few years, and were reused decades later, no longer to demonstrate species-specificity, but rather organ-specific cell sorting during vertebrate embryogenesis.

Sponge cell recognition is still one of the better known potential models for embryonal cell–cell recognition, primarily since some of the molecular parameters have been extensively studied. Such comparisons have, however, mainly heuristic value and should only be made with due caution, not only because species specificity must differ in biological function from organ specificity, but primarily because the phenomenon of cell–cell recognition in morphogenesis, let alone its mechanism, is still poorly defined.

This paper intends to analyse and characterize some of the morphogenetic processes in general in order to define the place of sponge cell recognition as a model for vertebrate embryonal cell recognition. It may

* Albert Einstein Medical Center, Northern Division Philadelphia, Pennsylvania, 19141, U.S.A.

[1]

serve as a fitting introductory chapter for the book, in that the problem is seen with a bias towards specific molecular mechanisms of adhesion, whereas the paper by Steinberg (this volume, pp. 25–49) puts more weight on the role of differential adhesion, and the paper by the editor (this volume, pp. 51–82) stresses control aspects. Since all these processes might have some validity in morphogenesis, the reader will eventually get a balanced picture, although emphasis and viewpoint might at a first glance be quite different.

Morphogenetic processes during embryogenesis may be divided into several subcomponents, of which cell–cell recognition is only one. Three such processes are depicted in Fig. 1:

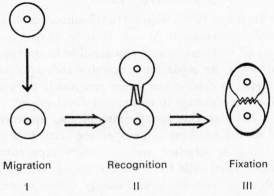

Fig. 1. Three stages in embryonic cell redistribution.

(I) Movement.
(II) Recognition of the correct location giving rise to the final cellular distribution.
(III) Fixation of the correct cellular distribution.

It should be noted that all three stages and processes, particularly movement, can apply to single cells as well as to groups of cells or tissues which can move together as a unit.

Such a dissection of morphogenetic processes *in vivo* not only applies to embryogenesis of higher vertebrates but is also relevant for Wilson's original experiment on species-specific sponge reconstitution. In these early experiments, the cells were allowed to move around on the glass surface before small sponges formed. As demonstrated later by Galtsoff (1925) such movement required Ca^{2+} and was essential for re-aggregation and re-organization of the sponge cells into functional tissues.

Movement

Cellular or tissue displacement, which accompanies most morphogenetic processes, can be either target-oriented or random. Random migration requires a recognition process in addition to the movement phase. Targeted migration does not necessarily require recognition at the end of the migratory period. For instance, if a chemotactic gradient was the cause of the targeted movement, then termination of the migration may simply be due to having reached the region in the embryo from which the chemotactic agent emanates. As chemotactic agents, both macromolecules as well as small molecules must be considered. Gradients (Wolpert, 1969; Wolpert, Hornbruch & Clarke, 1974) of such molecules are not necessarily extracellular, but may also be distributed in the interior of the cells of a tissue.

Information for the direction of the movement does not have to be in the form of soluble molecules, as in chemotaxis. It can also be provided by the immediate environment of a cell. Thus a cell may select its path of migration based on the best adhesions it can form. In some situations, such adhesions may be sufficient to create the movement (Carter, 1967) while in others they may merely guide the cellular machinery that provides the movement. Steinberg's differential adhesion hypothesis (1970 and this volume, pp. 25–49) can encompass both, provided one does not oversimplify these mechanisms and keeps in mind that the final, so-called equilibrium cell distribution is what is measured by most methods, and not movement per se. Such environmental adhesive guidance may be called contact guidance and has been observed in the formation of the cerebellum and the foetal neocortex (Racic, 1974). In some cases, this guidance can be provided by neighbouring cells where type and number of adhesions are presumably relevant (Levinthal, Macagno & Levinthal, 1976). In other cases extracellular structures and ground substance (fibres and proteoglycans) can provide the guidance. Contact guidance may explain why a given cell or tissue migrates in a given direction, but of course begs the question of how the guiding elements themselves found their position. Thus, it is only a mechanism for morphogenetic movement, but fails to explain how the re-distribution plan is set up, a question which has not been answered for chemotactic mechanisms either.

While adhesion, guidance and chemotaxis may be the most obvious principles for explaining movement, one should keep in mind that positive vectoral attractants are not the only possible mechanisms for directional movement. Cells may also move into areas where they encounter less tissue resistance, either because their own deformability is higher, or the extracellular space into which they penetrate or the tissue which they invade is softer. Tissue barriers, and particularly deformability of migrating cells or tissues, have not been sufficiently emphasized so far. Given a sufficient time interval, cells have been considered to be for all practical purposes fully deformable, an assumption not yet proven for the situation *in vivo* and deserving a good deal more attention.

Recognition

If a randomly migrating cell is to be integrated at a specific location according to a master plan, the migratory phase has to be terminated by a recognition process. As mentioned above, such a recognition step at the final location is not a necessary condition for those cells and tissues displaying targeted movement. However such theoretical considerations may not hold *in vivo*. Nature may, for security reasons, add a special recognition step, as is suggested by the following example from lower eukaryotes.

Although differentiating slime mould cells are attracted towards a centre by freely diffusing chemotactic agents, those cells that arrive at the centre develop species-specific surface adhesions (Beug, Katz, Stein & Gerisch, 1973) which are thought to involve newly appearing sugar-specific lectin-like macromolecules on the cell surface (Barondes & Rosen, 1976). It may later be discovered that such an aggregation system, evolved on top of a chemotactic migratory system, has additional functions. Similarly, mammalian morphogenesis may, for yet unknown reasons, require two systems simultaneously.

For the recognition process itself, we have pointed out two basically different mechanisms (Burger, Turner, Kuhns & Weinbaum, 1975): on one hand, the more popular type of recognition at the cell surface via recognizing molecules, and on the other hand, the poorly studied and often ignored recognition via the exchange of intracellular information between the two partner cells or tissues (Fig. 2).

Recognition molecules on the cell surface may – but need not – contribute directly towards the adhesion of the two cells or tissues. They may hold the two partners long enough in juxtaposition so as to enable the formation of secondary ties between the two. Or they can also act simply as a trigger for the establishment of secondary ties between partners or indirectly halt the migratory machinery of a cell.

The alternative mechanism for recognition – the exchange of information between neighbouring cells or tissues – is seldom mentioned in the literature. Cytoplasmic material can be exchanged through gap and perhaps other types of junctions. Such junctions can be formed quite quickly, at least within the time period during which two cells migrate past each

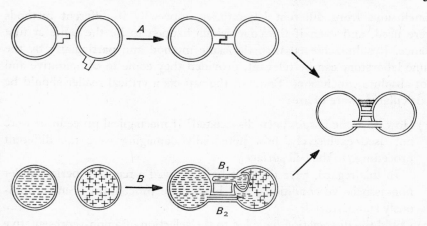

Fig. 2. Two possible mechanisms for recognition. The upper portion of the figure (*A*) shows recognition occurring via the interaction of specific surface molecules. In the lower portion (*B*), an alternative mechanism is depicted in which information signals are passed between cells via vesicular exchange (*B₁*) or via gap or other junctions (*B₂*). Such connections are likely to have a short life span and can be considered as mutual probing frequently seen at the ruffling edges of moving cells. Both types of *specific* interactions can lead either to secondary, stabilizing linkages (e.g. intercytoplasmic, intermembranous, or extracellular linkages as depicted by the cell doublet to the right), or to dissociation of the interacting cells (not shown). (From Burger *et al.*, 1975.)

other. The size of any exchanged molecular message would have to be small, however. Macromolecules will apparently not pass through gap junctions. They might still be exchanged, however, via pinocytosis or phagocytotic processes that have rarely been considered, but are clearly apparent in many in-vitro and in-vivo culture systems, particularly of embryonal cells (Fig. 2).

In such a discussion one should keep in mind that the above recognition mechanisms not only apply to morphogenesis, i.e. the positioning of cells, but to many other processes in embryogenesis, such as induction and other events in differentiation.

The definition of recognition *in vitro* is unfortunately flavoured by the assay used by the particular investigator. Thus, cells or groups of cells are separated from each other and then reassociation is monitored in an ever-increasing variety of manners. The rate of which aggregates are re-formed, the size that is reached, or the distribution of two cell types in mixed clumps are monitored. In some cases labelled cells are exposed to monolayers or clumps of cells and adhesion is recorded, and in most of these experiments the test partners are exposed to shear forces during the assay.

A novice to the field would be well advised not to attempt to compare conclusions from different laboratories, especially if different methods were used, and even if the conclusions happen to be the same at first glance. Results achieved with the same method and particularly by the same laboratory can be relevant, provided they come to comparative and not absolute conclusions. Some of the aspects a critical reader should be particularly aware of are:

(1) How have the tissues been dissociated? If mechanical procedures were not used exclusively, how potentially damaging were the different procedures to the cell surface?

In this regard, have the cells been allowed a recovery period under non-associative conditions, and have these procedures been scrupulously standardized?

(2) Could the dissociation have led to the selection of a non-representative cell population that is more resistant to the disruptive procedures?

(3) Did aggregation precede sorting out, or if separate clumps were found, did mixed aggregates precede the formation of separate clumps? In other words: have the investigators followed the composition and cell distribution of the aggregates right from the beginning of the assay?

This list of questions can easily be elongated to a sizeable catalogue which will no longer fit into a short review. We would like to stress, however, that the last question touches on important conceptual problems. Thus, if a mixed cell population gives rise, right from the beginning of the aggregation assay, to two distinct homotypic aggregates and essentially no mixed aggregates, then recognition must precede, or at least accompany, adhesion. Such recognition from the onset does not require sorting out and thus may not even require live cells, provided that recognition occurs by a purely molecular process that does not require ongoing energy metabolism or macromolecular synthesis. Sorting out, however, is thought to require active deformability and probably migratory capability of cells, both processes that seem to depend on the living state of a cell. In this case, a multitude of mechanisms could participate in formation of the final cell distribution. It is therefore not unexpected that comparisons of initial rates of aggregate formation of different cell types do not follow or parallel the hierarchy of final equilibrium distribution found for the same cell types after sorting out. Sorting out clearly requires the living state to achieve an equilibrium distribution, even if the equilibrium itself is not always influenced by living cell parameters such as cell deformability, microvilli formation, migration capability, etc. In any case, many of the assays in use are entirely different and, as pointed out above, we do not know exactly what physical and chemical parameters are being measured.

Fixation of the cell distribution

Following specific recognition steps the position of interacting cells may be fixed by series of secondary reactions. This halt in cell migration can occur in two ways – either by physical attachment (trapping) or by shutting down the migration machinery. In general, cell surface elements whose main function is in recognition, are likely to be readily reversible, and therefore unsuitable to provide permanent cohesion between cells. In addition, there is evidence that surface recognition elements have a rather short temporal existence during embryogenesis (Glaser, Santala, Gottlieb & Merrell, 1977). Other possible mechanisms for fixing the position of cells in the embryo are illustrated in Fig. 2. An obvious mechanism is the formation or stabilization of pre-existing gaps or other specialized junctions. In addition, intracellular filaments can aggregate into bundles, or so-called stress fibres, at zones of adhesion between neighbouring cells. Whether some filaments might actually pass between cells remains to be seen. Patches of membranes might become contiguous in the form of tight junctions (intramembranous bridges). Finally, other types of surface or extracellular material could provide the necessary cement. As examples collagen, elastin fibres or proteoglycans could not only create intercellular bridges, but surround and eventually solidly encapsulate whole groups of cells, tissues and organs.

SPONGE CELL RECOGNITION: SOME INSIGHTS INTO THE SPECIFICITY PROBLEM AT THE MOLECULAR LEVEL

Earlier studies at the cell and tissue level

In addition to Wilson's earliest experiments (1907) on species specificity, which utilized the disaggregation–reaggregation assay, several other test systems have been used to analyse species-specific recognition.

Species-specific fusion of free-swimming larvae was observed by Van de Vyver (1970). In certain species not all the larvae would fuse, suggesting specificity at the individual level, similar to allotype specificity in immunological terms.

In grafting experiments between species, heterospecific transplants were usually rejected while homospecific grafts survived (Moscona, 1968; McClay, 1974), although cases of heterospecific fusion (Paris, 1960) as well as intraspecific rejection in homospecific transplants have also been reported (Paris, 1960). All recent studies of homospecific transplants demonstrate uniformly successful graftings.

Yet another approach is shown in a study of explant fusion between spongelets. Simpson (1973) could observe in the same species (*Microciona prolifera*) the fusion of the canalicular systems of two explants brought into apposition. However, other explants, provided they did not come from the same mother sponge, were not capable of fusing their colonies into a common canalicular system. The latter observation lends support to some individuality of the allotype as mentioned above.

At a first glance, the results with explant fusion and with grafting experiments may seem contradictory. We have frequently observed that species which sort out in the re-aggregation assay may spontaneously grow on each other in the sea-water tank if they are inadvertently kept in the same compartment over the summer. Some naturally formed 'hybrids' of this type are occasionally collected by the supply department. In some cases a dark necrotic zone between the two sponge tissues was found, indicating that living tissues were no longer in apposition. One must be cautious in making hasty conclusions about compatibility between two cell types, particularly in grafting and regeneration experiments. A dead or inert contact zone between tissues may not be visible by eye, but only by the light- or even the electron microscope. One may ask whether, in some cases, the extracellular matrix may insulate different sponge individuals, and that intimate cell–cell surface contact is necessary not so much for recognition as for the rejection process. In the case of grafting experiments, the tissues are wounded to re-establish immediately artificial contact zones, while in the explant fusion experiments the tissues are permitted to approach each other after a certain time has elapsed without physical contact, and where selected cells grow out before they touch each other. A more detailed evaluation of these seemingly minor technical differences may be useful in resolving this apparent discrepancy and could lead to further insights into the question of whether a cell type or extracellular material is the carrier of the specificity.

Among the assays that have yielded information about the species-specificity of sponges at the cellular level is the collecting assay originally designed by Roth & Weston (1967) and applied to five species of Bermuda sponges by McClay (1971). In these experiments radioactively labelled cells were found to enter unlabelled aggregates only if they were of the same species. If they were of different species they either did not bind to the unlabelled clump of cells or they attached to the outside but did not mingle with the hetero-specific cells in the interior of the clump.

However, if two suspensions of sponge cells in the same heterospecific combination were mixed and rotated, the authors found non-specifically mingled clumps of cells at the very beginning of the experiment, before the two cell types sorted out. This observation stresses again what has been mentioned under 'Recognition'. On the one hand, the dissociation procedure may damage cell surfaces, which must then be regenerated in order to display recognition. Alternatively, the discrepancy in results between the two procedures could be interpreted on the basis of a multistep process of cell recognition and sorting. Thus, the cells may initially display strong, unspecific adhesion to one another, and only after being physically associated will they begin to probe each other and sort out.

Such problems of interpretation are common in the field, but can be solved if dissociated cells are permitted a full restoration of their surfaces under non-associative conditions. However, this prerequisite is not always easy to fulfill, since whatever steps are taken to prevent cell re-association, such as high shear forces or low Ca^{2+}, may at the same time impede full surface restoration. In order to approach these questions more directly, an analysis of the molecular components involved in the re-association process were begun by several laboratories.

The aggregation factor

Microciona prolifera or *parthena* cells dissociated into Ca^{2+}- and Mg^{2+}-free sea water were shown by Humphreys (1963) to release a high molecular weight 'factor' into the supernatant without which the cells could no longer re-aggregate at 4 °C. The *Microciona* factor was further purified and shown to be a 21×10^6 dalton proteoglycan complex (Cauldwell, Henkart & Humphreys, 1973; Henkart, Humphreys & Humphreys, 1973). Several such factors examined by Humphreys (1963) and by Moscona (1968) were found to be species-specific, in that they promoted the re-aggregation of the homospecific cells only. More recently, McClay (1974) confirmed that crude supernatants obtained from dissociated cells of five sponge species were species-specific in their aggregation-promoting activity, and that labelled supernatants appeared to bind to homospecific cells only.

Although species-specificity of sponge aggregation factor has been widely demonstrated, it is by no means universal. This is to be expected, since cells from some species of sponges sort out very poorly (Sarà, Liaci & Melone, 1966). We have concluded that the apparent specificity of factor–cell interactions is, in some cases, due to quantitative rather than qualitative differences. For example, at high doses the aggregation factor from *Haliclona occulata* was capable of aggregating *Microciona prolifera* cells, although its specific activity for homospecific versus heterospecific cell aggregation differed by an order of magnitude (Turner & Burger, 1973). This study utilized two sponge species belonging to separate orders. With

2

impure factors isolated from different families of sponges within one order, far less specificity was found (MacLennan & Dodd, 1967).

Not all specific surface 'factors' isolated from sponge cells directly enhance cell aggregation. McClay (1974) found, in addition to specific aggregation-promoting activities, inhibitory activity against the hetero-specific cells. After removal of the heterospecific inhibitor, the enhancing factor was still active in promoting homospecific aggregation. Curtis & Van de Vyver (1971) had earlier pointed out that a fresh-water sponge (*Ephydatia fluviatilis*) may produce a soluble factor that decreases the adhesiveness of heterologous cells.

Inhibition of aggregation can be interpreted in at least two ways. The problem of recognition of self versus non-self is discussed at several places in this volume (e.g. see articles by Curtis (pp. 51–82) or Heslop-Harrison (pp. 121–138)). Recognition of non-self could be an active process that involves a specific recognition of the heterologous cell type, leading to rejection. Alternatively, non-self 'recognition' and rejection could be a passive process that occurs *in the absence* of specific self-recognition. The inhibition of heterologous cell aggregation by a specific molecule separate from the homologous promoting factor is consistent with an active rejection process. In some cases, however, the homologous promoting factor may itself possess the inhibitory function. Just as antibodies and lectins can become inhibitory for the agglutination in the antibody/antigen excess range, so could sponge aggregation-promoting factors become, at least in theory, inhibitory at excessive concentrations. Since aggregation-promoting activities show some degree of cross-specificity (Turner & Burger, 1973), one can predict that inhibitory activities will be cross-specific as well. In situations where receptor sites for promoting factors have a range of K_m values, saturation of heterospecific binding sites could inhibit the homospecific cell aggregation. These are conjectural explanations of how some aggregation factors in certain concentration ranges may become inhibitory to heterospecific cells. It will be interesting to study such phenomena in more quantitative terms, once the factors can be obtained in sufficient purity with regard to affinities as well as the usual physical and chemical criteria.

The *Microciona* aggregation factor, as mentioned above, is at present the only such factor that has been biochemically characterized in detail. In the electron microscope, the 21×10^6 dalton molecule has a 'sunburst' configuration with a central 800 Å ring and about fifteen radiating 'arms', each about 1100 Å in length (Henkart *et al.*, 1973). Both 'ring' and 'arm' structures contain roughly equal proportions of protein and carbohydrate, including a substantial amount (10–20 %) of hexuronic acid. From our findings that glucuronic acid was a hapten-like inhibitor of factor-promoted aggregation of *Microciona* cells, and that a crude β-glucuronidase preparation could inactivate the factor, we suggested that a carbohydrate-specific receptor on the cell surface might act as an acceptor for the factor molecule. Since the carbohydrate–protein nature of the interaction was not rigorously established, we suggested the non-committal name of *baseplate*.

Baseplate: the second component

After removing most of the functional aggregation factor from *Microciona prolifera* cells, we tried to release a receptor-like activity by a variety of treatments, including proteolytic enzymes, butanol, octanol, guanidinium chloride, Triton X-100 and urea. These attempts at receptor solubilization were not successful, either because the conditions were insufficient for release, or the cells were destroyed or seriously damaged. We eventually utilized a carefully controlled hypotonic treatment, which had to be adjusted to different batches of sponges depending on the salt content of the harvest area, and which caused treated cells to lose their capability to respond to aggregation factor. The supernatant from the treated cells contained inhibitory activity if pre-incubated with aggregation factor followed by fresh, untreated cells (Weinbaum & Burger, 1973). In addition, partial reconstitution of the shocked cells was possible if they were incubated with this crude supernatant and washed. Such cells regained their responsiveness to aggregation factor. The active component released by the hypotonic treatment was termed *baseplate*.

Baseplate appeared to be a peripheral rather than an integral protein since it could be removed from the cell surface by a mild, non-peptic hypotonic treatment and since it could not be sedimented at 105 000 g for 90 min. It is heat sensitive (60 °C for 10 min) and non-dialysable. Unlike aggregation factor, it is stable towards EDTA, to a pH range of 3–12 as well as to freezing, thawing and lyophilization, suggesting that it is not simply a subunit of aggregation factor. A partial purification has been achieved as recently reported (Burger & Jumblatt, 1977).

Since the baseplate preparation did not promote, but rather inhibited aggregation of mechanically dissociated cells, and because it neither precipitated aggregation factor nor promoted aggregation of factor-coated beads, we concluded that it must be a functionally monovalent macromolecule. Some very preliminary immunofluorescence studies with W. J. Kuhns using antiserum against a baseplate preparation indicate that baseplate is present in sizeable amounts at the cell periphery.

Although many cell–cell recognition systems require dynamic properties of live cells, sponge cells can be re-aggregated specifically at 4 °C, or after

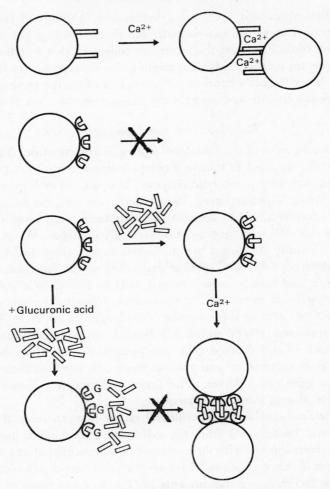

Fig. 3. Aggregation of agarose beads conjugated with baseplate or with aggregation factor. As depicted in the upper portion of the figure, beads conjugated with aggregation factor (bars) are found to aggregate rapidly in the presence of Ca^{2+}. This aggregation can occur even in the presence of glucuronic acid – a hapten sugar that specifically inhibits the interaction of the factor with the cell surface. Beads conjugated with baseplate (depicted as grooved, kidney-shaped receptors, (second row) do not aggregate spontaneously in the presence of Ca^{2+} but are efficiently aggregated when aggregation factor is added (rows 3 and 4, right). This factor-promoted aggregation of the beads can be prevented, however, by glucuronic acid (lines 3 and 4, left). (From Burger et al., 1975.)

aldehyde fixation (Moscona, 1968; Burger & Jumblatt, 1977). We asked ourselves a few years ago whether the sponge recognition system could be reconstituted in an acellular assay system. We coupled the crude baseplate preparation to Sepharose beads (Fig. 3) and found the beads as efficiently aggregated by factor as were the intact cells (Weinbaum & Burger, 1973). Further analysis revealed that this cell-free aggregation system shared several other qualitative aspects (effects of inhibitors, sequence of addition of factor or inhibitor, etc.) with the aggregation of live cells. Fig. 3 demonstrates schematically these experiments which utilized both insolubilized baseplate and insolubilized aggregation factor. Not shown is the finding that baseplate-coated beads were also capable of adhering to live cells that contained aggregation factor on their surfaces, but not to cells from which most of the aggregation factor had been removed by treatment with calcium–magnesium-free sea water.

Interaction of aggregation factor with the cell surface

Since cell aggregation is now considered to be a complex process involving a number of possible steps and different kinds of adhesive sites, it is important to know at what level the specificity of factor–cell interaction resides, and secondly, what role the essential divalent cations play in the re-aggregation process. To answer these questions, we have studied the binding of radioiodinated aggregation factor to dissociated cells from *Microciona prolifera*. The details of the labelling procedures, isolation of labelled aggregation factor and binding assays will be published elsewhere. The key results are summarized in the following paragraphs.

Binding of ^{125}I-labelled aggregation factor to cells in suspension did not appear to be saturable (Fig. 4) when binding was performed under our standard aggregation assay conditions – i.e. artificial sea water, 20 min shaking at room temperature. Based on separate studies of factor–factor interaction not presented here, we now believe such non-saturability to be caused by Ca^{2+}-dependent self aggregation of the factor molecules, continuously creating new sites and resulting in apparent co-operative binding. At the lowest factor concentration giving macroscopically visible cell aggregation (1 unit ml^{-1}), the number of factor molecules bound per cell is relatively low (< 1000) and corresponds to only a small fraction of potential 'sites'.

A more accurate quantitation of aggregation-factor binding at 1 unit ml^{-1} concentration is shown in Table 1. As we have shown previously (Burger & Jumblatt, 1977), glutaraldehyde fixation of cells resulted in no detectable loss of responsiveness to factor as measured by the endpoint assay. In accordance with this finding, the number of factor molecules

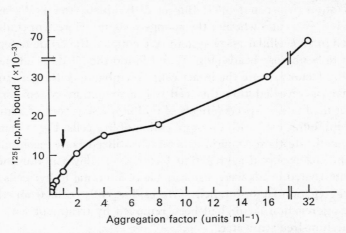

Fig. 4. Binding of ^{125}I-labelled aggregation factor from *Microciona prolifera* to dissociated cells. Chemically dissociated cells (1×10^7 ml^{-1}) from *Microciona prolifera* were incubated with increasing concentrations of ^{125}I-labelled aggregation factor on a rotary shaker at 20 °C. After 20 min incubation cells with bound factor were separated from unreacted factor by centrifugation through a cushion of sucrose/BSA (see Jumblatt & Burger, 1977, for details) and counted. The lowest concentration of ^{125}I-labelled aggregation factor which produced macroscopically visible cell aggregation (i.e. one unit ml^{-1}, by definition) is shown by the arrow.

Table 1. *Species-specificity of aggregation factor binding*

		Microciona ^{125}I-labelled aggregation factor bound*	
Cell type		Molecules per cell	Molecules per mg cell protein ($\times 10^{-10}$)
Microciona	CD (live)	516 ± 63	1.28 ± 0.11
Microciona	CD (fixed)	407 ± 67	1.36 ± 0.21
Halichondria	CD (live)	18	0.11
Cliona	CD (live)	28	0.11

* Live or 1 % glutaraldehyde fixed, chemically dissociated (CD) sponge cells (10^7 ml^{-1}) were incubated with ^{125}I-labelled aggregation factor from *Microciona prolifera* (1 unit ml^{-1}) for 20 min at room temperature. Cells were pelleted through a sucrose/BSA cushion to separate unbound factor, and cellular protein was determined by a modified Lowry procedure. Binding data for *Microciona* are the means of at least 4 determinations $\pm$ Standard deviation.

bound to fixed cells was not significantly different from live cells (Table 1). Furthermore, factor binding was clearly species-specific: binding of *Microciona* aggregation factor to either *Halichondria* or *Cliona* cells represented less than 10 % of amount bound to homologous cells (Table 1). Although the valency of aggregation factor molecule and actual number of

Table 2. *Effect of heat and EDTA inactivation of* ^{125}I-labelled aggregation factor on binding to cells

Experiment	Cell type		Treatment of aggregation factor†	Visible cell aggregation, 20 min	^{125}I-labelled aggregation factor binding* c.p.m. bound per mg cell protein ($\times 10^{-5}$)
1	*Microciona*	CD (fixed)	None	+	4.78
2	*Microciona*	CD (fixed)	EDTA (10 mM)	-	3.45
3	*Microciona*	CD (fixed)	60 °C, 20 min	-	7.46
4	*Haliclona*	CD (fixed)	None	-	0.26
5	*Haliclona*	CD (fixed)	EDTA (10 mM)	-	0.42
6	*Haliclona*	CD (fixed)	60 °C, 20 min	-	0.40
7	*Microciona*	CD (fixed)	Preincubation with 20 μl 'baseplate' (crude)	-	1.49

* 1 % glutaraldehyde fixed sponge cells (10^7 ml^{-1}) were incubated with native or denatured ^{125}I-labelled aggregation factor from *Microciona prolifera*.

† Where indicated, the factor was inactivated by preincubation with EDTA (10 mM final) at room temperature for 15 min, or by heating (60 °C for 20 min). In experiment 7 the 20 μl of crude 'baseplate' added contained approximately 1.3 μg of protein.

available surface sites have not yet been determined, the low number of bound factor molecules needed to produce cell aggregation and the high degree of binding specificity support Moscona's original concept of the factor as specific surface ligand (Moscona, 1963).

A number of early investigations showed Ca^{2+} to be essential for sponge cell aggregation. Since EDTA treatment of *Microciona* aggregation factor resulted in permanent loss of aggregation-promoting activity, it seemed likely that Ca^{2+} dependence might reflect the cation requirements for stability of the aggregation factor. In addition, it was predicted that Ca^{2+} would be required for aggregation-factor binding of factor to cells, since aggregation factor can be released from mechanically dissociated cells by soaking them in Ca^{2+}- Mg^{2+}-free sea water. We tested the effects of EDTA, low Ca^{2+} and heat denaturation of the factor molecule on subsequent binding to cells (Table 2). We found that although treated aggregation factor no longer produced cell aggregation, it still was able to bind to cells as well as untreated control factor. Moreover, further controls (Table 2) showed that the binding of EDTA- or heat-inactivated factor was still species-specific, thus ruling out a possible increase in non-specific 'stickiness' caused by the denaturing treatment. Subsequent studies (to be reported elsewhere) established that EDTA-inactivated aggregation factor has lost its capacity to interact or polymerize with other aggregation factor molecules – denatured or native – regardless of the cation concentration. We therefore conclude from these studies that both self-association between factor molecules *and* binding of aggregation factor to cells are necessary for cell aggregation, but only the former requires Ca^{2+}.

One might also predict that a soluble receptor, such as the baseplate component released from cell surfaces by hypotonic treatment, would competitively inhibit the binding of factor to cells. It was found that even a crude baseplate preparation at low concentrations was an effective inhibitor of binding (Table 2). We are presently investigating the chemical basis of baseplate–factor interaction.

Our current working model for factor-mediated cell aggregation in *Microciona prolifera* is shown in Fig. 5. We emphasize that it is only a working model, and will no doubt be revised and updated as further information on receptor component(s) and interacting sites on the aggregation factor becomes available. In accordance with our present state of knowledge, the factor molecule is depicted as a multivalent, multi-subunit ligand requiring Ca^{2+} ions for internal stability and association with other factor molecules. The baseplate (BP) component serves as a carbohydrate-specific attachment site for aggregation factor on the cell surface. Primary binding of the factor to surface receptors does not require Ca^{2+}. In addi-

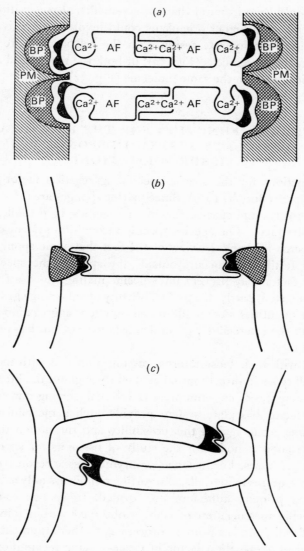

Fig. 5 (a) A tentative model for factor-mediated cell recognition in *Microciona prolifera*. AF (aggregation factor) molecules held in stable association by Ca^{2+} ions contain carbohydrate moieties (dark areas) that interact with surface sites, namely the baseplate component (BP). PM = plasma membranes. (b) An alternative model for ligand-mediated cell recognition. This model is distinguished from the *Microciona* model shown in (a) by its reversal polarity – i.e. the multivalent ligand lying between the cells is the recognizing component, and the antigenic or recognized component (black and hatched areas) is on the cell surface. (c) A third model for ligand-mediated recognition in which both recognized and recognizing components are present on the surface of the same cell. The complementary components can be distinct molecules or domains of the same molecule as depicted here. (From Burger *et al.*, 1975.)

tion, one should keep in mind that surface-mediated recognition in other sponge species or in higher organisms could involve alternative mechanisms in which carbohydrate residues on the cell surface are linked by protein ligands between cells (Fig. 5*b*); or both the recognized and recognizing entities reside in the same molecule (Fig. 5*c*).

IS THERE A FUNCTION FOR THE FACTOR AND BASEPLATE SYSTEM IN SPONGE TISSUE FORMATION?

Possible functions for the species-specific aggregation factor have been discussed earlier (Burger, 1977). Since marine sponges are colonial animals (Simpson, 1973), interspecies fusion of colonies is possible, although probably infrequent. The species-specific recognition system could serve to prevent such mishaps. In view of the fact that some sponges are oviparous and fertilization occurs outside the sponge, the species-specific aggregation factor may prevent interspecific mating. Such a function has not yet been tested, partly due to the difficulty of collecting eggs and sperm in sufficient quantities. As an additional function, aggregation factor may contribute to the intercellular ground substance, since it has proteoglycan character.

Sponges, with their loose internal organization and with most of their different cell types in direct mutual contact throughout the life span of the organism, may need an omnipresent internal sorting-out mechanism. Could the factor–baseplate system provide such a mechanism? Several considerations argue against this possibility and the notion that sponges might be as good a model for the study of tissue as for species-specific recognition processes. For instance, embryonic tissue reconstruction after dissociation requires living cells, and sorting out occurs only as a secondary process after primary adhesions are formed. In sponge re-aggregation species-specific reassociation of cells is observed within minutes, using live or fixed cells, and is a primary process. A further argument against the involvement of aggregation factor in tissue-specific recognition processes is the observation that aggregation-factor-promoted re-assembly seldom leaves any unaggregated cells in suspension. Essentially all dissociated cells are clumped, although very scanty information is available about possible selective losses of specific cell types during the dissociation procedures. Despite these arguments, the possibility remains that embryogenesis or cell sorting in the adult colonial organism are factor-dependent phenomena, not based on the strong forces observed in primary re-aggregation of cells by factor. Cell sorting and tissue formation *in vivo* could involve weaker, qualitatively different interactions with factor or possibly the modulation of the number or the distribution of factor molecules on the surfaces of distinct sponge cell types.

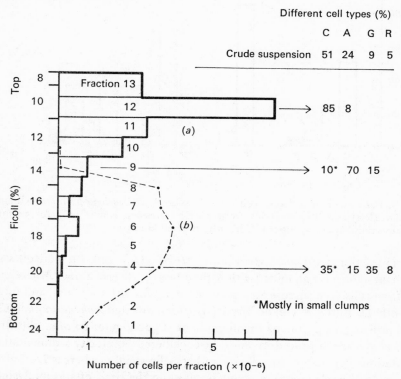

Fig. 6. Sponge cell distribution in a Ficoll density gradient ranging from 8 % to 24 % Ficoll (w/v) in 2 % steps. (a) 60 g for 50 min. (b) 500 g for 125 min. C, choanocytes; A, archeocytes; G, grey cells; R, rhabdiferous cells.

Almost all studies of cell recognition in sponges have utilized crude single-cell suspensions. The main cell types of *Microciona prolifera* that can be easily recognized in such suspensions are choanocytes, archeocytes, grey cells and rhabdiferous cells (Plate 1). Fig. 7 (left) shows the percentage of those four cell types in a fresh suspension from a sponge collected in the summer months (August). The relative numbers of these cells, as well as pinacocytes, globiferous cells, egg cells and spermatozoa show seasonal differences.

Several groups have attempted to fractionate such sponge-cell suspensions and to study the aggregative properties of the different cell types (John, Campo, Mackenzie & Kemp, 1971; Leith & Steinberg, 1972; Burkart & Burger, 1977; De Sutter & Van de Vyver, 1977). John *et al.*

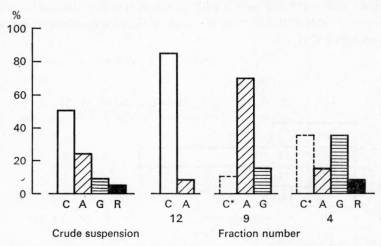

Fig. 7. Percentage of different cell types in a crude *Microciona prolifera* cell suspension (left) and in 3 fractions from a Ficoll gradient run at 60 *g* for 50 min (fraction numbers as in Fig. 6). * Mostly in small clumps.

(1971) fractionated *Ophlitaspongia* and *Halichondria* cells on a Ficoll step gradient at 60 *g* for 20 min; Leith & Steinberg (1971) separated *Microciona prolifera* cells on a sucrose step gradient at 1 *g* for several hours, and more recently De Sutter & Van de Vyver (1977) fractionated *Ephydatia fluviatilis* cells utilizing a continuous Ficoll gradient at 50 000 *g* for 60 min. The first two, as well as our procedure, must be considered as velocity sedimentation separations dependent on cell size and specific weight, whereas De Sutter & Van de Vyver's (1977) gradient resolves on the basis of specific weight only. We layered a crude cell suspension containing 10^7 cells ml^{-1} in 8 % Ficoll on a Ficoll gradient ranging from 10 % to 24 % in 2 % steps. Fig. 6 shows the result of a run at 60 *g* for 50 min at 0 °C. Fraction 12 (second from the top) contains around 85 % choanocytes, and fractions 9 and 10 about 70 % archeocytes. Fractions 3–7 contain most of the grey cells and the rhabdiferous cells together with some archeocytes and small clumps of choanocytes in varying amounts (Fig. 7). Recentrifugation of the bands with the highest enrichment in choanocytes or archeocytes on a second gradient resulted in a yet higher resolution of the peak fraction, although the yield was lower. The remaining contamination with other cell types, even after two consecutive centrifugations, is due to the wide fluctuations of cell size of discrete cell types of *Microciona prolifera*. Also, choanocytes, which as single cells remain near the top, form clumps of various sizes that can sometimes travel to the lowest fractions. Separation on the basis of specific weight alone was not successful, apparently because the dif-

PLATE I

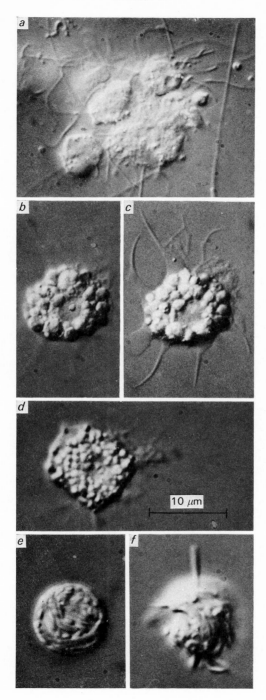

For explanation see p. 24

(facing p. 20)

ferences are too small. This is illustrated by the fact that longer centrifugation with stronger centrifugal forces brings almost all cells down to the
14–22% Ficoll layers (Fig. 6).

All fractions were observed to clump in the aggregation assay with
Microciona prolifera aggregation factor and Ca^{2+} (Leith & Steinberg,
1972; Burkart & Burger, 1977). Since the number of cells obtainable from
fractions is too small to allow aggregation assays at the standard cell concentrations, we tried to study the affinity of the different cell types to
Microciona aggregation factor by incubating the fractions with Sepharose
beads containing covalently bound *Microciona* aggregation factor on their
surface. After 30 min. shaking at room temperature at least 95% of all the
cells in all the tested fractions had bound to the beads. By visually examining
the cells which remain in the suspension and those attached to the beads,
there was no evidence that a distinct cell type would preferentially bind to
the factor beads. These results confirm that the main cell populations all
contain either factor or at least receptors for the factor on their surfaces.
Preliminary studies have not yet yielded any significant quantitative differences in factor-binding activities among different cell types.

Although sponges are relatively simple organisms, they provide a very
useful experimental system for studying specific cellular recognition of the
kind occurring during tissue and organ formation in higher animals. In
our present state of ignorance about molecular mechanisms for cell–cell
recognition, the sponge system offers considerable advantages – i.e. the
availability of cells of specific types in great quantities without resorting to
harsh disruptive treatments that might alter or obscure surface components involved in recognition. We and others have shown that the primary
mechanism for species-specific sponge cell recognition consists of a minimum of two macromolecular components present on all the cells, namely
the aggregation factor and the baseplate, and that this mechanism can be
studied in detail using fixed cells or a completely cell-free system.

We would like to acknowledge the continued help and interest of other
colleagues presently or formerly associated with the sponge project, which
is carried out mainly at the Marine Biological Laboratory in Woods Hole
(Drs R. Turner, W. Kuhns and Françoise Mir). In addition we would
like to express special thanks to Dr Tracy Simpson for his valuable advice
on identification of the different cell types, and to Ms Verena Schlup for
expert technical assistance. This work is supported by the Swiss National
Foundation, grant No. 3.721.76.

REFERENCES

BARONDES, S. H. & ROSEN, S. D. (1976). Cellular recognition in slime molds. Evidence for its mediation by cell surface species-specific lectins and complimentary oligosaccharides. In *Surface Membrane Receptors*, eds. R. P. Bradshaw, W. A. Frazier, R. C. Merrell, D. I. Gottlieb & R. A. Hogue-Angeletti, pp. 39–55. Plenum, New York.

BEUG, H., KATZ, F. E., STEIN, A. & GERISCH, G. (1973). Quantitation of membrane sites in aggregating *Dictyostelium* cells using tritiated univalent antibody. *Proceedings of the National Academy of Sciences, U.S.A.*, **70**, 3150–4.

BURGER, M. M. (1977). Mechanisms of cell–cell recognition, some comparisons between lower organisms and vertebrates. In *Cell Interactions in Differentiation*, eds. M. Karkinen, L. Saxén & L. Weiss, pp. 357–76. Academic, New York.

BURGER, M. M. & JUMBLATT, J. (1977). Membrane involvement in cell–cell interactions: a two-component model system for cellular recognition that does not require live cells. In *Cell and Tissue Interactions*, eds. J. W. Lash & M. M. Burger, pp. 155–72. Raven, New York.

BURGER, M. M., TURNER, R. S., KUHNS, W. J. & WEINBAUM, G. (1975). A possible model for cell–cell recognition via surface macromolecules. *Philosophical Transactions of the Royal Society, London*, B **271**, 379–93.

BURKART, W. & BURGER, M. M. (1977). Studies on cell populations from *Microciona prolifera* separated by Ficoll gradients. *Biological Bulletin* **153**, 417.

CARTER, S. B. (1967). Haptotaxis and the mechanism of cell motility. *Nature, London*, **213**, 256–60.

CAULDWELL, C., HENKART, P. & HUMPHREYS, T. (1973). Physical properties of sponge aggregation factor: a unique proteoglycan complex. *Biochemistry*, **12**, 3051–5.

CURTIS, A. S. G. & VAN DE VYVER, GYSELE (1971). The control of cell adhesion in a morphogenetic system. *Journal of Embryology and Experimental Morphology*, **26**, 295–312.

DE SUTTER, D. & VAN DE VYVER, G. (1977). Aggregative properties of different cell types of the fresh-water sponge *Ephydatia fluviatilis* isolated on Ficoll gradients. *Wilhelm Roux's Archives of Developmental Biology*, **181**, 151–61.

GALTSOFF, P. S. (1925). Regeneration after dissociation (an experimental study on sponges). I. Behaviour of dissociated cells of *Microciona prolifera* under normal and altered conditions. *Journal of Experimental Zoology*, **42**, 183–221.

GLASER, L., SANTALA, R., GOTTLIEB, A. I. & MERRELL, R. (1977). Cell–cell recognition in the embryonal nervous system. In *Cell and Tissue Interactions*, eds. J. W. Lash & M. M. Burger, pp. 197–208. Raven Press, New York.

HENKART, P. S., HUMPHREYS, S. & HUMPHREYS, T. (1973). Characterization of sponge aggregation factor: a unique proteoglycan complex. *Biochemistry*, **12**, 3045–50.

HUMPHREYS, T. (1963). Chemical and *in vitro* reconstruction of sponge cell adhesion. I. Isolation and functional demonstration of the components involved. *Developmental Biology*, **8**, 27–47.

JOHN, H. A., CAMPO, M. S., MACKENZIE, A. M. & KEMP, R. B. (1971). Role of different sponge cell types in species-specific cell aggregation. *Nature, London, New Biology*, **230**, 126–8.

JUMBLATT, J. & BURGER, M. M. (1977). Studies on the binding of *Microciona prolifera* aggregation factor to dissociated cells. *Biological Bulletin* **153**, 431.

LEITH, A. G. & STEINBERG, M. S. (1972). Velocity sedimentation, separation and aggregative specificity of discrete cell types. *Biological Bulletin*, **143**, 468.

LEVINTHAL, F., MACAGNO, E. R. & LEVINTHAL, C. (1976). Anatomy and development of identified cells in isogenic organisms. *Cold Spring Harbor Symposia on Quantitative Biology*, **40**, 321–33.

McCLAY, D. R. (1971). An autoradiographic analysis of the species specificity during sponge cell reaggregation. *Biological Bulletin*, **141**, 319–30.

McCLAY, D. R. (1974). Cell aggregation: Properties of cell surface factor from five species of sponge. *Journal of Experimental Zoology*, **188**, 89–102.

MACLENNAN, A. P. & DODD, R. Y. (1967). Promoting activity of extracellular materials on sponge cell reaggregation. *Journal of Embryology and Experimental Morphology*, **17**, 481–90.

MOSCONA, A. A. (1963). Studies on cell aggregation: Demonstration of materials with selective cell-binding activity. *Proceedings of the National Academy of Sciences, U.S.A.*, **49**, 742–7.

MOSCONA, A. A. (1968). Cell aggregation: Properties of specific cell ligands and their role in the formation of multi-cellular systems. *Developmental Biology*, **18**, 250–77.

PARIS, J. (1960). *Contribution à la biologie des éponges siliceuses Tethya lyncurium Linck et Suberites domuncula O: Histologie des greffes et sérologie.* Thèse, Montpellier. Causse, Graille, Castelnen, Imprimeurs.

RAKIC, P. (1974). Mode of cell migration to the superficial layers of fetal monkey neocortex. *Journal of Comparative Neurology*, **145**, 61–83.

ROTH, S. A. & WESTON, J. A. (1967). The measurement of intercellular adhesion. *Proceedings of the National Academy of Sciences, U.S.A.*, **58**, 974–80.

SARÀ, M., LIACI, L. & MELONE, N. (1966). Bispecific cell aggregation in sponges. *Nature, London*, **210**, 1167–8.

SIMPSON, T. L. (1973). Coloniality among porifera. In *Animal Colonies*, eds. Boardman, Cheetham & Oliver, pp. 549–65. Dowden, Hutchison and Ross, Strondsburg, Pa.

STEINBERG, M. S. (1970). Does differential adhesion govern self-assembly processes? Equilibrium configurations and the emergence of a hierarchy among populations of embryonic cells. *Journal of Experimental Zoology*, **173**, 395.

TURNER, R. S. & BURGER, M. M. (1973). Involvement of a carbohydrate group in the active site for surface guided reassociation. *Nature, London*, **244**, 509–10.

VAN DE VYVER, G. (1970). La non-confluence intraspecifique chez les spongiaires et la notion d'individu. *Morphogenèse*, **3**, 251–62.

WEINBAUM, G. & BURGER, M. M. (1973). A two-component system for surface guided reassociation of animal cells. *Nature, London*, **244**, 510–12.

WILSON, H. V. (1907). On some phenomena of coalescence and regeneration in sponges. *Journal of Experimental Zoology*, **5**, 245–58.

WOLPERT, L. (1969). Positional information and the spatial patterns of cellular differentiation. *Journal of Theoretical Biology*, **25**, 1–47.

WOLPERT, L., HORNBRUCH, A. & CLARKE, M. R. B. (1974). Positional information and positional signalling in *Hydra*. *American Zoologist*, **14**, 647–63.

EXPLANATION OF PLATE

Main cell types in *Microciona prolifera* cell suspensions. (*a*) Clump of choanocytes, note flagella. (*b*) Archeocyte characterized by prominent nucleolus and inclusions with wide variation in size. (*c*) Same archeocyte with pseudopods in focus. Picture was taken 8 min after deposition of cell suspension on a glass slide. Rapid attachment indicates that the cells are still viable after centrifugation in the Ficoll gradient. (*d*) Grey cell with vesicles of highly uniform size. (*e*), (*f*) Rhabdiferous cells: to the left intact cell which forms no processes, to the right partially disintegrated cell with 'cigar-like' structures. This transition is explosive and may occur within seconds.

Live, unstained cells viewed with Nomarski optics.

CELL–CELL RECOGNITION IN MULTICELLULAR ASSEMBLY: LEVELS OF SPECIFICITY

By MALCOLM S. STEINBERG

Department of Biology, Princeton University,
Princeton, N.J. 08540, U.S.A.

> ... since the production of *form* constitutes the essential feature of development, it is quite permissible to call the science of the causes of form developmental mechanics.
>
> Inasmuch as we call the causes of every phenomenon *forces* or *energies*, we may designate as the general problem of developmental mechanics *the ascertainment of the formative forces or energies*.
>
> Wilhelm Roux, 1894. *The Problems, Methods and Scope of Developmental Mechanics: An Introduction to the 'Archiv für Entwicklungsmechanik der Organismen'*. (Translated from the German by William Morton Wheeler.)

MORPHOGENETIC MECHANISMS AND 'RECOGNITION'

The various types of cells and tissues that constitute an animal do not all arise in their definitive locations. On the contrary, the period of cleavage typically gives rise to a simple ball or disc of cells. These cells and their progeny then engage in a series of movements, ranging from simple foldings to complex migrations, by means of which they take up their proper stations. Homogeneous-appearing cell masses partition into dissimilar layers. Layers split and diverge. Others flow together and combine in specific arrangements, sheath over core. Outer layers turn inward. Sheets roll up into tubes. Nodules vesiculate. And everywhere cells are deployed in specific locations, orientations and associations. By such a choreography, the embryo assembles itself.

As the opening quotation from the father of experimental embryology makes clear, the nature of the forces that direct these morphogenetic movements has been a central question for embryologists ever since their discovery. No single mechanism explains all such movements. In some cases, cells are attracted to their destinations individually or in groups by chemotaxis (e.g. Dubois, 1964). In other cases the movements involve not single cells but tissue masses or sheets. Herbert Phillips and I have recently divided these mass tissue movements into two sharply distinguished classes, set apart by the origin and location of the forces that govern them and by the mechanics of their operation (Phillips & Steinberg, 1978; Phillips, Steinberg & Lipton, 1977).

The first of these classes, which I will only mention here, consists of tissue shape changes that are the cumulative reflection of changes in the shapes of component cells typically organized as an epithelium. If cells arranged in a sheet flatten or palisade, the sheet as a whole expands and thins or contracts and thickens. If the cells constrict at their apices, the sheet curls or folds (Fig. 1a). The forces that produce these cell shape changes originate inside the cells and are considered to be pushing forces produced by the elongation of microtubules and pulling forces produced by the contraction of microfilament bundles (see Burnside & Jacobson, 1968; Wessells *et al.*, 1971; Burnside, 1973; Schroeder, 1973; Spooner, 1973; Karfunkel, 1974). These intracellular forces can be transmitted from cell to cell as supracellular tensions resulting in tissue shape changes only if cell slippage, which would dissipate these tensions, is prevented. Thus the operation of this morphogenetic mechanism precludes individual cell mobility within the epithelium.

(a)

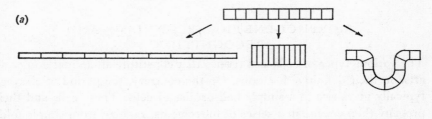

(b)

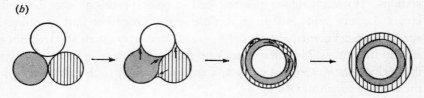

Fig. 1. (a) Cell shape changes can cause expansion, thickening and bending in epithelia. This requires that the cells be firmly linked. (b) Tissue flow requires that cells be able to shift their mutual positions, and thus cannot result from cell shape changes. Tissue systems that show this behaviour also show the other assembly properties of liquids (from Phillips, Steinberg & Lipton, 1977).

The second of these classes consists of tissue re-arrangements that involve a *flow* of cells from one position to another (Fig. 1b). In these cases the positions of cells relative to one another within the mass are not conserved, nor do the cells necessarily assume new and different shapes. Examples are the spreading of mesoderm between ectoderm and endoderm and the spreading of endoderm during early vertebrate development. This is the class of morphogenetic movements upon which I will focus, for the evidence strongly suggests that they are assembly processes that operate according to the same thermodynamic principles which govern the spreading behaviour and phase relations of liquids, the spontaneous assembly of viral subunits, alloy formation, crystal structure and the like.

The determination of macroscopic structure by the surface coding of inter-subunit interaction energies clearly requires some heterogeneity among the various surfaces if the structure so determined is to be anything but a sphere. It is commonly taken for granted that specific 'recognition' behaviour between cell populations is a more or less direct reflection of underlying molecular 'recognition' specificities at the surfaces of the participating cells. My purpose here is to explore the precise relationships between the multicellular, cellular and molecular levels of specificity.

MULTICELLULAR ASSEMBLY AS AN ANALOGUE OF LIQUID ASSEMBLY

The observations of Holtfreter (1939, 1944) and his student Townes (Townes & Holtfreter, 1955) dramatically demonstrated the abilities of amphibian embryonic tissues and cells to organize themselves by either tissue spreading or cell sorting from arbitrary arrangements into anatomically proper structures. These authors ultimately concluded that chemotaxis ('directed movements') determined the inside/outside distributions of cells by causing specific cells and cell groups to move inward or outward, while 'cell specificity of association' or 'selective adhesion' bound them to the appropriate neighbours after they were in position (Townes & Holtfreter, 1955).

Using mixtures of chick embryonic heart and retina cells bearing distinctive cell markers, I found the events of cell sorting within aggregates to differ significantly from expectations based upon a chemotactic mechanism. Rather, they agreed in detail with expectations based upon the hypothesis that cell sorting is an immiscibility phenomenon guided by differences in the intensities of the adhesions formed at the various kinds of cell interfaces (Steinberg, 1962a, b). It proceeded by a process of coalescence, with the formation of a discontinuous, internally segregating tissue phase and a continuous, externally segregating tissue phase, just as happens when oil and water de-mix from dispersion within a droplet. The same process was observed directly by Trinkaus & Lentz (1964) in living aggregates of intermixed heart and pigmented epithelial cells. Pursuing the analogy with liquid behaviour, I next sought to establish whether the tissue configuration generated by cell sorting was an equilibrium configuration – that is, whether this configuration would be approached not only by the de-mixing of scrambled cells but also along other pathways of rearrangement. This is the case with liquids. The principle that governs their rearrangements is that the subunits (molecules) continually shift their associations, stronger associations displacing weaker ones, until the energy of binding is maximal and the interfacial free energy is minimal, at which point the system's configuration is stable. To be sure, when pairs of intact fragments of a variety of embryonic tissues were allowed to adhere, a specific one of the two would spread around the other until the same configuration was reached that the same kinds of cells would arrive at through sorting out from a mixture (Steinberg, 1962c, 1963, 1964, 1970).

The similarities between the behaviour of embryonic tissues and liquids did not end here. The spreading relationships between immiscible liquids are determined by their relative surface tensions or specific interfacial free energies, the liquid of lower surface tension (intermolecular adhesiveness) always spreading over the liquid of higher surface tension. Thus if one has a series of mutually immiscible liquids, their mutual spreading tendencies will be transitively related, defining an inside/outside hierarchy. The same proved to be true of embryonic tissues. In an extensive series of embryonic tissue combinations, when tissue a was spread upon by tissue b, and b in turn was spread upon by tissue c, then tissue a was found in every case to be spread upon by tissue c (Steinberg, 1970). These liquid-like behaviours of embryonic tissues are summarized in Fig. 2.

Why should these embryonic tissues imitate liquids? The answer becomes apparent when one considers the circumstances that characterize the liquid state. To be a liquid, a body must first of all consist of a large number of subunits. These subunits must adhere to one another, comprising a coherent body of definite volume; and they must be mobile with respect to one another. If the subunits cohere but lack mobility, the body is a solid; if they are mobile but lack coherence, the body is a gas. The cell aggregates whose properties I have been discussing are bodies consisting of large numbers of subunits (cells) which can move with respect to one another while cohering. Thus these aggregates have the basic properties that characterize liquid droplets and should be expected to display the behaviour that results from these properties.

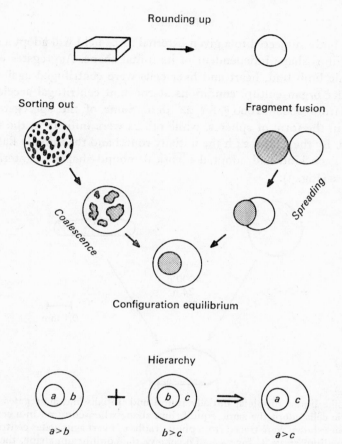

Fig. 2. Many of the behaviours well known in immiscible liquid systems are also displayed by combinations of embryonic cells and tissues. See text for details. (From Phillips, 1969.)

MECHANICAL PROPERTIES OF CELL AGGREGATES

I have reviewed above the *behavioural* evidence that embryonic cell populations sort out, spread over one another in specific orientations and adopt specific configurations because they share certain fundamental properties with liquids, which behave in the very same ways. These behaviours imply that the cell aggregates which display them must have the mechanical properties of liquids; and yet embryonic cell aggregates and tissue fragments have the feel of elastic solid bodies rather than liquid droplets when one handles them. In order to understand the behaviour of these embryonic tissues more fully, my colleagues and I have studied their mechanical properties directly.

Liquidity

A liquid body subjected to a given external force field will adopt a specific (equilibrium) shape independent of its initial shape. Aggregates of chick embryonic limb bud, heart and liver cells were centrifuged against hard agar under organ culture conditions at constant centrifugal accelerations ranging from 2000–12000 *g* for 24–48 h. Some of the aggregates were initially in the form of spheres, while others were initially in the form of flat discs. By the end of 24 h the initially round and the initially flat aggregates of each kind had adopted identical mound-shapes of intermediate roundness (Fig. 3).

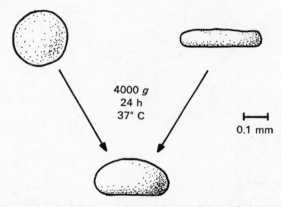

Fig. 3. Like liquid droplets, initially round and initially flat aggregates of chick embryonic cells adopt the same equilibrium shape when cultured in a centrifuge. The profiles shown were traced from photographs of heart aggregates centrifuged in culture medium at 37 °C for 24 h. To achieve the equilibrium shape, the initially flat aggregates must round up against the centrifugal force. (After Phillips & Steinberg, 1969.)

When a liquid body is at shape equilibrium, its interior is free of stress. Yet direct observations with the Harvey–Loomis microscope–centrifuge (Harvey & Loomis, 1930) showed that cell aggregates flattened and widened appreciably as soon as a centrifugal force was applied and returned promptly to their original shapes if the centrifugal force was quickly removed (Phillips & Steinberg, 1978). Since the interior of such aggregates must surely be put under tension when the centrifugal force is applied, the ultimate achievement of shape equilibrium during the succeeding hours implied that these internal tensions must come to be dissipated. These conclusions have been tested by observing, with the transmission electron microscope, the shapes of cells inside aggregates of chick embryonic liver cells before centrifugation, immediately after application of the centrifugal force and after 36 h of centrifugation, when shape equilibrium had been achieved (Phillips, Steinberg & Lipton, 1977). The internal cells were originally in a close-packed cuboidal (regular polygonal) configuration (Plate 1a). Upon application of the centrifugal force (2000 g) the aggregates immediately flattened and their internal cells were correspondingly distorted (Plate 1b). After 36 h of centrifugation the aggregates had flattened still more, but their internal cells, rather than being more distorted than before, had returned to their original, cuboidal shapes (Plate 2a). Thus it is seen, by both internal and external criteria, that these cell aggregates adopt equilibrium shapes and relax internal stresses as only liquids do.

Viscosity

Cell aggregates reach shape equilibrium slowly over the course of hours, sometimes requiring more than a day. Using preliminary experimental values for the specific interfacial free energies of chick embryonic tissues (Phillips, 1969, and personal communication) and on the rates of change of configuration in various circumstances (fusion of paired spherical aggregates; separation of an aggregate held in an elongated shape into several 'droplets'; rounding-up of an ellipsoidal tissue mass toward a spherical shape), Gordon has estimated their viscosities to be in the range of 10^6 to 10^8 poise (Gordon, Goel, Steinberg & Wiseman, 1972). He considers it likely that this extremely high viscosity results from high sliding friction between adhering cell surfaces.

Elasticity

It has been noted above that the living cell aggregates we have studied, while liquid, are extremely viscous; but that they nevertheless respond rapidly to both the imposition and the withdrawal of a centrifugal force by accommodating shape changes. These *rapid* responses to externally imposed forces are accompanied by cell stretching, which reverses if the force is quickly removed and which dissipates, if the force persists, as the aggregate approaches shape equilibrium. These and other observations show that the cell aggregates under consideration are *elasticoviscous liquid* bodies (Phillips & Steinberg, 1978; Phillips, Steinberg & Lipton, 1977) – viscous liquids composed of elastic subunits. The short-term *elastic* properties are doubtless contributed by the elasticity of the constituent cells and extracellular substances; the long-term liquidity by the ability of the cells to slip past one another; and the high viscosity probably by the sliding friction at cell surfaces due to their adhesion to one another and to extracellular structures.

Surface tension

The mutual attractions between neighbouring molecules in a liquid summate vectorially to produce forces which resist expansion of its surface. This 'surface tension', designated as σ, has a constant value, at a given temperature and pressure, for any liquid surface. Every liquid interface, whether it be with the air, with a solid, or with another liquid, is characterized by a value of σ. As I indicated earlier, these values are the principal determinants of the behaviour of liquids. If σ at the interface of two liquids (σ_{ab}) is zero or negative, the two liquids will intermix, whereas if σ is positive, they will be immiscible. Given that σ_{ab} is positive, if σ_{ao} and σ_{bo} (the σ values of liquids a and b against the external medium) are equal, the interface between an a and a b droplet of equal volume will be flat. If $\sigma_{ao} \neq \sigma_{bo}$, then the droplet of lower σ will envelop to some extent the droplet of higher σ. The degree of envelopment is determined by the precise *relative values* of σ_{ao}, σ_{bo} and σ_{ab}. This is illustrated in Fig. 4, which shows an equilibrium configuration for a pair of droplets characterized by the relationships $\sigma_{ao} + \sigma_{bo} > \sigma_{ab} > \sigma_{ao} - \sigma_{bo}$. This set of σ values determines an equilibrium configuration in which droplet a is partially enveloped by droplet b. The three values of σ are represented vectorially as tensions tangent to their respective interfaces at the juncture of phases a and b with the medium o. The lengths of the arrows represent the relative values of their respective surface tensions.

It can be seen that the set of values of σ characterizing the intensities of adhesion at the various interfaces of a liquid system will function as a set of *morphogenetic determinants*. These will cause the system to rearrange until it reaches configurational equilibrium, determined as that configuration in which the interfacial free energy of the system as a whole is minimized. The *differential adhesion hypothesis* (Steinberg, 1962c, 1963, 1964, 1970, 1975, 1978; Phillips, 1969) proposes that all of the liquid-like behavioural features of embryonic cell populations summarized in Fig. 2 are governed by tissue σ values that arise from the adhesive interactions among the constituent cells, just as σ for an ordinary liquid arises from the adhesive interactions among the constituent molecules. Tissue σ values would determine a specific anatomical configuration of cells as most stable and would also direct the constituent cells toward that configuration once they were in mutual contact. Thus the details of a morphogenetic movement governed by tissue σ values need not be separately programmed. It is sufficient to specify the values of σ and insure contact among the cells. Their inherent motility and the thermodynamic principles governing liquid

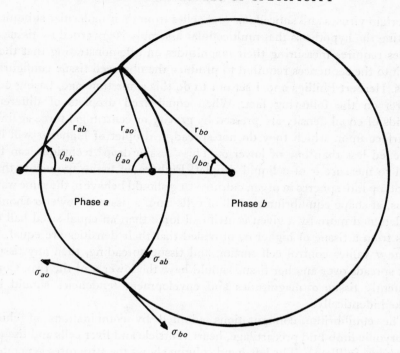

Medium *o*

Fig. 4. Continuous phases *a* and *b*, with spherical interfaces, in medium *o*. At equilibrium, forces of magnitudes σ_{ao}, σ_{bo} and σ_{ab}, tangent at their point of convergence to the *a–o*, *b–o* and *a–b* interfaces, respectively, will balance one another. From this balance of forces arise the adhesive relations and corresponding equilibrium configurations used in the differential adhesion hypothesis. (From Phillips, 1969.)

behaviour will then engender the morphogenetic movement necessary to produce the specified anatomy.

This method of programming structure would explain why the dissociated cells of a body part are able to reconstruct a semblance of that body part via an entirely abnormal morphogenetic pathway – sorting out from a random mixture. What is specified is not the pathway of morphogenesis but its end point. The pathway utilized is simply the thermodynamic downhill slide toward that end point from whatever the starting point happens to be. How much of the goal-directedness of embryonic development and tissue repair results from the use of this simple principle of structural regulation will be interesting to discover.

I take multicellular assembly to mean the spontaneous assembly of a cell population into a definite structure, parallel with the spontaneous assembly of certain viruses and subcellular organelles from their molecular subunits. Testing the hypothesis that multicellular assembly is governed by tissue σ values requires measuring their magnitudes and demonstrating that they fall into the sequences required to produce the observed tissue configurations. Herbert Phillips and I set out to do this some time ago, basing our efforts on the following fact. When equal-sized droplets of different liquids of equal density are pressed by gravity or centrifugal force against a surface upon which they do not spread, a droplet of higher σ will be flattened less than one of lower σ. This 'sessile drop' technique can be used to measure σ of a liquid droplet. Embryonic tissue fragments that round up into spheres in organ culture at $1\ g$ should behave in the same way. Thus, at shape equilibrium, a ball of cells from a tissue of lower σ should be flattened more by a given centrifugal force than an equal-sized ball of cells from a tissue of higher σ, provided that their densities are equal. If tissue σ values control cell sorting and tissue spreading, then any tissue that spreads over another tissue should have the lower σ of the two. Consequently tissue σ magnitudes and envelopment tendencies should be ranked identically.

The equilibrium configurations adopted by combinations of chick embryonic limb bud precartilage, heart ventricle and liver cells and tissues are shown in Plate 3. The left-hand column shows the structures generated by cell sorting, while the right-hand column shows the corresponding structures generated by tissue spreading. In each combination, the same (equilibrium) structure is produced through both processes. Heart ventricle envelops limb bud precartilage, liver envelops heart ventricle and (consistent with the transitivity principle) liver envelops precartilage (Steinberg, 1964). Fig. 5 shows the equilibrium profiles of aggregates of these same three tissues, of approximately equal volume, centrifuged for 24 or 48 h under organ-culture conditions. The sequence of their equilibrium shapes (limb bud considerably rounder than heart; heart slightly rounder than liver) corresponds exactly in the predicted way with their mutual spreading tendencies (Phillips & Steinberg, 1969).

If the densities of these three tissues were equal, we could immediately conclude that the three σ values also correspond as expected with the tissues' mutual spreading tendencies. As it happens, the three tissues' densities are not identical, and while the measured differences do not seem sufficient to throw the three values of σ into a different sequence (Phillips, 1969 and personal communication), the detailed measurements necessary for calculating the precise values of these three σ values are still in progress. Meanwhile, other experiments have permitted us to establish the correlation between tissue σ values and spreading behaviour.

Shapes of centrifuged aggregates

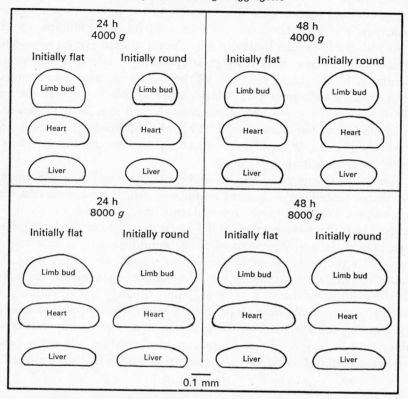

Fig. 5. Profiles of centrifuged aggregates traced from photographs. Each pair was derived from the same kind of primordium. The left-hand member of each pair was initially flat, while the right-hand member was initially round. *Upper left*: three pairs of aggregates spun at 4000 *g* for 24 h. *Upper right*: aggregates spun at 4000 *g* for 48 h. *Lower left*: aggregates spun at 8000 *g* for 24 h. *Lower right*: aggregates spun at 8000 *g* for 48 h. At the end of each run, the aggregates were fixed in the centrifuge before the speed of the rotor was reduced. (From Phillips & Steinberg, 1969.)

It was discovered in our laboratory that re-aggregates formed from chick embryonic heart ventricle fragments after tryptic dissociation and re-aggregation spread over the surfaces of certain control aggregates, whereas undissociated heart fragments were spread upon by control aggregates of the same kind. It was as though the dissociation–re-aggregation procedure had *lowered* the heart ventricle's σ. Moreover, when we examined the effects of varying the duration of organ culture upon the spreading properties of heart fragments, we discovered that fragments cultured in suspension for a longer time were spread upon by control aggregates of a kind which themselves would be spread upon by heart fragments cultured for a shorter time. Culturing in suspension appeared to *increase* σ of heart fragments (Wiseman, Steinberg & Phillips, 1972).

As with the limb bud, heart and liver tissues studied earlier, we had here three tissue 'phases' ranked by their spreading tendencies in a specific hierarchy. In this case, however, all three 'phases' were populations of heart ventricle cells from 5-day chick embryos, differing only in the manner of their pretreatment. Measurements showed that heart masses of all three kinds had the same density (Phillips, unpublished). Consequently their equilibrium shapes under centrifugation would be sufficient to reveal differences in their σ values.

Heart cell *re-aggregates* can easily be prepared as either spheroids or flat sheets containing a constant number of cells. One knows that shape equilibrium has been reached when both have achieved the same final shape. Heart *fragments* cannot readily be prepared as flat sheets. We therefore judged their achievement of shape equilibrium by a less desirable criterion – the cessation of further shape change with time. By 44 h of centrifugation at 1000 g against hard agar under organ-culture conditions, all three kinds of heart masses (of approximately equal volume) had reached shape equilibrium. The differences in their equilibrium shapes were quite pronounced (Fig. 6), in this case clearly corresponding with differences in their σ values. Moreover these σ values fell in the precise sequence deduced from the changes in spreading behaviour (Phillips,

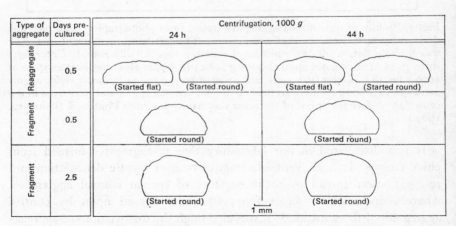

Fig. 6. Vertical profiles, traced from side-view photographs, of heart aggregates prepared in three different ways, as described in the figure, and cultured at 37 °C during centrifugation at 1000 g against hard agar for 24 or 44 h. Fixative was injected during centrifugation, 15–30 min before the rotor was stopped. (From Phillips, Wiseman & Steinberg, 1977.)

Wiseman & Steinberg, 1977). These constitute the first measurements of relative tissue σ values and the first critical test of the differential adhesion hypothesis.

TISSUE SURFACE TENSIONS AND
SELF-RECOGNITION

The discovery that an embryonic tissue's σ could be experimentally increased or decreased opened the way for a critical test of the role of tissue σ in morphogenetic self-recognition. The tissue assembly properties considered here constitute, after all, a form of self-recognition. Identical tissues coalesce and merge indistinguishably, whereas tissues of different kinds spread, one over the other, and their cells sort out from one another. As stated near the outset, a commonly held view has been that such specific recognition of self versus non-self between cells and tissues must directly reflect the molecular specificities of recognition molecules on the cell surfaces. According to this view, like or complementary surface recognition molecules would produce 'self-recognition' or affinity while dissimilar or non-matching surface recognition molecules would produce a 'foreign' reaction or disaffinity, more or less in the manner of immunological self-recognition. In the case of our heart aggregates, however, we had tissues of a single kind whose σ values had been experimentally raised or lowered. According to the differential adhesion hypothesis, these heart cell masses with different σ values should respond to one another like different tissues, and in a predictable way. Briefly cultured heart masses that had previously been dissociated and re-aggregated, possessing a lowered σ, should envelop undissociated heart masses cultured for the same short period of time. And the latter, in turn, should envelop undissociated heart masses cultured for a longer time. These experiments have been carried out (Phillips, Wiseman & Steinberg, 1977), with precisely the predicted results (Plate 4).

Not knowing the nature of the adhesion-mediating molecules in the heart cell surfaces, we cannot state with assurance that heart masses pretreated in these three ways utilize identical adhesion systems. Yet the fact that their σ values seem to be continuously variable within a limited range suggests that the main effects of these treatments has been to decrease or increase the number or area of adhesion sites. I have previously presented calculations showing that such purely quantitative differences in adhesion sites are sufficient to produce immiscibility and mutual spreading between two cell populations (Steinberg, 1963, 1964). These conclusions have been confirmed by computer-modelling (see Goel & Rogers, 1978; Rogers & Goel, 1978). And Overton (1977) has shown, by varying the numbers of desmosomes connecting isotypic and heterotypic cells in aggregates of mouse epidermal and chick corneal cells, that cells connected by different numbers of desmosomes sort out from one another, those with the larger number of desmosomes being enveloped by those with the smaller number. The number of desmosomes cross-connecting the two kinds of cells was equal to the geometric mean of the numbers of desmosomes connecting the two like kinds of cells. This numerical relationship would assure immiscibility of the two types of cells (Steinberg, 1963, 1964) if the desmosomes were energetically the most significant sites of adhesive interaction (Overton, 1977).

In the preceding section I described how the values of σ of liquid systems (depicted in Fig. 4) arise from the adhesive interactions between their constituent subunits and determine their mutual miscibility or immiscibility and spreading behaviour. In liquid *tissue* systems the subunits are *cells*, and the manner in which their adhesive interactions give rise to tissue surface tensions can be understood in the following way. Consider an irregular tissue fragment rounding up to become a sphere in organ culture. As it rounds up, its surface area decreases, cells that earlier were in the surface sinking to subsurface positions. The force driving this process arises from the adhesions made between these surface cells. This is particularly easy to visualize in the case of the zipping-together of cells lining surface crypts or valleys. Consider now that the tissue fragment is put under tension by an external force. Its surface area will increase by the separation of subsurface cells, which will be pulled back into the surface, as illustrated in Fig. 7. The greater the intensity of the adhesions between the cells, the greater will be the resistance to an external force acting to increase the aggregate's surface area. This resistance to surface expansion *at shape equilibrium* is measurable as the surface tension or specific interfacial free energy, σ, of the cell aggregate.

Pure liquids consist of myriads of identical molecules. But most tissues

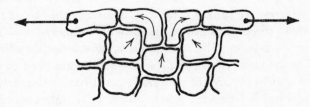

Fig. 7. Illustration of the manner in which the surface area of a liquid cell aggregate under tension is increased. Cells may at first be stretched, but originally subsurface cells continue to come to the aggregate's surface until the aggregate is again at shape equilibrium with the forces acting upon it. Because the surface of a liquid is exposed to anisotropic forces, surface cells may have a different shape than subsurface cells.

and organs are heterogeneous, consisting of a number of different types of cells deployed in characteristic associations and orientations. In attempting to explain what determines these arrangements, we must shift our attention from the aggregate surface to the aggregate interior – to the various possible pairs of apposable cell surfaces within the tissue or organ. In doing so, we must also shift our thermodynamic formulation away from the global or macroscopic surface tensions to the local interactions from which these surface tensions arise. These are conveniently dealt with in terms of the parameters called *reversible works of adhesion*, designated as W (Dupré, 1869; Steinberg, 1962c, 1963, 1964). W is the reversible work required to separate a column of liquid of unit cross-sectional area into two segments, with air being allowed to enter between them. Because *two* new surfaces of this area are thereby produced, W is related to σ in the following way (Dupré, 1869).

$$\sigma_{ao} = \frac{W_{a-a}}{2}; \quad W_{a-a} = 2\sigma_{ao}; \tag{1}$$

$$\sigma_{bo} = \frac{W_{b-b}}{2}; \quad W_{b-b} = 2\sigma_{bo}; \tag{2}$$

Also,

$$\sigma_{ab} = \frac{W_{a-a} + W_{b-b}}{2} - W_{a-b}; \quad W_{a-b} = \sigma_{ao} + \sigma_{bo} - \sigma_{ab}. \tag{3}$$

Although σ can be more rigorously formalized (Phillips, 1969) and more readily measured, W is in many ways easier to visualize, corresponding closely to one's intuitive impression of 'strength of adhesion' between two objects. Figure 8 depicts, for a two-phase liquid system, the equilibrium configurations determined by different sets of W. By convention, when the two phases differ in cohesiveness, the more cohesive phase is designated a and the less cohesive phase is designated b. In the figure, W_{a-a} and W_{b-b} are represented as having higher and lower arbitrary values represented by horizontal solid lines. The figure is divided into four shaded areas, each representing a range of values of W_{a-b} from infinity (column A) to zero (column D). A high value of W_{a-b} (higher than the average of W_{a-a} and W_{b-b}) causes the two phases to mix preferentially. Zero adhesion between a and b subunits (column D) causes phases a and b to round up into separate, isolated spheres. Low but positive values of W_{a-b} (between 0 and W_{b-b}) lead to a *partial* envelopment of phase a by phase b, while higher values of W_{a-b} (above W_{b-b} but below the average of W_{a-a} and W_{b-b}) lead to *complete* envelopment of phase a by phase b. In general, lowering the intensity of a–b adhesions relative to a–a and b–b adhesions decreases the number of a–b adhesions relative to a–a and b–b adhesions

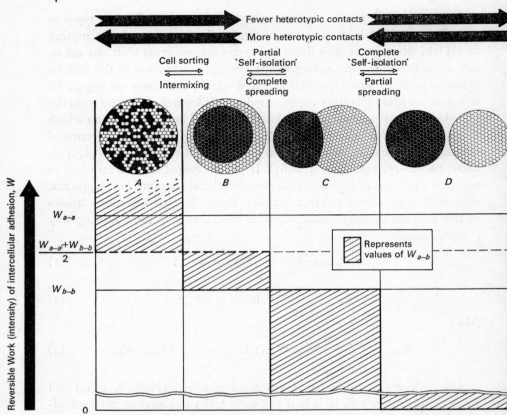

Fig. 8. Illustration of how the reversible works of cohesion (W_{a-a} and W_{b-b}) and adhesion (W_{a-b}) determine the most stable configuration of a liquid system. These relationships should apply to any multi-subunit system that adopts liquid-like equilibrium shapes, whether the subunits are molecules or cells. See text for full description.

in the most stable configuration, but not in a continuous way. Instead, there are critical transition points at which one type of phase distribution suddenly gives way to another. The most dramatic such point occurs when $W_{a-b} = (W_{a-a} + W_{b-b})/2$ (the transition between B and A). At precisely this point, phases a and b will be *randomly* intermixed except at the surface, which will be dominated by the less cohesive b subunits. If W_{a-b} is higher, mixing becomes more *preferential* (a alternating with b), while if W_{a-b} is lower, the two phases suddenly become immiscible, phase b totally enveloping phase a. As W_{a-b} drops still further, no change in the equilibrium configuration occurs until $W_{a-b} < W_{b-b}$. At this point, the coverage of phase a by phase b becomes incomplete, the degree of coverage (contact angle) being determined by the ratio W_{a-b}/W_{b-b}.

PLATE I

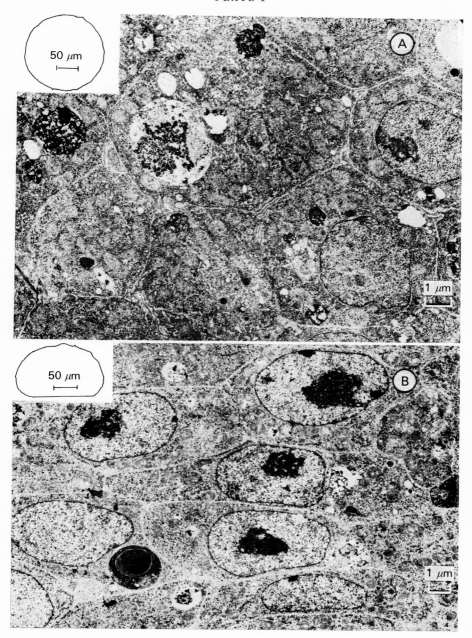

50 μm

50 μm

1 μm

1 μm

For explanation see p. 49

PLATE 2

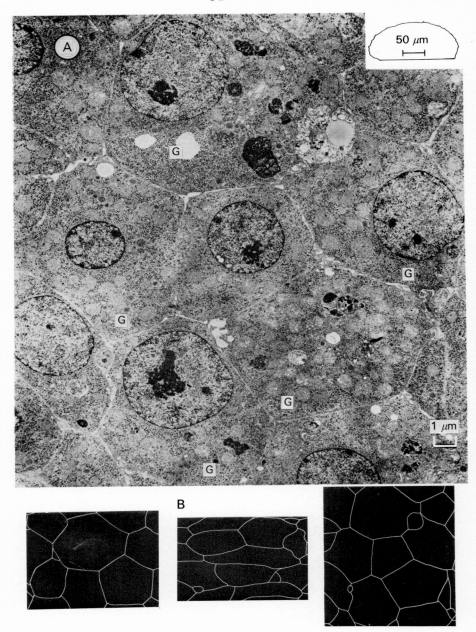

50 μm

A

G

G

G

G

G

1 μm

B

For explanation see p. 49

PLATE 3

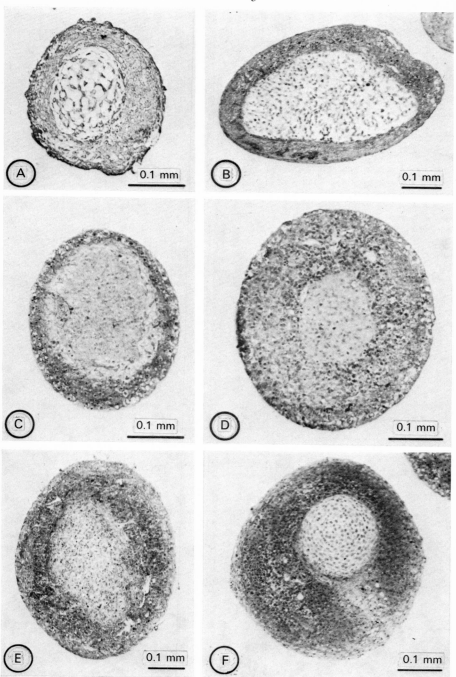

For explanation see p. 49

PLATE 4

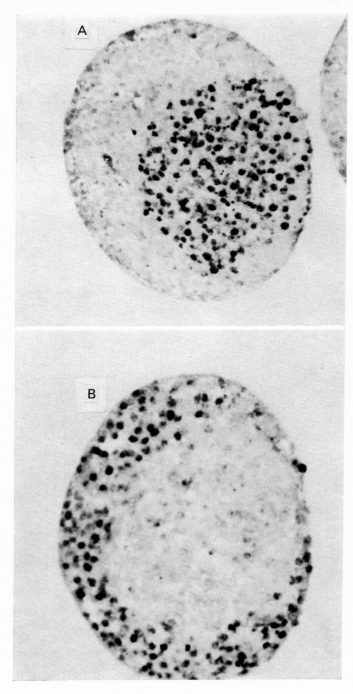

For explanation see p. 49

Tissue recognition versus cell–cell recognition

This brings us to perhaps the most crucial point in the consideration of how morphogenetic specificity and recognition at the cell *population* level arise from underlying interactions between cell *pairs*. Suppose *a* cells adhere to each other with an intensity we designate as 100 ($W_{a-a} = 100$) and *b* cells adhere to each other with an intensity of 50 ($W_{b-b} = 50$). If *a* cells adhere to *b* cells with an intensity (W_{a-b}) of 76, the two kinds of cells will preferentially mix up (Fig. 8*A*), while if they adhere with an intensity of 74, the two kinds of cells will preferentially sort out, and a *b* tissue mass will now also envelop an *a* mass instead of merging with it (Fig. 8*B*). Had we an instrument with which to measure these intensities of adhesion between cell pairs, we would have noted nothing special in the difference between these two situations. After all, in both cases the *b* cells adhere to the *a* cells with an intensity just about halfway between the intensities of *a–a* and *b–b* cell adhesion: just a bit higher in one case and just a bit lower in the other. There is nothing in the latter set of *W* values among and between these two kinds of cells that could be labelled 'specificity' or 'self-recognition'. The fact that these values determine that *groups* of *a* and *b* cells will be immiscible and spontaneously organize into a sheath-and-core configuration only emerges when *populations* of the cells come together. 'Tissue recognition' in this case is an emergent property that arises *de novo*, when cells congregate, out of underlying sets of cellular adhesive intensities that would not in themselves invite the label 'cell–cell recognition'.

I have chosen, in the above, the example that best illustrates this crucial point. Of course there are cases in which *a* and *b* cells, while adhering to others of their own kind perfectly well, adhere to each other very weakly if at all. This will be apparent not only when they are mixed in large numbers, in which case they will tend to form separate aggregates (Fig. 8*D*), but also when they are tested as pairs.

Ligand recognition versus cell–cell recognition

If 'specificity' and 'recognition' are not necessarily transferred bodily from cell pairs to cell populations, neither are they necessarily conveyed intact from cell-surface ligands or adhesion receptors to cell pairs. To explore this question, let us consider the series of situations modelled in Fig. 9. These are anything but comprehensive and have been selected only to illustrate my point. For the sake of illustration, I have utilized exclusively intercellular adhesion molecules of a lock-and-key nature embodying a high degree of recognition specificity. These have been distributed on the cell surfaces in different ways.

4

B cells differ from A cells in that the former have all the locks while the latter have all the keys. In this circumstance A and B cells adhere only to each other but not to 'self'. In terms of W, $W_{a-b} > 0 \simeq W_{a-a} \simeq W_{b-b}$. This set of W values will produce preferential intermixing if the adhesion sites occur on all parts of the cell surface. If the adhesion sites are all on one side of the cell, the result will be heterotypic pair formation. Both patterns are found in gametes, as for example in different yeasts (Crandall & Brock, 1968).

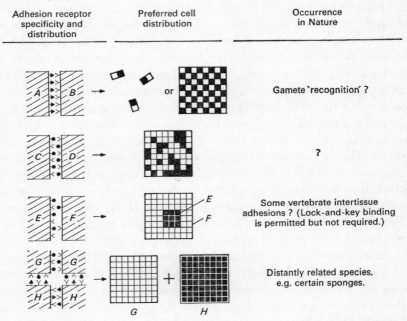

Fig. 9. Illustration of how the affinities, the abundance and the cellular distribution of adhesion receptor sites on cell surfaces determine particular sets of intercellular adhesive intensities which in turn (see Fig. 8) determine particular most-stable tissue configurations. Only 'lock-and-key' binding is illustrated, but many other possibilities exist. See text for details.

C and D cells also utilize a lock-and-key adhesion mechanism, but the same locks and keys both appear equally on both kinds of cells. The result is total interchangeability of the two kinds of cells: the adhesion molecules are stereospecific but adhesion is indiscriminate.

E and F cells utilize the same locks and keys as C and D cells, but F cells have fewer of them. If we do not permit these adhesion receptors to flow in the plane of the membrane, then E cells, able to make the largest number of cross-bonds, will cohere most strongly and F cells will cohere most weakly. E cells will adhere to F cells with intermediate intensity equal to the *geometric* mean of *E–E* and *F–F* adhesive intensities (see Steinberg, 1963, 1964, 1978). This being always less than the arithmetic mean, the adhesive relationships will fall into the set illustrated in Fig. 8*B*, determining immiscibility of the two cell types, a population of *F* cells completely enveloping a population of *E* cells at configurational equilibrium. This is a common result when different vertebrate embryonic cells and tissues are combined (see e.g. Plate 3). In this example, adhesion is stereospecific at the molecular level; it is *not* cell-type-specific, *F* cells even adhering to *E* cells more strongly than to each other; yet *E* and *F* cell populations show tissue-specific self-recognition by sorting out from a mixture and spreading, one over the other, rather than simply merging. And while the adhesion mechanism depends upon intermolecular recognition, the histotypic recognition displayed by the two cell populations does not depend upon this fact; for one could substitute for the specific locks and keys any other kind of attractive force and the histotypic recognition behaviour would remain exactly the same. In fact, Edwards (1978) has recently pointed out that the mere presence of different large polymers (e.g. glycoproteins) on the surfaces of different cells may be sufficient to cause them to be immiscible, as aqueous solutions of such polymers often are (Flory, 1953; Albertsson, 1960).

G and *H* cells both use locks and keys, but these, in addition to being stereospecific, are also cell-type-specific. Consequently *G* cells do not adhere to *H* cells. This is the case among certain distantly related species, the best known of which are certain sponges (Humphreys, 1967). Yet a different result could be obtained by allowing the *G* and *H* adhesion receptors to interact to a certain extent. If this interaction were relatively weak, a configuration like that shown in Fig. 8*C* would result. If it were stronger, configuration 8*B* or even 8*A* could result. In short, since all of the tissue configurations illustrated in Figs. 8 and 9 are shown to be capable of being determined by appropriate distributions of specific cell ligands, the generalized invocation of such specific ligands does not carry us very far toward understanding how a particular one of these configurations is determined in any specific case.

These examples serve to illustrate the following main facts.

(1) Morphogenetic tissue recognition reflects the number and topographic distribution of cell-surface ligands quite as much as it does their affinities.

(2) Tissue recognition does not require or depend upon (although it does, of course, permit) an underlying process of recognition at either the intercellular or the intermolecular level.

(3) Sharp transitions in cell population behaviour can arise from seemingly insignificant quantitative changes in intercellular adhesiveness.

Thermodynamic versus kinetic assays of cell–cell adhesive 'recognition'

In our analysis σ and W are thermodynamic parameters that measure the
reversible works required to increase cell population interfacial areas by
adding cells to them. In principle σ and W can be measured by any method
that measures liquid surface tensions; they can *not* be measured by determining *rates* of adhesion, e.g. between cells and aggregates, monolayers,
etc. This is because the rate at which a reaction proceeds is not a measure
of the change in free energy that accompanies the reaction. Catalysts, for
example, increase reaction rates without, of course, affecting free energies.

The rates of adhesion between cells of different kinds often show intriguing selectivity, and it is tempting to imagine that this may reveal the
associative 'specificities' that govern embryonic assembly processes (e.g.
Roth & Weston, 1967; Roth, 1968). In some cases this may indeed be so,
but the considerable body of evidence amassed to date clearly shows that
rates of intercellular adhesion can vary greatly, independently of changes
in the determinants of morphogenetic assembly processes. For example,
the aggregation rate of chick embryonic neural retina cells drops steadily
and markedly from 5 to 19 days of development (Moscona, 1961, 1962;
Gershman, 1970); yet the older and younger retinal cells do not sort out
when intermixed, nor do older and younger retinal fragments show preferential mutual spreading behaviour (Gershman, 1970). In another case,
Moyer & Steinberg (1976) found that the sequences of adhesion rates
between embryonic tissue masses of several kinds differed from the
sequences of W values required to generate the equilibrium configurations
that they adopted after fusion. The adhesion rate measured in assays of
the rate of adhesion of cells to aggregates or cell monolayers can be greatly
affected by the method of preparation of the cells and even by which of the
two cell populations provides the single cells and which provide the
monolayer (discussed in Moyer & Steinberg 1976). For example, chick
embryonic liver cells adhered at a significant rate to aggregates of pectoral
muscle, but pectoral muscle cells did not adhere at all to liver aggregates
(Roth, 1968). Similarly, chick embryonic heart cells adhered at a significant rate to mouse embryonic liver aggregates, but mouse liver cells did
not adhere at all to chick heart aggregates (Roth, 1968).

Rates of aggregation reflect the making of only those adhesions that
cause cells to stick initially, and do not reflect those adhesions that develop
subsequently. It is likely that these initial adhesions are not representative

of the cell binding mechanisms that dominate after cells have been in contact for many hours. Evidence that this is so has been provided by Umbreit & Roseman (1975), who found that initial adhesions between aggregating chick embryonic cells are made in the absence of metabolic energy production and are broken by gentle shearing, but that these early, weak adhesions are in time replaced by stronger bonds whose development requires metabolic energy production. This may be related to the fact that structural cell junctions such as desmosomes and adhaerens junctions take some time to develop after initial cell contact (Overton, 1962, 1977; Heaysman & Pegrum, 1973) but may contribute significantly to intercellular adhesive intensities after they have formed. The binding energy necessary to cause cell aggregation is extremely low, estimated at less than 10^{-4} erg cm^{-2} of apposition area (Edwards, 1978). This is in agreement with Curtis's (1969) measured values of about 10^{-6} to 10^{-4} erg cm^{-2} for the energy of initial adhesion between chick embryonic liver, limb bud and heart ventricle cells. By contrast, Phillips' measured values for σ of aggregates of these same three tissues at shape equilibrium after over a day in culture (Phillips, 1969) are on the order of about 10 erg cm^2, *one hundred thousand times the energy of initial adhesion that would be reflected in a kinetic assay!* That Phillips' values are reasonable is shown by calculating the equilibrium shapes that such cell aggregates would assume at 1 g for a range of assumed values of σ. At σ values below 10^{-1} erg cm^{-2}, gravity alone would significantly flatten aggregates that in fact adopt an almost spherical shape (Steinberg, unpublished data). Thus it appears highly likely that the adhesive selectivity observed in kinetic assays of cell adhesiveness often reflects the operation of adhesive determinants other than those that govern the assembly of anatomical structures.

SUMMARY

Vertebrate embryonic cells and tissues artificially combined in arbitrary configurations have the ability to rearrange themselves into semblances of normal embryonic structures. The configurations they thereby adopt have been shown to be most-stable or equilibrium configurations analogous to those adopted by multi-phase liquid systems. These tissue masses have the mechanical properties of elasticoviscous liquids – short-term elasticity, associated with cell deformation; long-term liquidity accompanied by cell rearrangement and relaxation; and high viscosity. Their specific interfacial free energies, insofar as they have been measured, fall into the sequences required to explain their assembly behaviour and presumably reflect the intensities of adhesion between their component cells. Thus the strategy of multicellular assembly, as of other self-assembly processes in Nature, is evidently to produce mobile subunits (cells) whose surfaces are designed to interact with one another and with the extracellular milieu in such a way as to specify the structure they are intended to form when the interfacial free energy of the system is minimized. The resulting system of cells is therefore goal-directed in its behaviour, healing wounds and correcting disturbances to its structure. A cell system designed in this way to produce a specific structure needs no separate programme to direct its morphogenesis. What is programmed is the end point, a thermo-dynamic valley into which the system as a whole will spontaneously slide. A separate programme to direct a structure's development may well be required in some cases if because of some impediment or barrier the initial configuration cannot spontaneously drift toward the final one. In such a case the adhesive coding of the cell surfaces will still serve to stabilize the specified structure.

The liquid nature of multicellular assembly systems determines that like tissue masses will coalesce and merge while dissimilar masses will often be immiscible, their cells sorting out when mixed and one tissue mass spreading over another. These behaviours represent a form of tissue self-recognition commonly ascribed to an underlying 'cell–cell recognition', in turn often attributed to 'molecular recognition' between cell-surface ligand molecules. We have here explored the precise relationships between the multicellular, cellular and molecular levels of specificity. We find that while multicellular assembly, differential cell adhesion and selective association between cell-surface ligands are indeed causally linked, 'recognition' is not necessarily transferred intact from one level of organization to the next but may arise *de novo* or be extinguished at each level of translation.

The research from my laboratory described in this article was carried out with the excellent technical assistance of Mr Edward Kennedy and Mrs Doris White. It was supported by grants no. G-5779, G-10896, G-21466, GB-2315 and GB-5759X from the National Science Foundation; grants no. BC-52B and P-532 from the American Cancer Society; and grant no. CA13605 from the National Cancer Institute, DEHW. We also benefited from the central equipment facilities in the Biology Department, Princeton University, supported by the Whitehall Foundation.

The later development and the physical testing of the differential adhesion hypothesis have been a joint effort with Dr Herbert M. Phillips, whose knowledge and judgment have been indispensable to the progress of this work.

REFERENCES

ALBERTSSON, P.-Å. (1960). *Partition of Cell Particles and Macromolecules*. Wiley, New York.

BURNSIDE, B. (1973). Microtubules and microfilaments in amphibian neurulation. *American Zoologist*, **13**, 989–1006.

BURNSIDE, B. & JACOBSON, A. G. (1968). Analysis of morphogenetic movements in the neural plate of the newt *Taricha torosa*. *Developmental Biology*, **18**, 537–52.

CRANDALL, M. A. & BROCK, T. D. (1968). Molecular basis of mating in the yeast *Hansenula wingii*. *Bacteriological Reviews*, **32**, 139–63.

CURTIS, A. S. G. (1969). The measurement of cell adhesiveness by an absolute method. *Journal of Embryology and Experimental Morphology*, **22**, 305–25.

DUBOIS, R. (1964). Sur l'attraction des cellules germinales primordiales par la jeune region gonadique chez l'embryon de Poulet. *Comptes rendus hebdomadaires des séances de l'Académie des Sciences, Paris*, **258**, 3904–7.

DUPRÉ, (1869). *Théorie Mechanique de la Chaleur*, 369. Cited in Davies, J. J. & Rideal, E. K., (1963), *Interfacial Phenomena*, 2nd edition, p. 19. Academic Press, New York and London.

EDWARDS, P. A. W. (1978). Is differential cell adhesion just a consequence of non-specific interactions between cell surface glyco-proteins? *Nature, London* **271**, 248–9.

FLORY, P. J. (1953). *Principles of Polymer Chemistry*. Cornell University Press, Ithaca, New York.

GERSHMAN, H. (1970). On the measurement of cell adhesiveness. *Journal of Experimental Zoology*, **174**, 391–406.

GOEL, N. S. & ROGERS, G. (1978). Computer simulation of engulfment and other movements of embryonic tissues. *Journal of Theoretical Biology*, **71**, 103–40.

GORDON, R., GOEL, N. S., STEINBERG, M. S. & WISEMAN, L. L. (1972). A rheological mechanism sufficient to explain the kinetics of cell sorting. *Journal of Theoretical Biology*, **37**, 43–73.

HARVEY, E. N. & LOOMIS, A. L. (1930). A microscopic centrifuge. *Science*, **72**, 42–4.

HEAYSMAN, J. E. M. & PEGRUM, S. M. (1973). Early contacts between fibroblasts: An ultrastructural study. *Experimental Cell Research*, **78**, 71–8.

HOLTFRETER, J. (1939). Gewebeaffinität, ein Mittel der embryonal Formbildung. *Archiv für Experimentelle Zellforschung Gewebezüchten*, **23**, 169–209.

HOLTFRETER, J. (1944). A study of the mechanics of gastrulation. *Journal of Experimental Zoology*, **95**, 171–212.

HUMPHREYS, J. (1967). The cell surface and specific cell aggregation. In *The Specificity of Cell Surfaces*, eds. B. D. Davis & L. Warren. Prentice Hall, Engelwood Cliffs, New Jersey.

KARFUNKEL, P. (1974). The mechanism of neural tube formation. *International Review of Cytology*, **38**, 245–271.

MOSCONA, A. (1961). Tissue reconstruction from cells. In *Growth in Living Systems*, ed. M. X. Zarrow, pp. 197–220. Basic Books, New York.

MOSCONA, A. (1962). Analysis of cell recombinations in experimental synthesis of tissues *in vitro*. *Journal of Cellular and Comparative Physiology*, **60**, (Suppl. 1) 65–80.

MOYER, W. A. & STEINBERG, M. S. (1976). Do rates of intercellular adhesion measure the cell affinities reflected in cell-sorting and tissue-spreading configurations? *Developmental Biology*, **52**, 246–62.

OVERTON, J. (1962). Desmosome development in normal and reassociating cells in the early chick blastoderm. *Developmental Biology*, 4, 532–48.

OVERTON, J. (1977). Formation of junctions and cell sorting in aggregates of chick and mouse cells. *Developmental Biology*, 55, 103–16.

PHILLIPS, H. M. (1969). 'Equilibrium measurements of embryonic cell adhesiveness: Physical formulation and testing of the differential adhesion hypothesis.' Ph.D. thesis. The Johns Hopkins University, Baltimore, Maryland.

PHILLIPS, H. M. & STEINBERG, M. S. (1969). Equilibrium measurements of embryonic chick cell adhesiveness: I. Shape equilibrium in centrifugal fields. *Proceedings of the National Academy of Sciences, U.S.A.*, 64, 121–7.

PHILLIPS, H. M. & STEINBERG, M. S. (1978). Embryonic tissues as elasticoviscous liquids. I. Rapid and slow shape changes in centrifuged cell aggregates. *Journal of Cell Science* 30, 1–20.

PHILLIPS, H. M., STEINBERG, M. S. & LIPTON, B. H. (1977). Embryonic tissues as elasticoviscous liquids. II. Direct evidence for cell slippage in centrifuged aggregates. *Developmental Biology*, 59, 124–34.

PHILLIPS, H. M., WISEMAN, L. L. & STEINBERG, M. S. (1977). Self vs. nonself in tissue assembly. Correlated changes in recognition behavior and tissue cohesiveness. *Developmental Biology*, 57, 150–9.

ROGERS, G. & GOEL, N. S. (1978). Computer simulation of cellular movements: Cell-sorting, cellular migration through a mass of cells and contact inhibition. *Journal of Theoretical Biology*, 71, 141–66.

ROTH, S. (1968). Studies on intercellular adhesive selectivity. *Developmental Biology*, 18, 602–31.

ROTH, S. A. & WESTON, J. A. (1967). The measurement of intercellular adhesion. *Proceedings of the National Academy of Sciences, U.S.A.*, 58, 974–80.

SCHROEDER, T. E. (1973). Cell constriction: Contractile role of microfilaments in division and development. *American Zoologist*, 13, 949–60.

SPOONER, B. S. (1973). Microfilaments, cell shape changes, and morphogenesis of salivary epithelium. *American Zoologist*, 13, 1007–22.

STEINBERG, M. S. (1962a). On the mechanism of tissue reconstruction by dissociated cells. I. Population kinetics, differential adhesiveness, and the absence of directed migration. *Proceedings of the National Academy of Sciences, U.S.A.*, 48, 1577–82.

STEINBERG, M. S. (1962b). Mechanisms of tissue reconstruction by dissociated cells. II. Time course of events. *Science*, 137, 762–3.

STEINBERG, M. S. (1962c). On the mechanism of tissue reconstruction by dissociated cells. III. Free energy relationships and the reorganization of fused, heteronomic tissue fragments. *Proceedings of the National Academy of Sciences, U.S.A.*, 48, 1769–76.

STEINBERG, M. S. (1963). Reconstruction of tissues by dissociated cells. *Science*, 141, 401–8.

STEINBERG, M. S. (1964). The problem of adhesive selectivity in cellular interactions. In *Cellular Membranes in Development*, ed. M. Locke, pp. 321–66. Academic Press, New York.

STEINBERG, M. S. (1970). Does differential adhesion govern self-assembly processes in histogenesis? Equilibrium configurations and the emergence of a hierarchy among populations of embryonic cells. *Journal of Experimental Zoology*, 173, 395–434.

STEINBERG, M. S. (1975). Adhesion-guided multicellular assembly: a commentary upon the postulates, real and imagined, of the differential adhesion hypothesis, with special attention to computer simulations of cell sorting. *Journal of Theoretical Biology*, 55, 431–44.

STEINBERG, M. S. (1978). Specific cell ligands and the differential adhesion hypothesis: How do they fit together? In *Specificity of Embryological Interactions*,
ed. D. Garrod. Chapman and Hall, London.
TOWNES, P. S. & HOLTFRETER, J. (1955). Directed movements and selective adhesion of embryonic amphibian cells. *Journal of Experimental Zoology*, **128**,
53–120.
TRINKAUS, J. P. & LENTZ, J. P. (1964). Direct observation of type-specific segregation in mixed cell aggregates. *Developmental Biology*, **9**, 115–36.
UMBREIT, J. & ROSEMAN, S. (1975). A requirement for reversible binding between
aggregating embryonic cells before stable adhesion. *Journal of Biological
Chemistry*, **250**, 9360–8.
WESSELLS, N. K., SPOONER, B. S., ASH, J. F., BRADLEY, M. O., LUDUEÑA, M. A.,
TAYLOR, E. L., WRENN, J. T. & YAMADA, K. M. (1971). Microfilaments in
cellular and developmental processes. *Science*, **171**, 135–43.
WISEMAN, L. L., STEINBERG, M. S. & PHILLIPS, H. M. (1972). Experimental
modulation of intercellular cohesiveness: Reversal of tissue assembly patterns. *Developmental Biology*, **28**, 498–517.

EXPLANATION OF PLATES

PLATES 1 AND 2

Electron micrographs of representative, approximately medial, cross-sections of
embryonic chick liver aggregates fixed ($1a$) before centrifugation; ($1b$) starting
after the fifth minute of 2000 g centrifugation and ($2a$) during the thirty-sixth hour
of 2000 g centrifugation. G indicates regions of densely packed glycogen granules.
Inserts: tracings of vertical profiles of each aggregate as seen in thick sections cut
adjacent to the thin sections shown in the electron micrographs. ($2b$) shows outlines of the cell boundaries in $1a$, $1b$ and $2a$. (From Phillips, Steinberg & Lipton,
1977.)

PLATE 3

Equilibrium configurations adopted by chick embryonic cell and tissue combinations through the sorting out of intermixed cells (A, C, E) and the spreading of
apposed, intact tissue fragments (B, D, F). Heart ventricle totally envelops limb
bud precartilage (A, B), liver totally envelops heart ventricle (C, D), and liver
totally envelops limb bud precartilage (E, F), illustrating the transitivity of inside–
outside positioning. According to the differential adhesion hypothesis, a cell population of lesser cohesiveness (σ) should tend to envelop one of higher σ. Thus the
σ values of the three tissues represented here should decline in the sequence
$\sigma_{limb\ bud} > \sigma_{heart} > \sigma_{liver}$. (From Steinberg, 1964.)

PLATE 4

Autoradiography of complete-envelopment configurations of heart aggregate pairs
cultured 48 h in centrifugation medium after fusion. A. An unlabelled, trypsin-
dissociated heart re-aggregate pre-cultured 0.5 d prior to fusion (top row in Fig. 6)
has enveloped a [^{3}H]thymidine-labelled (undissociated) heart fragment also pre-
cultured 0.5 d (middle row in Fig. 6). B. A [^{3}H]thymidine-labelled heart fragment
pre-cultured 0.5 d (middle-row in Fig. 6) has enveloped an unlabelled heart fragment cultured 2.5 d prior to fusion (bottom row in Fig. 6). (From Phillips
Wiseman & Steinberg, 1977.)

CELL–CELL RECOGNITION: POSITIONING AND PATTERNING SYSTEMS

By A. S. G. CURTIS

Department of Cell Biology, University of Glasgow,
Glasgow G12 8QQ, U.K.

One of the most important types of cell–cell recognition leads to the positioning and patterning of cells by controlling cell movement. Such positioning processes are seen extensively in embryonic development, in regeneration, in interactions between allogeneic organisms and sometimes in various xenogeneic combinations (see Table 1). Positioning establishes basic tissue structure and the geometrical relationships of one tissue to another. We tend to note these essential features of organisms without question as to the mechanism(s) that established them. An investigation of the cellular basis of positioning should provide us with a new insight into anatomy and histology – a newer anatomy in terms of the cell behaviour and interactions that establishes, maintains and remoulds body structure. Conversely an examination of the simple morphological features of positioning may provide useful clues as to the cellular mechanisms involved. I should emphasise that patterning can arise in two ways, either by the positioning of pre-differentiated cells or by the differentiation of cells that have already taken up their final positions. The latter process has of course been described by several generations of developmental biologists, using such terms 'fields', 'gradients' and 'positional information' (see Wolpert, 1969). Positioning processes apparently break down when malignant invasion or epizoism (see Rutzler, 1970) takes place.

The recognition processes seen in positioning normally result in the immobilisation of the cell at some particular position. It may immediately be questioned as to whether the locomotory mechanisms of each cell are paralysed or whether the cells stick so strongly that they cannot pull apart even though they attempt to do so. A third possibility is that the cells are able to move and do not adhere to one another over strongly, but are inhibited from moving from the correct environment to an incorrect one. This suggests that the control may lie in a recognition of unlike rather than one of like and also underlines the probability that recognition is a continuing process which not merely establishes the correct pattern of the body but also ensures that correctly positioned cells do not leave their normal site. Curiously, very little attention seems to have been given to

[51]

Table 1. *Positioning and patterning systems*

(1) Positioning alone: 1 cell type.
    e.g. Lens and lentoid structures.
        Tubule structures.

(2) Positioning and patterning of two or more cell types within one tissue.
    e.g. Structure of kidney, liver, pancreas etc.
        Epitheliomesenchymal interactions.
        Pigment cell positioning in dermis.

(3) Positioning and patterning of whole tissues.
    e.g. Embryogenesis, regeneration, graft acceptance.

(4) Allogeneic combinations.
    e.g. Non-coalescence between sponges: similar phenomena in botryllids,
        Hydractinia, Eunicella etc.
        Boundaries between lichen individuals.
        In general non-fusibility in sedentary organisms.
        Probably features of graft rejection in invertebrates.

(5) Xenogeneic combinations.
    Epizoism in sponges and other marine sedentary organisms.
    Oppositely non-fusibility between organisms.

those aspects of cell recognition that control cell locomotion though they may well be very important. Forms of cell behaviour which inhibit cell movement, such as contact inhibition of movement (Abercrombie & Heaysman, 1954) or contact paralysis, have been described but little attention has been given to the operation of such types of behaviour between normal cells of unlike type from the same species. The majority of combinations that have been made being between cell lines and malignant cells or between fibroblasts from different species. Chemotaxis or chemokinesis provides a mechanism by which positioning recognition might be achieved by control of movement but little evidence has yet been obtained for its operation between normal cells of a multicellular organism.

Our realisation that cells recognise one another is primarily based on observations on systems in which, when various types of cells are mixed together, they take up non-random combinations. This definition applies both to the mating type recognition described by other authors in this symposium, see Crandall (this volume, pp. 105–19) and Wiese & Wiese (pp. 83–103), to those systems in which two sponge species sort out (Burger *et al.*, this volume, pp. 1–24) and those systems in which positioning takes place.

Positioning mechanisms may operate at various different levels. For instance, some form of mechanism operates within a population of one type of cell to produce tubules or lentoid spheres etc. Recognition appears with those systems in which two or more cell types position themselves within one tissue. It operates at a larger level in the formation and separation of different tissues within an organism and may act to ensure that autografts fuse successfully with the surrounding tissue structure. It has been usual to investigate these systems by examining whether two different tissue types of cell will 'sort out' in mixed aggregates. This type of experiment establishes whether the positioning mechanism is an inherent property of the cells involved or whether it requires other factors and instructions. It is probably unfortunate that most studies on the assortment of cells have been carried out on pairs of cell types which never normally meet, for instance neural and liver, and thus the results may not be relevant to the formation of tissues or to the relative placing of tissue blocks. However the studies on retino-tectal recognition (see Barbera 1975, and Pierce, Marchase & Roth, this volume, pp. 261–74) and on the interactions of the choroid, sclera and pigmented retina of the avian eye (Büültjens & Edwards, 1977) though using slightly different test systems are clearly much more likely to provide information on positioning mechanism. One point of considerable interest discovered by Büültjens & Edwards is that unlike adhesions between pigmented retina and choroid are preferred to like adhesions between retina and retina or choroid and choroid.

The fourth level of positioning recognition involves the interaction of allogeneic cells. This of course occurs whenever a mating-type system operates as well as when sedentary organisms abut on one another or attempt to settle their larval stages on another member of the same species. It also may operate artificially when grafts are made, perhaps in both plants and animals. The final type of positioning occurs between xenogeneic organisms. This may be seen in the contact interactions between various sedentary organisms, in certain grafting situations and perhaps in experiments on the aggregation of mixed species of sponge.

The positioning seen in both the first (amongst one cell types) and the second (within one tissue) level is often circumferential or enclosing (see Fig. 1). The planes that separate the sets of assorting cells may appear as zones of apparently weak adhesion though they may be filled with collagen later. The nearly two-dimensional sorting out of xantho- and melanoblasts on the flanks of urodele amphibians (Twitty, 1954) is basically also very simple geometrically. These simple patterns can be explained as resulting from graded properties of cells or gradients of cell products.

Positioning between tissues (the third level) is often enclosing, as Steinberg (1963) noted, though it may also be circumferential. The patterning is rarely of a simple, geometric type though the precision of its repetition in different individual animals argues strongly for both precise geometric and

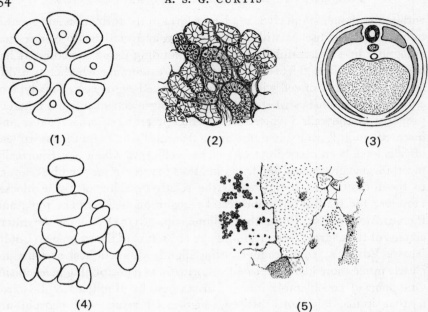

Fig. 1. Patterning and positioning. The five levels of positioning and four of patterning. (1) Positioning of cells of one type as in a lentoid. (2) Positioning and patterning of cells of two or more types in a tissue (drawn from a section of a salivary gland). (3) Positioning and patterning of tissues in an organism, illustrated by a simplified section of a somewhat idealised vertebrate. (4) Allogeneic positioning showing the pattern of allogeneic organisms, in this case a plan of *Hymeniacidon* sp. sponges at Linne Mhurich, Argyll, U.K. Each sponge is separated from the next by a zone of non-coalescence. (5) Xenogeneic positioning: the pattern of lichens of various species on a rock from Lephinchapel, Argyll, (*c.* ½ natural size).

genetic controls. The different tissues are often separated by a gap of 100–1000 μm width, though this may be bridged by connective tissue at places. This simple fact argues that there must be a cell type or types which is neutral as regards cell–cell recognition, possibly this is the fibroblast. The separate tissues are often lined with endo- or epithelia whose outer surfaces are non-adherent. It is unclear whether the separation is actively maintained by the epithelia, and whether the behavioural properties of these cells reported by Dipasquale & Bell (1974) and by Middleton (1973) which may tend to prevent adhesion of further cells, even of the same type, on their free surfaces, are of importance in such reactions. It is clear however from the work of Steinberg and others that differing tissues segregate (sort out) in mixed aggregates without the obvious intervention of epi- or endothelial cells, the segregation is usually enclosing or circumferential.

We do not know whether the maintenance of the separation of different tissues in an animal is due to the continued operation of positioning processes or whether the cells of the tissues are effectively 'frozen' in position, either by collagen fibres, endo- or epithelial cell linings or by irreversible mechanisms of adhesion. However there are two reasons for concluding that the positioning process normally operates continuously.

First, the relatively simple process of wounding or excising pieces of tissue allows cell emigration in cell culture. Second, the majority of solid tissues are accessible to infiltration by lymphocytes and leukocytes as well as by malignant cells, which suggests that emigration by the tissue cells is potentially possible at all times, because the various white blood cell types re-emigrate from tissues easily.

The patterns set up in allogeneic positioning are complicated by immunological reactions in vertebrates but in marine organisms such as sponges, (Curtis, 1978a), botryllids (Oka, 1970), bryozoans (Ryland, 1976) and perhaps in land plants such as lichens (Henssen & Jahns, 1974) the patterns are based initially on the vagaries of settlement of the founding cells for each organism. When the organisms have grown large enough to abut on other individuals, a zone about 1 mm wide, a no-man's land or zone of non-coalescence (van de Vyver 1970), remains between them. In sponges we know that the zone is probably due to the secretion of factors that diminish the adhesion of cells on the other side of the zone (Curtis, 1978a). The pattern that develops in such populations of allogeneic individuals is necessarily rather vague but it often consists of individuals of rather similar size forming a tesselation. If the substrate is confined, a chain of different organisms may develop; such chains have been achieved in experimental situations (Curtis, 1962). Xenogeneic positioning and patterning appears to have similar features. The constraints and vagaries of settlement in such situations are such that positioning in these situations must be rather irregular.

Thus there appears to be a progression of features through the five levels such that:

(i) There is a decreasing probability of the formation of long-lasting adhesions between unlike cells as they become less closely related. Gametes form a partial exception to this rule, since they must be allogeneic.

(ii) Decreasing regularity of patterning.

(iii) Increasing development of spacing planes and zones of non-coalescence as cells become less closely related.

In allo- and xenogeneic positioning the quality of positioning is usually of such low grade that it is uncertain as to whether any real positioning mechanism has operated other than the prevention of contact or overlap between the organisms. As an aside we may speculate as to whether graft rejection in invertebrates is anything more than such a reaction (see Dales, this volume, pp. 203–19). The question as to whether the five levels of recognition described above for positioning systems are expressions of a common mechanism or whether they are fundamentally different should remain open. Table 1 summarises the various types of positioning.

MECHANISMS FOR CELL POSITIONING

Basically four types of mechanism have been proposed in order to explain cell positioning.

(1) Specific adhesion of cells. It is hard to discover who originated this concept but it was clearly stated by Moscona (1962) and is re-stated in this symposium by Burger *et al.*, pp. 1–24. The concept usually appears as a self-recognition by complementary paired molecules though other systems are at least feasible in theory.

(2) Differential adhesion, a hypothesis which was first put forward by Steinberg (1962 *a, b, c*) and which is reviewed again by Steinberg in this symposium (pp. 25–49). In essence this theory dispenses with the need for any precise molecular interaction between recognising pairs of cells and explains the patterning and positioning of cells in terms of the quantitative differences in adhesion that they show. The theory has recently been criticised by Harris (1976).

(3) Chemotaxis or chemokinesis. This theory was proposed at least as early as 1955 by Townes & Holtfreter. Positioning of cells is supposed to occur in response to concentration gradients of chemotactic agents set up either by cells that have already acquired a definite position or in response to generalised differences in concentration of such substances as oxygen or carbon dioxide.

(4) The morphogen or interaction modulation factor theory. This was proposed in a simple form by Curtis & van de Vyver (1971) and restated by Curtis (1974, 1976). Dr de Sousa suggested to me the advantages of terming it the interaction modulation theory and it is thus described by Curtis (1978 *b*). This theory proposes that cells produce diffusible substances that diminish the adhesion of some unlike cell types so that they tend to allow unlike cell types to escape from their own environment. It is supposed that concentration gradients of these factors are set up and that these determine cell positioning. Chemotaxis or chemokinesis could be possible results of such interactions.

Requirements that must be met by explanations of cell positioning

There are three main types of positioning systems in organisms:

(1) Those which operate in arrangements of cells which are initially random and in which sorting-out mechanisms act to produce the whole pattern. This can be termed *de novo* positioning.

(2) Systems which use a pre-existing non-random arrangement of cells on which to build up a pattern. I shall refer to these as pre-positioned systems.

(3) Systems in which there is random accumulation of cells in various sites followed by selective survival or division of the cells at certain sites.

Do lymphocytes, for example, sort out within the lymph nodes into B- and T-dependent areas, according to their own type because of interactions with the stromal and other cells already in position (pre-positioning) or do they sort out because of properties inherent to the cells arriving in the node (*de novo*) or because they die or fail to multiply in incorrect sites?

Willis (1952) has argued that the positioning of secondary tumours in malignancy is due to selective survival and division. This is the 'nutritive' hypothesis. Similarly, Meyer (1964) has shown that the experimental and observational results on germ-cell localisation in birds are best explained in this manner.

It can be claimed that a great deal of the patterning seen in organisms is the result of differentiation *in situ* and that any subsequent positioning of migrating cells is simply due to the trapping of these cells onto pre-positioned 'target' cells. Nevertheless it is clear that *de novo* positioning can occur both *in vivo* and *in vitro*. The most complete example of *de novo* positioning in a normal organism probably takes place in the disaggregation and subsequent reaggregation of the embryo that occurs in the annual fish (Wourms, 1972). The sorting out or assortment of cells seen in aggregates of two or more cell types is a similarly complete *de novo* positioning in an experimental system.

Specific adhesion does not explain *de novo* positioning because the theory contains no element to direct cells into fixed relations with those of another type. The theory contains no patterning component. It might explain the trapping of cells onto a pre-existing pattern but cannot specify whether one cell type will surround another or take up any specified position when the cells are initially arranged randomly. On the other hand any mechanism that accounts for *de novo* positioning could also act in situations in which a degree of pre-positioning has taken place.

Positioning *de novo* or from a random start requires that a population of cells has a scalar or vectorial quantity graded or quantised in it which determines the direction in which a cell will probably move. The quantity may be smoothly graded or follow some complex relationship with position or may, as in the Steinberg hypothesis, be quantised according to cell type. Another possibility is that the quantity is the concentration of some diffusible chemical.

Steinberg suggested that the graded quantity was adhesiveness. Although this theory has been criticised by Harris (1976), Antonelli, Rogers & Willard (1973) and myself (Curtis 1967, 1978b) it should be appreciated that the criticisms have been directed at details of the theory, and at implications it has in other areas such as the strength and nature of cell contacts, rather than at the general proposition that a quantitative property is involved in sorting out. Steinberg (1962a, b, c, 1963) demonstrated this property in the 'hierarchy' of sorting out.

The differential adhesion hypothesis, as described by Steinberg (1962a, b, c, 1963, 1970, 1976) contains no explicit statement as to the means by which cells could change their surface properties. One mechanism would of course be differentiation, while another would be cell interaction. The concept that sorting out might be based on relatively rapid cell interactions altering surface properties was proposed in general terms by Curtis & van de Vyver (1971) as a consequence of the experimental results obtained on sorting out and non-coalescence in the various strain types of the sponge *Ephydatia fluviatilis*. This proposal was repeated more explicitly, including data from other systems, by Curtis (1974, 1976). Independently de Sousa (1973) suggested that lymphocyte recognitions might involve rapid interactions between lymphocyte classes and other cell types encountered by these cells.

On the other hand we must recognise that there is considerable evidence that very specific groupings involved in cell adhesion are to be found at the surface of cells. This evidence is reviewed in this volume by Burger *et al.*, pp. 1–24 and Wiese & Wiese, pp. 83–103, and elsewhere by Curtis 1978b. The differential adhesion theory does not require the presence of such groupings. Clearly it would be desirable if we could include both the evidence for quantitative properties involved in sorting out and the evidence for specific groupings in a single theory of positioning. A simple set of experimental findings illustrate this apparent dichotomy of evidence. I found (Curtis 1974) that well-washed chick embryonic neural retina and liver cells showed non-specific adhesion, while mixtures in the presence of a factor obtained from one of the cell types involved showed apparent specificity of adhesion, while yet again mixtures in the presence of factors from both cell types lost any specificity of adhesion.

Another pair of requirements which must be met by any satisfactory explanation of sorting out is the relatively high accuracy and rapid nature of sorting out. Both the specific adhesion and the differential adhesion theories can be criticised on these grounds. Antonelli *et al.* (1973) suggested, as a result of mathematical modelling, that the differential adhesion theory could not produce the accuracy of sorting-out required. Experimental work on microemulsions by Talmon & Prager (1977) implies that sorting-out due to surface free energy differences is not likely to take place in multi-component 'cellular' systems.

GENERAL FEATURES OF RECOGNITION SYSTEMS

The hormone signal which is received by a component of the cell surface forms a near-classical model recognition system. In other hormonal systems the receptor is soluble but is contained within some internal cell organelle. In mating-type recognition systems (see Crandall and Wiese & Wiese, this volume), both signal and receptor are insoluble, being bound to cell surfaces. Cell contact is required for the system to operate. I suggest that at least two classes which can be obtained if cell recognition is classified in this manner refer to cell–cell recognition (see Table 2).

Table 2. *Classes of cell recognition system and their possible involvement in cell–cell recognition*

Status of signal	Status of receptor	Example	Type of cell–cell recognition
Soluble	Cell surface bound	Insulin	Chemotaxis Chemokinesis Interaction modulation factors
Soluble	Internal receptor		?
Cell surface bound	Cell surface bound	*Hansenula* mating types	Specific adhesion

Specific adhesion theories propose that positioning must depend upon insoluble signal and insoluble receptor systems but we have seen that such explanations fail to account for the directionality seen in cell positioning. The differential adhesion theory provides directionality but does not require the operation of any form of signal-receptor system. Thus it is worth, for these reasons alone, reconsidering whether positioning may not

depend on soluble signal and insoluble receptor systems, as I have recently
suggested (Curtis, 1978 b), see Table 2.

An important attribute of soluble signal/insoluble receptor systems is
that they allow the existence of directionality. If concentration gradients
can be set up for the signal, then a whole range of types of positioning control can operate, for instance chemotaxis or chemokinesis. (The precise
use of these terms is somewhat in confusion, those investigating slime
mould behaviour (e.g. Garrod, Swan, Nicol & Forman, this volume, pp.
173–202) and those studying bacterial behaviour (Adler, 1966 for example)
using them in different ways. I shall use these terms in a provisional manner and simply refer to the change in cell motility and direction with concentration of a stimulating chemical, noting that response may be up-
gradient (positive) or down-gradient (negative).)

Other chemo-moto-responsive systems exist as well as chemotaxis or
chemokinesis – for instance, cell movement may be inhibited at a certain
concentration or in a certain gradient of the chemical, but in all such systems there is the potential of setting up directional cues to which the cells
can respond. Differences in diffusibility of two such substances (morpho-
gens if we follow the usage of Edelstein, 1971) could ensure that there are
differing relative concentrations of the two agents in different parts of an
aggregate and thus that different cell types would accumulate in differing
regions of the aggregate. The only requirements to establish such differential gradients are that there are differences in diffusibility and that the
agents are lost to the outside environment which acts as a sink. One further
important feature must be added if the system is to operate over long
periods: this is that there must be a turn-over of the saturated receptors at
the cell surface and their replacement by fresh receptors available to receive
new signal molecules. The level of response of a cell may be primarily
determined by the turn-over rate of such cell-surface components.

THE INTERACTION MODULATION HYPOTHESIS

Much of the background to this theory has already been examined above
and it can be seen that any explanation of cell–cell recognition in positioning and in related phenomena should:

(1) Account for the directionality of cell positioning, in other words for
    the patterning.
(2) Explain both the quantitative properties seen in sorting-out and the
    apparent specificity of molecular species in or on the cell surface
    which are involved in adhesion.

In addition the experimental evidence should throw light on whether:

(A) The system(s) involved is a soluble signal/surface receptor one or of some other type.
(B) Whether like or unlike are recognised.
(C) Whether the directionality results from the positive response to like–like recognition or to a negative response to unlike–unlike recognition.

The interaction modulation hypothesis is an attempt to marry these various considerations and much evidence on adhesion and positioning into a coherent whole that embraces the results of the differential and specific adhesion theories. In essence the interaction modulation hypothesis proposes that cell–cell recognition involves a soluble signal/insoluble receptor system. The recognition event is of like for like and prevents the signal molecule modifying the surface of the cell types that produce it. If unlike cells meet, their signals may modulate (modify) the surface properties, such as adhesion, of the opposite cell type.

In detail the theory suggests that all, or nearly all, cell types within an organism produce soluble diffusible substances – the interaction modulation factors (IMFs) that interact with unlike cell-types to reduce their adhesiveness in a relatively unspecific manner. The reduction in adhesion is a generalised one so that affected cells will be of low adhesiveness not merely for their own type but also for other cell types and for non-living surfaces. Gradients of these factors will be set up in cell populations so that the directionality required to establish pattern exists. Thus the theory includes the experimentally established feature of a quantitative relationship involved in positioning together with the possibility of very specific molecular species being required either to be the signals, the receptors, or the inhibitors of action that must be supposed to be present to prevent a cell being attacked by its own products.

Experimental evidence for the interaction modulation theory

The first prediction of the hypothesis is that actively metabolising cells will produce substances that diminish the adhesiveness of other cell types, not only for themselves but also for many other types of surface.

This has now been tested in five widely different groups of organisms. I and Gysele van de Vyver (Curtis & van de Vyver, 1971) found that α and δ strains of the sponge *Ephydatia fluviatilis* produce substances that diminish the adhesion of the opposite strain-type not merely for itself but also for such substrates as glass, cellulose acetate and the producer cell-type. α and δ strains are non-coalescent with each other. Later I (Curtis,

1974) demonstrated that embryonic chick neural retina and liver cells produced similar factors, while Maria de Sousa and I (Curtis & de Sousa, 1973, 1975 a, b) showed that mouse, rat and human B and T lymphocytes secrete factors that diminish the adhesiveness of the opposite cell type so that mouse T cells have their adhesion reduced by factors from mouse, rat or human B cells and vice-versa. Recently I have found that mouse liver also produces substances that affect lymphocyte adhesion and vice-versa. Kondo (1974) has described substances, possibly similar, in echinoderm embryos, as have Weinbaum & Burger (1973) in sponges, though these authors suggested that they were receptors for the cell ligands and not agents diminishing cell adhesiveness of other types of cell. Garrod *et al*, (this volume, pp. 173–202) report that pre-aggregation stages of slime moulds produce inhibitors for cell adhesion at later stages. Lomnitzer, Rabson & Koornhoff (1976) report the existence of lymphocyte products that reduce leukocyte adhesion.

A second important prediction of the hypothesis is that it should be possible to obtain apparent specific adhesion with the interaction modulation factors. For example, if you add IMF from neural retinal cells to a mixture of neural retinal and liver cells, the adhesion of the liver cells is reduced so that (Fig. 2) the aggregates formed will be composed predominantly of neural retinal cells. The result of such an experiment can be used to argue that you have specifically stimulated neural retinal cell adhesion, perhaps by adding a ligand to the system. However both measurements of the actual adhesiveness of the two cell types in presence and absence of added IMF and a second type of experiment show this to be an incorrect conclusion. The second experiment is to add IMFs from both liver and neural retinal cells to mixtures of these cells (see Fig. 2). The result is that though the overall adhesiveness of both cell types is diminished the aggregates are now composed equally of both cell types. In other words specificity of adhesion has now vanished. Consequently the IMFs cannot be effectors of specific cell adhesion and the experimental results argue strongly that an apparent specificity of adhesion can be produced by control of adhesive levels by IMFs. In other words we are seeing a specific control system for cell adhesion not specific mechanisms of adhesion. It would be interesting to repeat some of the experiments that have been claimed to demonstrate specificity of adhesion to discover whether the systems are in fact IMF controlled. Similar results have been obtained for mouse B and T lymphocytes, (Curtis & de Sousa, 1975 a).

However, the crucial test of any positioning mechanism is to demonstrate that interference in the mechanism produces predictable changes in positioning, preferably *in vivo* but at least in relatively well-known

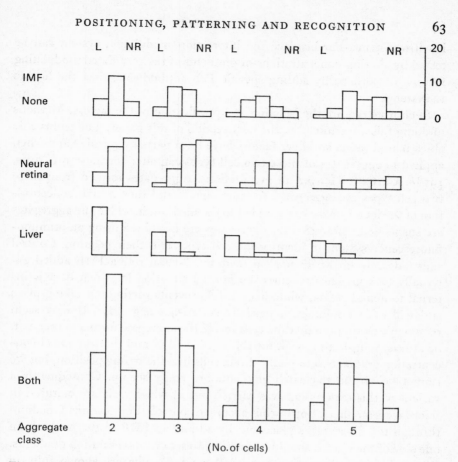

Fig. 2. The conditions for specific and non-specific aggregation. The composition of aggregates (2- to 5-cell aggregates) formed by 50:50 mixtures of chick embryonic neural retina and liver. The histograms give the proportion of aggregates containing various compositions of the cell types. For instance the 2-cell aggregates can be 100% neural retina, 50:50 liver and retina or 100% liver. In the absence of any added IMF these aggregates are most commonly composed of mixtures and the composition does not depart from that to be expected if there is no specificity of adhesion. When neural retina IMF is added the aggregates become biased in composition towards being rich in retinal cells, Row 2. When liver IMF is added the aggregates become rich in liver cells, Row 3. Thus these factors apparently stimulate specificity of adhesion. When both factors are added, Row 4, the aggregates become random in composition. The effect of applying both IMFs cannot be explained on the hypothesis that the factors produce specific adhesion but can be explained if it is proposed that they diminish the adhesion of the other cell type. The increased number of aggregates in each class of the histogram in Row 4 does not represent increased adhesiveness but merely the acquisition of a larger sample. Ordinate, number of aggregates in each aggregate composition class.

in-vitro systems. For instance the interaction modulation system can be tested by altering concentrations or gradients of the postulated modulating factors, or perhaps by adding specific Fab antibodies against the factors to systems.

Such tests require the isolation and purification of the factors. Methods of doing this are outlined in the next section of this paper. The embryonic chick neural retina and liver factors have been partially purified and when applied to aggregates of these two cell types can alter sorting-out (Curtis, 1978 b, Curtis & Hoover, in preparation). Aggregates formed from these two cell types are aggregated for some hours and then a high concentration of the factor or factors is added to the medium in which the aggregates are suspended. After 48 h the aggregates are examined using an immuno-fluorescence system to identify the cell types and their position. Control aggregates are grown in normal medium. Neural retina IMF added externally to aggregates reverses the normal situation in which liver is internal to neural retina, while liver IMF prevents sorting out taking place and will even randomise a partially sorted-out aggregate. It may seem reasonable to suppose that the cells inside the aggregate form a permanent, or at least long-lasting, sink for the applied IMF and that the novel concentration gradients set up now direct cells into the wrong position, but we cannot be certain that such gradients are really set up. Consequently a variant of this experiment was carried out in which cells are cultured in thick layers on filters. Sorting out occurs in controls. Permeation of medium through the filters takes place and by adding an IMF to the medium on one side of the filter it should be possible to set up concentration gradients. When this is done sorting out is reversed. Results are shown fully in Curtis (1978 b). Fig. 3 illustrates these experiments, while Fig. 4 shows an explanation of normal sorting out in terms of the interaction modulation hypothesis.

In-vivo experiments on IMFs have been carried out on lymphocyte positioning and these are described in the next section.

A number of other experiments can be conceived, which though not providing proof of the theory, could produce results consistent with it. For example, well-washed cells should be effectively free from IMFs until they have produced enough factor and should, in a good culture medium, recover their normal adhesiveness. Thus we should find that such cells should show a raised non-specific adhesiveness if they have been either well washed or cultured in the absence of other cell types. This has been described by myself for embryonic chick cells (Curtis 1970 a) and for the cells of various species of marine sponge (Curtis 1970 b).

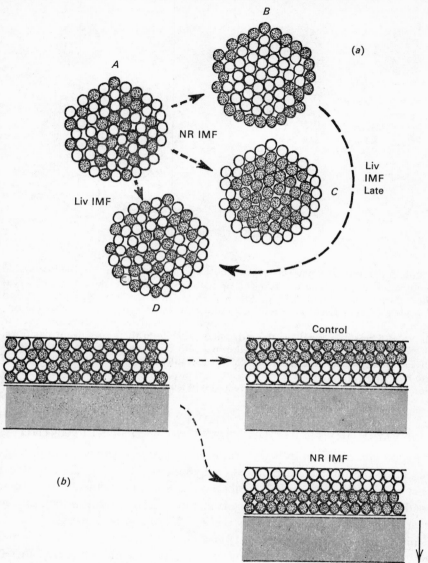

Fig. 3. (a) A diagram showing the results of applying embryonic IMFs on sorting out in aggregates, reported more fully by Curtis (1978b). A. Aggregates are prepared from liver cells (unshaded) and neural retina (shaded) in 50:50 proportions and have a nearly random arrangement of the two cell types at 4 to about 18 hours. Over the next 30 hours they normally sort out into the arrangement shown in B with liver internal to neural retina. If neural retina IMF is applied externally the arrangement of cells is reversed, as in C. If liver IMF is applied externally the aggregates do not sort out, see D, while sorted-out aggregates can be randomised by application of liver IMF after sorting-out was completed.

(b) Sorting-out of neural retina and liver grown in thick layers on Pellicon filter. Medium is permeated through the cell layer and subjacent filter. In controls liver (white) sorts out internally to neural retina but when neural retina IMF is permeated through the system from the top the positioning is reversed. Original results reported more fully in Curtis (1978b).

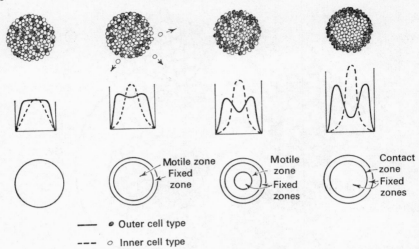

Fig. 4. A diagram to explain the sorting out of aggregates in terms of the interaction modulation theory. The top row shows a summary of the experimental observations during sorting-out. The middle row shows how the two cell types establish gradients of their IMFs whose initially different slopes are determined by the different diffusibilities of the IMFs. These gradients then show positive feedback to the final forms as a result of the movement of cells to their sorted out positions. Bottom row shows main areas of cell movement during sorting out on this hypothesis. In any region where there is a considerable difference in concentrations of the two IMFs, cells of the type opposite to that producing the greatest concentration of IMF will become less adhesive and will leave that region.

THE INTERACTION MODULATION SYSTEM IN LYMPHOCYTE RECIRCULATION AND POSITIONING

The lymphoid system in higher vertebrates forms an almost ideal testing-place for the interaction modulation hypothesis. There is the advantage that a considerable number of observations and experiments have been made on this system so that there is already a body of data which can be examined in the light of the hypothesis. The papers by Ford, Andrews & Smith (this volume, pp. 359–92) and de Sousa (1973 and this volume, pp. 393–409) provide many of the basic facts and some of the hypotheses about the system. Added to this is the fact that lymphocytes are available in relatively large quantities, that there are fairly easy methods for their characterisation and tracking and that they show several types of possible positioning reaction (see below).

Positioning selection might take place at or in any of the following situations.

(1) The selection of lymphocytes to attach to the specialised endothelium found in lymph nodes (high endothelium, or postcapillary

venule endothelium (a redundant name)) which is the main site at which lymphocytes emigrate from the blood stream. In addition the adhesion of lymphocytes to endothelium in other sites should be considered because this may show a selectivity.

(2) Release of lymphocytes from the outer side of the endothelium into surrounding organs. Surprisingly little thought seems to have been given to this as a possible source of selectivity.

(3) The sorting out of the lymphocytes into T- and B-dependent areas in spleen and lymph nodes, which may resemble the positioning found in sorting out in aggregates.

(4) The release of lymphocytes from the spleen, nodes and other organs, mainly into the lymph.

The sum total of these four processes in any one organ when compared with that in other organs can be looked at as organ routing, 'ecotaxis' or traffic, weakly termed 'homing' if you wish to be columbomorphic. It should be noted that organ routing is little understood even at the phenomenological level, for example there are many studies of the number of labelled cells which have been trapped by a given organ a stated number of hours after injection but we do not know whether this has been accompanied by any change in either the input or output of unlabelled cells from that organ nor do we know whether blood flow conditions may not have altered the chance that the cells will have of accumulating in that organ. Happily Ford, Sedgley, Sparshott & Smith (1976) have shown in some situations that treatments of lymphocytes may greatly alter blood flow.

Thus there are considerable opportunities for studies on lymphocyte positioning of which my colleagues and I have begun to take advantage. This section reviews this work and reports new results.

Perhaps one of the most important observations made in recent years on lymphocyte recirculation is that of Ford & Simmons (1972) who showed that the fast or slow recirculatory property of a lymphocyte is not an intrinsic property of that cell. This observation suggests that recirculation speed is either a random process or that it may be controlled by interaction with other cells. Thus it is worth considering whether lymphocytes interact by an interaction modulation system.

The initial findings made by Maria de Sousa and myself were that (i) mouse, rat and human thymocytes, and B cells produce factors that diminish the adhesiveness of the opposite cell type; (ii) peripheral T cells produce a factor probably identical with that produced by the thymocytes; (iii) B and T lymphocyte adhesion is non-specific when the cells are well

washed; (*iv*) addition of say B factor to a mixture of B and T cells produces
an apparent specificity of B cell adhesion; and (*v*) in the presence of both
factors adhesion is almost totally abolished so that the cells appear to be
non-adhesive (Curtis & de Sousa, 1973, 1975 *a*, *b*). It is well known that
lymphocytes prepared from lymph nodes by gentle dispersal are almost
totally non-adhesive even though they may localise and attach to organs
on re-injection. This suggests that the cells can easily recover some degree
of adhesiveness, even for other lymphocytes, when reinjected; for example
in the spleen or lymph nodes. We can now explain this finding by suggesting that when the node or spleen is teased, considerable quantities of
both B and T factors are released which diminish the adhesiveness of all
the cells.

In this initial work we also defined units of activity of the two factors
and showed that the factors were without effect on the adhesiveness of the
cell type from which they were derived. The mouse *CBA/ca* strain T
factor has an activity of 1×10^4 units if it just reduces the adhesiveness of
1×10^6 cells to zero, while the B factor has an activity of 2×10^4 units if it
reduces the adhesiveness of 1×10^6 T cells to zero. The dose response
curves can be defined by the relationships, *depression of adhesion is linearly
proportional to log activity of factor*, which when plotted linearly against
concentration are typically of the form of adsorption isotherms.

Biochemical properties of lymphocyte IMFs and site of action

Curtis & de Sousa (1975 *a*, *b*) also reported methods of preparing the
IMFs and some of their properties. They are prepared by incubating T
or B lymphocytes at relatively high population density (*c.* 1×10^7 ml^{-1})
in RPMI or Hank's Hepes 199 medium for 1 hour. The viability of the
cells at the start and end of the culture period is carefully checked and
should be above 85 %. At this point it is appropriate to digress to comment
that I and my colleagues feel that actively metabolising healthy cells are
more likely to produce products important in say positioning, than dying
cells such as those cultured in the absence of serum for prolonged periods,
which is the method of choice used by those preparing ligands (see Lilien,
1969). The cells are then removed by centrifugation and smaller particulate matter is removed by sterile filtration. These crude factors are
then purified by a molecular sieving process using Amicon filters so that the
active fraction contains species of molecular weight between 1000 and
12000 dalton, and is somewhat concentrated. These fractions are then run
on a P30 Biogel column in Hepes buffer, 0.01 M, pH 7.4. Using calibrated
columns the molecular weight of the B factor is 3000 dalton and the T
factor 9000 dalton.

Miles Davies and I (Curtis & Davies, in preparation) have shown that the activity of the T and B factors is destroyed by exposure to heat or to insolubilised trypsin or carboxypeptidase (Enzite forms of these enzymes), and that the T IMF fraction is mainly if not entirely protein as assayed by the Lowry method. The most active preparations at present have an activity of some 900 units per μg protein for the T IMF and 200 units per μg protein for B IMF.

Antibodies can be prepared against the factors in rabbits when injected subcutaneously with Freund's adjuvant. Use of these antibodies in a double antibody system with a Fluorochrome (tetramethyl rhodamine) attached to the second antibody, allows detection of cells which have bound the appropriate IMF. Using this system we can show that thymocytes have low levels of the T IMF at their surfaces and that target cells such as B lymphocytes exposed to T IMF bind this molecule. Prolonged washing removes the T IMF from other cell types. Certain types of cells, such as erythrocytes, appear to be unable to bind these modulation factors. Allogeneic or even xenogeneic cells will bind the T IMF from *CBA/ca* mice.

The observations described above on the loss of demonstrable IMF from cells on washing, and the recovery of adhesiveness on washing suggests either that the binding is weak and reversible or that the bound molecule is turned over or destroyed by the receptor cell. One matter of considerable speculative interest is the nature of the receptor. Possibly the receptors are the so-called ligands described for tissue cells by Lilien and others (see Lilien, 1969) though I shall present an alternative speculation later.

The molecular weight determinations, the sensitivity to heat inactivation and the origin from thymocytes and T cells argue that the T IMF is not thymosin. Dr D Katz was kind enough to supply samples of his AEF factor (see Katz & Armerding, 1975) which appears on test to lack any of the features of an interaction modulation factor.

Distribution and possible action of IMFs in vivo

The production of antibodies to IMFs has allowed mapping of the distribution of the T IMF in the mouse (see Table 3). I must point out that at the moment there are two ways of looking at this Table and that we have not fully resolved the answer to the two possibilities, which are (*i*) that any tissue which shows a positive reaction is a producer of T IMF, and (*ii*) that most tissues that show positive reactions simply have the receptors and the opportunity to be exposed to T IMF *in vivo*. Tracing the disappearance of T IMF in a mouse treated with ALS, to kill T cells, provides a test between the two possibilities. Incomplete results at present suggest that the thymocytes and peripheral T cells are the only producers of the factor and that other cells which appear positive in the normal mouse simply bind this interaction modulation factor.

Table 3. *Presence of T IMF antigen in* CBA/ca *mice*

Thymocytes	Weakly positive on their surface: strong internally.
Peripheral T cells	Weakly positive on their surface: strong internally.
Peripheral B cells	Weakly positive: negative internally.
Lymph nodes, spleen	All lymphocytes at least weakly positive. T-dependent areas strongly positive (in sectioned material). Stromal cells negative.
Macrophages	All positive.
Leucocytes	All positive.
Erythrocytes	All negative.
Cerebral hemispheres	Many non-glial cells positive, all glia negative.
Cerebellum	Glial cells negative, peduncular areas strongly positive.
Eye	Rods and cones positive, all other layers including sclera and choroid negative.
Liver	Very occasional positive cells (? lymphocytes, macrophages).
Kidney	Occasional positive cells in glomeruli.
Heart	All components negative.
Lungs	Alveolar surface positive, remainder negative.
Intestine	Villi weakly positive.
Skin	Epidermis negative, occasional positive cells in dermis.

The distribution shown in Table 3 has two important features. First, the number of cell types that show appreciable binding are few, and predominantly located in the nervous system or in possible sites for lymphocytes. Second, each of the other cell types that shows binding has been implicated in some way with lymphocyte distribution. There appears for example to be an association between thymocytes and degenerative diseases in the cerebellum. Again macrophages and lymphocytes are believed to be closely involved in interactions.

Both the small intestine and the lungs are common sites for lymphocyte routing. The rare occurrence of positive reactions in other tissues can probably be explained on the grounds that these are rare lymphocytes. Thus those tissues that show presence of the T IMF antigen are ones particularly associated with lymphocyte circulation and the density and ubiquity of labelling argues that all the cells in those sites carry the antigen. Preliminary experiments on the question of whether the T IMF antigen disappears from these sites in an ALS treated mouse suggest that this does indeed occur.

The sections of the thymus, spleen and lymph nodes (see Plate 1) are of interest in that they suggest that stromal components do not bind T IMF. Thus the question can be raised as to whether stromal–lymphocyte interactions are affected by the interaction modulation factors.

We have carried out two types of in-vivo experiment. In the first the release of lymphocytes into the blood, presumably mainly from spleen and lymph nodes, has been followed after an intravenous injection of T IMF. In the second the effect of T IMF on the distribution of injected labelled lymphocytes has been examined.

Miles Davies and I injected *CBA/ca* mice with 2×10^4 units of T IMF and collected blood samples at two hours subsequently, to discover lymphocyte proportions and total numbers in the blood: results are given in Table 4.

Table 4. *Effect of injection of T interaction modulation factor on blood lymphocytes and leukocytes 2 h after injection of factor*

		Control	Experimental	n	t-tests
White blood-cell count $\times 10^{-6}$	mean	2.92	4.03	10, 10	$t = 2.7$
	S.D.	0.71	1.07		$P < 0.02$
Percentage lymphocytes	mean	35.3	36.9	10, 8	0.03
	S.D.	5.7	8.1		
Percentage polymorphs	mean	63.5	63.3	10, 8	0.05
	S.D.	5.9	8.4		
Percentage IgG-positive lymphocytes		34.9	64.5	2*, 2*	

Column headed n gives number of individual animals sampled, control value first, experimental second, except for figure with * which is derived from samples of blood pooled from groups of five animals. Factor injected by tail vein. Saline injection in controls. Data will be described more fully by Davies & Curtis (in preparation).

This shows that T-factor injections increase the total load of polymorphs and lymphocytes in the blood, with a disproportionate increase in B cells. It could be argued that this result would be expected if the main effect of the factor was to reduce the adhesion of stromal cells for lymphocytes. A parallel experiment carried out by Curtis & de Sousa (1975a) showed that T-factor injections stimulated release of lymphocytes from the lymph nodes of *nu/nu* mice. Unfortunately the mice used were not of the same strain type as those which were the source of the T IMF. In the next section we shall see that there is an allogeneic effect superimposed on any intertissue effect. These experiments suggest that release of cells from lymph nodes or spleen may be due to the effect of T and presumably B

IMF, not only on each other's adhesiveness, but also on adhesion to the stroma of the nodes.

Bell & Shand (1975) reported that injections of T cells into a reconstituted rat in excess amounts tended to prevent T:B co-operation required for an antibody response to an antigen. It can now be seen that this result might be expected if the IMFs produced by the T cells suppressed the adhesion of themselves to the B cells. With this experiment in mind, Miles Davies and I (Davies & Curtis, unpublished observations) looked at the effects of injections of thymocytes on the localisation of labelled normal lymph node cells. Results are shown in Table 5.

Table 5. *Effects of lymphocyte traffic if excess thymocytes are added to the system*

CBA/ca mice. Results expressed as ↑ if a statistically significant increase, ↓ for a statistically significant decrease.

Tracer cells	Thymocyte injection	Organ	Effect
Lymph node 3×10^6 cells	3×10^7 cells	Spleen	↑
Spleen B cells, 3×10^6	2×10^7	Peripheral lymph nodes	↓

Tracer cells labelled with ^{51}Cr, injected by tail vein. Localisation after 4 h, $n = 3$. Other organs sampled were liver, lungs, blood, bone marrow, small intestine, kidneys.

When labelled B cells were injected with the thymocytes there was a significant decrease in lymph node routing by comparison with the B cell routing in control animals. Similarly, injection of about 7000 units of T IMF together with the B cells led to a similar change in routing (Table 6). If lymph-node cells (approximately 70 % T in type) were used as the labelled tracers, both thymocyte cell or T IMF injections raised spleen accumulation. These results are still very preliminary and have only been carried out for one time interval after injection, 4 h, but they clearly show that an interaction modulation factor can have considerable in-vivo effects and that the injection of the cell type that produces the factor has the same effects as factor injection. The changes in routing are easily explained if it is assumed that T IMF reduces stromal adhesion for lymphocytes in the lymph nodes.

Allogeneic effects: H-2 *control of interaction modulation factors*

Do allogeneic combinations of interaction modulation factors and cells have effects on adhesion? Simple experiments in which T or B cells from strains such as *Balb/c* was applied to B or T cells from *Cba/ca* mice

PLATE I

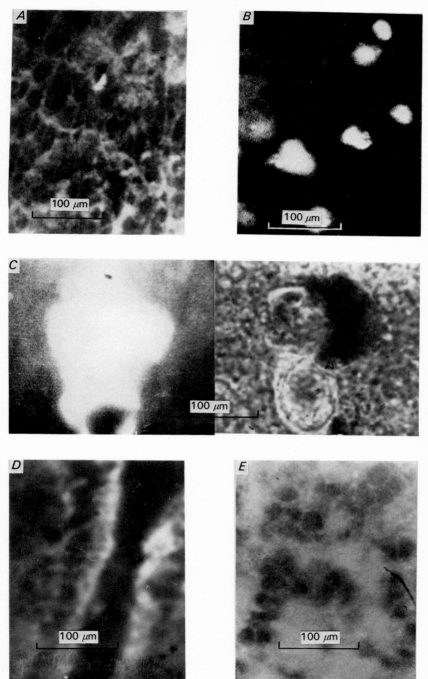

For explanation see p. 82

(*facing p.* 72)

Table 6. *Effects on lymphocyte traffic after injection of thymus IMF*

CBA/ca mice. Results expressed as ↑ for a statistically significant increase, and as ↓ for a statistically significant decrease.

Tracer cells	T factor injection	Organ	Effect
Lymph node cells	7000 units	Spleen	↑
		Liver	↑
		Peripheral lymph nodes	↓
		Mesenteric lymph nodes	↓
Spleen B cells	7350 units	Peripheral lymph nodes	↓
		Mesenteric lymph nodes	↓

Tracer cells labelled with ^{51}Cr and injected together with T IMF by tail vein. Localisation after 4 h, $n = 3$. Other organs sampled were lungs, blood, bone marrow, small intestine and kidneys.

The experiments on which this table and the preceding one are based, were primarily carried out by Miles Davies; a fuller report is in preparation (Davies & Curtis, in preparation).

showed that effects on adhesion were larger than expected from the activity of the factor assayed on its own strain type of cell. This led to a second question. Will an IMF derived from a given tissue of one strain type modify the adhesion of cells from the same tissue of a different strain type? Simple experiments show that this is so (see Table 7) provided that the target cells and those producing the IMF are of different histo-compatibility type. Although not enough different combinations of strain types have been tested, the effects seen between congenic strains show that this allogeneic effect is *H-2* based, and that it is probably linked to *H-2 D*.

The *H-2* basis of the T IMFs can be detected in another manner. If the lysis of thymocytes by antibodies against T IMFs in the presence of complement is examined (see Table 8) it is clear that lysis only takes place if the thymocytes are allelically identical at *H-2 D*. Again this work has been somewhat hampered by a lack of congenic strains but there is no combination of T IMF and target cell which shows lysis unless they are matched with respect to *H-2 D*.

It can of course be argued that the component responsible for the allogenic effect is different from that involved in the cross-tissue effect until it is established that IMFs have been purified to a single molecular species but it is of considerable interest that the cross-tissue type effects of IMFs may be linked to the major histocompatibility locus. Bennett (1975) has

6

Table 7. H-2 *relationship of IMFs; allogeneic effects of IMFs on adhesion of thymus cells*

Target cell strain type	IMF origin	Adhesion	H-2 mismatch	Congenic
Bᴛo.A,A	Bᴛo.A,A	13.8	None	—
	Bᴛo.A(2R)	0.0	D only	Yes
	Bᴛo.A(4R)	2.1	All except K & IA	Yes
	A/WySn	4.8	None	No
Bᴛo.A(2R)	Bᴛo.A(2R)	27.4	None	—
	Bᴛo.A,A	0.0	D only	Yes
	Bᴛo.A(4R)	13.9	IB, IJ, IE, IC, S & G	Yes
Bᴛo.A(4R)	Bᴛo.A(4R)	11.8	None	—
	Bᴛo.A(2R)	14.9	IB, IJ, IE, IC, S & G	Yes
	Bᴛo.A,A	0.0	All except K & IA	Yes
Bᴛo.S(7R)	Bᴛo.S(7R)	8.4	None	—
	Bᴛo.T(6R)	8.3	All except D	Yes
Bᴛo.G	Bᴛo.G	10.7	None	—
	Bᴛo.AKM	11.0	All except D	Yes
	AQR	0.0	All except K	No
A/WySn	A/WySn	9.4	None	—
	AQR	8.7	K	Yes
	Bᴛo.AKM	0.03	IC, S, G & D	No
	Bᴛo.A(5R)	9.0	K, IA & IB	No

Adhesion measured as collision efficiency percentages. Measurements made with IMF yields of *c*. 1×10^5 cells on 1×10^6 cells. Haplotype compositions compared at K, IA, IB, IJ, IE, IC, S, G and D.

speculated that the T-locus in the mouse may play a major role in cell–cell recognition at very early stages of development. The work reported above opens the possibility that cell–cell recognition involved in positioning may be *H-2* linked in the mouse. Katz (this volume, pp. 411–27) reviews some of the other cell–cell recognition phenomena which are related to the MHC locus (see also Katz & Armerding, 1975). McLaren (1976) discusses experiments on chimaerae constructed from mouse eggs of differing *H-2* type at blastocyst stages in which no particular evidence for allogeneic effects was found, but it can be argued that the two strain types would become tolerant to each other before the stage at which these allogeneic reactions might arise. Basically the problem of the intertissue effects of the IMFs within one genotype is the problem of how the prime differentiation of cells is specified. Technically it is the problem of determining which

Table 8. *Immune cytolysis of thymocytes by anti-T IMF antibodies*

Antibody against	Target cell strain	Cytolysis (%)	*H-2* mismatch
CBA/ca	CBA/ca	94.0	None
	B10.AKM	0.0	D
	B10.A(4R)	0.0	All except K & IA
	B10.A(2R)	1.8	IC, S, G & D
	AQR	21.0	K, IE, IC, S, G & D
C57 BL10/ScSn	C57 BL10/ScSn	69.0	None
	B10.AKM	11.0	All
	B10.A(2R)	77.2	All except D

Cytolysis expressed as maximum cytolysis with a range of antibody concentrations less background cytolysis. Background cytolysis in no case exceeded 12 % of the total number of cells.

tissue specific antigens are of importance in determining the separateness of the different tissues.

SPECULATIONS ON POSSIBLE FUNCTIONS OF IMFs

The experimental evidence described above shows that administration of T interaction modulation factor can alter the routing of lymphocyte traffic. It is possible to speculate that the natural production of this factor and of the B interaction modulation factor ensure both that there is a separation of B and T cells within nodes and spleen and that they reduce adhesion to stroma so that the cells are released from these organs into the lymph. We still do not know for example whether the drop in lymph node localisation after T IMF injection represents a reduced capture of cells or a much increased release. The discovery by Stamper & Woodruff (1977), that lightly fixed sections of lymph node will bind lymphocytes preferentially at the high endothelium could be explicable if other areas release IMFs or have had their adhesiveness lowered by them. Sprent (1973) found that lymphocyte recirculation in the B *nu/nu* mouse was some six times slower than that in the normal mouse: this could be explained on the grounds that the lack of T IMF in these animals will prevent reduction of B-cell adhesiveness so that they will tend to remain in the nodes. Cahill, Frost & Trnka (1976) report that antigen stimulation leads, after an initial fall, to a large increase in lymphocyte output from nodes in sheep. This might be explained on the interaction modulation hypothesis if antigen stimulation leads to increased production of IMFs. Curtis & de Sousa (1975 a, b) found evidence for such an effect in antigen stimulated mice.

It is also possible that T IMFs will tend to prevent T:B cell cooperation so that IMFs will tend to be suppressor agents. Partial doses of ALS which reduce the T cell population are known to increase T:B cooperation and would be expected to reduce T IMF levels if only because of the reduction in cell number.

Cahill, Poskitt, Frost & Trnka (1977) have shown that there are probably two separate populations of recirculating T lymphocytes in the sheep, one of them being mainly routed through the intestinal lymph organs. This type of result shows that the system, if based on IMFs, must be rather more complex than the relatively simple scheme outlined above, but Freitas, Rose & Parrott (1977) do not find such populations in the mouse. It is also of interest that Evans & Davies (1977) found subpopulations of thymocytes of differing adhesiveness which have different routing. This result underlines the probable importance of adhesive differences in routing and the possibility that these subpopulations have different responses to IMFs.

I have already suggested (see above) that one of the main roles of IMF production may be to prevent infiltration of one tissue by another. Thus we would expect to find IMFs in the body fluids of adult animals, and we have in fact detected them in the sera of mice and humans. It might be expected that there could be diseases in which cells are misplaced. Mycosis fungoides appears to be a disease in which T cells tend to localise in the dermis. It is of interest that patients suffering from this disease show, in double-blind tests, very high levels of T IMF in the sera. Again in chronic lymphatic leukemia there are abnormally low levels of both T and B IMFs in the sera (MacKie & Curtis, in preparation). Thus we can speculate that there may be mispositioning diseases, in some of which like CLL the ability to produce IMFs is lost, others in which perhaps the ability to respond to IMFs from other tissues vanishes, and some perhaps in which the diseased cells acquire the ability to produce IMFs typical of other tissues. Each of these classes of defect might correspond to some type of malignant invasiveness. Fidler (1973) was able to select clones of tumours with different metastatic siting: could this represent selection of tumours with different IMF production?

Again considering the apparent stability of cell position in the adult solid tissues, it is worth noting that the tissues are open to infiltration by white blood cells and since these appear to be producers of IMFs it is tempting to suggest that infiltration may depend on local reduction of adhesiveness by the IMFs. It is also tempting to suggest that the latent period observed before cells emigrate from explants of solid tissue in culture may simply represent the time required for penetration of IMFs present in the serum into the tissue in order to reduce adhesiveness and increase motility.

THE BASIC NATURE OF RECOGNITION PROCESSES

We can look at the question as to whether the recognition processes involved in positioning are of like by like or of unlike. In this symposium Katz argues cogently that the recognition processes involved in the immune response are a positive recognition of self. He instances the genetic restrictions on T cell cooperation with B lymphocytes as well as cytotoxicity reactions. These phenomena can be interpreted as suggesting that the reactions are of two like gene products on different cell types rather than the complementary ligand concept discussed by Burger in this symposium. Katz points out that it is possible to envisage histocompatibility molecules acting both as homologous as well as complementary reacting molecules. It should be pointed out that homologous pairing etc of molecules is of course widely present in biochemical systems, for instance those enzymes that assemble from monomers, e.g. lactic dehydrogenase. However the various cells that recognise each other in the immune system are differentiated as different cell types and so may have unlike surface components.

In the mating-type systems recognition is apparently of the complementary type but the nature of the process is more uncertain in other systems. Homologous recognition systems have not yet been clearly identified at the molecular level. Another system, which does not seem to have been suggested previously, combines features of complementary and of homologous mechanisms. Suppose that the recognition products produced by a given cell type interact with the surfaces of many other types of cell in a non-specific manner. The only method by which the cell escapes damage by its own products is to secrete the product with a masking sequence on it so that the compound is inactive. For example, many proteases, e.g. pepsin, are secreted with additional peptides attached which suppress the proteolytic activity. These peptides are removed, often in an autocatalytic manner, outside the cell (see Neurath (1975) for details of such enzymes). The relatively high concentration of masking peptide present near the producer cell tends to suppress enzyme activity. This model suggests that though the recognition effect is on unlike cells, the actual recognition event is the suppression of activity by the matching peptide in the vicinity of the producer cell.

In higher plants (Heslop-Harrison, 1975) the pollen–stigma reaction within one species permits fertilisation of allelically unlike individuals and prevents self-fertilisation. It also probably acts to prevent fertilisation by distant species. Does pollen-tube penetration of the allotypic stigma occur because a recognition reaction has taken place between unlike or because a reaction between like has failed to take place? Since a new mutant allelotype in the mating system, in for instance *Oenothera organesis* (see Lewis, 1952), is able to fertilise or be fertilised by all other allelotypes, it seems likely that recognition must be based on a self-inhibition system acting when like meets like. It is interesting to speculate as to whether the phenomena described by Katz for the immune system in this symposium are of the same type. In the immune system there are recognition reactions involving the question as to whether the cells come from different *H-2* genotypes, which is discussed by Katz later in this symposium volume. There are also I suggest, recognition reactions between different cell types within the same organism. This second type of recognition occurs predominantly in positioning reactions. This raises the question of the genetic basis of positioning reactions and of the possibility that the gene products of *H-2* involved are the prime ones by which tissue type is specified.

If recognition is of unlike, it is necessary to suppose that very large numbers of different complementary molecular pairs can be produced in different pairs of cell types. Not only does this require the production of very large numbers of alleles but it also requires the parallel selection of pairing molecules in genetically unlike strains. Even the theory that recognition is of like by the action of reciprocal receptor sites and 'ligands', present together on a given cell of one type, has considerable genetical problems. For instance if we explain the retino-tectal patterning in terms of a specific adhesion of complementary molecules which specifies each connection, an enormous number of alleles have to be postulated together with a stable geometrical control system for controlling the expression of these alleles. The explanation is considerably more complex than that required to account for the diversity of antibodies.

The mechanism I have described above, which proposes that the recognition event is the homologous suppression of a reaction by the inactivating peptide from the enzyme–inhibitor complex has several genetical advantages over rival proposals. First, selection has only to operate on one locus, which specifies the sequence of the effector molecule and the peptide which is attached to it in the apo-form. Second, if we suppose that the changes produced by the effector molecule are relatively non-specific, that they can be concentration dependent, and that gradients of these molecules can be set up then it is necessary only to postulate a small number of types of effector molecule to explain positioning.

If the action of such a system within an organism is considered, slight changes have to be made to the model. Each cell type which shows a definite positioning must express a different enzyme. In effect these substances will be one of the prime expressions of cell differentiation. Multiple allelic forms of enzymes are likely to occur and these will be seen to have effects when allogeneic combinations of tissue are made either naturally or experimentally.

REFERENCES

ABERCROMBIE, M. & HEAYSMAN, J. E. M. (1954). Observations on the social behaviour of cells in tissue culture. II. 'Monolayering' of fibroblasts. *Experimental Cell Research*, 6, 293–306.

ADLER, J. (1966). Chemotaxis in bacteria. *Science*, 153, 708–16.

ANTONELLI, P. L., ROGERS, T. D. & WILLARD, M. A. (1973). Geometry and the exchange principle in cell aggregation kinetics. *Journal of Theoretical Biology*, 41, 1–22.

BARBERA, A. J. (1975). Adhesive recognition between developing retinal cells and the optic tecta of the chick embryo. *Developmental Biology*, 46, 167–91.

BELL, E. B. & SHAND, F. L. (1975). Changes in lymphocyte recirculation and liberation of the adoptive memory response from cellular regulation in irradiated recipients. *European Journal of Immunology*, 5, 1–7.

BENNETT, D. (1975). The T-locus of the mouse. *Cell*, 6, 441–54.

BÜÜLTJENS, T. E. J. & EDWARDS, J. G. (1977). Adhesive selectivity is exhibited *in vitro* by cells from adjacent tissues of the embryonic chick retina. *Journal of Cell Science*, 23, 101–16.

CAHILL, R. N. P., FROST, H. & TRNKA, Z. (1976). The effects of antigen on the migration of recirculating lymphocytes through single lymph nodes. *Journal of Experimental Medicine*, 143, 870–88.

CAHILL, R. N. P., POSKITT, D. C., FROST, H. & TRNKA, Z. (1977). Two distinct pools of recirculating T lymphocytes: migratory characteristics of nodal and intestinal T lymphocytes. *Journal of Experimental Medicine*, 145, 420–8.

CURTIS, A. S. G. (1962). Pattern and mechanism in the reaggregation of sponges. *Nature, London*, 196, 245–8.

CURTIS, A. S. G. (1967). *The Cell Surface: its Molecular Role in Morphogenesis.* Logos Press, London.

CURTIS, A. S. G. (1970a). On the occurrence of specific adhesion between cells. *Journal of Embryology & Experimental Morphology*, 23, 253–72.

CURTIS, A. S. G. (1970b). Re-examination of a supposed case of specific cell adhesion. *Nature, London*, 266, 260–61.

CURTIS, A. S. G. (1974). The specific control of cell positioning. *Archives de Biologie*, 85, 105–21.

CURTIS, A. S. G. (1976). Le positionnement cellulaire et la morphogenèse. *Bulletin de la Société Zoologique de France*, 101, 1–9.

CURTIS, A. S. G. (1978a). Individuality and graft rejection in sponges: A cellular basis for individuality in sponges. In *Biology and Systematics cf Colonial Organisms*, ed. B. Rosen. Systematics Association (in press).

CURTIS, A. S. G. (1978b). Cell positioning. In *Receptors and Recognition*, vol. 3. Chapman & Hall, London.

CURTIS, A. S. G. & DE SOUSA, M. A. B. (1973). Factors influencing adhesion of lymphoid cells. *Nature, New Biology, London*, **244**, 45–7.

CURTIS, A. S. G. & DE SOUSA, M. A. B. (1975 a). Lymphocyte interactions and positioning. I. Adhesive interactions. *Cellular Immunology*, **19**, 282–97.

CURTIS, A. S. G. & DE SOUSA, M. A. B. (1975 b). Lymphocyte interactions and cell adhesion. *Behring Institute Mitteilungen*, **57**, 76–8.

CURTIS, A. S. G. & VAN DE VYVER, G. (1971). The control of cell adhesion in a morphogenetic system. *Journal of Embryological & Experimental Morphology*, **26**, 295–312.

DE SOUSA, M. (1973). The ecology of thymus-dependency. In *Contemporary Topics in Immunobiology*, vol. 2, eds. A. J. S. Davies & R. L. Carter, pp. 119–36. Plenum, New York.

DIPASQUALE, A. & BELL, P. R., jr. (1974). The upper cell surface: Its inability to support active cell movement in culture. *Journal of Cell Biology*, **62**, 198–214.

EDELSTEIN, B. B. (1971). Cell specific diffusion model of morphogenesis. *Journal of Theoretical Biology*, **30**, 515–32.

EVANS, C. W. & DAVIES, M. D. J. (1977). The influence of cell adhesiveness on the migratory behaviour of murine-thymocytes. *Cellular Immunology*, **33**, 211–18.

FIDLER, I. J. (1973). Selection of successive tumour lines for metastasis. *Nature, New Biology, London*, **242**, 148–9.

FORD, W. L. & SIMMONS, S. J. (1972). The tempo of lymphocyte recirculation from blood to lymph in the rat. *Cell & Tissue Kinetics*, **5**, 175–89.

FORD, W. L., SEDGLEY, M., SPARSHOTT, S. M. & SMITH, M. E. (1976). The migration of lymphocytes across specialized vascular endothelium. II. The contrasting consequences of treating lymphocytes with trypsin or neuraminidase. *Cell & Tissue Kinetics*, **9**, 351–61.

FREITAS, A. A., ROSE, M. L. & PARROTT, D. M. V. (1977). Murine mesenteric and peripheral lymph node; a common pool of small T cells. *Nature, London*, **270**, 731–3.

HARRIS, A. K. (1976). Is cell sorting caused by differences in the work of intercellular adhesion? A critique of the Steinberg hypothesis. *Journal of Theoretical Biology*, **61**, 267–85.

HENSSEN, A. & JAHNS, H. M. (1974). *Lichenes. Eine Einfuhrung in die Flechtenkunde*. Georg Thieme Verlag, Stuttgart.

HESLOP-HARRISON, J. (1975). Incompatibility and the pollen stigma interactions. *Annual Reviews of Plant Physiology*, **25**, 403–25.

KATZ, D. H. & ARMERDING, D. (1975). Evidence for the control of lymphocyte interactions by gene products of the *I* region of the *H-2* complex. In *Immune Recognition*, ed. A. S. Rosenthal, pp. 727–51. Academic Press, New York & London.

KONDO, K. (1974). Demonstration of a reaggregation inhibitor in sea urchin embryos. *Experimental Cell Research*, **86**, 178–81.

LEWIS, D. (1952). Serological reactions of pollen incompatibility substances. *Proceedings of Royal Society of London*, (B) **140**, 127–35.

LILIEN, S. (1969). Towards a molecular explanation for specific cell adhesion. *Current Topics in Developmental Biology*, **4**, 169–93.

LOMNITZER, R., RABSON, A. R. & KOORNHOF, H. J. (1976). Leucocyte capillary migration: an adherence-dependent phenomenon. *Clinical & Experimental Immunology*, **25**, 303–10.

MCLAREN, A. (1976). *Mammalian Chimaeras*. Cambridge University Press, Cambridge.

MEYER, D. B. (1964). The migration of primordial germ cells in the chick embryo. *Developmental Biology*, **10**, 154–90.

MIDDLETON, C. A. (1973). The control of epithelial cell locomotion in tissue culture. In *Locomotion of Tissue Cells*, Ciba Foundation Symposium, 14 (new ser.), eds. R. Porter & D. W. Fitzsimmons, pp. 251–70. Elsevier, Amsterdam, London, New York.

MOSCONA, A. A. (1962). Analysis of cell recombinations in experimental synthesis of tissues *in vitro*. *Journal of Cellular & Comparative Physiology*, Suppl. 1, 60, 65–80.

NEURATH, H. (1975). Limited proteolysis and zymogen activation. In *Proteases and Biological Control, Cold Spring Harbor Conference on Cell Proliferation*, vol. 2, eds. E. Reich, D. B. Rifkin & E. Shaw, pp. 51–64.

OKA, H. (1970). Colony specificity in compound ascidians. In *Profiles of Japanese Science and Scientists*, ed. H. Yukawa, pp. 195–206. Kodansha, Tokyo.

RUTZLER, K. (1970). Spatial competition among porifera: solution by epiziosm. *Oecologia*, 5, 85–95.

RYLAND, J. S. (1976). Physiology and ecology of marine bryozoans. *Advances in Marine Biology*, 14, 285–443.

SPRENT, J. (1973). Circulating T and B lymphocytes of the mouse. I. Migratory properties. *Cellular Immunology*, 7, 10–39.

STAMPER, H. B., jr. & WOODRUFF, J. J. (1977). An *in vitro* model of lymphocyte homing. I. Characterization of the interaction between thoracic duct lymphocytes and specialized high-endothelial venules of lymph nodes. *Journal of Immunology*, 119, 772–80.

STEINBERG, M. S. (1962a). On the mechanism of tissue reconstruction by dissociated cells. I. Population kinetics, differential adhesiveness, and the absence of directed migration. *Proceedings of the National Academy of Sciences, U.S.A.*, 48, 1577–82.

STEINBERG, M. S. (1962b). On the mechanism of tissue reconstruction by dissociated cells. II. Time-course of events. *Science*, 137, 762–763.

STEINBERG, M. S. (1962c). On the mechanism of tissue reconstruction by dissociated cells. III. Free energy relations and the reorganization of fused, heteronomic tissue fragments. *Proceedings of the National Academy of Sciences, U.S.A.*, 48, 1769–1776.

STEINBERG, M. S. (1963). Reconstruction of tissues by dissociated cells. *Science*, 141, 401–8.

STEINBERG, M. S. (1970). Does differential adhesion govern self-assembly processes in histogenesis? Equilibrium configurations and the emergence of a hierarchy among populations of embryonic cells. *Journal of Experimental Zoology*, 173, 395–434.

STEINBERG, M. S. (1976). Adhesion-guided multicellular assembly: a commentary upon the postulates, real and imagined of the differential adhesion hypothesis, with special attention to computer simulations of cell sorting. *Journal of Theoretical Biology*, 55, 431–43.

TALMON, Y. & PRAGER, S. (1977). Statistical mechanics of microemulsions. *Nature, London*, 267, 333–5.

TOWNES, P. L. & HOLTFRETER, J. (1955). Directed movements and selective adhesion of embryonic amphibian cells. *Journal of Experimental Zoology*, 128, 53–120.

TWITTY, V. C. (1945). The developmental analysis of specific pigment patterns. *Journal of Experimental Zoology*, 100, 141–8.

VAN DE VYVER, G. (1970). La non confluence intraspecifique chez les spongiaires et la notion d'individu. *Annales d'Embryologie et de Morphogenèse*, 3, 251–62.

WEINBAUM, G. & BURGER, M. M. (1973). Two component system for surface-guided reassociation of animal cells. *Nature, London*, 244, 510–12.

82 A. S. G. CURTIS

WILLIS, R. A. (1952). *The Spread of Tumours in the Human Body*, Butterworth, London.
WOLPERT, L. (1969). Positional information and the spatial pattern of cellular differentiation. *Journal of Theoretical Biology*, **25**, 1–47.
WOURMS, J. P. (1972). The developmental biology of annual fishes. II. Naturally occurring dispersion and reaggregation of blastomeres during the development of annual fish eggs. *Journal of Experimental Zoology*, **182**, 169–200.

EXPLANATION OF PLATE

Distribution of the T IMF antigen in tissues of *CBA/ca* mice. Antibody to *CBA/ca* thymus interaction modulation factor prepared in rabbits. 10 μm sections of mouse tissue unfixed but frozen in liquid nitrogen and cut on a cryotome. TRITC (Tetramethyl-rhodamine isothiocyanate) conjugated goat antibody to rabbit IgG used as the second antibody. Fluorescence microscopy with either UV or with green light incident illumination.

A, thymus; *B*, axillary lymph node, near periphery; *C*, spleen, showing *left*, periarteriolar area by fluorescence, *right*, phase contrast of frozen section; *D*, intestinal mucosa; *E*, cerebellum, peduncular region, which stains very strongly.

SEX CELL CONTACT IN *CHLAMYDOMONAS*, A MODEL FOR CELL RECOGNITION*

By LUTZ WIESE AND WALTRAUD WIESE

Department of Biological Science, Florida State University,
Tallahassee, Florida 32306, U.S.A.

Sex cell contact in *Chlamydomonas*, when studied comparatively with compatible and incompatible species, provides a sensitive and highly differentiating system for the analysis of cellular recognition and of its molecular basis (see Wiese, 1969, 1974). Cell recognition is expressed in the initial step of the copulation process, an instantaneous flagella adhesion between gametic cells at their random collision. The mechanism responsible for this mating type reaction (MTR) acts highly selectively. There is no reaction between gametes of identical sex, between cells of different sex in their vegetative phase, between gametes of one and vegetative cells of the other sex, or between gametes of different species (for review see Wiese, 1969). Sex contact must be based upon a species-typical complementarity with sex-specific components. The material basis for the MTR, hypothetical mating type substances are located at the flagella tips and are produced and exposed under gametogenic conditions only. The contact capacity and its specificity must reside in the functional structure of these mating type substances and involve special ligands in mutual receptor function. Sexual isolation between incompatible species is demonstrably caused by non-matching systems of mating type substances. Our comparative studies aim to identify the molecular basis of gamete contact and of the features which condition gamete recognition manifested in the bipolarity, species specificity and phase specificity of flagella adhesion. Not only because of the access to the problem of species specificity, but also by the nature of the complementarity, the contact system in *Chlamydomonas* appears more complex than the *Hansenula* system presented by Dr Crandall in this volume (pp. 105–19).

The research on sex cell contact in *Chlamydomonas* presently focusses on the nature of the contact-causing gametic complementarity (Witman, Carlson, Berliner & Rosenbaum, 1972; McLean, Laurendi & Brown, 1974; Wiese, 1974; Bosmann & McLean, 1975; McLean & Bosmann, 1975; Martin & Goodenough, 1975; Wiese & Wiese, 1975, 1977; Bergmann, Goodenough, Jawitz & Martin, 1976; Goodenough, Hwang & Martin, 1976; Snell, 1976 *a*, *b*):

*Dedicated to Professor Dr Lothar Geitler who initiated the research on algal gamete agglutinins.

[[83]]

(1) What is the chemical nature and the functional structure of the mating type substances?
(2) How are they associated with the functional structure of the flagella membrane?
(3) What is the nature of the actual ligands and of their combining sites?
(4) How do the ligands interact and what causes their species specificity?
(5) How complex is the flagella interaction?
(6) How is the molecular basis of the contact integrated into the gametic differentiation of the *Chlamydomonas* cell and into the total course of gamete copulation?

THE TEST SYSTEM

The experimental analysis of sex cell contact in these isogamous and dioecious model organisms is facilitated by specialties of their gamete copulation. The initial contact at the flagella tips is spatially and temporally isolated and separated from subsequent steps of the copulation event. The adhesion mechanism is extremely efficient, effecting contact in part of a second at collision of compatible (+) and (−) gametes and providing a test system easily to evaluate. A range of related sexually compatible and incompatible taxa exists and permits us to study the species-specificity of sex contact. In addition, sex cell contact in these isogamous species can be studied in two different assays, in the normal event as mating type reaction between (+) and (−) gametes and, in copy form with a great experimental potential, in the isoagglutination phenomenon. In each of the species investigated by us, both gamete types shed material which possesses their sex- and species-specific contact capacity (see Wiese, 1965, 1974). The contact capacity is bound to uniformous particles (10^8 daltons) which can be isolated from each gamete type's supernatant by high-speed centrifugations and salting-out procedures. Exerting its contact capacity, each such component – when added to the opposite sex – causes an instantaneous agglutination between sexually identical gametes (Plate 1). The isoagglutinin originating from (+) gametes agglutinates (−) gametes, the (−) isoagglutinin agglutinates (+) gametes. We explain these isoagglutinations, in analogy to a virus-, antibody-, or lectin-induced hemagglutination, with a functional bi- or multivalency of the isoagglutinating principles, their ligands interacting with receptors on several complementary gametes.

THE COMPATIBILITY RELATIONS

For the analysis of the molecular basis of gamete contact and of its specificity we chose a discriminating test system with compatible and incom-

Table 1. *Gametic isolation and compatibility between dioecious taxa of* Chlamydomonas

The specificity of gamete contact is expressed in the occurrence of the MTR between gametes (upper right half of the table) and of the isoagglutination of the gamete types at the left when exposed to the isoagglutinins of the gamete types above (lower left half of the table). A = agglutination in the MTR; I = isoagglutination; o = no reaction; X = not checked.

	C. moewusii eugametos		*C. moewusii syngen I*		*C. moewusii syngen II*		*C. moewusii f. rotunda*		*C. moewusii f. tenuichloris*		*C. mexicana*		*C. chlamydogama*		*C. reinhardti*	
	(+)	(−)	(+)	(−)	(+)	(−)	(+)	(−)	(+)	(−)	(+)	(−)	(+)	(−)	(+)	(−)
C. moewusii eugametos (+)	o	A	o	A	o	o	o	o	o	o	o	o	o	o	o	o
(−)	I	o	A	o	o	o	o	o	o	o	o	o	o	o	o	o
C. moewusii syngen I (+)	o	I	o	A	o	o	o	o	o	o	o	o	o	o	o	o
(−)	I	o	I	o	o	o	o	o	o	o	o	o	o	o	o	o
C. moewusii syngen II (+)	o	o	o	o	o	A	o	o	o	o	o	o	o	o	o	o
(−)	o	o	o	o	I	o	o	o	o	o	o	o	o	o	o	o
C. moewusii f. rotunda (+)	o	o	o	o	o	o	o	A	o	o	X	X	X	X	X	X
(−)	o	o	o	o	o	o	X	X	o	X	X	X	X	X	X	X
C. moewusii f. tenuichloris (+)	o	o	o	o	o	o	X	X	o	A	X	X	X	X	X	X
(−)	o	o	o	o	o	o	X	X	X	X	X	X	X	X	X	X
C. mexicana (+)	o	o	o	o	o	o	X	X	X	X	o	A	o	o	o	o
(−)	o	o	o	o	o	o	X	X	X	X	I	o	o	o	o	o
C. chlamydogama (+)	o	o	o	o	o	o	X	X	X	X	o	o	o	A	o	o
(−)	o	o	o	o	o	o	X	X	X	X	o	o	I	o	o	o
C. reinhardti (+)	o	o	o	o	o	o	X	X	X	X	o	o	o	o	o	A
(−)	o	o	o	o	o	o	X	X	X	X	o	o	o	o	I	o

patible dioecious taxa of the clade *C. moewusii* and extended these studies, in certain aspects, to some more unrelated dioecious species as *C. mexicana*, *chlamydogama*, and *reinhardtii*. The compatibility relations, the homology between sexes in these isogamous forms, and the identical action specificity of the isoagglutinins and of the mating type substances *in situ* are presented in Table 1. The data reported here deal preferentially with the first three taxa listed. *C. moewusii eugametos* and *C. moewusii syngen I* give a complete sexual cross-reaction and must possess an identical contact mechanism. Since their zygotes give regularly viable offspring the two taxa represent together only one true species in spite of their historically conditioned different names (Nybom, 1953; Bernstein & Jahn, 1955; Trainor, 1959; Gowans, 1963; Wiese & Wiese, 1977). Both taxa are absolutely isolated from *C. moewusii syngen II* which possesses a contact mechanism of its own, and from two more taxa, *C. moewusii f. rotunda* and *f. tenuichloris*, isolated and described by Dr Tsubo (1961). The latter two taxa could not be studied comparatively since the strains maintained in algal collections have meanwhile lost their competence for sexual reproduction. Both forms, however, have to be valued as individual syngens or species since Tsubo clearly established their sexual isolation from each other and from the other taxa of *C. moewusii*. The remaining species, *C. mexicana*, *chlamydogama* and *reinhardtii*, represent true species and are sexually isolated from each other as well as from the taxa of the *C. moewusii* group. Sexual isolation between the taxa mentioned is always caused by or connected with the absence of the mating type reaction.

THE SENSITIVITY OF SEX CONTACT TO SELECTED ENZYMES AND TO CONCANAVALIN A

The precise chemical characterization of the taxon-specific functional complementarity responsible for sex contact and its specificity, demands the analysis of the actual ligands of the various sexes and species. In order to isolate the ligands, the isoagglutinins with their high specific contact capacity are to be fractionated in defined manner by means of selected enzymes and by sodium dodecyl sulphate (SDS) with the aim to separate and identify components carrying the contact sites from material inert to the proper contact function.

The isoagglutinating principles produce on hydrolysis several sugars and a rather complete spectrum of amino acids rich in threonine, serine and aspartic acid and containing hydroxyproline (Wiese, 1974). The contact capacity is furthermore sensitive to periodate under conditions specified for oxidation of carbohydrates. Finally, the $(-)$ isoagglutinins of *C. moewusii eugametos* and of *C. moewusii syngen II* contain a considerable amount of sulphate. For the defined degradation of the isoagglutinins, therefore, the effect of various proteases, glycosidases, sulphatases, and of SDS is being studied. We first investigated the action of the selected enzymes upon the mating type activity of live gametes, i.e. on the mating type substances *in situ*. Individual gamete types were incubated with the particular enzyme, combined after washing with unincubated test gametes, and checked for the resistance of their mating type activity. For experi-

Table 2. *The sensitivity of the sex contact capacity of live gametes, i.e. of the mating type substances* in situ, *to various proteases*

R = Resistant, S = Sensitive.

Chlamydomonas species

Sex ...	eugametos		moewusii syngen I		moewusii syngen II		mexicana		reinhardtii		chlamydo-gama	
	+	−	+	−	+	−	+	−	+	−	+	−
Trypsin	R	S	R	S	R	S	R	S	S	S	S	S
Pronase	R	S	R	S	R	S	R	S	S	S	S	S
Subtilisin	R	S	R	S	R	S	R	S	S	S	S	S
Thermolysin	S	R	S	R	R	R	R	S	S	S	S	S
Chymo-trypsin	S	R	S	R	R	R	R	S	S	S	S	S
Papain	S	S	S	S	S	S	S	S	S	S	S	S
Proteinase K	S	S	S	S	S	S	S	S	S	S	S	S

mental details, enzyme concentrations, application conditions, etc. see the original papers (Wiese & Metz, 1969; Wiese & Shoemaker, 1970; Wiese & Hayward, 1972; Wiese & Wiese, 1975, 1977). In the species studied, a striking sensitivity pattern emerged (Table 2).

We detected, in *C. moewusii eugametos*, a unilateral sensitivity of the MTR to trypsin, the (+) sex being resistant, the (−) sex being sensitive (Wiese & Metz, 1969). The same pattern also appeared with proteolytic enzymes known for their broader action, such as pronase and subtilisin. This unilateral sensitivity of the contact mechanism to trypsin, pronase, and subtilisin was subsequently found to extend, understandably, to the compatible *C. moewusii syngen I* with its identical contact specificity. Remarkably, however, the same unilateral sensitivity was also detected in the incompatible *C. moewusii syngen II* with a different contact-causing complementarity. In all three taxa it is the (+) sex which is resistant and the (−) sex which is sensitive. The same pattern of a protease sensitivity was detected in *C. mexicana*, in which again one gamete type is resistant, the other sensitive to a series of proteases (see Table 2). Two more species, *C. reinhardtii* and *C. chlamydogama*, exhibit a different sensitivity pattern: the capacity of both gamete types to agglutinate is eliminated by trypsin, pronase, subtilisin and some other proteolytic enzymes checked (see Table 2). Finally, two proteolytic enzymes, papain and proteinase K, were shown to incapacitate both gamete types in all species investigated.

The enzymatic incapacitation of a given gamete type may be caused in very different manner. Its actual effect has to be elaborated for each enzyme individually. One may assume three basic types of an enzymatic incapacitation of the mating ability: (1) the enzyme action may directly involve a combining site essential for the contact; (2) the enzyme may detach side chains carrying the actual unaffected contact sites; (3) the enzyme may only indirectly affect the contact capacity by interfering with the meaningful conformation of the membrane components or by altering the functional structure of the membrane itself. We search especially for enzymes with the first two effects. The interpretation of the various enzyme effects is hampered by the present ignorance of the complexity of the functional membrane differentiation involved in the mating type reaction. Analysing the individual enzyme actions and the effects of other agents on the functional structure of the membrane, as masking functional groups, periodate oxidation, etc., are expected to provide the information needed.

The remarkable resistance of the (+) mating activity to broad-spectrum proteases in the species of the clade *C. moewusii* (and in *C. mexicana*) suggested the contact capacity to be inherent to components of carbohydrate nature. Verifying this assumption we proved the MTR sensitive to Concanavalin A (CONA) (Wiese & Shoemaker, 1970). Again, the sex contact sensitivity to CONA in the taxa of *C. moewusii*, not existent in *C. reinhardtii* and *chlamydogama* and also absent in *C. mexicana*, is unilateral, the trypsin-resistant (+) sexes being sensitive, the trypsin-sensitive (−) sexes resistant to CONA. Since the mating inhibition is completely reversible by D (+) mannose and α-methyl-D-mannoside, the CONA effect upon the MTR is caused by complex formation with appropriate sugars in the membrane and not on the basis of its binding potential to polyelectrolytes (Doyle, Woodside & Fishel, 1968). The elimination of the contact capacity is not indirectly caused due to a steric hindrance by the attached CONA molecule of a proper contact site with other than carbohydrate nature. The CONA effect could be pinpointed directly to the (+) contact site. Of the sugars known to interact with CONA only mannose

Table 3. *The sensitivity of the sex-contact capacity of live gametes, i.e. of the mating type substances* in situ, *to Concanavalin A and to α-exomannosidase, contrasted to the sensitivity pattern of trypsin*

R = Resistant, S = Sensitive.

Chlamydomonas species

	eugametos		moewusii syngen I		moewusii syngen II		mexicana		reinhardtii	
Sex ...	+	−	+	−	+	−	+	−	+	−
Concanavalin A	S	R	S	R	S	R	R	R	R	R
α-exomannosidase	S	R	S	R	S	R	R	R	R	R
Trypsin	R	S	R	S	R	S	R	S	S	S

is a component of the (+) isoagglutinin (Wiese, 1974). In the assumption then held that CONA reacts only with α-glycosidically linked mannose residues *in terminal position* we checked into the effect of an α-exomannosidase, first kindly provided by Dr Li (Tulane University).

α-mannosidase (Miles Research Products, 0.1 %, 0.0025 M acetate buffer, pH 5.4, with 0.001 M $CaCl_2$, 26 °C) inactivates the gametic contact capacity of the CONA-sensitive and trypsin-resistant (+) sexes in all three taxa (Table 3). Incubated (+) gametes no longer agglutinate with untreated (−) gametes. The contact capacity of the (−) gametes is not affected. α-mannosidase also inactivates the (+) but not the (−) isoagglutinin in those three taxa.

The sensitivity of the MTR to this glycosidase with its very specific action proves terminal mannose residues in α-glycosidic bond essential for the contact function of the (+) sexes.

The all-decisive dependence of the (+) contact capacity on terminal mannose residues demands an equally decisive sugar-binding potential of the (−) component.

This fundamental complementarity between a carbohydrate structure with terminal mannose and a carbohydrate-binding counterpart holds true for the three taxa studied despite of the incompatibility barrier between them. There must exist additional differentiating features on both sides of this complementarity to effect the absolute species specificity of the contact, i.e. a polymorphism of the demonstrated basic bipolarity must account for the absolute sexual isolation between incompatible taxa. It is presently unknown what these differentiating features are. Information is expected from the further analysis of the ligands and from a more comprehensive understanding of the complexity of the agglutinative cell adhesion.

The sensitivity of the carbohydrate-binding component to proteases does not imply that the contact function in the (−) sex is mediated by a protein structure. In *C. m. eugametos*, the (−) contact capacity is destroyed by sodium periodate under conditions which are considered to be specific for the oxidation of saccharides. The (−) component is also sensitive to a crude preparation of glycosidases isolated from *Charonia lampas* (Miles Research Laboratories; Elkhardt, Indiana). By incubation of the (−) isoagglutinin of this taxon with individual glycosidases present in the crude extract, the (−) contact capacity was found to be sensitive to α-*N*-acetylgalactosaminidase and to α-galactosidase and resistant to β-*N*-acetylhexosaminidase, β-galactosidase, and to α-mannosidase (see Table 4). Also the sensitivity of the (−) isoagglutinin to sulphatase probably indicates the involvement of a sulphate-bound component structure of carbohydrate nature in the contact function.

7

Table 4. *The sensitivity of the* (−) *isoagglutinin of* C. m. eugametos *to glycosidases, sulphatase and sodium periodate*

	Control	Experiment	Application
Charonia crude extract	128000	0	2 %, 0.0025 M Acetate buffer, pH 5.2, 26 °C
α-N-acetylgalactosaminidase	8000	0	0.1 %, 0.0025 M Acetate buffer, pH 5.2, 26 °C
β-N-acetylhexosaminidase	8000	8000	0.1 %, 0.0025 M Acetate buffer, pH 5.2, 26 °C
α-galactosidase	8000	0	0.1 %, 0.0025 M Acetate buffer, pH 5.2, 26 °C
β-galactosidase	8000	8000	0.1 %, 0.005 M Phosphate buffer, pH 7.0, 26 °C
α-mannosidase	8000	8000	0.1 %, 0.0025 M Acetate buffer, pH 5.2, 26 °C
Sulphatase (helix)	64000	0	160 units ml⁻¹, 0.0025 M Acetate buffer, pH 5.0, 26 °C
Sodium periodate	128000	0	1 mM, 0.0025 M Acetate buffer, pH 5.0, 0 °C

Isoagglutinin suspensions, adjusted for titres of 128000, 64000 and 8000, were incubated with the agents listed, and checked for their sensitivity in comparison with untreated controls.

THE DYNAMIC NATURE AND THE TRANSITORINESS
OF THE FLAGELLA ADHESION

The discussion of the nature of the contact-causing complementarity requires consideration of some more features characteristic for the MTR: the interacting components provide for a mechanism that causes a dynamic, exchangeable, and temporary flagella adhesion (Lewin, 1952; cf. Wiese, 1969). (1) The MTR does not result, *per se* and directly, in the permanent coalescence between the two gametes. Gamete copulation in these isogamous chlamydomonads consists of two spatially and temporally separated steps, the MTR and the proper cell fusion, the pairing. (2) The flagella adhesion, as readily observable under phase contrast, is of a kind which permits easy and repeated partner exchange within clusters of more than two gametes. The complementarity does not effect a permanently interlocking system but conditions a dynamic association. (3) The MTR is limited in time, i.e. a transient event. The MTR causes the initial contact between sexually different gametes and enables one (+) and one (−) gamete to pair. In the clade *C. moewusii*, pairing starts with the fusion of two papillae at the gametes' apices between the flagella bases and leads to membrane coalescence by establishing of a protoplasmic bridge. After pairing, the flagella of both partners disagglutinate and have altered properties: they no longer agglutinate with unmated gametes of either sex (Lewin, 1952, 1954).

Analysing these features which reflect upon the nature of the contact mechanism and its developmental control we proved that (1) the capacity to agglutinate *per se* is, under gametogenic conditions, potentially unlimited and (2) the ensued pairing terminates the flagella adhesion. This regulative correlation was studied in *C. moewusii syngen II* in which the MTR proved to be bilaterally resistant (see Wiese, 1974), the pairing sensitive to the proteolytic enzymes thermolysin and chymotrypsin. The differential sensitivity of both processes permits us to eliminate the pairing without interfering with the flagella agglutination and to document experimentally the interdependence between MTR and pairing, the existence of which had to be postulated from the course of gamete copulation. Many other proteolytic enzymes also interfere with pairing but they act, in addition, on the flagella agglutination (Wiese & Wiese, unpublished results).

Mixing of (+) and (−) gametes in equal numbers (by combining aliquots of cell suspensions with adjusted cell density) results, in presence or absence of either enzyme, in an instantaneous agglutination of all gametic cells (Fig. 1). In the normal course of copulation, the first pairs appear after 3–5 min, and full pairing with more than 85 % of the cells paired

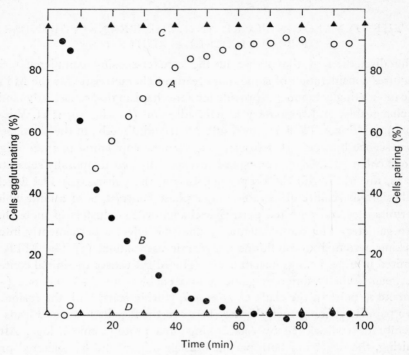

Fig. 1. Interdependence of MTR and pairing demonstrated in *C. m. syngen II*. Enzymatic elimination of the pairing by 0.026 % thermolysin causes unlimited permanence of the flagella agglutination. Curve *A* (○) illustrates pairing and curve *B* (●) agglutination in the control; curve *D* (△) pictures pairing and curve *C* (▲) agglutination in presence of thermolysin.

ensues in the next 30–45 min. The appearance of pairs is measured by counting paired against unpaired cells (curve *A*). Since flagella of paired cells no longer agglutinate and all unpaired cells remain in agglutination, the MTR declines to the degree as pairs arise (represented in the constructed curve *B*). Agglutination between sexually different cells ends after the last gamete of the sex which happened to be in minority has found its partner to pair. In presence of thermolysin or chymotrypsin, added at mixing or by mixing of preincubated gamete types, not one pair appears (curve *D*) and the entire gamete mass stays in uninterrupted agglutination (curve *C*) (see Wiese, 1969; Wiese, Wiese, & Cummings, in preparation). The enzymatic prevention of the second step, the pairing, causes an unlimited extension of the first one, the MTR, i.e. the capacity to agglutinate as such is unlimited and is, in the normal course of copulation, terminated upon a signal from the ensuing or ensued pairing. Enzyme-treated (+) and (−) gametes vehemently attempt to pair, their papillae gliding incessantly along each other. Under certain conditions just this intense trial, without actual permanent interconnection, results in the dispersion of the previously agglutinated gametes at loss of their agglutinability.

The unrestricted gametic potential to agglutinate could also be demonstrated in the isoagglutination phenomena. By constant adding of an isoagglutinin to the respectively opposite gamete type, gametes can be kept in uninterrupted isoagglutination, e.g. (+g) ametes of *C. moewusii eugametos* were isoagglutinated for 108 h when finally the assay was stopped.

The agglutinability of the gametes in *Chlamydomonas*, according to our concept anchored in a multitude of mutual receptors which by their interaction effect the dynamic contact and its specificity, expires in response to a signal produced or permitted by the papilla contact. We presently attempt to determine the nature of the perisemic effect (Reissig, 1974) and anticipate that knowledge of what reverses cell adhesion might inform on the adhesion mechanism *per se*.

The nature of flagella adhesion and the features responsible for its dynamic character are reflected in the course of the isoagglutinations. As mentioned at the outset, gametes of same sex agglutinate with each other by means of their flagella tips in reaction to the isolated contact principle, the isoagglutinin, of their opposite sex. These homotypic agglutinations possess the same specificity as the MTR (see Table 1). We study preferentially two systems: the isoagglutination of (+) gametes in *C. moewusii eugametos* and *C. moewusii syngen II* under the influence of their respective (−) isoagglutinin.

Upon addition of (−) isoagglutinin all (+) gametes immediately agglutinate. Even with high-titre preparations (64×10^6) the absolute sex-, species-, phase- and site-specificity of the causative principle is never lost. The isoagglutinations, just copying the initial contact, do not lead to pairing. After a certain time span, the homotypic agglutination ceases, the disagglutinated cells swimming around individually. In the supernatant of such disagglutinated gametes the added isoagglutinin is no longer demonstrable. Gametes emerging from isoagglutination can be re-agglutinated by newly added isoagglutinin. As mentioned before, this procedure can be repeated several times, i.e. the interaction with the isoagglutinin does not lead to saturation of the receptor sites on the part of the living cell: the responsiveness of the live cell is maintained, the added isoagglutinin, however, is inactivated in some manner.

This inactivation ensues in association with the matching complementary surface while causing isoagglutination. No inactivation results in combinations with incompatible gamete types. With complementary (+)

gametes, incapacitated by α-mannosidase, α-protease, or – in *C. moewusii syngen I* and *C. moewusii eugametos* – by thermolysin and α-chymotrypsin, no isoagglutination occurs and the (–) isoagglutinin added can be quantitatively rediscovered in the supernatant.

Preliminary experiments with ^{67}Ga-labelled (–) isoagglutinin have shown that the disappearance of the isoagglutinating activity is not due to its permanent adsorption to the (+) flagella. In suspensions of redispersed gametes the label is rediscovered in the supernatant, i.e. the inactivation of the isoagglutinating principle must have occurred during its temporary attachment to the functional complementary surface. The contact-causing molecular interaction is copied in the selective adsorption of the isoagglutinin to the complementary surface and in its subsequent membrane-bound inactivation. The contact made by carbohydrate binding is connected with the inactivation of the carbohydrate-binding component. With two live partners this inactivation does not appear, i.e. one has to assume that the capacity to agglutinate is constantly maintained as an essential part of the sexual differentiation of the gametic phase of the *Chlamydomonas* cell terminated only upon the signal by the ensued pairing (Forster & Wiese, 1954; Forster, 1957, 1959; Stifter, 1959; Hartmann, 1962).

DISCUSSION

The data presented reveal sex contact in the flagella adhesion of isogamous chlamydomonads as a rather special case. Some of the special features of the gametic differentiation are to be understood under the aspect that the gamete represents a functional state of the *Chlamydomonas* cell, inducible and reversed by environmental conditions and defined by its capacity to copulate, and that the gamete is not the endproduct of an irreversible differentiation as, for instance, egg and sperm. The molecular basis of the selective contact between compatible, sexually different gametes exhibits all the characteristics of a stereochemical complementarity some essential features of which could be elaborated in the enzyme sensitivity pattern, the carbohydrate-binding potential, the periodate sensitivity, and the function-essential role of amino groups (Wiese, 1974) and calcium ions (Lewin, 1954; Wiese & Jones, 1963).

The cell association effected is strikingly efficient, selective, and dynamic; its underlying bipolar basis provides for an unlimited agglutinability sustained under gametogenic conditions and programmed to be terminated by the ensuing papilla contact. The elaborated features of the contact mechanism do not yet permit classification of the complementarity as to its exact type of interaction. The most critical unresolved problem, we feel, is the unknown complexity of the reaction system, i.e. the lack of information on the size and nature of the hypothetical mating type substances. It is unknown as to whether they actually exist as individual molecules or represent, on both or eventually only on one gamete type, functional complexes of cooperating molecules. With one exception (Snell, 1976a) analytical separations of flagella homogenates by various teams have not yet produced evidence for sex and phase specific components.

The extraordinary dynamic nature of the flagella adhesion, manifested in the easy partner exchange, suggests a contact by easily reversible non-covalent bonds. Even though it is premature to make statements on the nature of the contact we would like to discuss how certain types of a specific bipolar complementarity fit the facts known for contact and recognition in *Chlamydomonas*.

Antigen–antibody concept

An interpretation of the demonstrated bipolar complementarity as an antigen–antibody-*like* relation between two membrane-integrated or membrane-associated components would readily explain the selectivity and efficacy of the MTR and could also, less easily, account for the constantly possible partner exchange during the flagella agglutination. The interpretation encounters two problems however. The two purified isoagglutinins of a species do not neutralize each other *in vitro* (see Wiese, 1969) and the mating type substances do not display a sex-specific antigenicity (Wiese & Baker, 1968). A causation of the intercellular adhesion – in the observed instantaneousness – by forces typical for an antigen–antibody interaction should require optimal exposure of the functional molecular groups responsible and suggest a distinct sex-specific antigenicity of the agglutinin. This is not the case. Injection of highly purified (−) isoagglutinin of *C. moewusii syngen I* into rabbits produces an antiserum which precipitates the antigen and agglutinates the (−) gametes by their flagella tips. However, the antiserum also precipitates the (+) isoagglutinin and agglutinates (+) gametes, trypsin-inactivated (−) gametes, and vegetative cells of both sexes. Cross-adsorption tests do not reveal any antigen-specific activity. The antiserum is species specific since it does not react with either gamete type of *C. reinhardtii* or *C. moewusii syngen II* (Wiese & Baker, 1968).

Both problems, non-neutralization and lacking antigenicity of the isoagglutinins, vanish if one assumes the sexual bipolarity in the MTR to be based upon a four-substance system with two pairs of complementary components. One component per gamete type may be inherent to a structure that is eventually detachable as isoagglutinin and has a receptor

of its own on the opposite sex. The combining sites of this isoagglutinin
would have to be comparable to those of an antibody with a selective contact capacity but not antigenic with respect to their specificity. Isoagglutinations would be possible by the action of only one pair of components,
the particular isoagglutinin and its complementary receptor. It still remains
the (not unsurmountable) problem that injections of gametic flagella homogenates do not produce a sex-specific antiserum either.

An interpretation of the MTR in *Chlamydomonas* by an antigen–
antibody-*like* relationship between the mating type substances should
imply that the selective adhesion is brought about by forces (electrostatic
attraction, hydrogen bonds, van der Waal forces) known to interact between
antigen and antibody, and it should not imply that one part of the complementarity is necessarily a protein. Presently, this latter relation is not
proven by the unilateral mating sensitivity to proteases like trypsin nor
disproven by the sensitivity of the $(-)$ sex to periodate and to α-galactosi-
dase and α-N-acetyl galactosaminidase. At the mentioned unknown complexity of the contact mechanism one may envisage carbohydrate–carbohydrate and carbohydrate–protein interactions, additional protein–protein
interactions being in no way excluded.

The termination of flagella adhesion after pairing is not simply due to a
dissociation between the complementary components analogous to the
splitting of an antigen–antibody complex since the disagglutinated flagella
no longer possess the capacity to agglutinate. In *Hansenula* (Crandall,
Lawrence & Saunders, 1974) and in *Saccharomyces cerevisiae* (Shimoda &
Yanagishima, 1975) the interacting agglutinins reappear in active form
after the (experimentally induced) splitting of their complex. The termination of the adhesion in *Chlamydomonas*, a regular, developmentally
controlled event, must be caused by or connected with an alteration of
the original complementarity or its exposure. The nature of the terminating
signal and the type of terminating the agglutinability is at present unknown.
The possible involvement of glycosyltransferases in this component process of gamete interaction is discussed below.

Glycosyltransferase–glycosylacceptor concept

The molecular forces causing contact in an antigen–antibody type of interaction are, in principle, not different from those effecting the binding of an
enzyme to its substrate. So far as the proper contact-making forces are
concerned, the concept of Dr Roseman (1970, 1974) that membrane-
bound glycosyltransferases of one cell interacting with glycosylacceptors
on another cell effect cell contact and recognition explains flagella adhesion
in *Chlamydomonas* as well as the antigen–antibody interpretation. The

establishment of flagella contact by formation of enzyme–substrate complexes and, moreover, the proper action of these enzymes to effect glycosylation of their receptors would suit several characteristic features of the MTR as its instantaneousness, its selectivity, its dynamic nature, and its reversibility. Even though, to our knowledge, intercellular contact by transglycosylation has not yet been reliably demonstrated in any system tested, we definitely envisage it as a possibility in *Chlamydomonas* where it could readily account for some of our data as for the decisive role of selective carbohydrate-binding, the dynamic nature of the contact, the membrane-bound incapacitation of the isoagglutinin, and for the induced termination of the MTR.

Several membrane-bound glycosyltransferase activities capable of binding added nucleotide monosaccharides have been demonstrated on vegetative cells and, in stronger concentrations, on gametes and in particles causing isoagglutination (McLean & Bosmann, 1975; Bosmann & McLean, 1975). In addition, a distinct enhancement of five of the six different enzymatic activities could be quantitatively demonstrated when (+) and (−) gametes or (+) and (−) isoagglutinins were mixed, with an especially dramatic increase of the sialyl transferase activity. No increase ensues if gametes of different species are mixed. On the basis of these data McLean & Bosmann claim that the mechanism for cell recognition–adhesion involves glycosyltransferase–glycosylacceptor systems and that the systems demonstrated act to this effect as proposed by Dr Roseman. Just for this decisive statement, however, whether flagella contact is made by the interaction of an enzyme of one with its receptor on the other gamete type, their reported data are inconclusive. The ambiguity resides in the speciality of the sexual differentiation in these chlamydomonads – that the gametic capacity to agglutinate is actively maintained, especially during the ensuing MTR. This active turnover is reflected in the enhanced appearance of isoagglutinins. (+) gametes in isoagglutination, caused by added (−) isoagglutinin, produce three to four times more (+) isoagglutinin than their untreated controls. At the demonstrated glycoprotein nature of the contact components, the maintenance of the agglutinating capacity will necessarily involve glycosyltransferases. The enhanced glycosyltransferase activities at mixing of the gametes or of the isoagglutinins may therefore not be the cause but the consequence of the contact. McLean & Bosmann failed to prove whether the enhanced enzymatic activity at the mixing of the isoagglutinins resulted in the bi- or unilateral loss of the capacity to cause isoagglutination, i.e. effected termination of the gametic agglutinability by glycosylation. At present the question as to whether glycosyltransferase–glycosylacceptor systems really effect intercellular contact in the MTR is entirely open. The strongest hint that such ecto-glycosyltransferases described by McLean & Bosmann actually may effect intergametic contact lies in our demonstration of the membrane-bound inactivation of the isolated contact component during isoagglutination. If this incapacitation of the isoagglutinin subsequent to its demon-strated sex- and species-specific adsorption to the opposite gamete type by means of sugar binding is connected with its glycosylation then, in-deed, the event copied in the isoagglutination, the sex contact in the MTR, is mediated by intercellularly acting glycosyltransferase systems.

In addition to their postulated collective role in gamete contact and recognition phenomena, individual glycosyltransferases might, according to McLean & Bosmann (1975), 'participate as messengers in initiating growth of the papilla which results in fusion of the gametes'.

Finally, there is the distinct possibility, envisaged by McLean & Bos-mann and by us, that glycosyltransferases may be responsible for the termination of the MTR. Induced by the signal from the ensuing pairing, the termination may occur directly by transglycosylation between flagella of sexually different partners or by glycosylation within each mating type or, indirectly, by alteration of the intricate functional structure of the flagella.

Lectin–substrate relation

With the demonstrated decisive role of selective carbohydrate-binding for the gamete contact, the components responsible may also be figured in a lectin–substrate interaction. Again, more detailed knowledge of the molecular basis of flagella adhesion should first be gathered in order to make such designation meaningful.

INCOMPATIBILITY AND SPECIATION CAUSED BY MUTATIONS IN THE GAMETIC RECOGNITION SYSTEMS

The taxon-specific contact mechanisms in the clade *C. moewusii* appear as modifications of a basic bipolarity common to all of them. Features speci-fying this fundamental complementarity provide for gamete recognition and account for the sexual isolation between related taxa by means of non-matching systems of mating types substances. In spite of their incomplete-ness our data suggest a simple mode of the evolution of such complex breeding systems, i.e. of the process of speciation and cladogenesis, by mutative alteration of the gametic recognition systems (Wiese & Wiese, 1977). The concept developed interprets (1) the existence of closely related but sexually isolated breeding units (syngens) within a nomenspecies, or of sibling species forming a clade; (2) the frequent occurrence of asexual or sterile strains in species which normally reproduce sexually in addition to their potential for asexual reproduction; and (3) the sporadic appearance of certain syngens or of individual sexual strains at locations geographically far apart (for details see Wiese, 1976).

According to the evidence presented, the gametic contact capacity and its specificity resides in the functional structure of membrane components which are of glycoprotein nature as expressed by their sensitivity to proteases, glycosidases, sulphatase, and periodate. Their biosynthesis and their specific molecular structure controlled by genes, such components are necessarily subjected to mutatively caused alterations. In each of the three taxa more closely investigated, gamete contact depends decisively on a carbohydrate component on the (+) gametes which, different between the syngens, is selectively recognized by a discriminating carbohydrate-binding counterpart on the (−) gametes. Any mutative alteration in the proper functional structure of either component which interferes with the matching complementarity responsible for gamete contact will result in a loss of the mating potential. At the capability of (+) and (−) gametes for parthenogenetic development and the unlimited potential of these unicells for asexual reproduction, such a mutant will not be eliminated but persists as a strain without sexual reproduction and will, upon isolation, be described as a sterile strain or as a strain without known sex partner (see Fig. 2).

Alterations which may affect the contact capacity on both sides of the complementarity are predetermined in their character by the chemical nature of the ligands. Assuming that a specific oligosaccharide chain codes for the specificity of the (+) contact function, the type, number, sequence, linkage type, and substitutions of the monosaccharides may be changed by mutation. Similar changes may alter the (−) contact capacity if its observed carbohydrate-binding potential should be a property of a carbohydrate compound (by establishing H-bonds). Should the carbohydrate binding of the (−) component involve a specific conformation of a protein or of the protein moiety of a glycoprotein, the type of possible alterations at the contact site would also be prescribed in their nature. Starting with a defined molecular complementarity, mutations will preferentially alter structural details but not the basic character of the interacting system. A mutative alteration of one side of the contact mechanism while causing incompatibility with the traditional mating partner, may create a functional structure capable of establishing a new complementarity when matched by an appropriate mutation of its erstwhile counterpart (Fig. 2; *D, G*).

A series of mutations bilaterally modifying a basic bipolar complementarity would logically account for a frequently observed incompati-

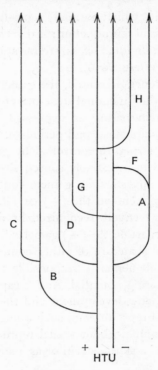

Fig. 2. Speciation by mutative alterations in the bipolar sex contact mechanism of a dioecious ancestor species (HTU = hypothetical taxonomic unit). Each sex may produce strains (A, G; B, D) incapable of reacting with their erstwhile partners. Cells of these strains may regain the original genotype by backmutations (F). Mutated strains may give rise to additional strains (C) not matching any partner, such strains will be detected only by chemical or eventually serological methods. Each sexual strain may produce strains (D, G) which, incompatible with their erstwhile partners, establish a new matching complementarity, thus creating a new species. A special case may be envisaged (H) in which a mutated strain is able to mate with two partner strains, his original complementary sex and its mutant (A). Such a situation fits reports in which two dioecious, geographically separated species can be crossbred in one but not in the reciprocal sexual combination.

bility phenomenon, as in *C. moewusii*, that a morphologically well-defined species at closer inspection appears composed of individual breeding units sexually isolated from each other. Once the gene flow within an ancestor species is interrupted by evolution of two systems of mating type substances, other genetic differences will appear, accumulate, and effect a stabilization of the individuality of the syngens produced. Individual recognition systems arising by complementary mutations on both sides of a bipolar contact mechanism are genetically and chemically well known, for instance, from the analysis of phage attachment by studying host–

range relations and phage-resistant mutants. Finally, since the character of the mutations possible is prescribed by the chemical nature of the contact components, identical mutations are bound to occur with a certain probability. The same mutations originating repeatedly in an ubiquitous ancestor species may be responsible for an incoherent geographic distribution of a syngen or for the existence of an individual sexual strain without a sex partner at its own location but compatible with strains from far distant areas. Sporadic appearances of syngens and of single sexual strains are traditionally explained by their secondary distribution having evolved just once (for review see Wiese, 1976).

It should be emphasized that this concept holds true for any macromolecular complementarity independent as to whether sex cell contact occurs by an antigen–antibody type of interaction, by a lectin–substrate relation or by glycosylacceptor–glycosyltransferase systems.

REFERENCES

BERGMAN, K., GOODENOUGH, U. W., JAWITZ, J. & MARTIN, H. (1975). Gametic differentiation in *Chlamydomonas reinhardtii*. II. Flagellar membranes and the agglutination reaction. *Journal of Cell Biology*, **67**, 606–22.

BERNSTEIN, E. & JAHN, T. L. (1955). Certain aspects of the sexuality of two species of *Chlamydomonas*. *Journal of Protozoology*, **2**, 81–5.

BOSMANN, H. B. & McLEAN, R. J. (1975). Gametic recognition: lack of enhanced glycosyltransferase ectoenzyme system activity of nonsexual cells and sexually incompatible gametes of *Chlamydomonas*. *Biochemical and Biophysical Research Communications*, **63**, 323–8.

CRANDALL, M., LAWRENCE, L. M. & SAUNDERS, R. M. (1974). Molecular complementarity of yeast glycoprotein mating factors. *Proceedings of the National Academy of Sciences, U.S.A.*, **71**, 26–9.

DOYLE, R. J., WOODSIDE, E. E. & FISHEL, C. W. (1968). Protein–polyelectrolyte interactions. The concanavalin A precipitin reaction with polyelectrolytes and polysaccharide derivates. *Biochemical Journal*, **106**, 35–40.

FORSTER, H. (1957). Das Wirkungsspektrum der Kopulation von *Chlamydomonas eugametos*. *Zeitschrift für Naturforschung*, **12b**, 765–70.

FORSTER, H. (1959). Die Wirkungsstarken einiger Wellenlangen zum Auslosen der Kopulation von *Chlamydomonas moewusii*. *Zeitschrift für Naturforschung*, **14b**, 479–80.

FORSTER, H. & WIESE, L. (1954). Gamonwirkungen bei *Chlamydomonas eugametos*. *Zeitschrift für Naturforschung*, **9b**, 548–50.

GOODENOUGH, U. W., HWANG, C., & MARTIN, H. (1976). Isolation and genetic analysis of mutant strains of *Chlamydomonas reinhardi* defective in gametic differentiation. *Genetics*, **82**, 169–86.

GOWANS, C. S. (1963). The conspecificity of *Chlamydomonas eugametos* and *Chlamydomonas moewusii*: an experimental approach. *Phycologia*, **3**, 37–44.

HARTMANN, K. M. (1962). 'Die Regulation der Gametogenese von *Chlamydomonas eugametos* und *C. moewusii* durch exogene und endogene Faktoren. Vergleichend morphologische, physiologische und biophysikalische Untersuchungen.' Dissertation, Universitat Tubingen.

LEWIN, R. A. (1952). Studies on the flagella of algae. I. General observations on *Chlamydomonas moewusii* Gerloff. *Biological Bulletin*, **102**, 74–9.

LEWIN, R. A. (1954). Sex in unicellular algae. In *Sex in Microorganisms*, ed. D. H. Wenrich, pp. 100–134. American Association for the Advancement of Science, Washington D.C.

MARTIN, N. C. & GOODENOUGH, U. W. (1975). Gametic differentiation in *Chlamydomonas reinhardtii*. 1. Production of gametes and their fine structure. *Journal of Cell Biology*, **67**, 587–605.

MCLEAN, R. J. & BOSMANN, H. B. (1975). Cell–cell interactions: Enhancement of glycosyltransferase ectoenzymesystems during *Chlamydomonas* gametic contact. *Proceedings of the National Academy of Sciences*, *U.S.A.*, **72**, 310–13.

MCLEAN, R. J., LAURENDI, C. J. & BROWN, R. M. (1974). The relationship of gamone to the mating reaction in *Chlamydomonas moewusii*. *Proceedings of the National Academy of Sciences*, *U.S.A.*, **71**, 2610–13.

NYBOM, N. (1953). Some experiences from mutation experiments in *Chlamydomonas*. *Hereditas*, **39**, 64–9.

REISSIG, J. L. (1974). Decoding of regulatory signals at the microbial surface. *Current Topics in Microbiology and Immunology*, **67**, 43–96.

ROSEMAN, S. (1970). The synthesis of complex carbohydrates by multiglycosyl-transferase systems and their potential function in intercellular adhesion. *Chemistry and Physics of Lipids*, **5**, 270–97.

ROSEMAN, S. (1974). Complex carbohydrates and intercellular adhesion. In *The Cell Surface in Development*, ed. A. A. Moscona, pp. 255–72. Wiley, New York.

SHIMODA, C. & YANAGISHIMA, N. (1975). Mating reaction in *Saccharomyces cerevisiae*. VIII. Mating-type specific substances responsible for sexual cell agglutination. *Antonie van Leeuwenhoek. Journal of Microbiology and Serology*, **41**, 521–32.

SNELL, W. J. (1976 a). Mating in *Chlamydomonas*: A system for the study of specific cell adhesion. I. Ultrastructural and electrophoretic analysis of flagella surface components involved in adhesion. *Journal of Cell Biology*, **68**, 48–69.

SNELL, W. J. (1976 b). Mating in *Chlamydomonas*: a system for the study of specific cell adhesion. II. A radioactive flagella-binding assay for quantitation of adhesion. *Journal of Cell Biology*, **68**, 70–79.

STIFTER, I. (1959). Untersuchungen uber einige Zusammenhange zwischen Stoffwechsel und Sexualphysiologie an dem Flagellaten *Chlamydomonas eugametos*. *Archiv für Protistenkunde*, **104**, 364–88.

TRAINOR, F. R. (1959). A comparative study of sexual reproduction in four species of *Chlamydomonas*. *American Journal of Botany*, **46**, 65–70.

TSUBO, Y. (1961). Chemotaxis and sexual behaviour in *Chlamydomonas*. *Journal of Protozoology*, **8**, 114–21.

WIESE, L. (1965). On sexual agglutination and mating type substances in isogamous dioecious Chlamydomonads. I. Evidence of the identity of the gamones with the surface components responsible for sexual flagella contact. *Journal of Phycology*, **1**, 49–54.

WIESE, L. (1969). Algae. In *Fertilization. Comparative Morphology, Biochemistry and Immunology*, eds. C. B. Metz & A. Monroy, pp. 135–88. Academic Press, New York.

WIESE, L. (1974). Nature of sex specific glycoprotein agglutinins in *Chlamydomonas*. *Annals of the New York Academy of Sciences*, **234**, 383–95.

WIESE, L. (1976). Genetic aspects of sexuality in *Volvocales*. In *Genetics of Algae*, ed. R. A. Lewin, pp. 174–97. Blackwell Scientific, Oxford.

WIESE, L. & JONES, R. F. (1963). Studies on gamete copulation in heterothallic Chlamydomonads. *Journal of Cellular and Comparative Physiology*, 61, 265–74.

WIESE, L. & BAKER, P. M. (1968). Immunological analysis of the mating type reaction in *Chlamydomonas*. *Journal of Phycology*, 4, Suppl. 3.

WIESE, L. & METZ, C. B. (1969). On the trypsin sensitivity of gamete contact at fertilization as studied with living gametes in *Chlamydomonas*. *Biological Bulletin*, 136, 483–93.

WIESE, L. & SHOEMAKER, D. W. (1970). On sexual agglutination and mating type substances (gamones) in isogamous heterothallic Chlamydomonads. II. The effect of Concanavalin A upon the mating type reaction. *Biological Bulletin*, 138, 88–95.

WIESE, L. & HAYWARD, P. C. (1972). On sexual agglutination and mating type substances in isogamous dioecious Chlamydomonads. III. The sensitivity of sex cell contact to various enzymes. *American Journal of Botany*, 59, 530–6.

WIESE, L. & WIESE, W. (1975). On sexual agglutination and mating type substances in isogamous dioecious Chlamydomonads. IV. Unilateral inactivation of the sex contact capacity in compatible and incompatible taxa by α-mannosidase and snake venome protease. *Developmental Biology*, 43, 264–76.

WIESE, L. & WIESE, W. (1977). On speciation by evolution of gametic incompatibility. A model case in *Chlamydomonas*. *American Naturalist*, 111, 733–42.

WITMAN, G. B., CARLSON, K., BERLINER, J. & ROSENBAUM, J. L. (1972). *Chlamydomonas* flagella. I. Isolation and electrophoretic analysis of microtubules, matrix, membranes, and mastigonemes. *Journal of Cell Biology*, 54 507–39.

EXPLANATION OF PLATE

Isoagglutination of (+) gametes of *C. m. eugametos* upon addition of (−) isoagglutinin. In the clusters the cells adhere to each other exclusively by means of their flagella tips, this adhesion being mediated by the isoagglutinin.

PLATE I

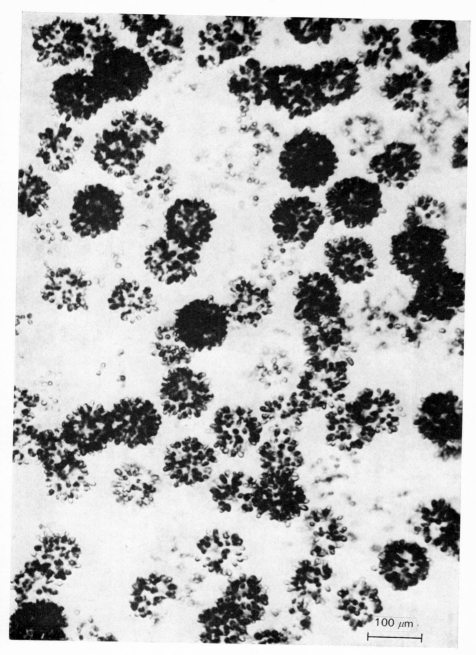

For explanation see p. 104

(*facing p.* 104)

MATING-TYPE INTERACTIONS IN YEASTS

By MARJORIE CRANDALL

Thomas Hunt Morgan School of Biological Sciences,
University of Kentucky, Lexington, Kentucky 40506, U.S.A.

RECEPTORS AND RECOGNITION

Cell–cell recognition during fertilization or tissue differentiation is a widespread phenomenon observed throughout the plant, animal and microbial kingdoms. Cell-surface receptor molecules responsible for specificity in these cellular recognition events are currently being sought and identified. In general, cell-surface substances from opposite sexes in microbial mating systems are complementary interacting macromolecules, whereas cell-surface substances from cells interacting to form a tissue are aggregation factors capable of recognizing and cross-linking like or similar cell types. A discussion of the types of biological assays to be used to isolate and characterize these different recognition or aggregation factor activities is presented in Crandall & Brock (1968a). Mating-type interactions in all microorganisms has been reviewed recently (Crandall, 1977). Recognition factors involved in mating in yeasts, algae, protozoa, fungi and bacteria have been identified. These recognition factors typically are high molecular weight proteins often conjugated with carbohydrate (summarized in Table 2.9 of Crandall, 1977). In this symposium, we will learn more about the details of recognition factors isolated from both microbial and mammalian sources. In this paper, I want to concentrate on the initial steps in cell fusion in yeasts, i.e. those steps involving mutual communication via pheromones and cellular recognition via glycoprotein agglutination factors and surface filaments.

SEX IN YEASTS

Sexual reproduction in ascomycetous yeasts involves fusion of haploid cells to form a diploid zygote followed by meiosis and ascospore formation thereby regenerating the haplophase (reviewed in Crandall, Egel & Mackay, 1977). Many of the medically important yeasts are classified in the Deuteromycetes because their perfect stage has not been found. However, the *Candida* species closely resemble the Ascomycetes in many features (Kreger-van Rij & Veenhuis, 1971) and, indeed, some of these species have recently been shown to belong to the genera *Kluyveromyces*, *Pichia*, *Hansenula*, etc. My recent research on the life cycle of *Candida albicans*

8

agrees with van der Walt's (1967) observation that this yeast exists primarily in the haplophase. Some strains of *C. albicans* become agglutinative and produce forms resembling zygotes in a medium containing yeast extract (M. Crandall & P. G. Hauck, unpublished observations). The zygotes and the large cells produced by the zygotes produce small, thick-walled (refractile) cells that give rise to the small-celled haplophase again. The relationship of these refractile cells to the sexual cycle is unclear at present but Mr Hauck and I are now attempting to induce ascosporogenesis in the large cell phase of *C. albicans*.

Opposite mating type strains exist in many yeast genera. In *Saccharomyces cerevisiae* these strains are called **a** and α; in *Hansenula wingei*, the two sexes are strains *5* and *21*.

The most studied yeast is *Saccharomyces cerevisiae*. Current laboratory strains are derivatives of 'baker's yeast'. *Sacch. cerevisiae* has a well developed genetic map, with 17 chromosomes defined by centromeres and linked markers (Mortimer & Hawthorne, 1975). The most exciting aspect of studies of *Sacch. cerevisiae* as far as this symposium is concerned is that mating is inducible. The induction process involves peptide hormones that reciprocally synchronize the sexual partners in G1 of the cell cycle, induce sexual agglutination and initiate cell-surface changes involved in conjugation tube formation (reviewed in Crandall *et al.*, 1977). (These sexual phenomena in *Sacch. cerevisiae* will be discussed under separate headings below.)

The yeast *Hansenula wingei* is a newcomer to this research field having been discovered as late as 1956 by Lynferd J. Wickerham (Wickerham, 1956). This yeast is constitutive for sexual agglutination forming huge cell clumps immediately upon mixing opposite mating types. This property was recognized by Thomas D. Brock as being valuable in the study of cell contact. Professor Brock (1961) studied the physiological conditions that maximized mating (late stationary phase cells of opposite sexes mixed in equal numbers in buffer containing only an energy source and counterions) and then went on to identify the two cell-surface macromolecules responsible for this sexual agglutination reaction (reviewed in Crandall & Brock, 1968*b*). Other laboratories then characterized the sexual agglutination factor from strain *5* in more detail (reviewed in Crandall, 1977; Crandall *et al.*, 1977). The two agglutination factors (5-factor and 21-factor) from the opposite sexes (strains *5* and *21*) form a neutralized complex *in vitro* analogous to the complementary interaction between an antigen and its specific antibody (Crandall, Lawrence & Saunders, 1974). This report is the first to demonstrate the formation of a complex between solubilized cell-surface recognition factors from complementary cell types.

(The agglutination factors from *H. wingei* will be discussed under a separate heading below.)

Cell recognition during mating in homothallic yeasts such as *Schizosaccharomyces pombe* probably proceeds by a somewhat different mechanism since sexual agglutination can occur in clonal cultures. Homothallic strains of yeast capable of recognizing 'self' and then undergoing sister cell fusion may be more interesting to the developmental biologist. (Studies of sexual flocculation in *Schizosacch. pombe* will be discussed under a separate heading below).

STEPS IN YEAST CELL FUSION

Heterothallic yeast strains are stable haploids capable of reproducing by budding indefinitely when cultured separately. However, when mixed together either on agar or in liquid, the two sexes initiate a series of steps that eventually lead to cell fusion. The steps in cell fusion are mutual communication via pheromones, cellular recognition, conjugation bridge formation, nuclear fusion and, finally, zygotic mitosis. The following discussion will focus on only the first two steps in cell fusion, i.e. those involved in pheromonal communication and cell–cell recognition. The mating habits of *Saccharomyces cerevisiae* and *H. wingei* will be presented in detail, since more information is available for these two yeasts than for any others. For reviews of steps in cell fusion in yeasts and other protists, see Crandall (1976), Crandall (1977) and Crandall *et al.* (1977).

MUTUAL COMMUNICATION *VIA* PHEROMONES

Saccharomyces cerevisiae strains **a** and α each secrete a variety of diffusible substances called pheromones that specifically act on the opposite sex in the absence of cell contact. Several such sex-specific responses have been reported.

Extracts of culture medium from strain **a** will specifically cause strain α to expand in cell volume; the same result obtains with the reciprocal experiment (Yanagishima, 1969). Initially it was reported that the substances in these extracts (called **a** and α hormones) were steroids (Takao, Shimoda & Yanagishima, 1970) but in a later publication, **a** hormone was reported to be *n*-octanoic acid (Sakurai, Tamura, Yanagishima & Shimoda, 1974). The relationship of these cell expansion hormones to the mating process remains unclear.

Sex-specific chemotropic responses have been observed in yeasts. When complementary mating types of *Sacch. cerevisiae* or *H. wingei* were placed about 15 μm apart on agar blocks, they budded toward each other (Herman, 1971). In nature, this sex-specific budding response may allow cells to grow to within touching distance where cell fusion could be initiated.

Pheromonally-induced cell elongation or protuberance formation occurs in both ascomycetous and basidiomycetous yeasts. Diffusible substances from one cell type induce the opposite cell type to undergo growth at one point on the cell resulting in a 'shmoo'-shaped cell as in *Sacch. cerevisiae* (Levi, 1956; Duntze, MacKay & Manney, 1970; MacKay & Manney, 1974) or in protuberances called conjugation tubes or copulatory processes in yeasts such as *Tremella mesenterica* (Bandoni, 1965; Reid, 1974), *Rhodosporidium toruloides* (Abe, Kusaka & Fukui, 1975) and *Sirobasidium magnum* (Flegel, personal communication). Pheromones have been searched for and not found in several well-studied yeast species (*H. wingei*, *Schizosaccharomyces pombe* and *Ustilago violacea*). In these cases, mating reactivity is induced by nutrient limitation and copulatory tubes are not elaborated until direct cell contact is established (see Table 2.4 in Crandall, 1977).

An important aspect of any process of cellular morphogenesis (such as conjugation in yeast) is the stage in the cell cycle. In general, cell fusion occurs between unbudded cells in the G1 phase. Cells may be synchronized at this point by nutrient limitation or by sex-specific pheromones (reviewed in Crandall, 1977). The pheromones from *Sacch. cerevisiae* strains **a** and α are called **a**-factor and α-factor, respectively. It now appears that **a**-factor is one molecular species with two activities, i.e. it causes G1 arrest in strain α and causes α cells to form shmoos (V. L. MacKay, personal communication). **a**-factor is a glycoprotein with a molecular weight greater than 10^6 daltons. In contrast, α-factor is actually a family of peptides that differ slightly in amino acid composition. The amino acids or derivatives that are sometimes missing are indicated inside parentheses in the following amino acid sequence of α-factor(s) reported by Stötzler, Kiltz & Duntze (1976): H_2N-(Trp)-His-Trp-Leu-Gln-Leu-Lys-Pro-Gly-Gln-Pro-Met(O)-Tyr-COOH. Since all of these related peptides are biologically active, the amino-terminal tryptophan residue and the penultimate carboxy-terminal methionine (sulphoxide) residues are probably not part of the active site (Stötzler & Duntze, 1976). These peptides cause **a** cells to cease budding, become sexually agglutinative and, in the absence of a receptive sexual partner, to form shmoos. Shmoos have small vesicles localized at the tip where cell wall and membrane synthesis is occurring (W. Duntze, personal communication). Research on these sex-specific pheromones from *Sacch. cerevisiae* is proceeding in about a dozen laboratories throughout the world and exciting new developments in this area of peptide hormone induction of cell differentiation should be forthcoming soon.

CELLULAR RECOGNITION FACTORS FROM HANSENULA WINGEI

The agglutination factor from mating type 5 (called 5-factor; abbreviated 5f) is detected by its agglutination of strain 21 cells (Taylor, 1964; Brock, 1965). The agglutination factor from mating type 21 (called 21-factor; 21f) is detected by its neutralization of the agglutination activity of 5f (Crandall & Brock, 1968b). Criteria for determining whether these isolated macromolecules are the mating-type specific cell-surface receptors are reviewed by Crandall (1977):

(1) 5f is obtained only from mating type 5; 21f is obtained only from mating type 21.

(2) 21f is released from the cell surface by the same treatment (trypsin digestion) that inactivates agglutination of strain 21 cells (Crandall & Brock, 1968b).

(3) 5f specifically agglutinates strain 21 cells but not cells of strain 5 or the 5 × 21 hybrid.

(4) 21f specifically inhibits 5f agglutination of strain 21 cells.

(5) Each mating factor is adsorbed only to the cell surface of the opposite mating type and not to its respective cell type or to the diploid.

(6) Cells with a large amount of adsorbed factor from the complementary type are inhibited from agglutinating with the opposite sex whereas treatment of the respective cell type with the same concentration of mating factor does not interfere with sexual agglutination.

(7) The isolated mating factors neutralize each other *in vitro* by forming a 5f–21f complex (Crandall *et al.*, 1974).

(8) The diploid hybrid normally does not produce either mating factor or the neutralized complex and is non-agglutinative and a non-mater with either mating type (Crandall & Brock, 1968c).

Cytoplasmic extracts of strain 5 cells yield 5f molecules of different sizes (Brock, 1965), with the number of combining sites roughly proportional to molecular weight (Crandall *et al.*, 1974). On the other hand, subtilisin digestion of strain 5 cells yields a homogeneous 5-agglutinin molecule of about 10^6 daltons that has 6 combining sites (Taylor & Orton, 1968). When released by subtilisin digestion and purified by affinity chromatography, 5-agglutinin was found to have a molecular weight of

9.6×10^5 daltons and to be composed of 85% carbohydrate (mostly mannose), 10% protein and 5% phosphate (Yen & Ballou, 1974). These workers proposed two possible structures for the 5-agglutinin based on their own results and those of Taylor, Tobin and Orton (Fig. 1).

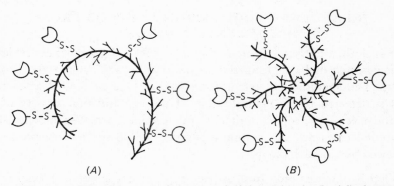

(A)   (B)

Fig. 1. Two possible structures of the 5-agglutinin molecule. In (A) the central core is composed of one polypeptide chain and in (B) it is composed of six polypeptide chains. The thick lines represent the polypeptide chains and the thin lines represent the carbohydrate chains. The active site is denoted by a tulip-shaped symbol connected to the core by a disulphide bond. The active site component is a glycopeptide similar in structure to the central core. (By kind permission of Dr C. E. Ballou and the American Chemical Society.)

Small binding fragments liberated from the large central core by reduction were estimated to be 12000 daltons by Taylor & Orton (1968). Yen & Ballou (1974) found these binding fragments to be composed of 28 amino acids and 60 mannose units, corresponding to a molecular weight of 12500 daltons. These two estimates for the molecular weight of the active site of 5f are certainly in excellent agreement. The central core of 5f is 10% protein and 90% mannose which is present in chains of eight sugar units linked to serine or threonine. These two hydroxy amino acids constitute 62% of the core protein. Since the molecule also contains phosphate, it should be called a phosphomannanprotein (Yen & Ballou, 1974). However, recent studies of the chemical nature of wild-type and mutant 5-agglutinin lacking phosphate indicate that the phosphate is not required for the agglutination reaction (Sing, Yeh & Ballou, 1976). The 5f remains active after boiling for 5 min (Brock, 1965); this heat stability of the 5f may be explained by a lack of α-helical structure in this branched, randomly-coiled molecule (Taylor & Tobin, 1966). The standard free energy of association of 5f with strain 21 cells is -14.5 kcal mol^{-1} which is high for reversible reactions and is due to cooperativity between the several combining sites on the 5f molecule (Taylor & Orton, 1970, 1971).

Cytoplasmic extracts or trypsin digestion of strain *21* cells yields 21f molecules that are univalent and homogeneous with respect to molecular weight (about 40000 daltons) (Crandall & Brock, 1968*b*; Crandall *et al.*, 1977). The 21f contains 25 to 35 % carbohydrate (mainly mannose) and is unstable to protein denaturants but stable to reducing agents.

Isolated 5f and 21f neutralize each other by forming a complex that is soluble probably because 21f is univalent and cannot form crosslinks as in the case of precipitating antibodies. The 5f–21f complex is stable presumably because of the large negative free energy of association (reported by Taylor & Orton, 1970). The neutralized complex is detected using a biological assay developed by Crandall & Brock (1968*b*) in which 21f is destroyed by alkali and 5f agglutination activity is recovered. Using this assay, it was found that the 5f–21f complex exists in multiple forms reflecting the molecular weight heterogeneity of the cytoplasmic 5f used to form the complex. Three peaks of complex were detected with estimated molecular weights of 0.5, 1.2 and 3.8×10^6 daltons; the corresponding number of combining sites on the 5f for these peaks were calculated to be roughly 6, 16 and 63 (Crandall *et al.*, 1974). The cytoplasmic 5f studied by Brock, Crandall and co-workers can vary in size from 15000 to 10^8 daltons whereas the cell-surface 5f studied by Taylor and co-workers and Ballou and co-workers is homogeneous (10^6 daltons). It seems reasonable that heterogeneous cytoplasmic 5f may represent both precursor molecules and aggregates thereof which will become cell-surface 5-agglutinin.

There is no enzymatic destruction of either factor during complex formation since the complex is stable and can be purified using conventional methods in protein chemistry. Furthermore, the activity of one of the factors can be recovered from the complex following specific chemical inactivation of the complementary factor. No attempts have been made yet to determine whether either factor has enzymatic activity.

Activity of the 5-agglutinin is destroyed by both pronase and exo-α-mannanase (Yen & Ballou, 1974) but it is not known at this time whether the specificity of interaction of the agglutinin with 21f is due to protein–protein or carbohydrate–protein interaction. Yen & Ballou (1974) presented additional results that suggest that the specificity of binding of 5f to strain *21* cells resides in the protein moiety but that the mannan serves to maintain the structural configuration of the 5f required for agglutination activity.

CELLULAR RECOGNITION FACTORS FROM
SACCHAROMYCES CEREVISIAE

The sex-specific substances responsible for sexual agglutination from strains **a** and α have recently been isolated. Both agglutination factors are univalent and may be measured only by inhibition of mixed cell agglutination. Shimoda & Yanagishima (1975) reported that both agglutination factors are about one million in molecular weight but recently Yoshida *et al* (1976) reported that **a** agglutination factor is 23 000 daltons and consists of two glycoprotein subunits.

Yanagishima and co-workers (Yoshida, Hagiya & Yanagishima, 1976; Hagiya, Yoshida & Yanagishima, 1977) report that the complementary agglutination factors from strains **a** and α form a neutralized complex *in vitro* as was reported for the recognition factors from *H. wingei* by Crandall *et al.* (1974).

CELLULAR RECOGNITION FACTORS FROM
SCHIZOSACCHAROMYCES POMBE

Conjugation in homothallic strains of *Schizosaccharomyces pombe* is also preceded by a sexual agglutination or flocculation reaction (reviewed in Crandall *et al.*, 1977). Using a floc-inhibition assay, G. B. Calleja (personal communication) has isolated from the culture filtrate a fraction which inhibits the reflocculation of heat-killed, induced cells of *Schizosacch. pombe*, but does not cause deflocculation. The fraction is heat stable and is not obtainable from uninoculated medium nor from media which do not support induction of flocculation. Passage of the fraction through Sephadex G25 suggests a molecular species smaller than 3000 daltons. This small molecule may be a binding fragment released into the culture medium from the larger aggregation factors that remain attached to the cell surface.

REGULATION OF RECOGNITION FACTORS

The diploid of *H. wingei* formed by mating between strains *5* and *21* has the genes for both agglutination factors and yet it is nonagglutinative. Neither 5f nor 21f nor the neutralized 5f–21f complex can be detected in normal diploid cultures (Crandall & Brock, 1968c). However, under certain extreme physiological conditions, the diploid can be induced to synthesize either 5f or 21f. Details of these inducing conditions have been summarized recently (Crandall & Caulton, 1975) and, therefore, will be reviewed here only briefly.

Induction of 5f occurs in the diploid when cells are grown: (*i*) to late stationary phase in a medium containing 0.67% yeast extract; or (*ii*) to late stationary phase in synthetic minimal medium containing 0.4 mM V^{4+} or V^{5+} (induction of 5f by vanadium is enhanced by oxygen limitation); or (*iii*) to limiting cell density in Ca^{2+}- and Mg^{2+}-free synthetic minimal medium; or (*iv*) to limiting cell density in vitamin-limited medium, then resuspended in a buffer containing phosphate, Mg^{2+}, glucose and pyridoxine and aerated for 2–4 h.

Induction of 21f occurs in the diploid when cells are grown: (*i*) to late exponential phase in a medium containing 0.67% yeast extract + 1 mM Na_2EDTA; or (*ii*) to stationary phase in a medium containing yeast extract, then washed, resuspended and grown in a synthetic minimal medium lacking trace metals.

It is apparent that the differential induction of 5f or 21f is controlled by the metal-ion concentration; high concentrations of metal ions induce synthesis of 5f whereas low concentrations induce synthesis of 21f. Since these inducing conditions have no effect on haploid agglutination (Crandall & Caulton, 1975), they must affect the diploid regulatory mechanisms rather than synthesis of the factors themselves. On the basis of these observations, the model proposed by Crandall & Brock (1968c) for the mutual repression of haploid genes in the diploid is still favoured. In this model, regulatory genes carried by each haploid genome operate in apposition to repress mating functions in the diploid. For example, strain *21* is visualized as carrying a structural gene for 21f and a regulatory gene for the repression of 5f synthesis and *vice versa*. This mutual repression in the diploid may be disrupted by the non-physiological conditions already listed in which growth-limitation induces either 5f or 21f in the diploid.

SURFACE FILAMENTS

Certain yeast-like fungi have recently been found to produce large numbers of external hairs or filaments similar to the fimbriae or pili found in Gram-negative bacteria. These filaments are about 0.007 μm in diameter and vary in length from about 0.5 μm in the Ascomycete *Sacch. cerevisiae* to over 10 μm in some species of the Basidiomycete genus *Ustilago* (Day & Poon, 1975). They are easily observed under the electron microscope after the cells have been thoroughly washed with water and ether and then shadowed with tungsten. The *Ustilago* filaments are composed of protein, but they may have carbohydrate molecules attached as part of a secondary structure. The filaments of *Ustilago violacea* have been demonstrated to be involved in conjugation and those of *Sacch. cerevisiae* to be important for flocculation (Day, Poon & Stewart, 1975). Similar filaments have been observed in strains *5* and *21* of the Ascomycete *H. wingei* (see Fig. 2a, b in Crandall & Caulton, 1975) using tungsten-shadowed cells prewashed with ether and water. In view of the results with *Saccharomyces* and *Ustilago*, it is possible that the filaments of *Hansenula* cells may be involved in sexual agglutination or conjugation (N. H. Poon & A. W. Day, personal communication). Fibrillar material external to the thick cell wall can be seen in all of the electron micrographs presented in the study of conjugation performed by Conti & Brock (1965). Cells were fixed in 1.5 % potassium permanganate and stained in 2 % osmic acid. The surface filaments are seen to overlap, forming a region of greater electron density between the opposite cell types during an early stage of agglutination (reviewed in Crandall *et al.*, 1977). This agglutination reaction is so strong that the cell walls are deformed along the length of contact even though the walls are not yet touching. Filaments on the cell surface of *H. wingei* have also been observed by K. Aufderheide (personal communication). Cells were prepared for electron microscopy using procedures similar to those of Conti & Brock (1965), but with minimal staining to increase visualization of the fibrous coat. The thickness of this coat on sexually agglutinative strains of *H. wingei* is approximately 0.2 μm. No filaments were observed on non-agglutinative strains of *S. cerevisiae*. When mating types *5* and *21* of *H. wingei* were mixed and allowed to agglutinate, only the outer fibrous coats were seen to be in contact, not the other cell wall layers. Furthermore, only the distal portions of the fibres appeared to be in contact, producing a region of greater electron density between the cell walls, perhaps as the result of an overlap of fibres. It is suggested by K. Aufderheide that the agglutination factors from strains *5* and *21* may be associated with these fibres and may possibly be located on their distal portions.

Recently, J. H. Caulton and I have performed an electron microscope study of the surface filaments on *H. wingei* using a Ruthenium Red staining procedure specific for acid-substituted polysaccharides or acidic polypeptides (Luft, 1966). We thought that this procedure would be suitable for staining the surface agglutination factors since 5f is a phosphomannan protein (Yen & Ballou, 1974) and 21f is also a mannanprotein (Crandall & Brock, 1968 *b*), probably also conjugated to phosphoric acid although this has not been studied. The purpose of this preliminary experiment was to compare strains *5*, *21* and the *5* × *21* hybrid and then to determine whether any change occurred when the hybrid was induced to synthesize one or the other agglutination factor. Cells were grown in yeast extract medium for one day under standard conditions (reported in Crandall & Caulton, 1975). Strain *5* agglutinated strongly with *21* tester, strain *21* agglutinated strongly with *5* tester and the diploid did not agglutinate with either tester as reported earlier (Crandall & Brock, 1968 *c*). Cells were fixed in 2 % glutaraldehyde containing 0.01 % Ruthenium Red, post-fixed and stained in 1 % osmium tetroxide containing 0.01 % Ruthenium Red followed by staining with 0.5 % uranyl acetate. Both strain *5* and strain *21* were seen to be covered with a fuzzy layer observed by earlier workers using different methods. When opposite sexes were mixed together, agglutination was immediate. The clumps were harvested after only 15 min of shaking in saline to prevent cell fusion. In these mixed-cell agglutinates, the fuzzy layers were seen to overlap (Plate 1) but no fine structure of the surface filaments could be seen. Most of the cells from a diploid culture lacked this fuzzy layer but appeared to have a double-layered membrane surrounding the cell wall (Plate 2). Less than 10 % of the diploid cells had a fuzzy surface resembling the haploids. However, in this one experiment, the diploid pellet was slightly larger than the pellets from the haploids and the possibility exists that the Ruthenium Red did not penetrate the pellet to stain all of the diploid cells. The unit membrane external to the cell wall of the diploid (Plate 2) looks very similar to the triple-track outer layer of the ascospore wall of *H. wingei* observed by Black & Gorman (1971). If the diploid is allowed to grow in yeast extract medium for 2 days instead of the standard time of growth (1 day), then it usually becomes agglutinative with strain *21* but this varies with the batch of yeast extract (Crandall & Brock, 1968*c*). In this experiment, unfortunately the diploid did not become induced to synthesize 5f after 2 days. In future experiments, we would study the kinetics of induction of 5f in the diploid using a defined medium containing vanadium that reliably induces 5f synthesis (Crandall & Caulton, 1973). In this Ruthenium Red experiment, cells were successfully induced to synthesize 21f (and, hence, become agglutinative with *5* tester) by growing the cells in yeast extract medium containing EDTA as reported by us earlier (Crandall & Caulton, 1973). These diploid → 21 cells were partially covered with a fuzzy layer (Plate 3) which correlated with a weaker agglutination reaction of the induced diploid → *21* cells compared with the corresponding haploid agglutination between strain *5* and *21* testers.

It is premature at this time to draw any conclusions about the role of surface filaments in yeast mating because of the preliminary nature of the electron microscopic studies done independently by the several groups of laboratory workers cited above. The facts that are certain, however, are as follows:

(1) surface fuzz external to the cell wall is observed on many different yeasts, grown in a variety of media;

(2) the sex-specific agglutination factors are on the surfaces of the complementary mating types because the factors may be removed by enzymatic digestion of whole cells.

It remains to be determined whether the agglutination factors are present on the fuzz or are bound to deeper layers of the wall. Use of the constitutively agglutinative strains *5* and *21* of *H. wingei* would facilitate answering this question because the factors are synthesized at maximal levels in all media in single cultures, i.e., no induction step in mixed cultures or by pheromones is necessary. Also, conditions are known for the specific removal of these factors from the respective cell surfaces (subtilisin specifically removes 5f and trypsin specifically removes 21f). The question of how sexual agglutination is regulated in the non-agglutinative, non-mating diploid hybrid may be approached by these methods as well. For example, the diploid may be grown in different media and induction of sexual agglutination correlated with appearance of surface filaments or changes in their texture or size. Then, the same treatments that specifically inactivate haploid agglutination may be used on the induced diploid to again determine what changes occur to the surface filaments.

Surface fuzz or a fluffy outer layer also may be seen on animal cells. At the point of cell contact between two endothelial cells in a capillary wall, tight junctions are observed which display a central dark element resulting from fused outer layers (see Fig. 3 in Luft, 1966). This material also stains with Ruthenium Red and is interpreted as being a fusion or condensation of the outer leaflets from each unit membrane that serves as a 'gasket' between the two 'mating' surfaces (Luft, 1966). A central dark element due to overlapping surface fuzz also occurs between agglutinated sexes of *H. wingei* (see Fig. 2*b* in Conti & Brock, 1965). It is interesting that such similar observations are made in such diverse biological systems as yeast and mammals.

Supported by U.S.P.H.S. Grant GM 21889 (Dr M. Crandall, Principal Investigator) and by an intramural grant from the Biomedical Research Support Grant Number 5 S05-RR 07114-08 (Dr Samuel F. Conti, Principal Investigator). Special thanks are extended to Mrs Joan Caulton for her dedication in performing the electron microscope studies and to Dr Anthony P. Mahowald for providing the electron microscope facilities.

REFERENCES

ABE, K., KUSAKA, I. & FUKUI, S. (1975). Morphological change in the early stages of the mating process of *Rhodosporidium toruloides*. *Journal of Bacteriology*, **122**, 710–18.

BANDONI, R. J. (1965). Secondary control of conjugation in *Tremella mesenterica*. *Canadian Journal of Botany*, **43**, 627–30.

BLACK, S. H. & GORMAN, C. (1971). The cytology of *Hansenula*. III. Nuclear segregation and envelopment during ascosporogenesis in *Hansenula wingei*. *Archiv für Mikrobiologie*, **79**, 231–48.

BROCK, T. D. (1961). Physiology of the conjugation process in the yeast *Hansenula wingei*. *Journal of General Microbiology*, **26**, 487–97.

BROCK, T. D. (1965). The purification and characterization of an intracellular sex-specific mannan protein from yeast. *Proceedings of the National Academy of Sciences, U.S.A.*, **54**, 1104–12.

CONTI, S. F. & BROCK, T. D. (1965). Electron microscopy of cell fusion in conjugating *Hansenula wingei*. *Journal of Bacteriology*, **90**, 524–33.

CRANDALL, M. (1976). Mechanisms of fusion in yeast cells. In *Microbial and Plant Protoplasts*, eds. J. F. Peberdy, H. J. Rogers, A. H. Rose & E. C. Cocking, pp. 161–75. Academic Press, New York and London.

CRANDALL, M. (1977). Mating-type interactions in microorganisms. In *Receptors and Recognition*, A, vol. 3, eds. P. Cuatrecasas & M. F. Greaves, pp. 45–100. Chapman and Hall, London.

CRANDALL, M. A. & BROCK, T. D. (1968a). Molecular aspects of specific cell contact. *Science*, **161**, 473–5.

CRANDALL, M. A. & BROCK, T. D. (1968b). Molecular basis of mating in the yeast *Hansenula wingei*. *Bacteriological Reviews*, **32**, 139–63.

CRANDALL, M. A. & BROCK, T. D. (1968c). Mutual repression of haploid genes in diploid yeast. *Nature, London*, **219**, 533–4.

CRANDALL, M. & CAULTON, J. H. (1973). Induction of glycoprotein mating factors in diploid yeast of *Hansenula wingei* by vanadium salts or chelating agents. *Experimental Cell Research*, **82**, 159–67.

CRANDALL, M. & CAULTON, J. H. (1975). Induction of haploid glycoprotein mating factors in diploid yeast. In *Methods in Cell Biology*, vol. 12, ed. D. M. Prescott, pp. 185–207. Academic Press, New York and London.

CRANDALL, M., LAWRENCE, L. M. & SAUNDERS, R. M. (1974). Molecular complementarity of yeast glycoprotein mating factors. *Proceedings of the National Academy of Sciences, U.S.A.*, **71**, 26–9.

CRANDALL, M., EGEL, R. & MACKAY, V. L. (1977). Physiology of mating in three yeasts. In *Advances in Microbial Physiology*, vol. 15, eds. A. H. Rose & D. W. Tempest, pp. 307–98. Academic Press, London and New York.

DAY, A. W. & POON, N. H. (1975). Fungal fimbriae. II. Their role in conjugation in *Ustilago violacea*. *Canadian Journal of Microbiology*, **21**, 547–57.

DAY, A. W., POON, N. H. & STEWART, G. G. (1975). Fungal fimbriae. III. The effect on flocculation in *Saccharomyces*. *Canadian Journal of Microbiology*, **21**, 558–64.

DUNTZE, W., MACKAY, V. & MANNEY, T. R. (1970). *Saccharomyces cerevisiae*: A diffusible sex factor. *Science*, **168**, 1472–3.

HAGIYA, M., YOSHIDA, K. & YANAGISHIMA, N. (1977). The release of sex-specific substances responsible for sexual agglutination from haploid cells of *Saccharomyces cerevisiae*. *Experimental Cell Research*, **104**, 263–72.

HERMAN, A. I. (1971). Sex-specific growth responses in yeast. *Antonie van Leeuwenhoek. Journal of Microbiology and Serology*, **34**, 379–84.

KREGER VAN RIJ, N. J. W. & VEENHUIS, M. (1971). A comparative study of the cell wall structure of basidiomycetous and related yeasts. *Journal of General Microbiology*, **68**, 87–95.

LEVI, J. D. (1956). Mating reaction in yeast. *Nature, London*, **177**, 753–4.

LUFT, J. H. (1966). Fine structure of capillary and endocapillary layer as revealed by ruthenium red. *Federation Proceedings*, **25**, 1773–83.

MACKAY, V. & MANNEY, T. R. (1974). Mutations affecting sexual conjugation and related processes in *Saccharomyces cerevisiae*. I. Isolation and phenotypic characterization of nonmating mutants. *Genetics*, **76**, 255–71.

MERCER, E. H. & BIRBECK, M. S. (1972). *Electron Microscopy*: *A Handbook for Biologists*, 3rd edition. Blackwell Scientific, Oxford.

MORTIMER, R. K. & HAWTHORNE, D. C. (1975). Genetic mapping in yeast. In *Methods in Cell Biology*, vol. XI, ed. D. M. Prescott, pp. 221–34. Academic Press, London and New York.

REID, I. D. (1974). Properties of conjugation hormones (erogens) from the basidiomycete *Tremella mesenterica*. *Canadian Journal of Botany*, **52**, 521–4.

SAKURAI, A., TAMURA, S., YANAGISHIMA, N. & SHIMODA, C. (1974). Isolation and identification of a sexual hormone in yeast. *Agricultural and Biological Chemistry*, **38**, 231–2.

SHIMODA, C. & YANAGISHIMA, N. (1975). Mating reaction in *Saccharomyces cerevisiae* VIII. Mating-type-specific substances responsible for sexual cell agglutination. *Antonie van Leeuwenhoek. Journal of Microbiology and Serology*, **41**, 521–32.

SING, V., YEH, Y.-F. & BALLOU, C. E. (1976). Isolation of a *Hansenula wingei* mutant with an altered sexual agglutinin. In *Surface Membrane Receptors*, eds. R. A. Bradshaw, W. A. Frasier, R. C. Merrell, D. I. Gottlieb, R. A. Hogue-Angeletti, pp. 87–97. Plenum, New York.

STÖTZLER, D. & DUNTZE, W. (1976). Isolation and characterization of four related peptides exhibiting α factor activity from *Saccharomyces cerevisiae*. *European Journal of Biochemistry*, **65**, 257–62.

STÖTZLER, D., KILTZ, H.-H. & DUNTZE, W. (1976). Primary structure of α-factor peptides from *Saccharomyces cerevisiae*. *European Journal of Biochemistry*, **69**, 397–400.

TAKAO, N., SHIMODA, C. & YANAGISHIMA, N. (1970). Chemical nature of yeast sexual hormones. *Development, Growth and Differentiation*, **12**, 199–205.

TAYLOR, N. W. (1964). Specific soluble factor involved in sexual agglutination of the yeast *Hansenula wingei*. *Journal of Bacteriology*, **87**, 863–6.

TAYLOR, N. W. & ORTON, W. L. (1968). Sexual agglutination in yeast. VII. Significance of the 1.7S component from reduced 5-agglutinin. *Archives of Biochemistry and Biophysics*, **126**, 912–21.

TAYLOR, N. W. & ORTON, W. L. (1970). Association constant of the sex-specific agglutinin in the yeast, *Hansenula wingei*. *Biochemistry*, **9**, 2931–4.

TAYLOR, N. W. & ORTON, W. L. (1971). Cooperation among the active binding sites in the sex-specific agglutinin from the yeast, *Hansenula wingei*. *Biochemistry*, **10**, 2043–9.

TAYLOR, N. W. & TOBIN, R. (1966). Sexual agglutination in yeast. IV. Minimum particle count per cell for fast-sedimenting 5-agglutinin. *Archives of Biochemistry and Biophysics*, **115**, 271–6.

VAN DER WALT, J. P. (1967). Sexually active strains of *Candida albicans* and *Cryptococcus albidus*. *Antonie van Leeuwenhoek. Journal of Microbiology and Serology*, **33**, 246–56.

WICKERHAM, L. J. (1956). Influence of agglutination on zygote formation in *Hansenula wingei*, a new species of yeast. *Comptes rendus des Travaux du Laboratoire de Carlsberg* (*serie Physiologie*), **26**, 423–43.

YANAGISHIMA, N. (1969). Sexual hormones in *Saccharomyces cerevisiae*. *Antonie van Leeuwenhoek. Journal of Microbiology and Serology*, **35** (suppl.), C9–C10.

YEN, P. H. & BALLOU, C. B. (1974). Partial characterization of the sexual agglutination factor from *Hansenula wingei* Y-2340 type 5 cells. *Biochemistry*, **13**, 2428–37.

YOSHIDA, K., HAGIYA, M. & YANAGISHIMA, N. (1976). Isolation and purification of the sexual agglutination substance of mating type *a* cells in *Saccharomyces cerevisiae*. *Biochemical and Biophysical Research Communications*, **71**, 1085–94.

EXPLANATION OF PLATES

PLATE 1

Cell contact between strain 5 and strain 21 of H. wingei. Yeast stocks 5 *cyh-1 lys-1* (resistant to 100 µg/ml of cycloheximide and requiring lysine) and 21 *ade-1 his-1* (requiring adenine and histidine) were inoculated 0.50 ml per 25 ml of a medium containing 0.7 % yeast extract (Difco), 0.5 % KH_2PO_4 and 2 % glucose. Cultures were aerated in a 125 ml Erlenmeyer flask on a rotary shaker at 250 r.p.m. for 24 h at 30 °C. Cells were harvested, washed twice in sterile 0.9 % saline and then resuspended in half the volume of culture (2 × cell density). Equal volumes (12.5 ml) of each mating type were added to a 125 ml Erlenmeyer flask, shaken at 250 r.p.m. for 15 min, then harvested with the control cultures. The pellets were removed with a spatula and mixed into sterile, dried silica gel to desiccate the cells for storage. Cells were rehydrated in sterile distilled water for 1 h, harvested and then fixed with 2 % glutaraldehyde (Electron Microscopy Sciences) containing 0.01 % Ruthenium Red (Polysciences, Inc.) for 30 min at room temperature followed by 1.5 h on ice. The following steps were performed at 0 °C. Cells were washed three times for 30 min each time in 0.2 M Sucrose (Mallinckrodt) plus 0.1M sodium cacodylate (K & K Laboratories) buffer; post-fixed for 2 h in 1 % osmium tetroxide (Electron Microscopy Sciences) Zetterquist solution containing 0.01 % Ruthenium Red according to the procedures in Mercer & Birbeck (1972); washed for 10 min in water; stained in 0.5 % uranyl magnesium acetate (Electron Microscopy Sciences) overnight; washed for 10 min in water; dehydrated in ethanol going from 25 % to 50 % to 75 % for 10 min each time; and finally placed in 95 % ethanol and allowed to warm to room temperature. Then cells were soaked in absolute alcohol for 15 min and added to propylene oxide twice for 15 min each time. One part of this cell suspension in propylene oxide was added to one part of DER embedding resin (Electron Microscopy Sciences) for 1 h at room temperature followed by 37 °C overnight and then 60 °C for 24–36 h until hard. Thin sections were made with a diamond knife following standard procedures for electron microscopy.

PLATE 2

Absence of surface filaments on the uninduced diploid of H. wingei. Yeast stock D44 which is a hybrid of 5 *cyh-1 lys-1* × 21 *ade-1 his-1* was grown and prepared for electron microscopy as described in the legend of Plate 1 for the haploids.

PLATE 3

Presence of irregular surface fuzz on diploid cells induced to synthesize 21-factor. Yeast stock D44 was grown as described in the legend of Plate 1. Induction of 21f was achieved by adding 1 mM Na_2EDTA to the yeast extract medium and growing the cells for 44 h.

PLATE I

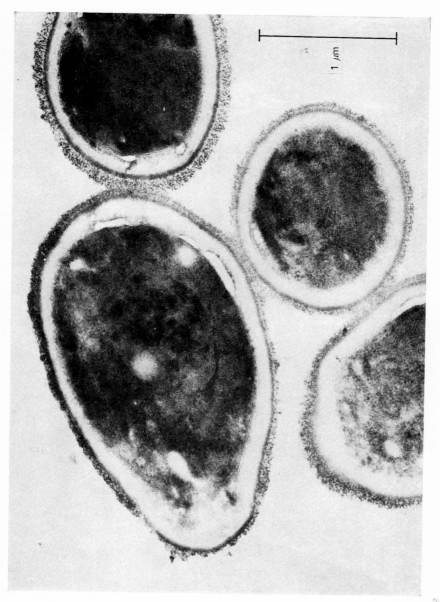

1 μm

For explanation see p. 119

PLATE 2

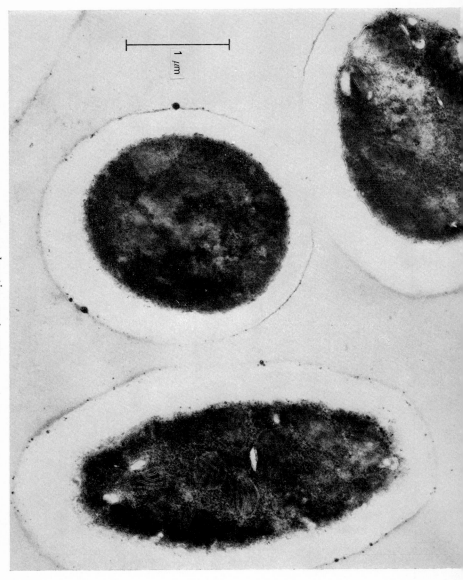

For explanation see p. 119

PLATE 3

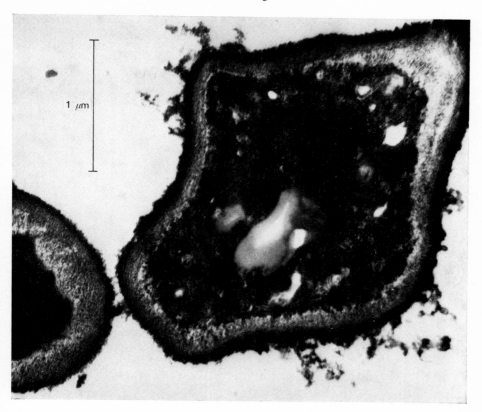

For explanation see p. 119

RECOGNITION AND RESPONSE IN THE POLLEN–STIGMA INTERACTION

BY J. HESLOP-HARRISON

Welsh Plant Breeding Station, Plas Gogerddan,
nr Aberystwyth, U.K.

INTRODUCTION

The angiosperms share with the higher gymnosperms a siphonogamous mode of fertilisation in which gametes are delivered to the ovule through a pollen tube. The evolution of this form of gamete transfer evidently conferred numerous advantages, not the least being a new potentiality for the control of breeding behaviour through the regulation of pollen germination and tube growth by the tissues of the diploid sporophytic parent of the female gametophyte (J. Heslop-Harrison, 1975*c*, 1976*a*, 1976*b*). The stigma and style can provide barriers to the invasion of foreign pollen and so contribute to the isolation of species, or, within species, can select between different genetical classes of pollen and so control the breeding system.

In this paper we will be concerned mainly with controls of the second type, specifically those which favour outcrossing by discriminating against self-pollen – the so-called self-incompatibility systems now known in nearly one hundred families of flowering plants. The genetics of angiosperm incompatibility systems have been treated recently in two excellent monographic reviews (Galun, 1976; Nettancourt, 1977), and here we need note only the principal features. Control in all of the investigated species is vested in one or a very few loci, and in species where there is no morphological distinction between the breeding groups numerous alleles occur. Where the behaviour of each haploid pollen grain is determined by its own genotype – so-called gametophytic systems – rejection takes place when the allele carried by the pollen matches one of those in the diploid pistil, or, where more than one locus is concerned, when the alleles at each are matched. In sporophytic systems, the behaviour of the pollen grain is determined by its parent, again with the same basic rules, identity of alleles in the two diploid parents ensuring rejection. These mechanisms, were they to work perfectly, would necessarily enforce outbreeding, generating heterozygosity at the incompatibility loci in each generation.

The most striking feature of self-incompatibility systems is undoubtedly the high level of specificity shown in their functioning. The numbers of

9

[121]

incompatibility ('S') alleles known in some groups is astonishingly high: for example, upwards of fifty have been identified in the sporophytic system of *Brassica* (Cruciferae) (Ockenden, 1975), and comparable numbers are known in *Trifolium* (Leguminosae) with a gametophytic system. The specificities must be imposed on both pollen and pistil, and rejection or acceptance must result from an interaction at the time of pollination or soon thereafter, when mutual recognition must take place. This much was clear to East & Mangelsdorf (1925) and East (1934), pioneers of work on the genetics of self-incompatibility systems; and it is noteworthy that East drew an analogy between such systems and the antigen–antibody reaction of vertebrate immune responses. The recognition event must be followed by secondary processes that discriminate between male gametophytes, either by blocking the growth of the incompatible, or promoting that of the compatible. We will take the two parts of the overall response in turn, beginning with the primary interaction.

THE RECOGNITION STEP

Stigmatic surfaces are invariably glandular in character, and, after the capture of acceptable pollen, they provide the conditions for germination and growth of the tube, which reaches the ovules either by penetrating the stigma surface and growing through the intercellular spaces of a specialised transmitting tract, or by entering a pre-existing canal leading into the cavity of the ovary and growing through this to the ovules. In all sporophytic systems and certain gametophytic systems the incompatible tubes are inhibited on the stigma surface or very shortly after the tube penetrates. In most gametophytic systems, pollen germination and tube growth proceeds in much the same manner in compatible and incompatible combinations, but incompatible tubes are retarded or arrested during their growth through the transmitting tissue of the style or in the stylar canal.

In those systems where the inhibition is at or near the stigma surface it seems probable that the incompatibility factors are synthesised in *anticipation* of the actual event of pollination, so that they are part of 'preset' systems in both pollen and pistil at the time of anthesis (Lewis, 1965; Linskens & Kroh, 1967). The timing of the response favours this conclusion. In the Cruciferae, a family with a sporophytic system and surface inhibition, the recognition event is probably completed in under three minutes (J. Heslop-Harrison, Y. Heslop-Harrison & Barber, 1975a), and in the gametophytic system of the grasses, considered in more detail later, the interaction can be completed in under ninety seconds from the moment of first contact of pollen and stigma.

This situation cannot be assumed in those gametophytic systems where the response is delayed until the tube is traversing the style. Here there are three principal possibilities, (a) that the incompatibility factors are synthesised in both pollen and pistil before the time of pollination, but those on the female side are localised in certain zones of the style, as supposed by East (1934), to be encountered by the tubes when this part is reached during growth; (b) that the factors on the male side are not synthesised until the tube is actively growing, when the interaction with pre-existing recognition molecules in the style begins, or (c) that the stylar factors are not synthesised until after pollination, perhaps in consequence of induction by the pollen, again leading to a delayed reaction. These alternatives have different implications, and these have been considered in various recent papers and reviews, for example by Donk (1974, 1975), Ascher (1975) and Nettancourt (1977).

Pollen and stigma-surface proteins

Where the rejection of incompatible pollen occurs at or in the stigma surface, there is a strong presumption that the recognition reaction must be mediated through factors carried on or near the surfaces of the interacting partners, pollen and stigma. Since the early work of Lewis (1952) and Linskens (1960) it has been generally accepted that the incompatibility factors are proteins or glycoproteins, and much recent work has been concerned with these constituents in the surface materials of pollen and stigma.

The pollen wall is structurally and chemically complex. The two main layers are the outer exine, composed of sporopollenin, a unique polymer of carotenoids and carotenoid esters (Brooks & Shaw, 1971), and the inner, the intine, a polysaccharide wall with cellulose, hemicellulose and pectic components. Each wall layers carries protein fractions (reviews, J. Heslop-Harrison, 1975a, 1975c).

The *intine* seems invariably to incorporate proteins during its development, the product first of the young spore and then of the vegetative cell of the male gametophyte (Knox & J. Heslop-Harrison, 1969, 1970; Knox, 1971). These include a range of acid hydrolases and also non-enzymic proteins and glycoproteins. The inclusions are ultimately separated from the plasmalemma and sealed in the wall during the final stages of thickening. In this respect the intine forms a gametophytic domain adjacent to the cell surface of the pollen grain. The *exine* is often deeply sculptured, and this layer of the wall receives proteins and lipids from the nurse tissue of the anther, the tapetum, during the final stages of pollen maturation (J. Heslop-Harrison, 1968a, 1968b; J. Heslop-Harrison, Y. Heslop-Harrison,

Knox & Howlett, 1973 *b*; Dickinson & Lewis, 1973 *b*). The exine proteins are synthesised in diploid parental tissue, so this layer forms a sporophytic domain of the pollen wall.

The proteinaceous constituents of the wall, although generally sealed in by lipidic or polysaccharide layers, are not bound tenaciously, and move out rapidly on moistening, with the exine fractions, when these are present, diffusing first (J. Heslop-Harrison, Knox, Y. Heslop-Harrison & Mattsson, 1975 *b*). In species of families like the Cruciferae and Malvaceae the out-flow from the exine begins within a few seconds in media of suitable tonicity, while loss from the intine begins later and continues for a longer period. The different spectra in short- and long-term diffusates of the pollen of *Althaea rosea* (Malvaceae) are shown in the electrofocussing microgels illustrated in Plate 1. Ultimately the grains autolyse, and the intracellular proteins are then released. A comparison of the wall diffusates of pollen of *Cosmos bipinnatus* (Compositae) and an extract from the same pollen sample is given in Fig. 1. The residual wall proteins are still evident in the extract, but the bulk of the material at the high-pH end of the scan is derived from the cytoplasm.

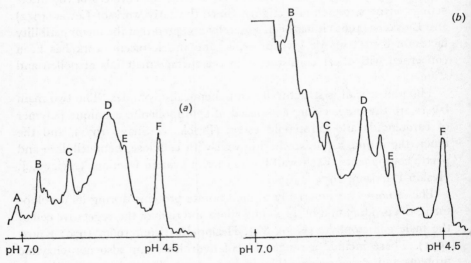

Fig. 1. Densitometric scans of Coomassie Blue-stained microelectrofocussing gels of proteins of pollen of *Cosmos bipinnatus* (Compositae). Gel volume, *c.* 5 μl; LKB ampholytes. (*a*) Pollen surface eluate, equivalent to *c.* 50 grains, in 12 % sucrose with 0.5 % Triton × 100 for 4 min. (*b*) Macerate of the same sample eluate for a further 4 min. The main peaks visible in the scan of the surface eluate are present, but much additional protein has now been released from the interior of the grains. I am grateful to Dr Reinhard Rüchel for preparing these gels, making the scans and for instructing me in the art of microgel electrophoresis.

The great heterogeneity of the pollen-wall proteins and their taxonomic diversity has been demonstrated in several recent studies. In *Cosmos* with a self-incompatibility system of the sporophytic type, the wall diffusates show a minimum of fourteen components on SDS–polyacrylamide gels, two of which contain considerable carbohydrate, suggesting that they are glycoproteins (Knox, J. Heslop-Harrison & Y. Heslop-Harrison, 1975; Howlett, Knox, Paxton & J. Heslop-Harrison, 1975). The spectrum is dominated by components with molecular weights of 11 500, 14 000, 26 000 and 30 000 daltons. Among the antigens released from the pollen wall of *Cosmos* is a fraction cross-reacting with Antigen-E, one of the principal allergens held in the pollen wall of *Ambrosia* spp., the ragweeds notorious as a cause of hayfever in the US (Knox & J. Heslop-Harrison, 1971). Interestingly, the spectrum of proteins present in diffusates from *Brassica* also includes a dominant low-molecular-weight fraction, seemingly derived from the exine domain, and therefore of sporophytic origin (Knox *et al.*, 1975).

In many families the pollen receives no appreciable contribution from the parent sporophyte, the protein load of the wall being mainly in the intine and so of gametophytic origin. This is true for many Iridaceae (Knox, 1971; Y. Heslop-Harrison, 1977), and for the Gramineae. In the latter family, generally regarded as advanced among the monocotyledons, the pollen has a single germination aperture, and the exine is porous. The emission on hydration is readily observed, both in artificial media and on the stigma (Plate 2). The fractions emitted include the allergens responsible for human hayfever, and at least three glycoproteins. The complexity of the, proteins held in the intine is indeed astonishing; thus in rye (*Secale cereale*) gel electrofocussing reveals some 30 components (Plate 1 *c*).

The taxonomic diversity of the pollen-wall materials is shown in the protein and antigen spectra (Howlett *et al.*, 1975) and in the capacity for binding lectins (Watson, Knox & Creaser, 1974). In general, so far as the immunological differences are concerned, the closer the taxonomic relationship of the species, the greater the degree of cross-reactivity. Nevertheless, the diversity of the wall proteins within a family may be quite considerable, as shown by comparison of the short-term exine diffusates of two species of different genera of the Malvaceae in Plate 1 *a* & *b*. Exploration of the *intraspecific* diversity of the wall-held constituents have been limited as yet, but the important observations of Lewis and collaborators on the antigens of the pollen of *Oenothera organensis* (Onagraceae, gametophytic system) are significant in considering which of the pollen-borne fractions might be concerned in the recognition reactions of the self-incompatibility system. Lewis (1952) and Mäkinen & Lewis

(1962) showed that pollen extracts of *O. organensis* contain antigens specific for the *S*-allele, and Lewis, Burrage & Walls (1967) found that the *S*-genotype of individual, intact pollen grains could be identified by precipitin reactions between diffusate and specific antibodies held in the hydrating medium. The nature of the genetical control in *O. organensis* indicates that the recognition factors on the male side should be of gametophytic origin. In an incompatible match in this species the pollen tubes are arrested in the surface layer of the stigma (Emerson, 1940; Hecht, 1966), and this fact, together with the observation that *S*-gene specific antigens are released from non-germinated pollen into artificial media, strongly suggests that the incompatibility factors are held in the intine, although a direct demonstration of this is lacking. The poral intine of the Onagraceae is the repository of massive amounts of protein, and the early diffusion is exclusively from this source.

Protein or antigen diversity of a kind that might be associated with the self-incompatibility system has yet to be demonstrated in the pollens of species with sporophytic control of pollen behaviour. Such trials as have been made – all with species of Cruciferae – have produced uniformly negative results (Nasrallah & Wallace, 1967a, b; Knox et al., 1975).

The pollen-receiving surfaces of angiosperm *stigmas* vary in the amount of secretion they carry at maturity, and this variation is broadly related to the type of incompatibility system (J. Heslop-Harrison *et al.*, 1975a; Y. Heslop-Harrison & Shivanna, 1977). All species with known sporophytic systems have slight amounts of secretion, evincing the 'dry' type of stigma. Most species with gametophytic systems carry a more or less copious fluid secretion, notable exceptions being the grasses, with dry plumose stigmas.

Characteristics of the receptive surfaces of stigma papillae of the dry type have now been examined in several families. In all cases the surface material is mainly proteinaceous, although lipids are usually also present. The layer lies outside of the cuticle, which is invariably discontinuous, with channels from the underlying pectocellulosic wall through which the secretion reaches the surface (Heslop-Harrison *et al.*, 1975b). The protein layer characteristically shows non-specific esterase activity (Mattsson, Knox, J. Heslop-Harrison & Y. Heslop-Harrison, 1974). The layer is readily disrupted and dispersed by proteolytic enzymes, but stripped only slowly by water or buffers. Detergents quickly remove it from the cuticle, and an earlier attempt to characterise the components released by SDS from the stigmas of *Hibiscus* (Malvaceae) revealed the presence of seven fractions separable by polyacrylamide gel electrophoresis, with two esterase isozymes (J. Heslop-Harrison *et al.*, 1975a), and two glycoproteins.

Microgel electrofocussing with the surface eluate of the equivalent of ten stigmas of rye shows about 24 components (Plate 1 d), and in *Gladiolus* (Iridaceae, no self-incompatibility system) Knox, Clarke, Harrison, Smith & Marchalonis (1976) have found that at least 17 fractions are present in the stigma surface pellicle, including 3 esterase isozymes and 3 glycoproteins.

The stigma surface layer has obviously an important claim to be regarded as the receptor site in the pollen recognition system, since it receives the early outflow from the pollen wall during the initial stages of hydration. Significantly, the extracuticular layer binds lectins such as concanavalin A (Gramineae, Y. Heslop-Harrison, 1976; Iridaceae, Knox *et al.*, 1976), and the binding is blocked by the appropriate sugars. Should the layer be truly the recognition site, it might be expected that it will carry constituents bearing *S*-allele specificity in those species where the inhibition in incompatible combinations is at the stigma surface. Ten years ago, Nasrallah & Wallace provided the first evidence that the stigmas of *Brassica* (Cruciferae, sporophytic system) do indeed contain *S*-gene specific antigens (Nasrallah & Wallace, 1967a; 1967b; also Nasrallah, Barber & Wallace, 1970), and showed furthermore that these diffuse slowly from the surfaces of intact stigmas in veronal buffer. We have shown that the antigens present in the stigma diffusate of *Brassica* are derived mainly from the surfaces of the papillae (J. Heslop-Harrison *et al.*, 1975a), strongly suggesting that it is the extracuticular layer that carries the *S*-gene specific fractions, although these are no doubt reinforced continuously throughout the functional life of the stigma by further secretion from within the papillae.

Binding reactions on the stigma surface

Upon contact with the stigma surface, pollen of the same species will normally hydrate and then release the wall-held constituents, which pass out onto the stigma surface (Knox, 1973; Knox & J. Heslop-Harrison, 1973; J. Heslop-Harrison, Knox & Y. Heslop-Harrison, 1974). The outflow begins before germination, but may continue for some considerable time after the emergence of the tube. In species with the 'dry' type of stigma the pollen protein exudates bind rapidly to the surface proteins of the papilla in the meniscus formed between the grain and the papilla surface. Some of the association is no doubt non-specific, but the binding is partly species-specific, even in genera like *Gladiolus*, where there is no self-incompatibility system (Knox *et al.*, 1976; Knox, personal communication).

The speed of the passage of the wall materials onto the stigma is graphically illustrated in the grasses (Table 1). The interface between exine and stigma surface in *Gaudinia fragilis* is seen in the electron micrograph of Plate 3*c* which may be interpreted as showing the outflow of the intine-held materials through the micropores of the exine and their subsequent association with the extracuticular surface proteins of the stigma (Plate 3*a*). The pollen contribution, originating as it does in the intine, is here from the gametophytic domain. With the emergence of the tube, a further contribution is made to the meniscus from the apertural intine. This outflow corresponds to that seen in Plate 2*b* and 2*c*, where the grain is in an artificial medium. The transfer of antigens from this site to the stigma surface can be followed using the immunofluorescence method (Knox & Heslop-Harrison, 1973; Plate 2*d*).

Table 1. *Initial behaviour of pollen following self- and cross-pollination in* Secale cereale

| | Meniscus appears (s) | Exine exudate appears (s) | Tube growth rate, μm s^{-1} | |
			to stigma contact	through stigma
Self	23.5 ± 1.6	43.5 ± 2.9	1.55 ± 0.11	arrested
Cross	21.7 ± 1.6	36.2 ± 2.6	1.35 ± 0.21	2.23

Observations made by Dr K. R. Shivanna and Dr Y. Heslop-Harrison, June 1975.

Several events follow upon the passage of pollen-borne materials onto the stigma surface, many not at all associated with the incompatibility response. In species with dry stigmas the pollen tube cannot enter until the cuticular layer is penetrated, and this depends on enzyme activation at the stigma surface near the contact face with the emerging tube tip (J. Heslop-Harrison *et al.*, 1975*b*; Y. Heslop-Harrison, 1977). Again taking a grass species as an example, we see in Plate 3*b* the intense surface esterase activity in the vicinity of the contact face between the tube tip and the stigma surface in *Alopecurus pratensis*. While it is not at all certain what this cytochemically detectable activity might betoken, it is possible that cutin-esterase activity is indicated. In any event, the erosion of the cuticle begins in the vicinity of this surface activity. The disruption at the tube tip in *Gaudinia fragilis* is seen in the electron micrograph of Plate 3*d*. The intine exudates themselves are incapable of lysing the cuticle (Mattsson, unpublished results; Y. Heslop-Harrison, 1977), suggesting that activation on the stigma surface is obligatory. That the surface materials of the stigma are concerned in this has now been shown in several experiments. If the superficial protein layer is dispersed enzymically, the tube tip cannot penetrate (Caryophyllaceae: J. Heslop-Harrison & Y. Heslop-Harrison,

1975; Cruciferae: J. Heslop-Harrison, 1975b; Iridaceae: Y. Heslop-Harrison, 1977). Furthermore, a similar effect is achieved if lectins are bound to the stigma (Knox et al., 1976). Complementations of this kind have the potentiality of forming interspecific compatibility barriers (J. Heslop-Harrison, 1976a, 1976b), but there is no reason to suppose that they are concerned in the operation of the self-incompatibility system.

Surface materials and the incompatibility response

The evidence bearing on the recognition reaction of the self-incompatibility system remains fragmentary and also largely circumstantial. Most is derived from work on the sporophytic systems of the Cruciferae and Compositae. In these families the expectation is that the recognition event occurs on the stigma surface, since it is here that the inhibition is imposed. In each family, rejection of incompatible pollen is accompanied by the development of callose bodies in the contiguous stigma papillae (Knox, 1973; Dickinson & Lewis, 1973a; J. Heslop-Harrison, Y. Heslop-Harrison & Knox, 1973a). The stigma thus signals the advent of an incompatible grain on its surface in a quickly recognisable way, since callose is readily located cytochemically. Employing this bioassay for the recognition reaction the involvement of the pollen surface fractions in the process has now been demonstrated in the Cruciferae: diffusates of incompatible pollen induce the stigma rejection response; those from compatible do not (Dickinson & Lewis, 1973b; J. Heslop-Harrison et al., 1973b; J. Heslop-Harrison et al., 1974). The genetical control of pollen behaviour in the Cruciferae is sporophytic, a fact which suggests control on the pollen side by the exine materials derived from the parental tapetum (Heslop-Harrison, 1968a, 1968b). It is therefore a satisfactory circumstance that the response is indeed activated by the exine fractions, and even by the tapetal material alone in the absence of pollen, provided that the combination is an incompatible one.

Specific effects on pollen germination by aqueous stigma diffusates have recently been demonstrated by Ferrari & Wallace (1975, 1976) in Brassica. So far as the initial recognition response is concerned, the important feature of these experiments seems to be that low concentrations of self-stigma diffusate inhibit pollen germination, whereas diffusates from compatible stigmas do not. The factors present in the diffusates responsible for these effects are yet to be characterised, but Ferrari & Wallace note the relationship with the earlier work in which S-genotype specific proteins were detected in stigma diffusates of Brassica. Clearly the possible involvement of these proteins in the response now requires urgent attention.

Circumstantial evidence for the involvement of the stigma surface proteins in the recognition response in the Cruciferae comes from experiments in which the surface layer has been disrupted enzymically. Removal of the surface proteins of *Raphanus* stigmas by mild pronase digestion does not kill the papillae, but, as in the Caryophyllaceae (J. Heslop-Harrison & Y. Heslop-Harrison, 1975) and Iridaceae (Y. Heslop-Harrison, 1977), prevents the penetration of the pollen tube tip in both compatible and incompatible combinations. The percentage of germination of incompatible grains on the stigmas may increase, and at the same time the intensity of the callose reaction in adjacent papillae is often reduced, the deposition becoming general. The result is consistent with the view that the surface pellicle carries the female-side recognition factors as well as those concerned with such enzyme activations as follow upon the advent of the pollen, and that their removal disrupts the chain of events that would normally lead to the rejection responses in an incompatible mating. The implication is, of course, that the factors involved are themselves proteins or glycoproteins, corresponding to those detected by Wallace and colleagues in *Brassica*; but again we are at present without direct proof.

THE REJECTION STEP

The recognition event of the self-incompatibility response is followed by processes which lead ultimately to a discrimination between compatible and incompatible pollen, the former being allowed to convey its gametes to the ovule and the latter not. This could be achieved in several different ways: for example, (a) stigma and style may provide all the requirements for germination and tube growth, incompatible tubes being actively inhibited; (b) stigma and style may supply activators for germination and growth to compatible grains while incompatible grains are not activated, or (c) both inhibition and activation may be involved, simultaneously or sequentially.

Alternative (a) has an intuitive appeal, and was that favoured by earlier workers in the field, such as East (1934) and Sears (1937). It poses the question of the nature of the inhibition and the relationship with the recognition event determining the specificity of the rejection. We shall return to it.

Alternatives (b) and (c) embrace several possibilities, and versions have been favoured in various theories of the incompatibility response. Ascher (1966) proposed that a regulatory molecule arises from the interaction of the pollen and stylar recognition factors which, in an incompatible pollination, prevents the switching on of metabolic pathways essential for rapid tube growth, with a consequential discrimination in favour of compatible tubes that do make the transition to rapid growth. Donk (1975) assumes that 'rejection of pollen tubes is the "normal" reaction of the style', and that pollen tubes can only reach the ovary if a special set of genes is switched on in the style, the products of which in turn activate the pollen genome. In incompatible combinations this activation is blocked, so that tube growth slows and stops. These two models have been based mainly on the responses of plants with gametophytic self-incompatibility systems, in which inhibition of incompatible tubes occurs in the style. The model proposed by Ferrari & Wallace (1976 and *in litt.*) relates to the response in *Brassica*, a genus with a sporophytic self-incompatibility system and surface inhibition. This model assumes that compatibility relationships are regulated by proteins synthesised in the pollen soon after hydration, one an activator, the other an inhibitor. In compatible combinations, the activator prevails and the effect of the inhibitor is nullified, while in incompatible combinations the production of the activator is blocked, so that the inhibitor alone is effective.

In a system with so many elements there are almost limitless possibilities for model building, and those mentioned clearly do not exhaust the field. However, I would like to urge here the merits of the 'primitive' hypothesis, namely that incompatibility results from a specific growth inhibition imposed on the tube following the initial recognition reaction. We may take the grasses as a model.

As we have seen, the grasses have a gametophytic self-incompatibility system. The early observations of Hayman (1956) and others established that the block to pollen tube growth in incompatible combinations occurs at or very near the stigma surface; in this respect the response is more akin to that seen in sporophytic systems, and not like that in the Solanaceae, Liliaceae and various other families with gametophytic control of pollen behaviour where the tube is arrested in the style. Moreover, the grasses share with families like the Cruciferae and Compositae, which have sporophytic systems, the possession of a 'dry' stigma surface.

The grasses are especially informative because of the extraordinary speed of the whole sequence of events following upon pollination. After the first attachment the pollen grain becomes hydrated, in rye increasing in volume up to 20 % in under thirty seconds. Thereafter exudation begins through the micropores of the exine, as first recorded by Watanabe (1955). Germination follows immediately, the tube growing directly towards the adjacent stigma papilla. Detailed observations on the events following compatible and incompatible pollinations in rye were made in 1975, and those pertaining to the rates of various processes are summarised in Table 1. The notable feature is that until the moment of contact with the stigma surface there is no difference in the behaviour of self and foreign pollen tubes. The first manifestation of incompatibility is the arrest of self-tubes after their tips make contact with the stigma papilla, when growth ceases in a matter of seconds. Shivanna has studied the response in various other grasses, and has found that, where a self-incompatibility system exists, inhibition is invariably delayed in a self-pollination until the tube tip makes its initial contact, or in certain cases until it penetrates and makes a limited amount of growth between the papillar cells.

It seems, then, that the key interaction is between the tube tip and the stigma surface. The recognition event occurring then is followed in an incompatible pollination by a dramatically rapid arrest of growth. In the case of rye, the whole response may be completed in as short a period as $1\frac{1}{2}$ minutes after the first capture of the grain and the beginning of hydration. It is wholly improbable that within this time there should be a complete sequence of events involving gene activation or repression with consequent changes in protein spectra; or indeed that there should be protein synthesis mediated by pre-existing mRNA, as proposed in some published models for the self-incompatibility response. In the grasses, the recognition factors are clearly present in pollen and stigma at the time of pollination, and we have seen already where they are likely to be situated. Furthermore, the speed with which the inhibition is imposed on incompatible tubes strongly suggests a trigger-like response, again not requiring further protein synthesis.

It is informative to compare the fates of compatible and incompatible pollen after contact with the stigma. In the compatible match in rye, the tube maintains its average growth rate of 2–3 μm s^{-1} through the stigma for a distance of 2–4 mm until the tip reaches the ovule. The starch reserves of the grain are transferred to the tube to contribute to the polysaccharide of the tube wall. Ultimately the intine is also scavenged, leaving the empty exine on the stigma surface. Growth of the grass pollen tube, like that in many other genera, proceeds through the apposition of vesicles containing precursor materials to the plasmalemma immediately behind the tip. The precursors are processed on the outer face of the plasmalemma, forming the outer cellulosic and then the inner callosic layers of the wall. In an *incompatible* match, the storage starch is mobilised in a normal fashion, and the production of the precursor wall materials is *not* arrested when the inhibition is imposed. The vesicles continue to circulate up to the tip for up to two hours after the arrest, and cyclosis continues in the grain itself. The synthesis of callose continues, and the short tube then becomes partly occluded. There is therefore no general metabolic arrest; the inhibitory event specifically effects tube growth.

We may also consider the behaviour following germination on artificial media. On suitable semi-solid media, the pollen grains of rye behave essentially as on the stigma, hydrating and germinating with the same precipitate rapidity, and the tube maintains an initial growth rate of c. 1.5μm s^{-1} – comparable, that is, to the rate on the stigma. The maximum length attained in the most favourable cultures is about one quarter of the mean length achieved in the pistil. At this time, however, the starch reserves of the grain are used up, and erosion of the intine has begun; we may suppose accordingly that the arrest occurs when endogenous reserves are exhausted. The structure of the tube wall is quite normal *in vitro*, and callose is deposited regularly as an inner lining up to the moment when growth fails. Significantly, however, various features of the incompatibility response *can* be simulated in pollen tubes in culture. When growth is brought to a stop by transfer to a hypotonic medium, callose accumulates in the tip and ultimately in the grain itself, much as in incompatible grains arrested on the stigma.

In the light of the above and other evidence I do not see that it is necessary to hypothesise any form of control in the grass self-incompatibility system other than one which directly and immediately affects the tip growth mechanism of the incompatible tube following upon the recognition event. The immediate target is probably some component of the enzyme systems concerned with cellulose synthesis, which are likely to be located in or on the plasmalemma (Mueller, Brown & Scott, 1977).

What, then, lies between the target and the site of the initial recognition reaction? There are two main possibilities, namely that some product of the binding itself moves to the plasmalemma where it acts as an inhibitor, or that the interaction of pollen- and stigma-borne factors mobilises an intermediate, highly diffusible, 'messenger', of the nature of an allosteric inhibitor, which passes to the target enzymes in the plasmalemma. Currently I favour the second possibility, which is embodied also in a scheme proposed for the incompatibility response of the Cruciferae (J. Heslop-Harrison *et al.*, 1975 b). The Cruciferae provide supporting evidence of a different kind for the idea of a mobile intermediary. In the crucifers the stigma also reacts in an incompatible pollination, the papillae producing a callose lenticule adjacent to the contact face with the grain. Dickinson & Lewis (1973 a), in a fine-structural study of *Raphanus*, showed that this response is initiated at the plasmalemma of the papilla, the callose being laid down in waves on the inner face of the pectocellulosic wall where none would otherwise accumulate during the normal life of the cell. The callose synthetase system, like that for cellulose, is held at the plasmalemma. The fact that it is activated in a specific and strictly localised manner in the stigma rejection response shows very clearly that an early link in the chain of events following upon the recognition reaction involves the re-programming of enzymes in this site.

From the earliest period of research on self-incompatibility in flowering-plants those who have studied the cytological aspects of tube inhibition have been struck by evidences of abnormal tip growth (e.g., Sears, 1937; review, Linskens & Kroh, 1967). The fine-structural observations of Nettancourt and collaborators (Nettancourt et al., 1973) on *Lycopersicon* (Solanaceae, gametophytic system) have shown that here, as in the Gramineae, the arrest of a pollen tube involves an accumulation of wall-precursor vesicles in the tip zone, accompanied eventually by bursting or callose occlusion. In *Lycopersicon* the inhibition is imposed during growth through the style, but as in the Gramineae with surface inhibition the events are fully compatible with the conclusion that the effect is on an enzyme system in the plasmalemma of the tip zone.

We do not therefore require hypotheses involving gene action or necessarily protein synthesis after the primary recognition event to explain the inhibitory responses of the incompatibility reaction. This is not to say, however, that the active promotion of germination and tube growth may not also be a factor in the overall control. Tubes penetrating the stigma surface eventually become dependent nutrionally on the female tissue, and it is entirely possible that the pollen itself might initiate metabolic changes in preparation for the advent of compatible tubes. A scheme of this kind could be applied also to those gametophytic systems where the inhibition is in the style, taking into account that in such instances the *S*-gene products may be synthesised during the growth of the pollen tube, so making the response a cumulative one, with recognition and rejection phases overlapping.

REFERENCES

ASCHER, P. D. (1966). A gene action model to explain gametophytic self-incompatibility. *Euphytica*, 15, 170–83.

ASCHER, P. D. (1975). Special stylar property required for compatible pollen-tube growth in *Lilium longiflorum* Thunb. *Botanical Gazette*, 136, 317–21.

BROOKS, J. & SHAW, G. (1971). Recent developments in the chemistry, biochemistry, geochemistry and post-tetrad ontogeny of sporopollenin derived from pollen and spore exines. In *Pollen: Development and Physiology*, ed. J. Heslop-Harrison, pp. 99–114. Butterworths, London.

DICKINSON, H. G. & LEWIS, D. (1973a). Cytochemical and ultrastructural differences between intraspecific compatible and incompatible pollinations in *Raphanus*. *Proceedings of the Royal Society, London*, B 183, 21–38.

DICKINSON, H. G. & LEWIS, D. (1973 b). The formation of the tryphine coating the pollen grains of *Raphanus*, and its properties relating to the self-incompatibility system. *Proceedings of the Royal Society, London*, B, **184**, 149–65.

DONK, J. A. W. M. VAN DER (1974). Synthesis of RNA and protein as a function of time and type of pollen-tube-style interaction in *Petunia hydrida* L. *Molecular and General Genetics*, **134**, 93–8.

DONK, J. A. W. M. VAN DER (1975). Recognition and gene expression during the incompatibility reaction in *Petunia hybrida* L. *Molecular and General Genetics*, **141**, 305–16.

EAST, E. M. (1934). Norms of pollen tube growth in incompatible matings of self-sterile plants. *Proceedings of the National Academy of Sciences, U.S.A.*, **20**, 225–30.

EAST, E. M. & MANGELSDORF, A. J. (1925). A new interpretation of the hereditary behaviour of self-sterile plants. *Proceedings of the National Academy of Sciences, U.S.A.*, **11**, 166–71.

EMERSON, S. H. (1940). Growth of incompatible pollen tubes in *Oenothera organesis*. *Botanical Gazette*, **101**, 890–911.

FERRARI, T. E. & WALLACE, D. H. (1975). Germination of *Brassica* pollen and expression of incompatibility *in vitro*. *Euphytica*, **24**, 757–65.

FERRARI, T. E. & WALLACE, D. H. (1976). Pollen protein synthesis and control of incompatibility in *Brassica*. *Theoretical and Applied Genetics*, **48**, 243–9.

GALUN, E. (1977). Chapters on Incompatibility. In *Pollination Mechanisms, Reproduction and Plant Breeding*, by R. Frankel & E. Galun. Springer, Berlin, Heidelberg, New York.

HAYMAN, D. L. (1956). The genetical control of incompatibility in *Phalaris coerulescens*. *Australian Journal of Biological Sciences*, **9**, 321–31.

HECHT, A. (1966). Growth of pollen tubes of *Oenothera organensis* through otherwise incompatible styles. *American Journal of Botany*, **47**, 32–6.

HESLOP-HARRISON, J. (1968 a). Ribosome sites and *S*-gene action. *Nature, London*, **218**, 90–91.

HESLOP-HARRISON, J. (1968 b). Tapetal origin of pollen-coat substances in *Lilium*. *New Phytologist*, **67**, 779–86.

HESLOP-HARRISON, J. (1975 a). Incompatibility and the pollen–stigma interaction. *Annual Reviews in Plant Physiology*, **26**, 403–25.

HESLOP-HARRISON, J. (1975 b). The physiology of the incompatibility reaction in the Cruciferae. *Proceedings of the Eucarpia Meeting on the Cruciferae*, eds. E. B. Wills & C. North, pp. 14–19. Scottish Horticultural Research Institute.

HESLOP-HARRISON, J. (1975 c). The physiology of the pollen-grain surface. *Proceedings of the Royal Society, London*, B **190**, 275–99.

HESLOP-HARRISON, J. (1976 a). A new look at pollination. *Annual Report of the East Malling Research Station for 1975*, pp. 141–57.

HESLOP-HARRISON, J. (1976 b). Male gametophyte selection and the pollen–stigma interaction. In *Gamete Competition in Plants and Animals*, ed. D. L. Mulcahy, pp. 177–90. North Holland, Amsterdam.

HESLOP-HARRISON, J. & HESLOP-HARRISON, Y. (1975). Enzymic removal of the proteinaceous pellicle of the stigma papilla prevents pollen tube entry in the Caryophyllaceae. *Annals of Botany*, **39**, 163–5.

HESLOP-HARRISON, J., HESLOP-HARRISON, Y. & KNOX, R. B. (1973 a). The callose rejection reaction: a new bioassay for incompatibility in Cruciferae and Compositae. *Incompatibility Newsletter*, **3**, 75–6.

HESLOP-HARRISON, J., HESLOP-HARRISON, Y., KNOX, R. B. & HOWLETT, B. (1973 b). Pollen-wall proteins: 'Gametophytic' and 'Sporophytic' fractions in the pollen walls of the Malvaceae. *Annals of Botany*, **39**, 403–12.

HESLOP-HARRISON, J., KNOX, R. B. & HESLOP-HARRISON, Y. (1974). Pollen-wall proteins: exine-held fractions associated with the incompatibility response in the Cruciferae. *Theoretical and Applied Genetics*, **44**, 133–7.

HESLOP-HARRISON, J., HESLOP-HARRISON, Y. & BARBER, J. T. (1975a). The stigma surface in incompatibility responses. *Proceedings of the Royal Society, London*, B **188**, 287–7.

HESLOP-HARRISON, J., KNOX, R. B., HESLOP-HARRISON, Y. & MATTSSON, O. (1975b). Pollen-wall proteins: emission and role in incompatibility responses. In *The Biology of the Male Gamete*, eds. J. G. Duckett & P. A. Racey. *Biological Journal of the Linnean Society*, **9** (Suppl. 1), 189–202.

HESLOP-HARRISON, Y. (1976). Localisation of concanavalin A binding sites on the stigma surface of a grass species. *Micron*, **7**, 33–6.

HESLOP-HARRISON, Y. (1977). The pollen–stigma interaction: pollen-tube penetration in *Crocus*. *Annals of Botany*, **41**, 913–22.

HESLOP-HARRISON, Y. & SHIVANNA, K. R. (1977). The receptive surface of the angiosperm stigma. *Annals of Botany*, **41**, 1233–58.

HOWLETT, B., KNOX, R. B., PAXTON, J. H. & HESLOP-HARRISON, J. (1975). Pollen-wall proteins: physicochemical characterisation and role in self-incompatibility in *Cosmos bipinnatus*. *Proceedings of the Royal Society, London*, B **188**, 167–82.

KNOX, R. B. (1971). Pollen-wall proteins: localisation, enzymic and antigenic activity during development in *Gladiolus*. *Journal of Cell Science*, **9**, 209–37.

KNOX, R. B. (1973). Pollen-wall proteins: pollen-stigma interactions in ragweeds and *Cosmos* (Compositae). *Journal of Cell Science*, **12**, 421–43.

KNOX, R. B. & HESLOP-HARRISON, J. (1969). Cytochemical localisation of enzymes in the wall of the pollen grain. *Nature, London*, **223**, 92–4.

KNOX, R. B. & HESLOP-HARRISON, J. (1970). Pollen-wall proteins: localisation and enzymic activity. *Journal of Cell Science*, **6**, 1–27.

KNOX, R. B. & HESLOP-HARRISON, J. (1971). Pollen-wall proteins: localisation of antigenic and allergenic fractions in the pollen grain walls of *Ambrosia* spp. (ragweeds). *Cytobios*, **4**, 49–54.

KNOX, R. B. & HESLOP-HARRISON, J. (1973). Pollen-wall proteins: fate of intine-held antigens on the stigma in compatible and incompatible pollinations of *Phalaris tuberosa* L. *Journal of Cell Science*, **9**, 239–52.

KNOX, R. B., HESLOP-HARRISON, J. & HESLOP-HARRISON, Y. (1975). Pollen-wall proteins: localisation and characterisation of gametophytic and sporophytic fractions. In *The Biology of the Male Gamete*, eds. J. G. Duckett & P. A. Racey. *Biological Journal of the Linnean Society*, **9** (Suppl. 1), 177–87.

KNOX, R. B., CLARKE, A., HARRISON, S., SMITH, P. & MARCHALONIS, J. J. (1976). Cell recognition in plants: determinants of the stigma surface and their pollen interactions. *Proceedings of the National Academy of Sciences, U.S.A.*, **73**, 2788–92.

LEWIS, D. (1952). Serological reactions of pollen incompatibility substances. *Proceedings of the Royal Society, London*, B **140**, 127–35.

LEWIS, D. (1965). Incompatibility: a protein dimer hypothesis. *Genetics Today*. *Proceedings XI International Congress on Genetics, The Hague*, 1963, ed. S. J. Geerts, **3**, 1729–35.

LEWIS, D., BURRAGE, S. & WALLS, D. (1967). Immunological reactions of single pollen grains. Electrophoresis and enzymology of pollen protein exudates. *Journal of Experimental Botany*, **18**, 371–8.

LINSKENS, H. F. (1960). Zür Frage der Entstehung der Abwehr-Körpen in der Inkompatibilitäts-reaktion von *Petunia*. III. *Zeitschrift für Botanik*, **48**, 126–35.

LINSKENS, H. F. & KROH, M. (1967). Inkompatibilität der Phanerogamen. *Encyclopaedia of Plant Physiology*, ed. W. Ruhland, vol. 18, pp. 506–30. Springer, Berlin, Heidelberg, New York.

PLATE I

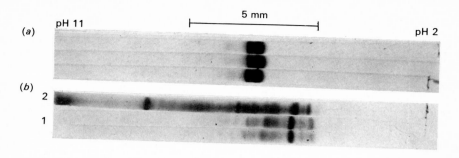

(a) pH 11   5 mm   pH 2

(b)   2

       1

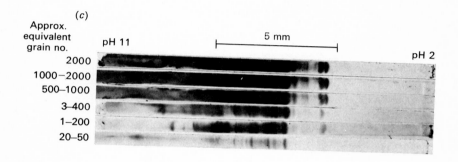

(c)

Approx.
equivalent
grain no.   pH 11   5 mm   pH 2

2000
1000–2000
500–1000
3–400
1–200
20–50

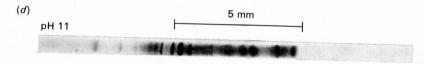

(d)

pH 11   5 mm

For explanation see p. 137

PLATE 2

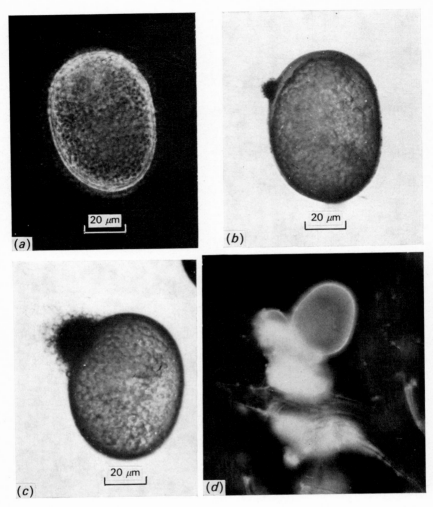

For explanation see p. 138

PLATE 3

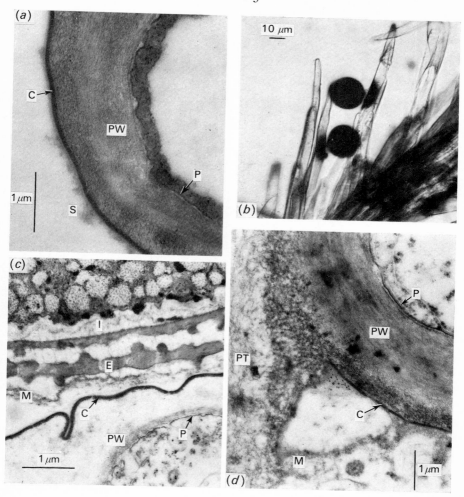

For explanation see p. 138

Mäkinen, Y. L. A. & Lewis, D. (1962). Immunological analysis of incompatibility (S) proteins and of cross-reacting material in a self-compatible mutant of *Oenothera organensis*. *Genetical Research*, **3**, 352–62.

Mattsson, O., Knox, R. B., Heslop-Harrison, J. & Heslop-Harrison, Y. (1974). Protein pellicle of stigma papilla as a probable recognition site in incompatibility reactions. *Nature, London*, **247**, 298–300.

Mueller, S. C., Brown, R. M. & Scott, T. K. (1977). Cellulose microfibrils: nascent stages of synthesis in a higher plant cell. *Science*, **194**, 949–51.

Nasrallah, M. E. & Wallace, D. H. (1967a). Immunochemical detection of antigens in self-incompatibility genotypes of cabbage. *Nature, London*, **213**, 700–701.

Nasrallah, M. E. & Wallace, D. H. (1967b). Immunogenetics of self-incompatibility in *Brassica oleracea* L. *Heredity*, **22**, 519–27.

Nasrallah, M. E., Barber, J. T. & Wallace, D. H. (1970). Self-incompatibility proteins in plants: detection, genetics, and possible mode of action. *Heredity*, **25**, 23–7.

Nettancourt, D. de (1977). *Incompatibility in Angiosperms*. Springer, Berlin, Heidelberg, New York.

Nettancourt, D. de, Devreux, M., Bozzini, A., Cresti, M., Pacini, E. & Sarfatti, G. (1973). Ultrastructural aspects of the self-incompatibility mechanism in *Lyocopersicum peruvianum* Mill. *Journal of Cell Science*, **12**, 403–419.

Ockenden, D. J. (1975). Dominance relationships between S-alleles in the stigmas of Brussels sprouts (*Brassica oleracea* var. *gemmifera*). *Euphytica*, **24**, 165–72.

Sears, E. R. (1937). Cytological phenomena associated with self-sterility in the flowering plants. *Genetics*, **22**, 130–51.

Watanabe, K. (1955). Studies on the germination of grass pollen. 1. Liquid exudation of the pollen on the stigma before germination. *Botanical Magazine*, **68**, 40–44.

Watson, L., Knox, R. B. & Creaser, E. H. (1974). Con A differentiates among grass pollens by binding specifically to wall glycoproteins and carbohydrates. *Nature, London*, **249**, 574–6.

EXPLANATION OF PLATES

PLATE I

Electrofocussing of pollen-wall and stigma eluates in $4 \mu l$ poly-acrylamide gels. Serva ampholytes, pH 2–11; 14 V mm^{-1}.

(a) Short-term (5–10 min) eluates of pollen of *Malva sylvestris* (Malvaceae) in 25 % sucrose, equivalent of c. 20 grains per gel. The protein yield is mainly from the exine domain, and is therefore derived from the parent sporophyte.

(b) 1. Replicate gels of short-term (5–10 min) eluates of *Althaea rosea* (Malvaceae) pollen in 25 % sucrose, c. 20 grains per gel. Protein mainly from the exine domain. 2. Gel of long-term (30–40 min) eluate in 25 % sucrose, equivalent to c. 15 grains. There is now a considerable contribution from the intine (gametophytic) domain.

(c) Pollen wall eluates of rye (*Secale cereale*, Gramineae). Elution for 30–60 min in 25 % sucrose, under which conditions the protein yield is mainly from the intine.

(d) Stigma surface eluate of rye. Elution for 1–2 h in 12 % sucrose with 0.005 % Triton X 100. Most of the protein is probably from the surface of the papillae. but there will be some contamination from cut cells.

PLATE 2

(*a*) Emission of proteins from the pollen wall of rye (*Secale cereale*) immersed in a stain–fixing medium (o.oo5 % Coomassie Blue, 2 % acetic acid, 10 % methanol); early passage of intine material through the micropores of the exine. Print in negative contrast from a positive colour original.

(*b, c*) Emission of proteins from the apertural site of the pollen of rye in the stain fixing medium. (*b*) 90 s; (*c*) 3 min. The outflow at this time is almost exclusively from the apertural intine, and these proteins are those mainly contributing to the separations of Plate 1.

(*d*) Fluorescent antibody localisation of antigens released during the early germination of pollen on the stigma of *Phalaris tuberosa* (Gramineae). The outflow is mainly from the apertural intine (details of method in Knox & Heslop-Harrison, 1973).

PLATE 3

(*a*) Electron micrograph of the tip of a stigma papilla of *Gaudinia fragilis* (Gramineae). Glutaraldehyde fixation, OsO_4 post-fixation, followed by extended uranyl acetate staining of the thin section. The surface protein layer (S) may be seen lying outside of the cuticle (C), separated from the plasmalemma (P) by the pectocellulosic wall (PW), which in this tip zone is 1–2 μm thick.

(*b*) Esterase activity in the vicinity of the contact face between pollen-tube and stigma papilla shortly after pollination in *Alopecurus pratensis* (Gramineae). α-naphthyl acetate substrate in a coupling reaction with Fast Blue B salt.

(*c*) Electron micrograph of the contact face between an ungerminated pollen grain and the stigma surface of *Gaudinia fragilis* (Gramineae). Glutaraldehyde fixation, OsO_4 post-fixation; uranyl acetate and lead citrate post-staining. The material held in the meniscus (M) includes contributions from both pollen grain and stigma surface. E, Exine; I, intine; C, cuticle of stigma papilla, separated somewhat from the pectocellulosic wall (PW), which is faced by the plasmalemma (P).

(*d*) Contact face between pollen tube tip (PT) and wall of stigma papilla (PW) in *Gaudinia fragilis*, preparation as in Plate 3(*c*). Immediately under the tube tip the cuticle (C) of the stigma papilla is completely eroded. Meniscus material is seen at M; P, plasmalemma.

CELLULAR INTERACTIONS DURING GRAFT FORMATION IN PLANTS, A RECOGNITION PHENOMENON?

BY M. M. YEOMAN, D. C. KILPATRICK,
M. B. MIEDZYBRODZKA AND A. R. GOULD*

Department of Botany, University of Edinburgh,
Mayfield Road, Edinburgh EH9 3JH, U.K.

INTRODUCTION

The joining together of parts of different plants by man to form a single viable entity has been an important technique in vegetative propagation through the ages, and is universally known as grafting (Roberts, 1949; Rogers & Beakbane, 1957). In recent times the major botanical interest has been focussed on the use of graft assemblages in studies on the movement of secondary metabolites between stock and scion (Dawson, 1944), translocation of assimilates across the graft union (de Stigter, 1961), polarity (Roberts, 1949), transfer of the stimulus to flower (Zeevaart, 1958), the ability of 'graft hybrids' to form an integrated unit as an indicator of taxonomic similarity or diversity (Kloz, 1971), and cellular interactions between the regenerating surfaces of stock and scion (Lindsay, Yeoman & Brown, 1974; Yeoman & Brown, 1976). A fact which has emerged from both the horticultural and botanical aspects of this work, is that not all grafts are successful. Some combinations of stock and scion are compatible and a successful union results, while others are incompatible and fail to produce an integrated unit. The basis of compatibility has received considerable attention but is not completely understood, particularly in a grafting context. Clearly the terms compatible and incompatible are interpreted differently by different workers. It appears that compatibility in a strictly horticultural sense (Garner, 1970) refers to a graft union which survives throughout the life of the plant and any breakdown of the graft prior to the death of the plant indicates that the combination was incompatible. Roberts (1949) states, 'By compatibility it seems best at present to refer to the long time success of a graft for economic, esthetic or scientific purposes. Anything less is incompatibility, partial compatibility, delayed incompatibility or some similarly expressed condition. This view-point

---

* Present address: Department of Biochemistry, University of Adelaide.

10-2

considers compatibility as any inter-influence between stock and scion and not merely effects arising directly from the union.' The investigations carried out on herbaceous plants in our laboratory have provided a rather different definition of compatibility which is based on the early developmental changes which occur during the formation of the graft union. In our view, the critical structural event in the formation of a successful graft is when the vascular elements of stock and scion become united (Fig. 1). Subsequently, such plants can survive, grow, flower and fruit under normal greenhouse conditions. Incompatible graft combinations do not display vascular continuity and can only survive in a very humid environment for a limited period (Yeoman & Brown, 1976). In addition, when the breaking weight of the graft, determined using the technique of Roberts & Brown (1961), is plotted against time of development, the slope of the curve obtained is indicative of whether the graft is compatible or not (Yeoman & Brown, 1976).

A major reason for studying the structural and physiological changes

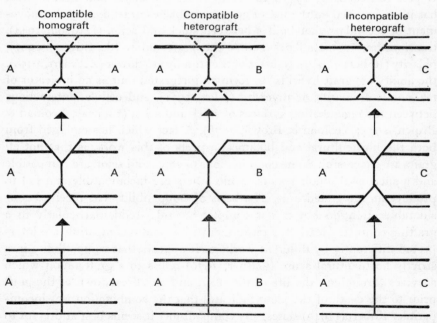

Fig. 1. A diagram illustrating the development of the graft union between the stock and scion of (1) a compatible homograft, (2) a compatible heterograft and (3) an incompatible heterograft. A = *Lycopersicon esculentum*, B = *Datura stramonium* and C = *Nicandra physaloides*. The vascular system is shown as a thick horizontal black line, the area enclosed by a dotted line is the new regenerated tissue.

which occur during the development of compatible and incompatible grafts is to discover the cause of incompatibility and to establish whether it is due to structural rather than physiological differences. If the latter, then the possibility of a recognition system operating must be considered. Our investigations have been made with precision under strictly controlled environmental conditions generally within one taxonomic group. The results and ideas which have emerged during the course of these investigations form the basis of this article and an attempt is made to relate our work to the relevant studies that have preceded it.

THE EXPERIMENTAL SYSTEM

Members of the family Solanaceae are amenable to experimental manipulation and have been used repeatedly in grafting studies. Working with the tomato, *Lycopersicon esculentum*, Roberts & Brown (1961) invented a quantitative technique to follow the development of the graft union. The technique was based on the use of a machine which could measure the mechanical strength of a graft, and therefore provided for the first time a basic parameter to compare the development of different graft combinations. Subsequently, this method was exploited by Lindsay *et al.* (1974) and Yeoman & Brown (1976) in studies on the physiology of grafting. For these investigations large numbers of very uniform young plants of a variety of Solanaceous species were manipulated and grafted under strictly controlled conditions. 'Graft hybrids' were only made between plants in which the diameter and shape of the opposing stock and scion were similar. A cut was made at right angles to the main axis of the plant midway between the cotyledons and the first leaf. The graft was assembled immediately and all subsequent procedures were performed as described by Lindsay *et al.* (1974).

STRUCTURAL EVENTS OCCURRING AT THE INTERFACE BETWEEN STOCK AND SCION

The freshly cut surfaces of stock and scion are held together tightly at the initiation of the graft. Initially, contact is established over the complete area of both faces but within a few hours shrinkage and collapse of the cells occurs. There is however, a greater degree of shrinkage in the peripheral region of the stem, in the cortex and around the vascular bundles, than in the pith (Fig. 1). This results in a partial separation of the opposing surfaces in stock and scion and this cavity is filled with exudate from the plant. Throughout the first two days only the cells of the piths (medullas) are in contact, and the cohesion between stock and scion, and therefore the strength of the graft, depends on this event. This initial cohesion between opposing pith cells occurs whether the combination is compatible or incompatible. Even widely incompatible 'graft hybrids' such as cucumber (*Cucumis sativus*) and runner bean (*Phaseolus coccineus*) produce an initial union with significant mechanical strength.

The initial cohesion between the pith cells of stock and scion involves

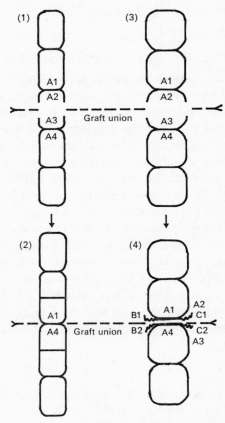

Fig. 2. A diagram illustrating the spatial relationships between the walls of cells at the graft union before and after a homograft of tomato has been established. Files of cells (1) and (2) are within the cortex in association with the vascular strands, while files (3) and (4) are in the medulla or pith. The letters designate individual cell walls or parts of cell walls, A_1 and A_4 are the walls of the last row of intact cells in the scion and stock respectively, A_2 and A_3 are the walls of the remnants of the cut cells which abut onto A_1 and A_4; B_1, B_2, C_1 and C_2 are the remains of the vertical cell walls remaining after the graft has been initiated. In part (4) the graft union has been 'exploded' to show the relationship between the various walls and parts of walls. (Redrawn from Yeoman & Brown, 1976.)

collapse of the cut cells and the accumulation of pectic materials leading to a substantial thickening of the cell walls forming the sandwich (Fig. 2). This series of changes takes place regardless of whether the two parts are compatible or not. If this initial cohesion is prevented by the insertion of an inert layer of Millipore filter (0.45 μm pores), or polythene, or if the pith is removed to a depth of 2 mm (on both sides) prior to the assembly of the graft, graft development will be retarded somewhat but ultimately a successful fusion of stock and scion will occur in compatible homografts and heterografts. However, it is in the outer regions of the stem, in the cortex and around the vascular bundles, that the critical changes occur which lead to the establishment of a successful graft. In parallel with the events which lead to the cohesion of the opposing pith regions, cells in and around the vascular bundles are induced to divide (Plate 1a) and the products of these divisions expand into the space filled with fluid between stock and scion (Plate 1a). Eventually cells growing from opposing surfaces meet and fuse, subsequently differentiating into elements which join together the vascular bundles of stock and scion (Plate 2a). In incompatible situations cells divide, expand and meet in a similar fashion but do not differentiate into xylem and phloem cells to give vascular continuity (Plate 2b).

EVIDENCE FOR A RECOGNITION SYSTEM

The simple fact that incompatibilities exist even within a family suggests that some form of recognition system is present in higher plants (Table 1). Of course, it is possible that the inability of two separated parts to achieve vascular continuity could be due to a structural or major biochemical difference between stock and scion (Table 2). Gross differences in anatomy could presumably prevent the formation of a successful union, or toxic materials released from one part could prevent, retard or modify the grafting process. Certainly within the groups of herbaceous plants explored in our investigations (which are probably virus-free) such structural or biochemical incompatibilities are only remote possibilities. In tomato homografts, rotation of the scion, so that the vascular bundles are put out of register with those of the stock, can slow down the formation of the graft to some slight extent but cannot prevent the attainment of a successful union. Indeed, at least within the species tested in the Solanaceae, there is no correlation between the anatomy of the two components of a heterograft and the ability to form a successful union. There are reports of biochemical incompatibilities between pear scions and quince rootstocks (Gur, Samish & Lifshitz, 1968) and between peach scions and almond

Table 1. *Grafting relationships within the Solanaceae*

	Scion						
	L. esculentum	*D. stramonium*	*N. physaloides*	*N. tabacum*	*S. melongena*	*P. hybrida*	*C. frutescens*
Lycopersicon esculentum	+	+	−	+	+	+	−
Datura stramonium	+	+	+	+	+	+	+
Nicandra physaloides	−	+	+	+	+	+	NT
Nicotiana tabacum	+	+	+	+	+	+	NT
Solanum melongena	+	+	+	+	+	+	NT
Petunia hybrida	+	+	+	+	+	+	NT
Capsicum frutescens	−	+	NT	NT	NT	NT	+

+, compatible; −, incompatible; NT, not tested.

Table 2. *Possible causes of incompatibility in the establishment of the graft union*

1. Differences in anatomy between stock and scion
2. Presence of a virus in stock or scion
3. Presence of toxic substances in stock/scion affecting the establishment of the graft
4. Release of a toxic substance from stock/scion by an enzyme in the scion/stock
5. Inability to achieve vascular continuity consequent on the recognition of stock/scion by scion/stock
6. Inability to achieve vascular continuity consequent on the non-recognition of stock/scion by scion/stock

rootstocks (Gur & Blum, 1973) where a cyanogenic glycoside descending from scion to stock is hydrolysed by a β-glycosidase, releasing cyanide, killing the cells in the union and preventing or disrupting graft formation. It is probable that gross biochemical incompatibilities do not occur within the group of plants studied in our laboratory, because such differences would promote necrosis of cells within the graft union and that has not been observed in our histological preparations. Therefore the available evidence is consistent with the view that incompatibility occurs for some other reason, at least for the species used in our investigations. It is suggested that some form of cell recognition occurs as a result of cell interaction and that the most likely site of recognition is the point of contact between the cells growing out from around and within the vascular bundles.

Preliminary studies with protoplasts have provided some limited circumstantial evidence which is consistent with the presence of a recognition capability associated with the outer surfaces of plant cells. Protoplasts carefully prepared from the leaves of *Lycopersicon*, *Datura* and *Nicandra* spp. provide a useful tool for the study of interactions between plasmalemmas. The separation of the cells and the removal of the cellulose wall is achieved enzymically (Cocking, 1974) and the protoplasts are kept intact in a solution of sucrose to prevent osmotic rupture.

It was found that a saline extract of *Lycopersicon* seeds would agglutinate protoplasts prepared from leaves of *Lycopersicon* or *Datura* at a lower concentration than the minimum required to agglutinate protoplasts from *Nicandra* leaves. Conversely, a saline extract of *Nicandra* seeds agglutinated *Nicandra* and *Datura* protoplasts more readily than those of *Lycopersicon*. There is therefore an apparent correlation between compatibility in a grafting context and agglutination. The possibility that this relationship is purely coincidental should be seriously considered, especially in the light of the report of Glimelius, Wallin & Eriksson (1974) that protoplasts prepared from cultured carrot cells can be readily agglutinated by concanavalin A, a major component of saline extract of the jack bean. The carrot and the jack bean are universally regarded as being only distantly related!

A different approach has been to look for grafting-induced protein synthesis. Briefly this was carried out as follows. A cut was made in a stem of *Lycopersicon esculentum*, then the two surfaces were grafted together again. After three days the internode (the stem region immediately below and above the graft union) was excised and incubated for twenty-four hours in a medium containing [^{3}H]methionine. It was then ground in buffered sucrose solution using a mortar and pestle. The resulting homogenate was centrifuged (10000 g; 10 min) to remove large particles like cell walls, nuclei and chloroplasts. The supernatant (containing various membranous organelles and soluble proteins) was treated with sodium dodecyl sulphate and subjected to polyacrylamide slab gel electrophoresis. Finally, the gel was cut into 1 mm sections each of which was counted for radioactivity. Identical treatment was given to a similar piece of stem which had not been cut and grafted together again, except that it was incubated in the presence of [^{35}S]methionine. The corresponding segments labelled with ^{3}H and ^{35}S were homogenised together before application to the gel. Any difference in the spectrum of protein synthesis between the grafted and ungrafted stem sections should be apparent as a variation in the ratio of ^{3}H c.p.m. to ^{35}S c.p.m. across the gel. Such a variation was indeed observed (Fig. 3*a*). To try to eliminate the possibility that the observed difference in protein syntheses was simply a response to wounding, a stem

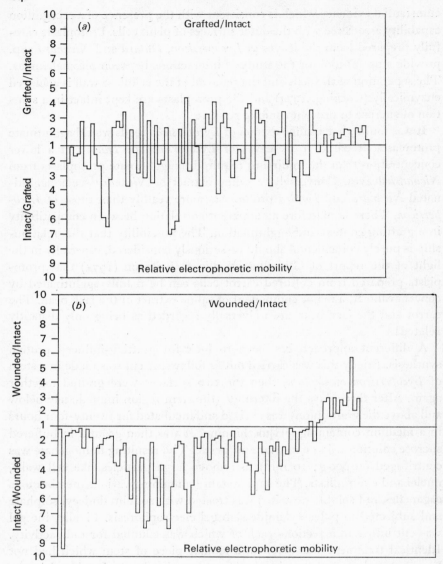

Fig. 3. Comparison of patterns of radioactive methionine incorporation into proteins during grafting and wounding.

(a) Ratios of ^{3}H c.p.m. (grafted internode) to ^{35}S c.p.m. (intact internode) of proteins separated by electrophoresis of a supernatant of an homogenate of (i) a region 1 mm on either side of the graft and (ii) an equivalent region on the intact internode taken from internodes of *Lycopersicon esculentum* labelled for 24 h, 3 days after grafting.

(b) Ratios of ^{3}H c.p.m. (wounded internode) to ^{35}S c.p.m. (intact internode) of protein from (i) a region 1 mm below the wounded surface and (ii) an equivalent region on the intact internode taken from internodes of *Lycopersicon esculentum* labelled for 24 h, 3 days after wounding.

Values for ^{3}H:^{35}S less than unity have been expressed as reciprocals on a displaced scale, making differences in the patterns of (a) and (b) more readily apparent.

PLATE I

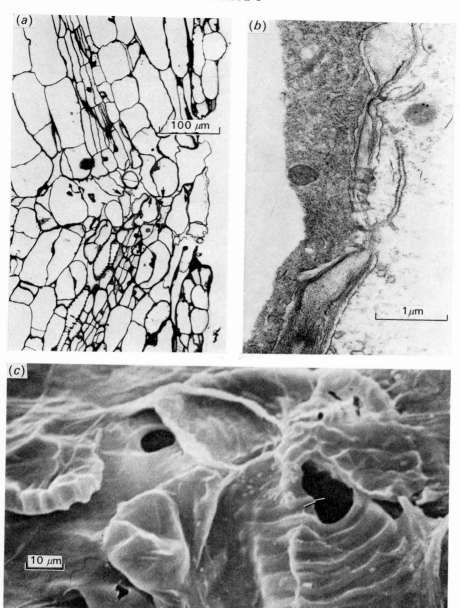

For explanation see p. 160

PLATE 2

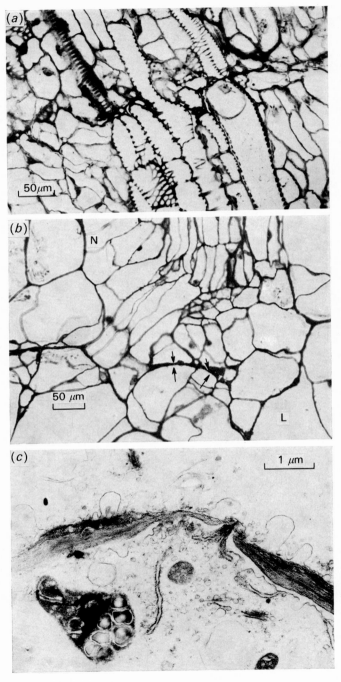

For explanation see p. 160

was cut as though for grafting, but the stem was capped by a layer of PTFE (a presumably inert material) instead of grafting on the scion. After this treatment, differences in protein synthesis were still apparent compared with an uncut stem section but the pattern of synthesis differed also from that of the genuine graft (Fig. 3b). The tentative conclusion to be drawn from these experiments is that some proteins are preferentially synthesised in response to wounding but when a graft is forming, additional proteins are also preferentially synthesised. It is tempting to speculate that the graft-specific protein synthesis is directed towards the production of proteins involved in the compatibility mechanism.

LOCATION OF THE RECOGNITION SYSTEM

It is clear that the events which finally give rise to a complete functioning graft union or an unsuccessful graft are preceded by specific cellular interactions. As the cells from opposing surfaces touch, the dissolution of the opposing walls is initiated, a hole appears rapidly and the plasmalemmas come into contact (Plate 1b). This sequence occurs whether the parts are compatible or incompatible (Plate 2c). Yeoman & Brown (1976) have suggested that the primary recognition event occurs at the point the cells touch and that the basis of the system is that protein molecules released from the plasmalemma move into and across the wall, forming a complex with catalytic activity which subsequently initiates a sequence in development which results in the formation of a successful graft. A contrary view is that recognition occurs only when the plasmalemmas come into contact after the cell walls have been dissolved. Both interpretations implicate a major role taken by the plasmalemma.

In addition there is a large volume of circumstantial evidence implicating the plasma membrane of animal cells in a variety of recognition functions (Cook & Stoddart, 1973; Hughes, 1976) and it is reasonable to suppose that the plasmalemma of plant cells may also house a recognition system. It is therefore important to devise a technique to enable plasmalemmas from graft unions to be prepared in high yield and of high purity.

Problems of the isolation and identification of plant plasmalemma

A number of reports have appeared during the last few years in which it is claimed that membrane preparations containing 50% or more plasmalemma have been obtained from a variety of plant tissues. Although details may vary, all these preparations have relied on differential centrifugation followed by density gradient centrifugation. The fractionation scheme we used was that of Hodges, Leonard, Bracker & Keenan (1972) which is typical of these methods. After grinding with a mortar and pestle, the homogenate was subjected first to moderate (13 000 g for 15 min), then to heavy (80 000 g for 30 min) centrifugation in order to obtain mitochondrial and microsomal pellets, respectively. The microsomal pellet was then layered onto a discontinuous sucrose gradient, and it was estimated that the vesicles sedimenting at the interfaces between the densest layers (34/38 % and 38/45 % sucrose) contained around 75 % plasma membrane. All putative plasmalemma preparations isolated in other laboratories have a similar density, but several other workers (Hardin, Cherry, Morré & Lembi, 1972; Lee, 1974; Sinensky & Strobel, 1976) have layered onto their sucrose gradients particulate matter sedimenting much more readily than microsomes.

A major problem which has not yet been satisfactorily overcome is the lack of a suitable marker enzyme by which the plasmalemma may be identified. The phosphohydrolases associated with the plasma membranes of many animal cells, 5′-nucleotidase, alkaline phosphatase and Na^+, K^+-dependent ATPase, have not been shown to be exclusively associated with the plasmalemma of plant cells, and indeed may possess negligible activity in plant homogenates (Morré, Roland & Lembi, 1970; Hodges & Leonard, 1974). A glycosyl transferase system, 'glucan synthetase', has been associated with putative plasmalemma fractions from various sources including onion stem (Van der Woude, Lembi & Morré, 1972), oat roots (Hodges *et al.*, 1972), sugar cane leaves (Thom, Laetsch & Maretzki, 1975) and soybean hypocotyls (Hardin *et al.*, 1972). This enzyme system may be of general occurrence in plasmalemma from plants but is also associated with the Golgi apparatus and possibly other organelles, so is of rather limited value as a plasmalemma marker. A Mg^{2+}-dependent, K^+-stimulated ATPase is associated with the putative plasmalemma fractions from oat roots (Hodges *et al.*, 1972) and also from sugar cane leaves (Sinensky & Strobel, 1976), but the possibility of its occurrence elsewhere in the cell has not been eliminated.

In the absence of suitable enzyme markers, most procedures for the isolation of plasmalemma from plants have relied on the carbohydrate-sensitive phosphotungstic acid–chromic acid (PACP) stain (Morré *et al.*, 1970) for identification and assessment of purity, although no unequivocal evidence has been presented that the PACP staining is specific. Indeed, it is inherently unlikely that the PACP staining affects only the plasmalemma since carbohydrate is known to be associated with various intracellular membranes of animal cells (Hughes, 1976) including lysosomes, nuclei and mitochondria, and it is possible that some carbohydrate is generally associated with the biological membranes of all higher life forms. It has already been demonstrated (Thom *et al.*, 1975) that intracellular membranes of sugar cane cells react positively when subjected to the PACP-staining procedure. Experiments in this laboratory on the use of the PACP stain suggest that staining is more intense with some tissues than with others, and that the conditions under which staining is carried out are critical. At best, the PACP stain is not quantitative and estimates of the enrichment of a PACP-staining fraction over the homogenate can only be very approximate.

It would clearly be desirable to introduce an external marker to intact tissue before homogenisation and subsequent fractionation of the homogenate. In general, lactoperoxidase-catalysed iodination of externally exposed proteins (Morrison, 1974) has emerged as the method of choice for this purpose on account of its specificity, relatively high degree of radioisotope incorporation and absence of damage to the membrane or its proteins. This method has been used to label the cell surface of a wide variety of animal cell types including erythrocytes (Phillips & Morrison, 1971; Hubbard & Cohn, 1972), lymphocytes (Marchalonis, 1971), fibroblasts (Hynes, 1973) and slime-mould amoebae (Green & Newell, 1974). An encouraging application of this technique to coleoptiles of *Zea mays* has been reported (Hendriks, 1976). The pellets obtained by differential centrifugation of that tissue displayed a significant increase in specific radioactivity over the homogenate while the soluble fraction displayed a corresponding decrease. When the mitochondrial pellet was subjected to sucrose density-gradient centrifugation, a peak of radioactivity could be obtained largely separated from peaks of NADPH–cytochrome c reductase (endoplasmic reticulum?) and cytochrome c oxidase (mitochondrial membrane?). Hall & Roberts (1975), however, found that lactoperoxidase-mediated iodination of tissue from maize roots was non-selective. The moderately high degree of incorporation they obtained in the absence of lactoperoxidase may have been due to the action of endogenous peroxidases, and since peroxidases are of widespread occurrence in plant tissues, it is possible that this method may be of only limited value for plant cells in general.

In the absence of any single way of unequivocally identifying the plasmalemma at present, it is clearly desirable to use several different criteria, and if they are consistent, identification may be made with some confidence. Sinensky & Strobel (1976) identified plasma membranes of sugar cane leaves by means of their ability to bind the toxin, helminthosporoside. This was justified by the previous demonstration of the location of the toxin-binding protein on the external surface of the plasmalemma (Strobel & Hess, 1974) by a variety of means including the use of specific antibodies to the protein, surface labelling with pyridoxyl phosphate/Na B[^{3}H$_4$] and PACP staining. Although no single experiment alone would have been adequate, several independent tests taken together constituted compelling evidence that the membrane with which the toxin-binding protein was associated was plasmalemma.

No laboratory has claimed the preparation of plant plasmalemma of really high purity, and it is doubtful whether existing methods are adequate for obtaining very pure preparations in reasonable yields. The association of the helminthosporoside-binding protein with the plasmalemma suggests a possible way of improving on existing technology. Specific antibodies to the toxin-binding protein could be covalently attached to an insoluble support and used for affinity chromatography. If and when cell surface antigens are isolated from other plant sources, antibodies raised against such antigens could be used in a similar way.

In conclusion, it is clear that the existing methods for both the isolation of plant plasma membranes and their identification are inadequate. This prevents the elaboration of a line of investigation in which the composition and properties of the plasmalemma would be studied to discover if proteins are freely exchanged between adjacent cells. Such an approach must await the discovery of a suitable means of labelling, and therefore identifying, the organelle, which would subsequently assist the development of improved methods of isolation.

Location of proteins synthesised in response to grafting

A dual-labelling experiment similar to those previously described provided results consistent with the involvement of the plasmalemma in a recognition role. A homograft was labelled with [^{35}S]methionine and an ungrafted control with [^{3}H]methionine. After homogenisation together the material was fractionated by the scheme of Hodges et al. (1972) to obtain membrane fractions of different density, each of which was then subjected to electrophoresis. It was found that the differences in the ratio of ^{3}H to ^{35}S were most marked in the higher density fractions which were also those reacting most strongly with the PACP strain. For reasons discussed above, no definite conclusion can be drawn from such results, but a failure to correlate differential protein synthesis with the putative plasmalemma fraction would certainly have constituted evidence against a recognition role for the plasmalemma, and such was not obtained.

POSSIBLE INVOLVEMENT OF LECTINS IN
CELL INTERACTIONS

Saline extracts of plant seeds often possess haemagglutinating activity which can usually be inhibited by simple sugars or glycosides (Sharon & Lis, 1972). Some phytohaemagglutinins or lectins (Boyd & Shapleigh, 1954) are superficially similar to vertebrate immunoglobulins in being di- or multivalent and able to recognise specific saccharides or saccharide sequences. The interaction of lectins with animal surfaces has been studied intensively (Nicolson, 1974) and it is well known that they can bring about major changes in cell-surface architecture and/or initiate important cellular events such as mitosis.

A protein with these properties is an obvious possible candidate for a recognition role in the formation of grafts in higher plants. The lectin could either be a soluble messenger which travels across the graft (Yeoman & Brown, 1976) conveying information by binding to a glycoprotein receptor on the plasmalemma; or, the lectin itself could be a component of the plasmalemma, able to recognise and specifically bind to a diffusible carbohydrate-containing messenger substance. The physiological function of lectins in general is not known although various suggestions have been made (Table 3), for example that they are storage proteins or part of a defence mechanism similar to the role played by antibodies in vertebrates (Mirelman, Galun, Sharon & Lotan, 1975). The best-substantiated suggestion, however, is that the lectins of legumes are responsible for the recognition event essential to the establishment of the symbiotic relationship between *Rhizobium* species and various members of the Leguminosae (Hamblin & Kent, 1973; Bohlool & Schmidt, 1974; Albersheim & Wolpert,

Table 3. *Possible functions of lectins in plants*

1. Defence against bacterial or fungal attack	Albersheim & Anderson (1971); Mirelman et al. (1975)
2. Control of germination	Howard et al. (1972); Southworth (1975)
3. To act as storage protein	Liener (1976)
4. To facilitate storage or transport of carbohydrates	Liener (1976)
5. Control of cell division	Liener (1976)
6. Enzymes of unknown activity with carbohydrate specificity	Liener (1976)
7. Attachment of glycoprotein enzymes to multi-enzyme complexes	Liener (1976)
8. Recognition in pollen–stigma interactions	Heslop-Harrison et al. (1975)
9. Specific binding of *Rhizobium* spp. to roots of legumes	Hamblin & Kent (1973); Bohlool & Schmidt (1974)

1976; Bhuvaneswari, Pueppke & Bauer, 1976). The binding of the bacteria to the root is an essential preliminary to the formation of the nitrogen fixing nodule. The mechanism of *Rhizobium*–legume root specificity and that of compatible or incompatible graft development might well have a similar molecular basis.

When considering the function(s) of lectins it is important to know the location of such proteins, both within the plant and within the cell. Lectins are usually prepared from mature seeds which are a readily available source and one in which the lectin constitutes a significant and easily extractable proportion of the total protein. There is reason, however, to believe that lectins generally have a much wider distribution in the plant. They have been detected, for example, in the emerged cotyledons and roots of several species including lentil (Howard, Sage & Horton, 1972), kidney bean (Mialonier *et al.*, 1973) and soybean (Pueppke & Bauer, 1976) and to a lesser extent in other tissues of these species. We have dissected *Datura, Lycopersicon* and *Nicandra* into stem, leaf, root and emerged cotyledons and examined saline extracts of these parts for their phytohaem-agglutinating activity against fixed erythrocytes. All extracts were active but showed no blood group specificity when A, B or O group cells were used in agglutination assays. A detailed study on *Datura* has been carried out in which the plant was dissected into its component parts, and examined for the presence of an agglutinin occurring in *Datura* seeds. Saline homogenates were tested for agglutinating activity against rabbit erythrocytes and also for immunological activity with antibodies raised against the purified agglutinin. By this means, the *Datura* lectin (Table 4) was readily detected in homogenates prepared from roots, cotyledons and flowers. When the flowers were further dissected into their component parts the lectin was detected in ovaries, stamens and petals.

Table 4. *Properties of the* Datura *lectin**

Activity	Will agglutinate rabbit or human erythrocytes.
Inhibitors	The major erythrocyte glycoprotein (glycophorin); trypsinate of fresh erythrocytes.
Monosaccharide inhibitors	None known.
Molecular weight (subunits)	40 000 and 45 000.
Other properties	Activity is lost after extensive heating at 100 °C or trypsinisation; can be precipitated with ammonium sulphate; displays a very broad profile on elution from Sephadex G-100, indicative of molecular aggregation.

* A highly purified preparation of the *Datura* lectin was obtained by treating glutaraldehyde-fixed human erythrocytes with a saline extract of *Datura stramonium* seeds. After washing the cells, some of the bound protein was removed by warming the cells to 56 °C.

Even less information is available on the subcellular location of lectins in plants. Lectin activity has been detected in various membrane fractions (dictyosomes, endoplasmic reticulum, plasma membrane and mitochondria) isolated from mung bean hypocotyls (Bowles & Kauss, 1975) and has been positively identified in the inner membrane of mitochondria prepared from the endosperm of *Ricinus communis* (Bowles, Schnarrenberger & Kauss, 1976). Using fluorescein-labelled globulins, Clarke, Knox & Hermyn (1975) have found concanavalin A and PHA to be mainly located in the cytoplasm of cells in sections cut from jack beans and red kidney beans, respectively. We have used FITC-conjugated antibody to rabbit IgG in conjugation with rabbit anti-lectin IgG to detect the positions occupied by the lectin in *Datura* stem. Most of the specific fluorescence was found in the cytoplasm, apparently associated with membrane organelles including the plasmalemma. The fluorescent antibody technique, however, does not permit the exact location to be determined and we are now using ferritin-conjugated antibodies combined with electron microscopy to improve the resolution. Hopefully, these experiments will yield an unequivocal answer to the precise intracellular location of the *Datura* lectin.

A quite different approach to the involvement of lectins in recognition phenomena has been initiated by Jermyn (1975) who observed that precipitation lines sometimes formed on gel-diffusion plates between pairs of seed extracts. The active factor from extracts of jack beans appeared to be identical to concanavalin A, and since methyl-α-mannoside acted as a general inhibitor of these reactions, it is possible that certain lectins are generally involved in a mechanism responsible for distinguishing self from non-self. The possible relevance of these findings to the grafting situation was tested with phosphate-buffered saline extracts from species which form compatible grafts (e.g. *Lycopersicon* and *Datura*) and those which form incompatible grafts (*Lycopersicon* and *Nicandra*). The expectation was that extracts from compatible species would not be expected to form precipitation lines in double diffusion tests, while extracts from species which form incompatible heterografts (e.g. *Lycopersicon* and *Nicandra*) should react to give an insoluble complex. No precipitation reaction was observed between any of the extracts, which might be an indication that a different recognition mechanism is operative in grafting.

Jermyn & Yeow (1975) have found that seed extracts from many plants will form an insoluble complex with any one of a selection of chemically synthesised 'antigens', and have claimed that the glycoproteins which were found to be responsible comprise a distinct class of lectins. These have been termed 'all-β-lectins' since the specificity of the reactions resides in the β-glycosidic linkage and not in the nature of the saccharide. The 'all-β-lectins' from jack beans and kidney beans are located between plasmalemma and cell wall (Clarke *et al.*, 1975) and so could be involved in recognition in general and/or grafting in particular.

POSSIBLE MOLECULAR BASIS OF A RECOGNITION SYSTEM IN PLANTS

The existence of incompatibility in plant grafting has been demonstrated and it has been argued that a recognition capability is probably responsible. No concrete evidence is available regarding the biochemical basis of plant grafting, but it is relevant within this context to consider what possibilities exist, and to relate them to recognition phenomena displayed outside the plant kingdom.

Assuming a recognition system exists, it could be either negative or positive in operation. That is to say, the stock or scion of one plant and the stock or scion of another plant taken at random will tend to graft together unless there is a specific signal exchanged to indicate incompatibility; or, such stock and scion tissues will each tend to regard the other as foreign until a positive signal occurs enabling them to recognise each other as compatible. It is our opinion that present knowledge favours the negative mechanism since all stock/scion combinations form an immediate, non-specific union irrespective of potential compatibility, and the number of compatible combinations greatly exceeds the number of incompatible combinations in the examples we have studied (Table 1). Reasoning by analogy with animal grafting, however, favours the positive recognition mechanism, since tissue grafts even from a member of the same (e.g. human) species are positively recognised by T-lymphocytes as foreign.

There is also considerable doubt concerning the molecular nature of the recognition capability in plants. A low-molecular weight factor, such as nucleotide, peptide or steroid, would presumably have no difficulty crossing the graft from stock to scion or vice-versa and could rapidly elicit a response either by binding to a receptor on the plasmalemma or by passing through it to enter the cell. Such messenger substances that are known in plants, for example the various plant hormones, are all of low molecular weight. On the other hand, macromolecules are responsible for many recognition events in the animal kingdom including the species-specific sorting-out of dispersed sponge cells (Henkart, Humphreys & Humphreys, 1973), the identification of damaged serum glycoproteins by the liver (Hudgin *et al.*, 1974; Stockert, Morell & Scheinberg, 1974) and the recognition of lymphoid tissue by circulating lymphocytes (Gesner & Ginsburg,

1964). The dual protein-labelling experiments we have described earlier give limited encouragement to the view that proteins perhaps associated with the plasmalemma are synthesised specifically in response to the formation of a graft union, and, if so, such proteins could form part of the recognition system. The observation that opposing cell walls dissolve during the second phase of graft union development is difficult to interpret but is consistent with protein involvement since its purpose might be to facilitate further an exchange of macromolecules between stock and scion cells, or to permit contact between macromolecules attached to the surfaces of the opposing plasmalemmas, or to allow the deposition of an extracellular catalytic complex. On the other hand it is difficult to see how this structural event could assist a mechanism mediated by a low molecular weight factor(s).

Proteins found outside the cell are usually glycoproteins (Eylar, 1966). The majority of lectins that have been characterised are also glycoproteins (Sharon & Lis, 1972). Certain carbohydrate-binding proteins (or animal lectins) have been implicated in specific recognition mechanisms like the binding of asialoglycoproteins by the liver (Stockert et al., 1974) or the aggregation of the slime moulds *Dictyostelium discoideum* (Simpson, Rosen & Barondes, 1974) and *Polysphondylium pallidum* (Simpson, Rosen & Barondes, 1975). Against this background, it is not surprising that glycoproteins and lectins appear to be involved in the recognition events leading to compatible pollination, the best-studied recognition phenomenon in the plant kingdom. For example it has been shown that glycoproteins (lectin receptors?) are present on the stigma surface of *Gladiolus gandavensis* and when these receptors were blocked by treating the stigma surface with concanavalin A, pollen tubes failed to penetrate the cuticle, preventing pollination (Knox et al., 1976). Working with Cyads, Pettitt (1977) has found lectin-receptor sites on the magaspore wall, and also that a saline extract of megagametophyte and megaspore wall was able to agglutinate erythrocytes. Clearly, the involvement of a lectin in grafting compatibility is a possibility worthy of continued experimental investigation.

WIDER SIGNIFICANCE OF GRAFTING IN PLANTS

Grafting does occur naturally in the wild. Roots of trees in the substratum of a wood are forced into close contact with one another and fusion can be observed. However, it is difficult to appreciate how the ability to graft, which can be detected in almost all species tested, can confer a selective advantage upon an individual. If this reasoning is correct then grafting is a reflection of other processes which contribute to the growth, development and survival of the plant. Clearly, regeneration is one of these processes but this alone cannot account for all of the events involved in the fusion or non-fusion of stock and scion in compatible or incompatible combinations respectively. It is the cellular interactions which occur during graft formation which must provide a clue to this mystery. It has been suggested by Yeoman & Brown (1976) that the interactions between the two surfaces of the graft union are regulated by polar gradients within the cells. This indicates that corresponding gradients regulate the interaction of cells in a meristematic tissue and that the cells in a tissue cohere as a result of the interactions of compatibility substances released from surfaces with opposing polar properties. The resistance to diffusion is probably less into the cell wall and this will be enhanced by the formation of a compatibility complex at the outside surface of the plasmalemma. The formation of this compatibility complex is presumably the basis of the recognition system in the grafting situation and initiates a developmental sequence which includes the formation of a bridge between the opposing vascular systems, the event which promotes the major part of the mechanical strength of the graft. The movement of proteins from opposing plasmalemmas into the wall and across the wall could, in addition to promoting and maintaining cohesion between cells, provide a mechanism for producing one cell different from those surrounding it. The cessation or alteration of the turnover of these proteins could produce changes only in a particular cell and lead to the initiation of a different pathway of development in that cell. The nature of the agency which modulates the turnover of these recognition proteins is unknown but plant hormones are an obvious candidate. From all of this it is readily apparent that the results of these studies on grafting may provide a better understanding of the mechanisms involved in the development and organisation of plants.

The authors wish to express their thanks to Mrs E. Mills, Miss M. McCagie and Mrs E. Raeburn for assistance with the preparation of the manuscript and to Mr W. Foster and Mr D. Denham for the provision of photographic material. We are also indebted to the Agricultural Research Council for a grant in support of this research.

REFERENCES

ALBERSHEIM, P. & ANDERSON, A. J. (1971). Proteins from plant cell walls inhibit polygalacturonases secreted by plant pathogens. *Proceedings of the National Academy of Sciences, U.S.A.*, **68**, 1815–19.

ALBERSHEIM, P. & WOLPERT, J. S. (1976). The lectins of Legumes are enzymes which degrade the lipopolysaccharides of their symbiont *Rhizobia*. *Plant Physiology, Lancaster*, **57**, Suppl., 79.

BHUVANESWARI, T. V., PUEPPKE, S. G. & BAUER, W. D. (1976). Differential binding of the Soybean lectin to *Rhizobium* species. *Plant Physiology, Lancaster*, **57**, Suppl., 80.

BOHLOOL, B. B. & SCHMIDT, E. L. (1974). Lectins: a possible basis for specificity in the Rhizobium-Legume root nodule symbiosis. *Science*, **185**, 269–71.

BOWLES, D. J. & KAUSS, H. (1975). Carbohydrate-binding protein from cellular membranes of plant tissue. *Plant Science Letters*, **4**, 411–18.

BOWLES, D. J., SCHNARRENBERGER, C. & KAUSS, H. (1976). Lectins˜as membrane components of mitochondria from *Ricinus communis*. *Biochemical Journal*, **160**, 375–82.

BOYD, W. C. & SHAPLEIGH, E. (1954). Specific precipitating activity of plant agglutinins (lectins). *Science*, **119**, 419.

CLARKE, A. E., KNOX, R. B. & JERMYN, M. A. (1975). Localisation of lectins in legume cotyledons. *Journal of Cell Science*, **19**, 157–67.

COCKING, E. C. (1974). The isolation of plant protoplasts. *Methods in Enzymology*, **31**, 578–83.

COOK, G. M. W. & STODDART, R. W. (1973). *Surface Carbohydrates of the Eukaryotic Cell*. Academic Press, London and New York.

DAWSON, R. F. (1944). Accumulation of anabasine in reciprocal grafts of *Nicotiana glauca* and Tomato. *American Journal of Botany*, **31**, 351–5.

EYLAR, E. H. (1966). On the biological role of glycoproteins. *Journal of Theoretical Biology*, **10**, 89–113.

GARNER, R. J. (1970). *The Grafter's Handbook*, 3rd edition, pp. 36–45. Faber and Faber, London.

GESNER, B. M. & GINSBURG, V. (1964). Effect of glycosidases on the fate of transfused lymphocytes. *Proceedings of the National Academy of Sciences, U.S.A.*, **52**, 750–5.

GLIMELIUS, K., WALLIN, A. & ERIKSSON, T. (1974). Agglutinating effects of Concanavalin A on isolated protoplasts of *Daucus carota*. *Physiologia Plantarum*, **31**, 225–30.

GREEN, A. A. & NEWELL, P. C. (1974). The isolation and subfractionation of plasma membrane from the cellular slime mould *Dictyostelium discoideum*. *Biochemical Journal*, **140**, 313–22.

GUR, A. & BLUM, A. (1973). The role of cyanogenic glycoside in incompatibility between peach scions and almond rootstocks. *Horticultural Research*, **13**, 1–10.

GUR, A., SAMISH, R. M. & LIFSHITZ, E. (1968). The role of cyanogenic glycoside of the quince in the incompatibility response between cultivars and quince rootstocks. *Horticultural Research*, **8**, 113–34.

HALL, J. L. & ROBERTS, R. M. (1975). Biochemical characteristics of membrane fractions isolated from Maize (*Zea mays*) roots. *Annals of Botany*, **39**, 983–93.

HAMBLIN, J. & KENT, S. P. (1973). Possible role of phytohaemagglutinins in *Phaseolus vulgaris* L. *Nature, New Biology, London*, **245**, 28–30.

HARDIN, J. W., CHERRY, J. H., MORRÉ, D. J. & LEMBI, C. A. (1972). Enhancement of RNA polymerase activity by a factor released by auxin from plasma membrane. *Proceedings of the National Academy of Sciences, U.S.A.*, **69**, 3146–50.

HENDRIKS, T. (1976). Iodination of Maize coleoptiles: a possible method for identifying plant plasma membranes. *Plant Science Letters*, **7**, 347–57.

HENKART, P., HUMPHREYS, S. & HUMPHREYS, T. (1973). Characterisation of sponge aggregation factor. A unique proteoglycan complex. *Biochemistry*, **12**, 3045–50.

HESLOP-HARRISON, J., HESLOP-HARRISON, Y. & BARBER, J. (1975). The stigma surface in incompatibility responses. *Proceedings of the Royal Society, London*, B, **188**, 282–97.

HODGES, T. K. & LEONARD, R. T. (1974). Purification of a plasma membrane-bound adenosine triphosphatase from plant roots. *Methods in Enzymology*, **32**, 392–3.

HODGES, T. K., LEONARD, R. T., BRACKER, C. E. & KEENAN, T. W. (1972). Purification of an ion-stimulated ATPase from plant roots: association with plasma membranes. *Proceedings of the National Academy of Sciences, U.S.A.*, **69**, 3307.

HOWARD, I. K., SAGE, H. J. & HORTON, C. B. (1972). Studies on the appearance and location of hemagglutinins from a common lentil during the life cycle of the plant. *Archives of Biochemistry and Biophysics*, **149**, 323–6.

HUBBARD, A. L. & COHN, Z. A. (1972). The enzymatic iodination of the red cell membrane. *Journal of Cell Biology*, **55**, 390–405.

HUDGIN, R. L., PRICER, W. E., ASHWELL, G., STOCKERT, R. J. & MORELL, A. G. (1974). The isolation and properties of a rabbit liver binding protein specific for asialoglycoproteins. *Journal of Biological Chemistry*, **249**, 5536–43.

HUGHES, R. C. (1976). *Membrane Glycoproteins: A Review of Structure and Function*. Butterworth, London and Boston.

HYNES, R. O. (1973). Alteration of cell surface proteins by viral transformation and by proteolysis. *Proceedings of the National Academy of Science, U.S.A.*, **70**, 3170–74.

JERMYN, M. A. (1975). Precipitation reactions between components of plant tissue extracts. *Australian Journal of Plant Physiology*, **2**, 533–42.

JERMYN, M. A. & YEOW, Y. M. (1975). A class of lectins present in the tissues of seed plants. *Australian Journal of Plant Physiology*, **2**, 501–31.

KLOZ, J. (1971). Serology of the Leguminosae. In *Chemotaxonomy of the Leguminosae*, eds. J. B. Harborne, D. Boulter & B. L. Turner, pp. 309–65. Academic Press, London and New York.

KNOX, R. B., CLARKE, A., HARRISON, S., SMITH, P. & MARCHALONIS, J. J. (1976). Cell recognition in plants: determinants of the stigma surface and their pollen interactions. *Proceedings of the National Academy of Sciences, U.S.A.*, **73**, 2788–92.

LEE, T. T. (1974). Cytokinin control in subcellular localisation of indoleacetic acid oxidase and peroxidase. *Phytochemistry*, **13**, 2445–53.

LIENER, I. E. (1976). Phytohemagglutinins (phytolectins). *Annual Review of Plant Physiology*, **27**, 291–319.

LINDSAY, D. W., YEOMAN, M. M. & BROWN, R. (1974). An analysis of the development of the graft union in *Lycopersicon esculentum*. *Annals of Botany*, **38**, 639–46.

MARCHALONIS, J. J., CONE, R. E. & SANTER, V. (1971). Enzymic Iodination. A probe for accessible surface protein of normal and neoplastic lymphocytes. *Biochemical Journal*, **124**, 921–7.

MIALONIER, G., PRIVAT, J.-P., MONSIGNY, M., KAHLEM, G. & DURAND, R. (1973). Isolement, propriétés physico-chimiques et localisation *in vivo* d'une phytohémagglutine (lectre) de *Phaseolus vulgaris* L. (Var. rouge). *Physiologie Végétale*, **11**, 519–37.

MIRELMAN, D., GALUN, E., SHARON, N. & LOTAN, R. (1975). Inhibition of fungal growth by wheat germ agglutinin. *Nature, London*, **256**, 414–16.

MORRÉ, D. J., ROLAND, J.-C. & LEMBI, C. A. (1970). Comparisons of isolated plasma membranes from plant stems and rat liver. *Proceedings of the Indiana Academy of Science*, **79**, 96–106.

MORRISON, M. (1974). The determination of the exposed proteins on membranes by the use of lactoperoxidase. *Methods in Enzymology*, **32**, 103–9.

NICOLSON, G. L. (1974). Interactions of lectins with animal cell surfaces. *International Review of Cytology*, **39**, 89–190.

PETTITT, J. M. (1977). Detection in primitive gymnosperms of proteins and glycoproteins of possible significance in reproduction. *Nature, London*, **266**, 530–2.

PHILLIPS, D. R. & MORRISON, M. (1971). Exposed protein on the intact human erythrocyte. *Biochemistry*, **10**, 1766–71.

PUEPPKE, S. G. & BAUER, W. D. (1976). Distribution and localisation of the soybean lectin in *Glycine Max*. *Plant Physiology, Lancaster*, **57**, Suppl., 80.

ROBERTS, J. R. & BROWN, R. (1961). The development of the graft union. *Journal of Experimental Botany*, **12**, 294–302.

ROBERTS, R. H. (1949). Theoretical aspects of graftage. *Botanical Reviews*, **15**, 423–63.

ROGERS, W. S. & BEAKBANE, A. B. (1957). Stock and scion relations. *Annual Review of Plant Physiology*, **8**, 217–36.

SHARON, N. & LIS, H. (1972). Lectins: Cell-agglutinating and sugar-specific proteins. *Science*, **177**, 949–59.

SIMPSON, D. L., ROSEN, S. D. & BARONDES, S. H. (1974). Discoidin, a developmentally regulated carbohydrate-binding protein from *Dictyostelium discoideum*. Purification and characterisation. *Biochemistry*, **13**, 3487–93.

SIMPSON, D. L., ROSEN, S. D. & BARONDES, S. H. (1975). Pallidin. Purification and characterisation of a carbohydrate-binding protein from *Polysphondylium pallidum* implicated in intercellular adhesion. *Biochemica et Biophysica Acta*, **412**, 109–19.

SINENSKY, M. & STROBEL, G. (1976). Chemical composition of a cellular fraction enriched in plasma membranes from sugarcane. *Plant Science Letters*, **6**, 209–14.

SOUTHWORTH, D. (1975). Lectins stimulate pollen germination. *Nature, London*, **258**, 600–601.

DE STIGTER, H. C. M. (1961). Translocation of ^{14}C photosynthates in the graft Musk Melon *Cucurbita ficifolia*. *Acta botanica Neerlandica*, **10**, 466–73.

STOCKERT, R. J., MORELL, A. G. & SCHEINBERG, I. H. (1974). Mammalian hepatic lectin. *Science*, **186**, 365–6.

STROBEL, G. A. & HESS, W. M. (1974). Evidence for the presence of the toxin binding protein on the plasma membrane of sugarcane cells. *Proceedings of the National Academy of Sciences, U.S.A.*, **71**, 1413–17.

THOM, M., LAETSCH, W. M. & MARETZKI, A. (1975). Isolation of membranes from sugarcane cell suspensions. Evidence for a plasma membrane enriched fraction. *Plant Science Letters*, **5**, 245–53.

VAN DER WOUDE, W. J., LEMBI, C. A. & MORRÉ, D. J. (1972). Auxin (2,4D) stimulation (*in vivo* and *in vitro*) of polysaccharide synthesis in plasma membrane fragments isolated from onion stems. *Biochemical and Biophysical Research Communications*, **46**, 245–53.

YEOMAN, M. M. & BROWN, R. (1976). Implications of the formation of the graft union for organisation in the intact plant. *Annals of Botany*, **40**, 1265–76.

ZEEVART, J. A. D. (1958). Flower formation as studied by grafting. *Mededeelingen van de Landbouwhoogeschool te Wageningen*, **58** (3), 1–88.

EXPLANATION OF PLATES

All plant material for light and transmission electron microscopy (TEM) was fixed in 3 % buffered glutaraldehyde and osmium tetroxide. Araldite sections (*c.* 0.5 μm) used for light microscopy were stained in 1 % toluidine blue at 50 °C for 20 s. Sections for TEM were stained in lead citrate and uranyl acetate. Material for scanning electron microscopy (SEM) was fixed in 3 % buffered glutaraldehyde, freeze dried, placed on a stub and coated with gold.

PLATE 1

Homografts of tomato (*Lycopersicon esculentum*)

(*a*) A vertical section through a 3-day-old graft, showing cellular contact between stock and scion of the products of division from cells in and around the vascular bundles. Points of contact between cells from opposing surfaces are arrowed. Section thickness *c.* 0.5 μm.

(*b*) An electron micrograph of a 'thin' section (*c.* 60 nm) next to the 'thick' section shown in Plate 1 *a*. The area shown is the point of contact between two cells which have grown out from the opposing surfaces of the stock and scion. The intervening cell walls have been partially dissolved away and a pit is forming which will allow contact between the protoplasts of the two cells.

(*c*) A scanning electron micrograph of the scion surface of a graft, 4 days after grafting. The stock and scion have been pulled apart by applying a force along the axis of the stem. The disruption of the graft has separated the stock and scion exposing the recently formed connection (arrowed) between the opposing xylem elements (which originated as a pit), (see Plate 1 *b*).

PLATE 2

(*a*) A vertical section through a tomato homograft 5 days after grafting. The xylem elements of stock and scion are connected by a bridge of new xylem cells.

(*b*) A vertical section through an incompatible heterograft between tomato (*Lycopersicon esculentum*) and *Nicandra physaloides*. The graft is 5 days old. Differentiated xylem elements have not appeared in the regenerated tissue between the peripheral parts of stock and scion (compare with Plate 2 *a*). Points of contact between cells from opposing surfaces are arrowed. (N = *Nicandra*; L = *Lycopersicon*). Section thickness *c.* 0.5 μm.

(*c*) An electron micrograph of a 'thin' section (*c.* 60 nm) next to the 'thick' section shown in Plate 2 *b*. The area shown is a point of contact between the two cells which have grown out from the opposing surfaces of stock and scion. The intervening walls have been partially dissolved away and a pit is forming which will allow physical contact between the protoplasts of the two cells. (Compare with Plate 1 *b*).

CELL-SURFACE cAMP RECEPTORS
IN *DICTYOSTELIUM*

By P. C. NEWELL and I. A. MULLENS

Department of Biochemistry, University of Oxford,
South Parks Road, Oxford OX1 3QU, U.K.

During the life cycle of the cellular slime mould *Dictyostelium discoideum* (Fig. 1) the organism passes through a phase during which populations of amoebae aggregate toward collecting centres. The resulting aggregates produce small multicellular organisms that then develop the ability to migrate over surfaces and eventually differentiate into fruiting bodies composed of spore and stalk cells.

The system of cellular communication used during the aggregation phase is particularly amenable to biochemical study as the amoebae can be induced to start signalling in a synchronous manner and because amoebae exhibit visible changes in their optical properties during the time that they are responding to the chemotactic stimuli. Light and dark bands of cells can be seen emanating from the aggregation centres of dense cultures of amoebae on agar surfaces using dark field optics, or fluctuations of the optical density of bubbled suspensions of amoebae can be observed with a spectrophotometer. (Gerisch, 1968; Alcântara & Monk, 1974; Gerisch & Malchow, 1976).

The trigger for aggregation is starvation and this in some way initiates the release of pulses of an attractant by some of the starving cells. In the case of *D. discoideum* this attractant has been found to be cyclic AMP (cAMP) (Konijn, van de Meene, Chang & Barkley, 1969).

The cAMP is not released continuously by the attracting centres but in pulses every 5–10 min. The pulses of cAMP from the centres are soon broken down by phosphodiesterase enzymes also produced by the starving cells (Chassy, 1972; Pannbacker & Bravard, 1972; Malkinson & Ashworth, 1973) but before this happens, some of the cAMP binds to the surface of amoebae in the vicinity of the attractive centre. This binding initially triggers two major events: firstly the amoebae respond by moving toward the source of the pulse (and continue to do so for about 100 s) and secondly they themselves produce and liberate a new pulse of cAMP that diffuses away for a short distance and attracts amoebae further out from the centre. The signal is in this way relayed outward as the amoebae move periodically inward toward the centre. Details of this process and references may be

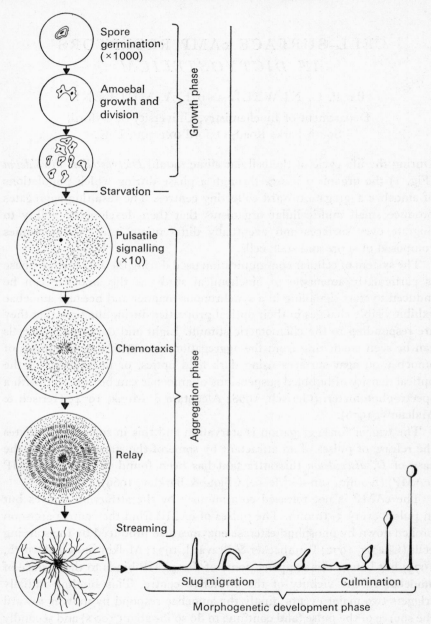

Fig. 1. The life cycle of *Dictyostelium discoideum* showing the growth phase, aggregation phase and morphogenetic development phase (from Newell, 1977*b*).

found in several recent reviews (Gerisch & Malchow, 1976; Bonner, 1977; Newell, 1977 a, b). This paper will be concerned with one aspect of this cell communication system, namely the cell-surface receptors for binding the cAMP. Recent studies have shown these receptors possess equilibrium and kinetic binding properties strikingly similar to certain hormone receptors in higher systems.

DISCOVERY OF cAMP RECEPTORS

The ability of aggregating *D. discoideum* amoebae to bind cAMP on their cell surfaces was first found by Malchow & Gerisch (1974) and has since been studied by Green & Newell (1975), Henderson (1975) and Mato & Konijn (1975) using different types of binding assay. One of the problems in assaying the binding was the presense of the phosphodiesterase enzyme that rapidly destroyed the added (radioactive) cAMP. This problem was overcome by Malchow & Gerisch by using a vast excess of cGMP that transiently preoccupied the phosphodiesterases (whose cyclic nucleotide specificity is low) but that did not greatly inhibit the highly specific cAMP receptors. In studies by Green & Newell (1975) and Mullens & Newell (1978) the phosphodiesterases were inhibited with the disulphide reducing agent dithiothreitol. Mato & Konijn carried out aggregation at pH 4.6, a value that they reported inhibited the phosphodiesterases but not aggregation or cAMP binding.

The cell surface cAMP binding of several species and genera has been investigated (Fig. 2). Considerable cAMP binding was observed in aggregating amoebae of the species *D. discoideum*, *D. purpureum* and *D. mucoroides*, but little or no binding to *D. minutum*, *P. pallidum* or *P. violaceum* could be detected. (Mullens & Newell, 1978). The binding of cAMP to *D. purpureum* and *D. mucoroides* has also been reported by Mato & Konijn (1975) using a different method of assay. The difference in binding activity between these genera and species is in accord with the chemotaxis data of Bonner *et al.* (1972), indicating that cAMP is an attractant for *D. discoideum*, *D. purpureum* and *D. mucoroides* but not for *D. minutum*, *P. pallidum* and *P. violaceum*.

CURVILINEARITY OF SCATCHARD PLOTS

Malchow & Gerisch calculated from Scatchard plots of equilibrium binding data that *D. discoideum* had 5×10^5 cAMP receptors per cell formed during aggregation and these had an apparent dissociation constant of 100–200 nM. The Scatchard plots were analysed as being linear but a

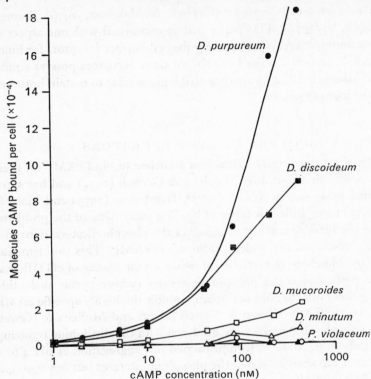

Fig. 2. The binding of cAMP to aggregating amoebae of various species and genera of cellular slime mould. ●, *Dictyostelium purpureum*; ■, *D. discoideum*; □, *D. mucoroides*; △, *D. minutum*; ▲, *Polysphondylium violaceum*; ○, *P. pallidum*. (Redrawn from Mullens & Newell, 1978.)

degree of curvilinearity was noticeable at low cAMP concentrations. Scatchard plots of the binding observed by Green & Newell with dithiothreitol in the absence of cGMP (which may compete with cAMP to some extent at low cAMP concentrations) were found to be strongly curvilinear and were concave upwards. Such curvilinearity has more recently been found by Mullens & Newell (1978) in other species of cellular slime mould such as *D. purpureum* and *D. mucoroides* (Fig. 3). The apparent relative affinities (calculated as equilibrium dissociation constants, K_d) varied between these species; *D. purpureum* showed much more lower affinity binding than *D. discoideum* while *D. mucoroides* showed only a very small (but reproducible) degree of curvilinearity.

Curvilinear Scatchard plots have also been observed by a number of workers for hormone binding to target cells of higher systems, but (as will be discussed below) the implications of such binding characteristics are

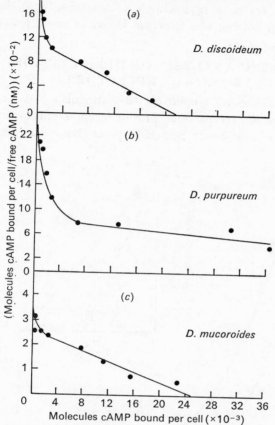

Fig. 3. The binding of cAMP to three species of *Dictyostelium* plotted by the method of Scatchard. The receptor affinities (measured as the equilibrium dissociation constants (K_d) from the slopes of the curves) are seen to vary with the cAMP concentration. (a) *D. discoideum*, K_d = 10–100 nM. (b) *D. purpureum*, K_d = 10–580 nM; (c) *D. mucoroides*, K_d < 10–95 nM. (Redrawn from Mullens & Newell, 1978.)

currently incompletely understood. In general, however, they are thought to indicate either the presence of two or more types of receptor molecules with different affinities or the presence of receptors that show variable affinity in response to cAMP induced receptor–receptor (negative cooperativity) or receptor–effector interactions.

The possibility of more than one type of cAMP receptor being present is rendered plausible by the fact that aggregating amoebae show at least two responses that require the ability to perceive the cAMP signals, namely the chemotactic (movement) response and the signal relay response. The two responses appear at different stages of the aggregation process

(Robertson, Drage & Cohen, 1972; Gingle & Robertson, 1976), the ability to relay appearing an hour or two after the ability to move toward a local source of cAMP.

KINETIC ANALYSIS OF DISSOCIATION
FROM cAMP RECEPTORS

Equilibrium studies alone are unable to distinguish between multiple forms of receptors and receptor–receptor or receptor–effector interactions being the cause of curvilinear Scatchard plots (Boeynaems & Dumont, 1975).

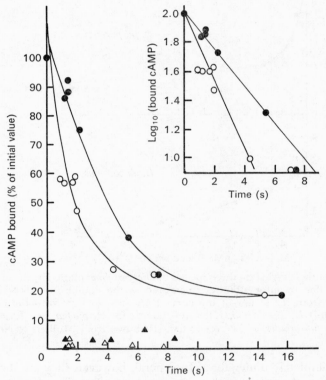

Fig. 4. The dissociation of [³H]cAMP from cAMP receptors in the presence and absence of excess unlabelled cAMP. The dissociation of 1 nM [³H]cAMP from aggregating amoebae was induced by rapid 80-fold dilution of the amoebal suspension into buffer (●) or buffer containing excess (100 μM) unlabelled cAMP (○). The extent of dissociation was determined at the times shown, by rapidly filtering the diluted suspension on to membrane filters and scintillation counting of the filters and attached amoebae. The inset shows a semilogarithmic plot of the dissociation data. The triangles show the rate of association of 0.012 nM [³H]cAMP with diluted cells in the presence (△) and absence (▲) of 100 μM unlabelled cAMP as a control test of the extent of reassociation under the conditions used. (Redrawn from Mullens & Newell, 1978.)

Kinetic analysis of the dissociation of hormones from their receptors has been employed, however, to help distinguish between such models (see review by Bradshaw & Frazier, 1977). The technique of measuring the rate of loss of small amounts of labelled ligand from receptors after dilution into buffer with or without an excess of unlabelled ligand (a method that was first used for insulin binding to animal cells), (DeMeyts, Roth, Neville & Gavin, 1973) has recently been employed for studying cAMP binding to *D. discoideum* (Mullens & Newell, 1978). Aggregating amoebae were incubated with small amounts of tritium-labelled cAMP and were then diluted into buffer to give about an 80-fold dilution. This displaces the equilibrium and rapid dissociation of the cAMP ensues. When the rate of dissociation was measured using a rapid filtration system, it was found that dissociation was almost complete in under 10 s except for approximately 20 % of the bound radioactivity which dissociated extremely slowly (10 % still remaining after a minute). Significantly, the rate of dissociation was considerably accelerated by the presence of excess unlabelled cAMP in the dilution buffer (Fig. 4). (Control experiments showed that this finding could not be attributed to competition for rebinding of dissociated labelled cAMP by the excess unlabelled ligand.) It was concluded that the affinity of the cAMP receptors was decreased at high cAMP concentrations, indicating that the curvilinear Scatchard plots could be due to receptors with variable affinity rather than to the presence of two types of receptor molecule with fixed but different affinities.

Very recent studies on the insulin receptor by Pollet, Standaert & Haase (1977), however, have questioned the interpretation of all such dissociation experiments, as they find that for the insulin receptor system, the degree of receptor occupancy cannot be correlated with the increase in rate of dissociation caused by unbound insulin that would be expected on the basis of negatively cooperative receptor interactions. Further studies on other receptor systems are clearly essential to clarify the interpretation of these kinetic experiments.

MODELS OF RECEPTOR SYSTEMS

One of the striking features of cAMP binding to species of *Dictyostelium* is its similarity to the binding of certain peptide hormones to their plasma membrane receptors. For example, curvilinear Scatchard plots have been seen for the binding of insulin to target tissues of animal and human origin (De Meyts *et al.*, 1973); nerve growth factor to sympathetic and dorsal root ganglia (Frazier, Boyd & Bradshaw, 1974), thyrotropin releasing factor (TRF) to anterior pituitary membranes (Boss, Vale & Grant, 1975), alprenolol to β-adrenergic receptors in frog erythrocyte membranes

(Limbird, De Meyts & Lefkowitz, 1975), prostaglandin $F_2\alpha$ to bovine corpora lutea cell membranes (Rao, 1976) and vasoactive intestinal peptide (VIP) to corpora lutea, (Christophe, Robberecht, Conlon & Gardner 1976). Moreover, the increase in dissociation rate on dilution in the presence of excess hormone (compared to its absence) is also a characteristic of all but the last two of these examples.

The similarity of binding characteristics of cAMP receptors and hormone receptors also extends to 'desensitization' observed for hormone binding to higher cells (Gavin *et al.*, 1974; Lesniak & Roth, 1976) as Klein & Juliani (1977) have recently reported a similar decrease in ligand binding with *D. discoideum* when cAMP is pre-incubated with amoebae for prolonged periods of time.

Models that could account for the behaviour of the cAMP receptor (based on models of hormone binding) are shown in Fig. 5. The simplest model with two independent molecular types of receptor could account for the curvilinear Scatchard plots but fails to account for the kinetic data. If a second species of receptor is present, it must be of a very low capacity, very high affinity type.

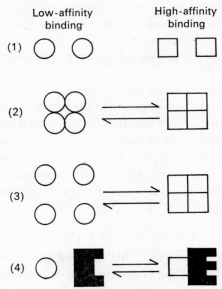

Fig. 5. Models of cAMP receptors. (1) Two distinct types of receptor protein with different, fixed affinities. (2) Negatively cooperative subunit interactions (arbitrarily shown as four subunits). (3) The affinity of the receptor is affected by association or dissociation of the subunits in the plane of the membrane lipid layer. (4) Interaction of receptor with its membrane effector molecule activates the effector and simultaneously converts the receptor to the high-affinity form. (The hypothetical effector molecule is shown in black).

Models that postulate negatively cooperative interactions between subunits can satisfactorily account for the equilibrium data although interpretation of the kinetic dissociation experiments in terms of negatively cooperative interactions has recently been questioned. With such models the dissociation constants measured at high and low cAMP concentrations would represent the affinities of the sites when fully cooperating and when non-cooperating, respectively. The subunit interaction need not be of the classical form seen with soluble enzymes but may be, for example, association or dissociation in the plane of the membrane lipid layer. Such ideas are strengthened by the finding that solubilized insulin receptors exist as tetramers which dissociate in the presence of high insulin concentrations to a monomeric form (Ginsberg, Cohen & Kahn, 1976).

Models have also been put forward recently in which (hormone) receptors are homogeneous and interact with their membrane effector molecules such that curvilinear Scatchard plots and ligand induced dissociation occur without the need for receptor–receptor interactions (Boeynaems & Dumont, 1975; Jacobs & Cuatrecasas, 1976). In the case of *D. discoideum*, if the receptor–effector complex had a higher affinity for cAMP than did the receptor on its own, while conversely the binding of cAMP to the receptor increased the affinity of the receptor for the effector, then the observed equilibrium and kinetic binding characteristics of cAMP to *Dictyostelium* membranes could be theoretically generated. From the recent work of Wurster, Shubiger, Wick & Gerisch (1977) and Mato *et al.* (1977), showing very rapid cyclic GMP synthesis after cAMP binding to aggregating amoebae, it is an interesting possibility that the effector molecule with which the cAMP receptor interacts in *D. discoideum* could be the enzyme guanylate cyclase.

We thank the S.R.C. for a research studentship (to I.A.M.) Mrs Stephanie J. Rogers for excellent technical assistance and Mr Frank Caddick for drawing the figures.

REFERENCES

ALCÂNTARA, F. & MONK, M. (1974). Signal propagation during aggregation in the slime mould *Dictyostelium discoideum*. *Journal of General Microbiology*, **85**, 321–34.

BONNER, J. T. (1977). Some aspects of chemotaxis using the cellular slime mold as an example. *Mycologia*, **69**, 443–59.

BONNER, J. T., HALL, E. M., NOLLER, S., OLESON, F. B. & ROBERTS, A. B. (1972). Synthesis of cyclic AMP and phosphodiesterase in various species of cellular slime molds and its bearing on chemotaxis and differentiation. *Developmental Biology*, **29**, 402–9.

BOEYNAEMS, J. M. & DUMONT, J. E. (1975). Quantitative analysis of the binding of ligands to their receptors. *Journal of Cyclic Nucleotide Research*, **1**, 123–42.

12

Boss, B., Vale, W. & Grant, G. (1975). Hypothalamic hormones. In *Biochemical Actions of Hormones*, vol. 3, ed. G. Litwack, pp. 87–118. Academic Press, New York.

Bradshaw, R. A. & Frazier, W. A. (1977). Hormone receptors as regulators of hormone action. *Current Topics in Cellular Regulation*, **12**, 1–37.

Chassy, B. M. (1972). Cyclic nucleotide phosphodiesterase in *Dictyostelium discoideum*: Interconversion of two enzyme forms. *Science*, **175**, 1016–18.

Christophe, J., Robberecht, P., Conlon, T. P. & Gardner, J. D. (1976). Specific binding and ability of vasoactive intestinal octacosapeptide (VIP) to activate adenylate cyclase in isolated acinar cells from the Guinea pig. In *Surface Membrane Receptors*, eds. R. Bradshaw, W. Frazier, R. Merrell, D. Gottlieb & R. Angeletti, pp. 269–89. Plenum, New York.

DeMeyts, P., Roth, J., Neville, D. M. Jr. & Gavin, J. R. iii (1973). Insulin interactions with its receptors: Experimental evidence for negative cooperativity. *Biochemical and Biophysical Research Communications*, **55**, 154–61.

Frazier, W. A., Boyd, L. F. & Bradshaw, R. A. (1974). Properties of the specific binding of ^{125}I-Nerve growth factor to responsive peripheral neurones. *Journal of Biological Chemistry*, **249**, 5513–19.

Gavin, J. R. iii, Roth, J., Neville, D. M. Jr., DeMeyts, P. & Buell, D. N. (1974). Insulin-dependent regulation of insulin receptor concentrations. A direct demonstration in cell culture. *Proceedings of the National Academy of Sciences, U.S.A.*, **71**, 84–8.

Gerisch, G. (1968). Cell aggregation and differentiation in *Dictyostelium*. *Current Topics in Developmental Biology*, **3**, 157–97.

Gerisch, G. & Malchow, D. (1976). Cyclic AMP receptors and the control of cell aggregation in *Dictyostelium*. *Advances in Cyclic Nucleotide Research*, **7**, 49–68.

Gingle, A. R. & Robertson, A. (1976). The development of the relaying competence in *Dictyostelium discoideum*. *Journal of Cell Science*, **20**, 21–7.

Ginsberg, B. H., Cohen, R. M. & Kahn, C. R. (1976). Insulin-induced dissociation of receptors into subunits: Possible molecular concomitant of negative cooperativity. *Proceedings of the American Diabetes Association*, Abstract **5**, 322.

Green, A. A. & Newell, P. C. (1975). Evidence for the existence of two types of cAMP binding sites in aggregating cells of *Dictyostelium discoideum*. *Cell*, **6**, 129–36.

Henderson, E. J. (1975). The cyclic adenosine 3′:5′-monophosphate receptor of *Dictyostelium discoideum*. *Journal of Biological Chemistry*, **250**, 4730–6.

Jacobs, S. & Cuatrecasas, P. (1976). The mobile receptor hypothesis and 'cooperativity' of hormone binding: Application to insulin. *Biochimica et Biophysica Acta*, **433**, 482–95.

Klein, C. & Juliani, M. H. (1977). cAMP-induced changes in cAMP-binding sites on *Dictyostelium discoideum* amebae. *Cell*, **10**, 329–35.

Konijn, T. M., Van de Meene, J. G. C., Chang, Y. Y. & Barkley, D. S. (1969). Identification of adenosine-3′,5′-monophosphate as the bacterial attractant for myxamoebae of *Dictyostelium discoideum*. *Journal of Bacteriology*, **99**, 510–12.

Lesniak, M. A. & Roth, J. (1976). Regulation of receptor concentration by homologous hormone. *Journal of Biological Chemistry*, **251**, 3720–29.

Limbird, L. E., DeMeyts, P. & Lefkowitz, R. J. (1975). β-adrenergic receptors: Evidence for negative cooperativity. *Biochemical and Biophysical Research Communications*, **64**, 1160–68.

Malchow, D. & Gerisch, G. (1974). Short-term binding and hydrolysis of cyclic 3′:5′-adenosine monophosphate by aggregating *Dictyostelium* cells. *Proceedings of the National Academy of Sciences, U.S.A.*, **71**, 2423–7.

MALKINSON, A. M. & ASHWORTH, J. M. (1973). Adenosine 3',5' cyclic mono-phosphate concentrations and phosphodiesterase activities during axenic growth and differentiation of cells of the cellular slime mould *Dictyostelium discoideum. Biochemical Journal*, **134**, 311–19.

MATO, J. M. & KONIJN, T. M. (1975). Chemotaxis and binding of cyclic AMP in cellular slime moulds. *Biochemica et Biophysica Acta*, **385**, 173–9.

MATO, J. M., VAN HAASTERT, P. J. M., KRENS, F. A., RHIJNSBURGER, E. H., DOBBE, F. C. P. M. & KONIJN, T. M. (1977). Cyclic AMP and folic acid mediated cyclic GMP accumulation in *Dictyostelium discoideum. F.E.B.S. Letters*, **79**, 331–6.

MULLENS, I. A. & NEWELL, P. C. (1978). Kinetics of cAMP binding to cell surface receptors of *Dictyostelium. Differentiation*, (in press).

NEWELL, P. C. (1977a). Aggregation and cell surface receptors in cellular slime moulds. In *Microbial Interactions*, ed. J. L. Reissig, pp. 1–57. (Receptors and Recognition, Series B). Chapman & Hall, London.

NEWELL, P. C. (1977b). How cells communicate: the system used by slime moulds. *Endeavour*, **1**, (new series), 63–8.

PANNBACKER, R. G. & BRAVARD, L. J. (1972). Phosphodiesterase in *Dictyostelium discoideum* and the chemotactic response to cyclic adenosine monophosphate. *Science*, **175**, 1014–15.

POLLET, R. J., STANDAERT, M. L. & HAASE, B. A. (1977). Insulin binding to the human Lymphocyte receptor: Evaluation of the negative cooperativity model. *Journal of Biological Chemistry*, **252**, 5828–34.

RAO, CH. V. (1976). Lack of negative cooperativity among binding sites for prostaglandin $F_2\alpha$ in bovine corpus luteum cell membranes. *Life Sciences*, **18**, 499–506.

ROBERTSON, A., DRAGE, D. J. & COHEN, M. H. (1972). Control of aggregation in *Dictyostelium discoideum* by an external periodic pulse of cyclic adenosine monophosphate. *Science*, **175**, 333–5.

WURSTER, B., SCHUBIGER, K., WICK, U. & GERISCH, G. (1977). Cyclic GMP in *Dictyostelium discoideum*. Oscillations and pulses in response to folic acid and cyclic AMP signals. *F.E.B.S. Letters*, **76**, 141–4.

CELLULAR RECOGNITION IN SLIME MOULD DEVELOPMENT

By D. R. GARROD, ALMA P. SWAN,*
A. NICOL and D. FORMAN†

Medical Oncology Unit, Centre Block, Southampton General Hospital,
Southampton, SO9 4XY U.K.

The cellular slime moulds, especially *Dictyostelium discoideum*, are widely regarded as a model system for development. During the developmental sequence which is their life cycle (Fig. 1) they undergo processes of differentiation, pattern formation and morphogenetic movement, analogous in many respects to similar processes in animal development. Thus, as well as being fascinating in its own right, the study of the development of these simple eukaryotes should provide essential clues for the experimental analysis of many fundamental aspects of development in general.

Nowhere does this seem truer than in the field of cellular recognition and selective cell adhesion. In fact, Garrod (1974) pointed out that the cellular slime moulds show the same types of adhesive selectivities as do embryonic cells, particularly the embryonic cells of vertebrates. Firstly, they show changes in adhesion properties with time during development, particularly during the events leading to the chemotactic aggregation stage of the life cycle. During embryonic development, changes in cellular adhesiveness at particular stages are believed to be crucial in controlling morphogenetic cell movements (see Trinkaus, 1963, and Gustafson & Wolpert, 1967). Garrod (1974) coined the term 'chrono-specific' adhesion to denote such changes in adhesiveness as are specific to certain stages of development. Secondly, the slime moulds appear to show species-specific recognition which may be akin to species-specific cell adhesion demonstrated for vertebrate embryonic cells by Burdick & Steinberg (1969) and Burdick (1970, 1972). Thirdly it was suggested somewhat tenuously that *D. discoideum* showed a type of tissue-specific cell adhesion, not unlike the tissue-specific adhesion of vertebrate embryonic cells (Steinberg, 1964). Thus, the sorting-out of potential spore and stalk cells during the early post-aggregation phase of the slime mould life cycle may play an important part in the formation of the spore-stalk pattern of the final fruiting body.

* Department of Structural Biology, St George's Hospital Medical School, Cranmer Terrace, London SW7, U.K.

† Imperial Cancer Research Fund Laboratories, Burtonhole Lane, London, NW7 1AD, U.K.

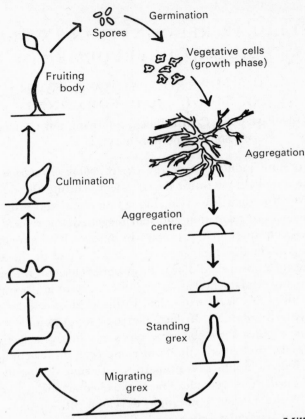

Fig. 1. The life cycle of *Dictyostelium discoideum*. Development on Millipore filters from the time of plating to fruiting body formation takes 24 h.

During the past four years, work in our laboratory has concentrated on these aspects of cell adhesion in the slime moulds. Our philosophy has been that it is essential to understand the descriptive aspects of developmental processes to complement studies trying to determine their biochemical and molecular basis. The story of primary embryonic induction, which was first reported in 1918 by Spemann and which is still not understood, is surely adequate justification for this belief. It is fundamental to obtain as much descriptive information as possible about the phenomenon one is trying to explain before proceeding logically to define the questions which need to be asked at the molecular level.

We have pursued this philosophy in descriptive work on both pattern formation and species-specific cell adhesion in the slime moulds. In addition we have developed some more biochemical work which may eventually help to elucidate the mechanism of cell adhesion in these organisms. Before reporting and discussing this work, we review briefly what is known about the mechanism of slime mould cell adhesion.

CONTACT SITES AND CARBOHYDRATE-BINDING PROTEINS IN SLIME MOULD CELL ADHESION

It has been known for some time that the cohesive properties of *D. discoideum* cells alter during the pre-aggregation phase of the life cycle (Gerisch, 1961; Born & Garrod, 1968; Garrod & Gingell, 1970; Garrod, 1972). A variety of surface differences between pre-aggregation and aggregation-competent cells have been reported recently. Of these, two main lines of study have related particularly to cellular cohesiveness, the work of Gerisch and colleagues and of Rosen, Barondes and their group.

The approach of Gerisch and coworkers has been to prepare antibodies against sites on the surfaces of pre-aggregation and aggregation-competent cells (i.e. cells which readily show chemotactic aggregation when placed on a solid substratum (Beug *et al.*, 1970; Gerisch *et al.*, 1974). Univalent (Fab) fragments of these antibodies have highly specific inhibitory effects on cell cohesion and adhesion (from here on we use the word 'cohesion' to refer to the sticking of cell to cell and the word 'adhesion' to refer to the sticking of cell to substratum). From this work, the possibility has emerged that two types of contact sites, A and B, are involved in cell cohesion (and adhesion). 'Contact sites A' are present on the surface of aggregation-competent cells. Beginning at an almost zero level on vegetative cells, their quantity rises simultaneously with the acquisition of aggregation-competence and EDTA-insensitive cohesion. 'Contact sites B' are present on both pre-aggregation and aggregation-competent cells. Blocking contact sites A with Fab fragments prevents the end-to-end cohesion characteristic of aggregating cells, whereas blocking contact sites B prevents side-to-side cohesion. Blocking both contact sites A and B prevents adhesion. Fab-binding to the contact sites is inactivated by periodate and is heat sensitive, suggesting that the sites are carbohydrate-associated proteins. Contact sites A have been purified and shown to be glycoproteins of molecular weight 120000–130000. There appear to be about 3×10^5 contact sites A per cell as is shown by the number of Fab molecules required to block them completely. Finally, there is some evidence that contact sites A may in fact be a mixture of two complementary sites, as would seem necessary to allow binding between cell surfaces.

The approach of Rosen & Barondes and their colleagues has led to the discovery of carbohydrate-binding proteins (lectins) on the surfaces of slime mould cells (Rosen, Kafka, Simpson & Barondes, 1973). The carbohydrate-binding protein from *D. discoideum* has been called 'discoidin' and that from *Polysphondylium pallidum*, 'pallidin' (Rosen, Simpson, Rose & Barondes, 1974). It has been shown that discoidin appears on the cells 6–8 h after food deprivation (i.e. the beginning of the developmental phase of the life cycle). The appearance of discoidin parallels the development of EDTA-insensitive cohesiveness by the cells prior to chemotactic aggregation. Anti-pallidin Fab prevents cell cohesion (Rosen, Haywood & Barondes, 1976). Apparently discoidin continues to increase up to 12 h of development (i.e. beyond the aggregation stage) to a level 400-fold greater than at the time of its first appearance. Receptors for the lectins also appear on the cells at the time of aggregation-competence (Chang, Reitherman, Rosen & Barondes, 1975; Reitherman, Rosen, Frazier & Barondes, 1975, Chang, Rosen & Barondes, 1977). Discoidin can be fractionated into discoidins I and II each having a molecular weight of 100000 to 110000.

Carbohydrate-binding proteins have been isolated from four other slime mould species as well as from *D. discoideum* and *P. pallidum*, (Rosen, Reitherman & Barondes, 1975). These lectins have discriminable carbohydrate-binding properties as determined by sugar inhibition of erythrocyte agglutination, though there is overlap of their binding specificities, (Rosen *et al.*, 1975).

Despite many apparent similarities in their properties, contact sites A and discoidin seem to be different molecules (Heusgen & Gerisch, 1975). A possibility which has not been excluded, however, is that contact sites A represent discoidin plus its receptor. Because of this discrepancy, there is still considerable doubt about the molecular mechanism of cell adhesion in slime moulds.

A LOW MOLECULAR WEIGHT INHIBITOR OF CELL COHESION FROM AXENIC CULTURES

We had shown that populations of *D. discoideum* cells grown axenically under slightly different conditions would sort out from each other during the life cycle (Leach, Ashworth & Garrod, 1973). For example, when cells grown in axenic medium plus 86 mM glucose to the log or stationary phases were mixed and allowed to develop together, they formed mixed aggregates and subsequently sorted out during development, the log-phase cells predominating in the sporeheads and the stationary-phase cells predominating in the stalks of the final fruiting bodies. Since sorting out of many different types of cells has been attributed to differences in cohesive properties, we began an investigation of the cohesiveness of these two cell populations (Swan & Garrod, 1975). The main result of this investigation so far has been the discovery of a low molecular weight inhibitor of cell adhesion which accumulates in axenic medium during growth of a cell population. This inhibitor has a variety of properties which lead us to believe that it may be important in elucidating the mechanism of cohesion of slime mould cells.

Log phase glucose cells rapidly cohere when shaken in phosphate buffer and in fresh axenic medium (Fig. 2). In stationary phase medium (SPM) however, cohesion is completely inhibited (Fig. 2). The inhibitory activity is removed by dialysis of the SPM against phosphate buffer or distilled water (Fig. 2).

The inhibitor has been purified from SPM by dialysis, column chromatography, ion exchange chromatography and thin layer chromatography (Swan, Garrod & Morris, 1977).

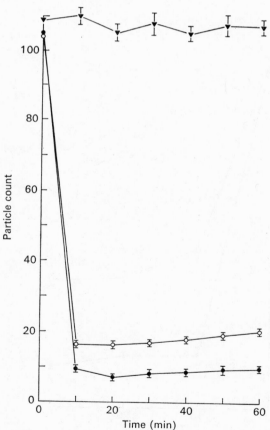

Fig. 2. Cohesion of log phase cells in shaken suspension in phosphate buffer (●); stationary phase medium (▼); and dialysed stationary phase medium (○).

The molecular weight of the inhibitor was estimated by the method of Andrews (1964) on a 40 × 2.6 cm chromatography column packed with Sephadex G-50 and eluted with 10 mM phosphate buffer, pH 7.0. The column was calibrated using a range of water-soluble standards and the molecular weight of the inhibitor was found to be between 500 and 700.

Properties of the inhibitor have been investigated using a partially purified fraction. The most important of these properties are as follows:

(1) A concentration of the inhibitor which completely inhibits the cohesion of log phase cells of *D. discoideum* (*Dd*) Ax-2 glucose cells only partially inhibits cohesion of aggregation-competent cells of this species (Fig. 3).

(2) The effect of the inhibitor is not species-specific. Vegetative cells of

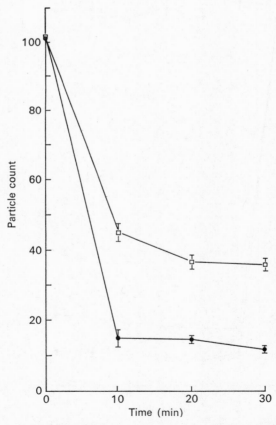

Fig. 3. Cohesion of aggregation-competent Ax-2G cells in phosphate buffer (●); and in the inhibitor solution (□).

three other species, *D. purpureum (Dp)*, *D. mucoroides (Dm)* and *Polysphondylium violaceum (Pv)* are also prevented from cohering by the inhibitor. Complete inhibition of *Dp* and *Dm* occurred as with *Dd* Ax-2 log cells, but *Pv* was only partially inhibited at the same inhibitor concentration (Fig. 4).

(3) As well as preventing cohesion, the inhibitor diminishes adhesion of cells of all four species to glass (Table 1).

(4) The development through the life cycle of *Dd* Ax-2 cells on Millipore filters (Sussman, 1966) is also reversibly blocked when the inhibitor is applied prior to the aggregation stage. After aggregation, the inhibitor has no effect on development.

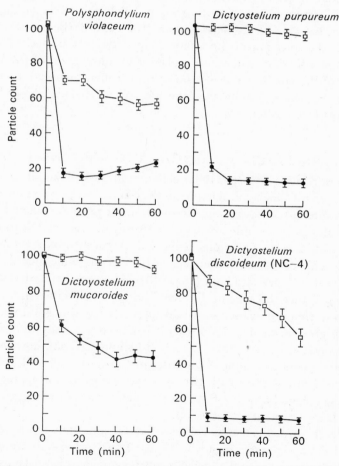

Fig. 4. Cohesion of pre-aggregation cells of four species of cellular slime moulds in phosphate buffer (●); and in inhibitor solution (□).

These results have implications in relation to several aspects of cellular recognition in slime moulds. Firstly, sensitivity to the inhibitor is developmentally regulated as are both the contact sites mechanisms and the carbohydrate-binding proteins already discussed. There is a possibility that the inhibitor is specifically active against the cohesive mechanism of early pre-aggregation cells (contact sites B). Since we have preliminary evidence that the inhibitor binds to the cells it may be that it binds specifically to contact sites B (Swan *et al.*, 1977).

The effect of the inhibitor on different species is interesting with respect to our findings on interspecific cohesive relationships. It suggests that the cohesive mechanisms employed by vegetative cells of these species may have a similar or even identical molecular basis (see below).

The diminution of adhesion to glass parallels the effect of the anti-homogenate Fab (directed against contact sites A and B) (Beug *et al.*, 1970). This raises the question of the relationship between the mechanisms of cohesion and adhesion. Adhesion to glass is an illustration of the lack of specificity of the adhesive mechanism of these cells which may be related to the palpable lack of specificity of the cohesive mechanism shown by our other results (see below).

INTERSPECIFIC COHESION AND SORTING OUT

Recent work relating to intercellular cohesion in the cellular slime moulds has centred on the biochemical nature of the cell surface (see section above). Regarding interspecific cohesion, Rosen *et al.* (1975) have isolated lectin-like molecules from six different species. However, it is difficult to understand the significance of this biochemical work because little is known of the interspecific cohesive relationships of the cellular slime moulds. It has been shown by Raper & Thom (1941) Shaffer (1957 *a*, *b*) and Bonner & Adams (1958) that the cells of different species would either not stick to each other during aggregation or would aggregate together and then separate during grex formation. From this it has been generally inferred that cell adhesion is species-specific. Garrod (1974) pointed out, however, that these experiments are difficult to interpret in terms of cell cohesion. Specific recognition shown by two species which did not aggregate together might have been due to different chemotactic mechanisms rather than to a lack of mutual cell cohesion.

In order to investigate the cohesive interactions between species we have carried out experiments in which cells of different species were mixed in binary combinations and allowed to form aggregates in shaken suspension. This provided a situation in which chemotactic mechanisms were

Table 1. *Effect of inhibitor on adhesion of slime mould cells of different species to glass*

Species	Test solution	Mean % decrease in adhering cells	S.D.	t	P	No. of observations
D. discoideum (NC-4)	Buffer	2.58	2.01	3.94	0.02	60
	Inhibitor	19.34	6.90			60
D. discoideum (NC-4) (aggregation-competent)	Buffer	− 1.03	7.34	8.73	0.02	90
	Inhibitor	45.44	1.12			90
P. violaceum	Buffer	2.28	1.05	7.66	0.001	80
	Inhibitor	22.72	5.24			80
D. purpureum	Buffer	2.75	0.60	2.81	0.05	80
	Inhibitor	16.89	10.06			80
D. mucoroides	Buffer	2.85	4.04	3.37	0.02	60
	Inhibitor	22.52	10.97			60

The mean % decrease of adhering cells after 30 min is shown (see text for details). The two treatments (buffer and inhibitor) for each species were compared by Student's t test. Differences with $P \leqslant 0.05$ are considered significant. Results from three or more experiments.

negated so that mutual adhesion alone could be studied. Aggregation stage cells which had been mechanically dissociated by trituration in distilled water were used and in each mixture one of the species was labelled with [^{3}H]thymidine to allow identification by autoradiography. The species used were *Dictyostelium discoideum* (*Dd*), *D. mucoroides* (*Dm*), *D. purpureum* (*Dp*) and *Polysphondylium violaceum* (*Pv*). The following binary combinations were tested: *Dd/Dm*, *Dd/Dp*, *Dd/Pv* and *Dm/Pv*. The mixed suspensions were shaken for a maximum of 24 h. In all cases the cells of the different species were found to be mutually cohesive. There were some differences of detail between the results obtained with different mixtures, as follows:

Dd/Dm: At times up to 4 h both cell types were present in aggregates and were not arranged in any discernible pattern (Fig. 5). In 8 h and 24 h aggregates the cells were distinctly localised according to species (Fig. 5). The pattern of localisation was inconsistent.

Dd/Pv: At 1 h aggregates were small and loosely packed and the cell types did not appear localised (Fig. 6). At 2 h aggregates were larger and more compact and in some the cells were localised according to species (Fig. 6). At 4 h and 8 h localisation according to species was distinct and, although the aggregates were irregularly shaped there was a definite tendency for *Dd* to surround *Pv* (Fig. 6). At this stage a few aggregates appeared to consist of one species only. By 24 h some mixed aggregates were present similar in size and appearance to those found at 8 h. However, the majority of aggregates consisted entirely or almost entirely of cells of one species only (Fig. 6).

Dd/Dp: At 1 h, 8 h and 24 h aggregates all shared distinct localisation of cell types. Again, the pattern of localisation was inconsistent (Fig. 7).

Dm/Pv: At 1 h there were three types of aggregates, those consisting of cells of either species alone, and mixed aggregates in which there appeared to be localisation according to species (Fig. 7). 8 h aggregates were larger and more regular in shape. Again three types of aggregates were present. The mixed aggregates showed localisation according to species and a tendency for *Dm* to surround *Pv* (Fig. 7). At 24 h there were again three types of aggregates (Fig. 7).

These results show that cohesion between slime mould cells is not completely species-specific. Rather, cells of different species readily adhere to

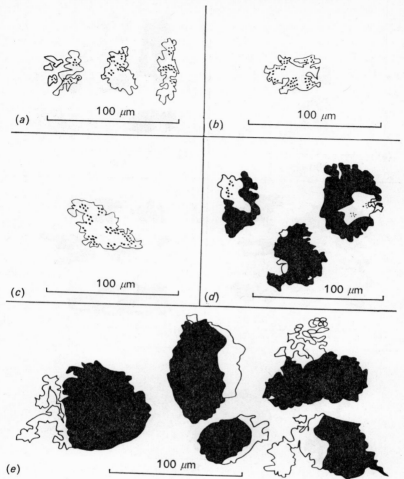

Fig. 5. Cohesion and sorting out between *Dm* and *Dd*. The diagrams are tracings taken from autoradiographs of aggregate sections. The black areas represent regions of labelled cells (in this case *Dd*) and the white areas regions of unlabelled cells (*Dm*). Black dots indicate the presence of labelled cells. *Dm* and *Dd* cells were mixed in a ratio of 1:1. Aggregates fixed after shaking for (*a*) 1 h, (*b*) 2 h, (*c*) 4 h, (*d*) 8 hr and (*e*) 24 h.

each other, forming mixed aggregates. They then segregate according to species by a process of sorting out, and, except in the case of *Dd/Pv*, remain mutually cohesive within aggregates.

It is important to discuss what these results indicate about the molecular mechanisms of cell adhesion in the different species of slime moulds.

It is now widely believed that cell–cell adhesion occurs because of

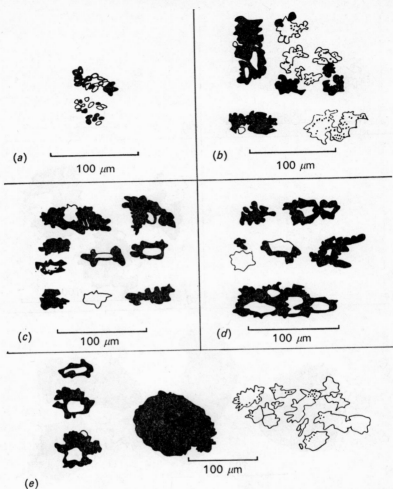

Fig. 6. Cohesion and sorting out between *Pv* and *Dd*. *Dd* labelled with [³H]thymi-dine. Cell ratio 1-1. Aggregates fixed after (*a*) 1 h, *b*) 2 h, (*c*) 4 h, (*d*) 8 h, (*e*) 24 h. Other details as for Fig. 5.

binding between complementary molecules on the apposed membranes between adhering cells, akin to the binding between antigen and antibody or between enzyme and substrate. Such a mechanism was first proposed by Tyler (1946) and Weiss (1947) and some reasonable evidence in its favour exists at least in the case of the sponge *Microciona prolifera* (Burger, Turner, Kuhns & Weinbaum, 1975; Turner, 1977) and the cellular slime mould *Dictyostelium discoideum* (see above section).

The evidence from our experiments strongly suggests that such a

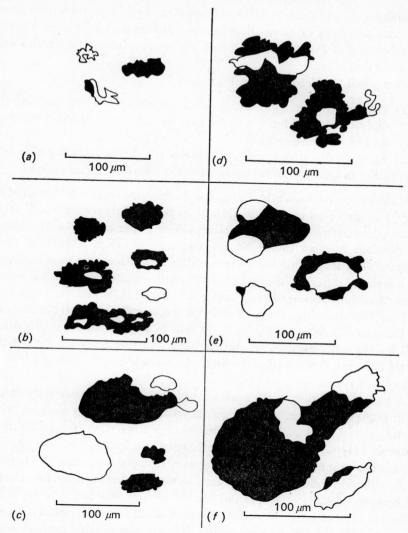

Fig. 7. Cohesion and sorting-out between *Pv* and *Dm*, (*a*, *b*, and *c*) and between *Dp* and *Dd* (*d*, *e* and *f*). *Dm* and *Dd* respectively were [³H]thymidine labelled. Cell mixtures are 1:1 ratios except in (*e*) which is 3:1 *Dp*:*Dd*. Aggregates fixed after (*a* and *d*) 1 h, (*b* and *e*) 8 h and (*c* and *f*) 24 hr. Other details as for Fig. 5.

mechanism in the slime mould is unlikely to be completely specific to individual species, because the cells of all four species readily cohere with each other. For example, because *Dm* and *Pv* cells cohere with each other, we cannot suppose that there is one set of cohesive sites for *Dm*, which are completely specific for *Dm*, and another set for *Pv*, which are completely

13

specific for *Pv* and entirely different from those for *Dm*. Rather, there must be some degree of cross reaction between *Dm* sites and *Pv* sites, i.e. the cohesive sites of *Dm* must be able to form binding interactions with those for *Pv* and vice versa.

Two possible mechanisms are:

(1) The cohesive sites of different slime mould species are qualitatively slightly different, but not so different as to preclude binding between sites of different species.
(2) The cohesive sites of different species are qualitatively the same but differ in quantity on the surfaces of the cells of different species.

If the lectin-like molecules isolated from different species by Rosen *et al.* (1975) are involved in cell adhesion, their binding properties would seem to support the first of these possibilities. This paper reports on the ability of different hexoses to inhibit the haemagglutinating activity of their lectins. They found that lectins from different species were each inhibited by a series of different hexoses, usually to different degrees. Thus it would seem that these lectins are not highly specific in their binding properties but can interact with a range of different sugar groups. A cell bearing one of these lectins on its surfaces should thus be able to adhere to any other cell having some combination on its surface of the hexoses to which the lectin can bind. Such a cohesive mechanism could in principle give rise to the type of results we have obtained.

A second possibility is the same as the suggestion put forward by Steinberg (1964) to explain the configurations achieved during tissue-specific sorting out of chick embryo cells. Steinberg's suggestion (the Differential Adhesion Hypothesis) is the only satisfactory explanation of how one cell type can surround another in sorting out. It necessitates that the heterotypic adhesion should be intermediate in strength between the two classes of homotypic adhesions. It is difficult to see how this could be so, if the adhesive sites of different cell types are qualitatively different. (It would mean that the sites of the surrounding cell type should bind more strongly to the sites of the other cell type than to their own). On the other hand if the different cell types possessed different numbers of identical sites it is quite easy to see how the heterotypic adhesions could be of intermediate strength.

Only in two of our combinations have we found a regular tendency for one of the cell types to surround the other, and even here other configurations were present simultaneously. These other configurations were similar to Steinberg's case 3 (Steinberg, 1964, p. 328) for which it was suggested that the heterotypic adhesions were weaker than either of the two types of homotypic adhesion. However, except in the case of Dd/Pv at 24 h, the strength of the heterotypic adhesion was clearly not zero. We feel that this may be compatible with our first suggestion of adhesive sites which are qualitatively slightly different. Our work with the inhibitor of cell adhesion from *D. discoideum* axenic cultures would be compatible with either possibility. If cells of each different species have more than one type of contact site (e.g. contact sites A and B of *D. discoideum*) the situation would be more complicated but the same arguments with respect to the binding site specificities should apply. We should also point out that once the cells of different species have adhered to form aggregates in suspension, the different chemotactic mechanisms of different species might give rise to sorting out within aggregates.

CELLULAR RECOGNITION IN PATTERN FORMATION

By 'pattern formation' we mean the spatial organisation of cellular differentiation.

The study of pattern formation in general has benefited in recent years from the concept of 'positional information' which has been developed by Wolpert (1969, 1971). Wolpert has suggested that cells can determine their position within a morphogenetic field by reference to some pervasive property of that field such as a chemical gradient and that the genome of each individual cell responds to this 'positional information' by instigating a type of differentiation appropriate to its position. An important contribution of this concept has been in successfully providing a framework for the experimental analysis of pattern formation in a variety of organisms.

The fruiting body of *Dictyostelium discoideum* has a very simple linear pattern consisting of a prolate mass of spores on top of a narrow, conical cellular stalk and thus ostensibly provides a good system for studying the mechanism of pattern formation, many of the patterns of differentiation in animal development being structurally more complex and therefore more difficult to analyse.

The spore–stalk pattern of the fruiting body is preceded during the migrating and early culminating grex stages of the life cycle by a pattern of pre-spore and pre-stalk cells which are present at the back and front of the grex respectively (Fig. 8). These cells are the precursors of the fully differentiated cells of the fruiting body (see reviews by Bonner, 1967; Garrod & Ashworth, 1973; and Loomis, 1975; and recent work of Hayashi & Takeuchi, 1976; and Forman & Garrod, 1977a).

There are essentially two schools of thought about the mechanism of pattern formation in *D. discoideum* (see Forman & Garrod, 1977b; Garrod

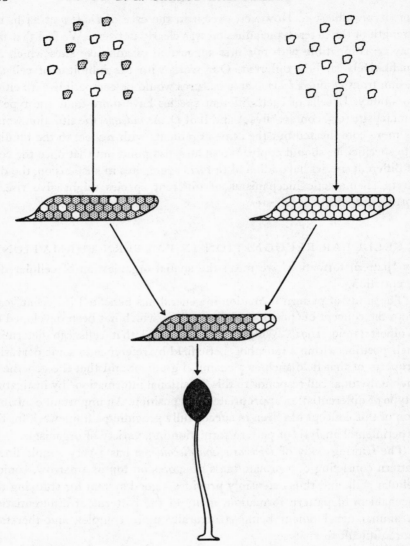

Fig. 8. Diagram summarising two alternative views of pattern formation in *Dictyostelium discoideum*. On the right-hand side the view that cells differentiate in position in the grex giving rise to pre-stalk cells at the front and pre-spore cells at the back. These then form the stalk and spores of the fruiting body respectively. On the left-hand side, the view that pattern formation occurs by sorting out of cells which are initially randomly arranged in the cell mass and which may be predisposed to form either spore or stalk even before aggregation.

& Forman, 1977). These are summarised in Fig. 8. One view (represented on the right of the figure) is that pre-stalk and pre-spore cells differentiate in position in the grex in response to some type of positional information. The polarity of the grex tip and the signalling or organising properties of the grex tip are in some way essential components of theories suggesting this kind of mechanism (McMahon, 1973; Pan, Bonner, Wedner & Parker, 1974; Rubin & Robertson, 1975). The other view is that pattern formation occurs by sorting out within the grex of cells which have become predisposed to form spore or stalk at an earlier stage of the life cycle, perhaps before the aggregation stage (Takeuchi, 1969; Bonner, Sieja & Hall, 1971; Garrod & Ashworth, 1973; Leach *et al.*, 1973; Garrod, 1974; Maeda & Maeda, 1974).

It is important to distinguish between these alternative views, not least because *D. discoideum* is widely regarded as a model for the development of other organisms. Is it a model for pattern formation by positional information or is it not? If not, does the sorting out of cells with tendencies to differentiate as either spore or stalk provide a model for a process akin to the tissue-specific sorting out of vertebrate embryo cells (Garrod, 1974)?

We have begun to try to decide which, if either, of these mechanisms is operative in the following way.

It has long been known that vegetative cells of *D. discoideum* cohere to form spherical aggregates if maintained in shaken suspension in phosphate buffer (Gerisch, 1968). Such cells are capable of differentiating to the chemotactic aggregation stage but were thought to develop no further unless brought to an air–water interface at which they would develop a tip and form normal fruiting bodies (Gerisch, 1968). We decided to determine whether cells in aggregates which were maintained in shaken suspension were capable of differentiating into pre-spore and pre-stalk cells and of forming a pre-spore–pre-stalk pattern. Because aggregates formed in this way are spherical (Plate 1a) they possess neither the morphological polarity not the tip of the grex stages of the normal life cycle. These experiments therefore enabled us to ask whether polarity and the grex tip, crucial ingredients of theories dependent on positional information, were necessary for differentiation and pattern formation. In order to test for the formation of pre-spore cells, we stained sections of aggregates with a fluorescein-labelled antiserum prepared against the spores of another species, *D. mucoroides*, which specifically stains pre-spore vesicles (PSV) in the cytoplasm of pre-spore cells of Dd (Takeuchi, 1963) (Full details of the experimental procedures used are given by Forman & Garrod (1977a, b).)

Our first finding was that pre-spore cells developed within spherical aggregates at about 12–13 h after the beginning of shaking. This corresponds to the time of appearance of pre-spore cells during the normal life cycle of *D. discoideum*, about 4 h after the chemotactic aggregation stage and 12 h after the beginning of development (Hayashi & Takeuchi, 1976; Forman & Garrod, 1977 *a*). However, between 12 and 18 h, although pre-spore differentiation was apparent in suspension aggregates, no pattern of pre-spore and pre-stalk cells was present. Instead, the pre-spore cells appeared to be essentially randomly arranged and mingled with other cells which did not show specific staining for PSVs (Plate 1*c*). Sections of 24 h aggregates on the other hand showed the presence of a pattern of specific staining very similar to that found in the migrating grex of the normal life-cycle. Pre-spore cells were localised in one region of the spherical aggregates and were partially surrounded by a cap of non-staining cells, presumably pre-stalk cells (Plate 1*d*). Thus although being still completely spherical and showing no overt polarity nor any sign of tip development, the aggregates had become internally polarised as a consequence of pattern formation.

Since these aggregates were initially unpolarised by the criteria normally taken to indicate polarity of the grex, namely an elongated shape and the presence of a tip, we are tempted to conclude that these features are not essential for pattern formation. We must, however, be cautious of the interpretation of these results for a number of reasons. Firstly, we cannot completely exclude the possibility of a tip region being present in these aggregates, even though a tip was not morphologically recognisable. Even if a clandestine tip was present, however, our results would seem inconsistent with the simplest view of differentiation of cell types in position. Illustrated in Fig. 9*a*, this view would be that a tip region gives rise to positional information within the aggregate which defines a boundary within the aggregate. Cells on one side of this boundary then differentiate into pre-stalk cells and cells on the other side into pre-spore cells. We never observe within our aggregates during the earliest stages of pre-spore differentiation (12 to 18 h) a region comparable in size to the non-staining region of 24 h aggregates, so that simple differentiation in position can be excluded. Another possibility based on positional information is still possible, however. This is that cells initially differentiate into pre-spore cells in a spatially random way. At a later stage, a boundary is defined, possibly in relation to a tip region, and the cells on one side of the boundary then redifferentiate forming pre-stalk cells (Fig. 9*b*). Cells of a mutant strain (V-12 M 2), when spread at low density in the presence of 10^{-3} M cyclic AMP have been shown to develop into pre-spore cells and then revert to stalk cells (Kay, Garrod & Tilly, 1978) though this does not happen with

PLATE I

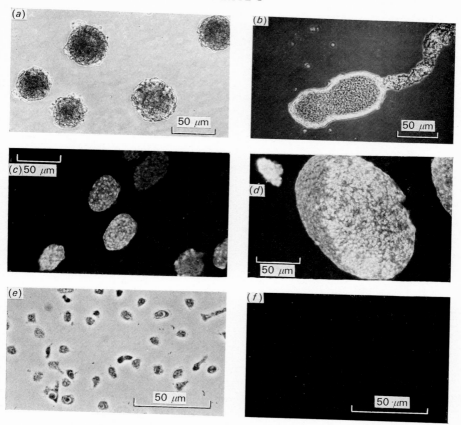

For explanation see p. 202

Ax-2, the strain used in our experiments. Because the surfaces of our 24 h patterned aggregates are not smooth, unlike the surface of the normal grex, but instead allow the outlines of individual cells to be seen, we believe that the aggregates are not covered by a slime sheath. We therefore question the view (Ashworth, 1971; Farnsworth & Loomis, 1975) that the pattern may be formed in response to positional information generated by the slime sheath of the grex.

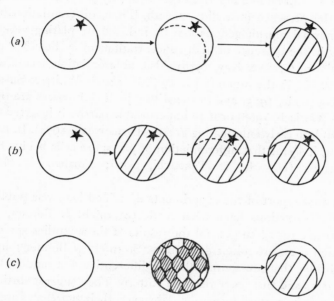

Fig. 9. Alternative possibilities for pattern formation in spherical aggregates. (a) A tip region, though not visible, is present at some point in the aggregate (represented by a star). A boundary line is specified in relation to the tip. Cells differentiate in position within the aggregate, pre-spore cells (cross-hatched) on one side of the boundary and pre-stalk cells (unshaded) on the other. (b) An invisible tip region is present. All the cells differentiate as pre-spore cells. A boundary line is then specified. Cells on the tip side of the boundary line change into pre-stalk cells. (c) No tip region is present. Pre-spore and pre-stalk cells differentiate at random within the aggregate and sort out to give a pre-stalk–pre-spore pattern.

The interpretation of our results which we favour at present is that pattern formation occurs by the sorting out within these aggregates of pre-spore and pre-stalk cells which initially differentiate with spatial randomness (Fig. 9c). The results though certainly consistent with this opinion, are open to at least one other interpretation. In addition there are other reasons for favouring sorting out rather than differentiation in position as a mechanism for pattern formation. Firstly, there is considerable evidence from other work to demonstrate that sorting out of cells can and does occur in the normal development of *D. discoideum* (Bonner, 1959; Takeuchi, 1963, 1969; Bonner *et al.*, 1971; Francis & O'Day, 1971; Garrod & Ashworth, 1973; Leach *et al.*, 1973; Muller & Hohl, 1973; Garrod, 1974; Maeda & Maeda, 1974; Takeuchi, Hayachi & Tasaka, 1977). Secondly, there is much evidence that differentiation of cell types does not necessitate (*i*) that the cells should be in position in the grex, (*ii*) that the cells should be in a polarised cell mass (*iii*) the presence of a grex tip, or (*iv*) the presence of a slime sheath. Thus, differentiation in the absence of normal morphogenesis can be induced by plating vegetative cells on agar in the presence of high concentrations of cyclic AMP (Bonner, 1970; Town, Gross & Kay, 1976). Also, differentiation occurs within tipless cell masses in the mutants FR-17 (Sonneborn, White & Sussman, 1963) and P-4 (Chia, 1975) and in the wild type if cell masses are treated with EDTA (Gerisch, 1968) or if an impermeable barrier is inserted to the correct depth into early culminating grexes (Farnsworth, 1974). In normal development, it seems that differentiation of pre-spore cells begins before a tip develops (Hayashi & Takeuchi, 1976; Forman & Garrod, 1977*a*).

Since the first report of the experiments described here was published, two similar observations have been made (Sternfeld & Bonner, 1977; Takeuchi, *et al.*, 1977). In general the results of these studies agree with ours, although there are minor differences. Sternfeld & Bonner found, in common with us, that pre-spore cells first differentiated at random within their aggregates and then gave rise to a pattern. They suggest sorting out as a mechanism for pattern formation. However, their aggregates appeared to possess both polarity and slime sheath. This difference between our results and those of Sternfeld & Bonner arises because different strains of *D. discoideum* were used, Ax-2 glucose in our case and the wild type, NC-4, in theirs and not because the method used for forming and maintaining aggregates were slightly different. Under our conditions NC-4 cells also form sausage-shaped (polarised) masses which appear to possess a slime sheath (Plate 1*d*). However, since spherical Ax-2 glucose aggregates form a pre-stalk–pre-spore pattern, we conclude that polarity is not essential for pattern formation. Takeuchi *et al.* (1977) obtained spherical aggregates in rolled test tubes, but reports that his aggregates were already patterned at the earliest time of appearance of pre-spore cells. He believes that sorting out precedes pre-spore differentiation in that PSVs do not appear in the cells until they have reached the appropriate position in the aggregate. This interpretation would permit a mechanism of pattern formation involving both sorting out and differentiation in position. However, we have presented above our reasons for believing that cells do not need to be in the correct position within a cellular mass in order to differentiate.

As a result of our experiments on pattern formaton in spherical aggregates and the experimental work of others we have made a series of suggestions about the mechanism of pattern formation and morphogenesis of *D. discoideum* during normal development (Forman & Garrod, 1977*b*). Four of these suggestions are as follows:

(1) The two cell types differentiate at random within the early grex and then sort out to give the pre-spore–pre-stalk pattern.

(2) Differentiation, sorting out and pattern formation begin before the formation of the grex tip. Thus the grex tip does not play an organising role in any of these processes.

(3) The early pre-stalk region itself gives rise to the grex tip. It is the formation of the pattern which generates the polarity of the grex tip rather than the polarity which gives rise to the pattern.

(4) Control of the ratio of pre-stalk to pre-spore cells may be independent of the mechanism which determines the spatial arrangement of cell types. This suggestion is presented in diagrammatic form in Fig. 10.

In many respects these suggestions are original, and hopefully, provocative. We do not regard any of them as proven. However, we have presented them as a basis for discussion and as a stimulus to experimental work in the field. Even if they can be conclusively shown to be incorrect, our understanding of pattern formation in the slime mould will have advanced. Some of these suggestions will now be examined in more detail.

How can cells differentiate at random within a population? Firstly, it has been shown that during aggregation of the species *Polysphondylium violaceum*, founder cells, the cells which originate or found aggregation centres, appear to differentiate at random within the population (Shaffer, 1961). Once an aggregation centre has been set up, it inhibits the formation of other aggregation centres in its vicinity (Shaffer 1963). Also spatially random differentiation of aggregation centres occurs within uniform fields of *D. discoideum* amoebae (see Gerisch, Hulser, Malchow & Wick, 1975). The idea of spatially random cell differentiation in slime moulds is not new therefore. In the case of pre-spore and pre-stalk cell differentiation, we suggest that cells within the population may be continuously variable with regard to some metabolic parameter, such as rate of differentiation into either pre-stalk or pre-spore cells. Once within the cell mass, the cells which can differentiate fastest, say into pre-spore cells, would begin differentiation. This differentiation would be on a spatially random basis

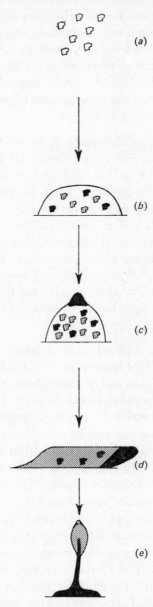

Fig. 10. Proposed model for pattern formation. (*a*) Prior to aggregation cells un-differentiated, though there may be tendencies within the population. (*b*) Following chemotactic aggregation, cells start to differentiate at random within the aggregate (pre-stalk cells black; pre-spore cells cross-hatched). (*c*) The pre-stalk cells sort out from the pre-spore cells forming the tip of the aggregate. (*d*) The grex, led by the pre-stalk cells, migrates with the pattern fully formed, though there may be some residual sorting out. (*e*) Fruiting body. (From Forman, 1977).

because cells capable of different rates of differentiation would be incorporated into the cell mass at random during aggregation. Such a mechanism would probably necessitate some type of end-product inhibition such that differentiated pre-spore cells would inhibit the formation of further pre-spore cells. When inhibition reached a certain threshold level due to the accumulation of many pre-spore cells, those cells which had not differentiated, the slowest cells with regard to differentiation, would be prevented from forming pre-spore cells, and would become pre-stalk cells. A continuous variation in cellular properties could be suggested by the sorting-out experiments of Takeuchi (1969), and Takeuchi feels that a continuous variation in cellular properties may be important with regard to differentiation (Takeuchi, personal communication).

It should be pointed out that a mechanism based on differential rate of differentiation and inhibition could give rise to regulation of the proportions of pre-stalk and pre-spore cells. It would probably be necessary for inhibition to be mediated by a diffusible substance which is released by the differentiating cell type, and to assume that the threshold concentration of the substance be regulated to approximately the same value irrespective of the size of the cell mass. This could be done in a variety of ways which it would be inappropriate to develop here. The important thing is that such a mechanism for regulating proportions would not be dependent on cells being in position in the grex (suggestion 4, p. 193). Proportions could be regulated with a random distribution or a patterned distribution of pre-spore and pre-stalk cells. It would be expected, however, that the proportions would vary slightly with the manner in which the cells were distributed, because the spatial distribution of the inhibitor would vary. It should be stressed that there is no evidence for such an inhibitor of differentiation. The fact that regulation of both fragments occurs when the grex is cut to separate the pre-stalk and pre-spore cells (Raper, 1940; Bonner, Chiquoine & Kolderie, 1955; Gregg, 1965; Sampson, 1976) could be explained on the basis of this type of inhibition as well as a number of other mechanisms.

CELLULAR RECOGNITION IN DIFFERENTIATION

There is now increasing evidence that some type of contact-mediated cell–cell recognition is required for the continued development and differentiation of *D. discoideum* cells beyond the aggregation stage. Our own work has shown that cell contact is required for the formation of PSVs and the differentiation of pre-spore cells among cells maintained in shaken suspension. Thus cells which were shaken so fast that they were prevented from forming contacts were capable of becoming aggregation-competent but did not form PSVs (Plate 1 e & g) (Forman & Garrod, 1977 b; Garrod & Forman, 1977). It has previously been shown that if pre-spore cells were dissociated from the grex they would dedifferentiate, losing their PSVs (Sakai & Takeuchi, 1971).

Several other pieces of recent evidence stress the need for cell contact in differentiation beyond the aggregation stage. Thus, cell contact is required for the accumulation of certain developmentally regulated enzymes (threonine deaminase and tyrosine transaminase) after a certain point in development around the aggregation stage (Grabel & Loomis, 1977). When cell contacts are formed around the time of aggregation competence, about thirty new proteins begin to be synthesised and the synthesis of about six proteins which began previously ceases. (Lodish & Alton, 1977). The synthesis of new proteins is accompanied by the appearance of homologous translatable mRNAs. If contact formation is prevented, these changes in protein synthesis do not occur. It has been shown by Newell, Longlands & Sussman (1971) that if developing cells of D. discoideum are disaggregated at any time before the formation of spore walls, and replated, they would recapitulate the normal developmental sequence from aggregation onwards. The cells also recapitulate the accumulation of certain enzymes so that repeated disaggregation and re-aggregation of the cells resulted in accumulation of enzymes to two or three times the normal level. These accumulations require additional RNA synthesis (Newell, Franke & Sussman, 1972).

Further, it has been shown that the period of propagation of chemotactic waves through fields of aggregating cell changes abruptly from about 4 min to 1.5–2.5 min during aggregation. At the same time there is a cessation of accumulation of cell-bound phosphodiesterase and the synthesis of uridine diphosphate glucose pyrophosphorylase begins (Gross et al., 1977). Observations on a mutant which does not undergo these changes suggests that the formation of end-to-end contacts may be the crucial event which triggers the changes.

It seems probable that the cohesion of cells triggers a sequence of differentiative changes possibly in a manner analogous to the changes brought about by the binding of hormones to cell surfaces. Thus the interaction of receptors on apposed cell surfaces may trigger cytoplasmic events leading to differentiation. Is this trigger provided by binding between cohesive sites such as contact sites A, as may be suggested from the work of Gross et al. (1977) or are other components of the cell surface involved? It has been found that some enzymes, such as alkaline phosphatase, which normally accumulate late in development, can be synthesised by cells which are kept apart in suspension and pulsed with cyclic AMP (Rickenberg, Tihon & Guzel, 1977). It may be therefore, that there are at least two aspects to the control of differentiation after aggregation: a combination of cell-contact formation and cyclic-AMP pulsing may be required for further development.

CONCLUSION

Work on the molecular basis of cell cohesion in the cellular slime moulds has made good progress with the discovery of contact sites A and B, and of carbohydrate-binding proteins such as discoidin. At present a major problem is that of the relationship between these two possible cohesive mechanisms, and it has yet to be shown precisely how either or both are involved in cell cohesion. The low molecular weight inhibitor of cell cohesion which we have found has a range of extremely interesting biological effects. As yet we know little about its action at the molecular level, but there exists the possibility that it may help in elucidating the mechanism of cohesion. In particular it is more effective against pre-aggregation cells than against aggregation-competent cells, and may therefore enable us to investigate the mechanisms of the earliest cohesions about which little is known.

The inhibitor is effective against cohesion of cells of at least four species cellular slime moulds. Further, our studies on inter-specific cell cohesion immediately tell us something about the molecular basis of cohesion – whatever the mechanism is, it cannot be completely specific to cell type.

With regard to pattern formation in *D. discoideum* we can put forward a reasonable argument that sorting-out is involved. However, we cannot yet prove it. We need to obtain separate populations of pre-spore and pre-stalk cells, to label them and to mix them in aggregates. We can then ask whether they move from an initially random arrangement to a patterned one. If they do, then sorting-out as a mechanism for pattern formation will have been demonstrated.

It seems quite clear that some type of cell-contact-mediated recognition is required for differentiation beyond the aggregation stage. The molecular basis for this recognition is a fundamental problem of immediate importance. Here again our inhibitor, which inhibits development probably because it inhibits cell cohesion, may prove extremely valuable.

The financial support of the Science Research Council is gratefully acknowledged. We thank Susanna Johns, Sherilee Taylor and Hilary Nossiter for technical assistance.

REFERENCES

ANDREWS, P. (1964). Estimation of molecular weights of proteins by Sephadex gel filtration. *Biochemical Journal*, **91**, 222–8.

ASHWORTH, J. M. (1971). Cell development in the cellular slime mould *Dictyostelium discoideum*. In *Control Mechanisms of Growth and Differentiation*, SEB Symposium XXV, eds. D. D. Davies & M. Balls, pp. 27–49. Cambridge University Press.

BEUG, H., GERISCH, G., KEMPFF, S., RIEDEL, N. & CREMER, G. (1970). Specific inhibition of cell contact formation in *Dictyostelium* by univalent antibodies. *Experimental Cell Research*, **63**, 147–58.

BONNER, J. T. (1959). Evidence for the sorting out of cells in the development of the cellular slime moulds. *Proceedings of the National Academy of Sciences, U.S.A.*, **45**, 379–84.

BONNER, J. T. (1967). *The Cellular Slime Moulds*. Princeton University Press, Princeton, N.J.

BONNER, J. T. (1970). Induction of stalk cell differentiation by cyclic-AMP in the cellular slime mould *Dictyostelium discoideum*. *Proceedings of the National Academy of Sciences, U.S.A.*, **65**, 110–3.

BONNER, J. T. & ADAMS, M. S. (1958). Cell mixtures of different species and strains of cellular slime moulds. *Journal of Embryology and Experimental Morphology*, **6**, 346–56.

BONNER, J. T., CHIQUOINE, A. D. & KOLDERIE, M. Q. (1955). A histochemical study of differentiation in the cellular slime moulds. *Journal of Experimental Zoology*, **130**, 133–57.

BONNER, J. T., SIEJA, T. W. & HALL, E. M. (1971). Further evidence for the sorting out of cells in the differentiation of the cellular slime mould *Dictyostelium discoideum*. *Journal of Experimental Zoology*, **25**, 457–65.

BORN, G. V. R. & GARROD, D. R. (1968). Photometric demonstration of aggregation of slime mould cells showing effects of temperature and ionic strength. *Nature, London*, **220**, 616–18.

BURDICK, M. L. (1970). Cell sorting out according to species in aggregates containing mouse and chick embryonic limb mesoblast cells. *Journal of Experimental Zoology*, **175**, 357–68.

BURDICK, M. L. (1972). Differences in the morphogenetic properties of mouse and chick embryonic liver cells. *Journal of Experimental Zoology*, **180**, 117–25.

BURDICK, M. L. & STEINBERG, M. S. (1969). Embryonic cell adhesiveness: do species differences exist among warm-blooded vertebrates? *Proceedings of the National Academy of Sciences, U.S.A.*, **63**, 1169–73.

BURGER, M. M., TURNER, R. S., KUHNS, W. J. & WEINBAUM, G. (1975). A possible model for cell–cell recognition via surface molecules. *Philosophical Transactions of the Royal Society, London*, B, **271**, 379–93.

CHANG, C. M., REITHERMAN, R. W., ROSEN, S. E. & BARONDES, S. H. (1975). Cell surface location of discoidin, a developmentally regulated carbohydrate-binding protein from *Dictyostelium discoideum*. *Experimental Cell Research*, **95**, 136–42.

CHANG, C. M., ROSEN, S. D. & BARONDES, S. H. (1977). Cell surface location of an endogenous lectin and its receptor in *Polysphondylium pallidum*. *Experimental Cell Research*, **104**, 101–9.

CHIA, W. K. (1975). Induction of stalk cell differentiation by cyclic-AMP in a susceptible variant of *Dictyostelium discoideum*. *Developmental Biology*, **44**, 239–2.

FARNSWORTH, P. A. (1974). Experimentally induced aberrations in the pattern of differentiation in *Dictyostelium discoideum*. *Developmental Biology*, **33**, 869–77.

FARNSWORTH, P. A. & LOOMIS, W. F. (1975). A gradient in thickness of the surface sheath in pseudoplasmodia of *Dictyostelium discoideum*. *Developmental Biology*, **46**, 349–57.

FORMAN, D. (1977). 'Pattern formation in the cellular slime mould *Dictyostelium discoideum*.' Ph.D. Thesis, University of Southampton.

FORMAN, D. & GARROD, D. R. (1977 a). Pattern formation in *Dictyostelium discoideum*. I. Development of prespore cells and its relationship to the pattern of the fruiting body. *Journal of Embryology and Experimental Morphology*, **40**, 215–28.

FORMAN, D. & GARROD, D. R. (1977 b). Pattern formation in *Dictyostelium discoideum*. II. Differentiation and pattern formation in non-polar aggregates. *Journal of Embryology and Experimental Morphology*, **40**, 229–43.

FRANCIS, D. W. & O'DAY, D. H. (1971). Sorting out in pseudoplasmodia of *Dictyostelium discoideum*. *Journal of Experimental Zoology*, **176**, 265–72.

GARROD, D. R. (1972). Acquisition of cohesiveness by slime mould cells prior to morphogenesis. *Experimental Cell Research*, **72**, 588–91.

GARROD, D. R. (1974). Cellular recognition and specific cell adhesion in cellular slime mould development. *Archives de Biologie (Liege)*, **85**, 7–31.

GARROD, D. R. & ASHWORTH, J. M. (1973). Development of the cellular slime mould *Dictyostelium discoideum*. In *Microbial Differentiation*, SGM Symposium 23, eds. J. M. Ashworth & J. E. Smith, pp. 407–35. Cambridge University Press.

GARROD, D. R. & FORMAN, D. (1977). Pattern formation in the absence of polarity in *Dictyostelium discoideum*. *Nature, London*, **265**, 144–6.

GARROD, D. R. & GINGELL, D. (1970). A progressive change in electrophoretic mobility of preaggregation cells of the slime mould, *Dictyostelium discoideum*. *Journal of Cell Science*, **6**, 277–84.

GERISCH, G. (1961). Zellfunktionen und Zellfunktionswechsel in der Entwicklung von *Dictyostelium discoideum*. V. Stadienspecifishe Zelkontaktbildung und ihre quantitative Erfassung. *Experimental Cell Research*, **25**, 535–54.

GERISCH, G. (1968). Cell aggregation and differentiation in *Dictyostelium discoideum*. In *Current Topics in Developmental Biology*, vol. 3, eds. A. A. Moscona & A. Monroy, pp. 157–197. Academic Press, New York and London.

GERISCH, G., BEUG, D., SCHWARZ, H. & VON STEIN, A. (1974). Receptors for intercellular signals in aggregating cells of the slime mould, *Dictyostelium discoideum*. In *Biology and Chemistry of Eukaryotic Cell Surfaces*. Miami Winter Symposium, volume 7, pp. 49–66. Academic Press, New York and London.

GERISCH, G., HULSER, D., MALCHOW & WICK, U. (1975). Cell communication by periodic cyclic-AMP pulses. *Philosophical Transactions of the Royal Society, London*, B, **272**, 181–92.

GRABEL, L. & LOOMIS, W. F. (1977). Cellular interactions regulating early biochemical differentiation in *Dictostelium*. In *Development and Differentiation in the Cellular Slime Moulds*, eds. P. Cappuccinelli & J. M. Ashworth, pp. 189–99. Elsevier/North Holland, Amsterdam.

GREGG, J. H. (1965). Regulation in the cellular slime moulds. *Developmental Biology*, **12**, 377–93.

GREGG, J. H. (1971). Developmental potential of isolated *Dictyostelium* amebae. *Developmental Biology*, **26**, 478–85.

GREGG, J. H. & BADMAN, W. S. (1970). Morphogenesis and ultrastructure in *Dictyostelium*. *Developmental Biology*, **22**, 96–111.

GROSS, J., KAY, R., LAX, A., PEACY, M., TOWN, C. & TREVAN, D. (1977). Cell contact, signalling, and gene expression in *Dictyostelium discoideum*. In *Development and Differentiation of the Cellular Slime Moulds*, eds. P. Cappuccinelli & J. M. Ashworth, pp. 135–47. Elsevier/North Holland, Amsterdam.

GUSTAFSON, T. & WOLPERT, L. (1967). Cellular contact and movement in sea urchin morphogenesis. *Biological Reviews*, **42**, 442–98.

HAYASHI, M. & TAKEUCHI, I. (1976). Quantitative studies on cell differentiation during morphogenesis of the cellular slime mould, *Dictyostelium discoideum*. *Developmental Biology*, **50**, 302–9.

HEUSGEN, A. & GERISCH, G. (1975). Solubilized contact sites A from cell membranes of *Dictyostelium discoideum*. *F.E.B.S. Letters*, **56**, 46–9.

KAY, R. R., GARROD, D. R. & TILLY, R. (1977). Requirements for cell differentiation in *Dictyostelium discoideum*. *Nature, London*, **271**, 58–60.

LEACH, C. K., ASHWORTH, J. M. & GARROD, D. R. (1973). Cell sorting-out during differentiation of mixtures of metabolically distinct populations of *Dictyostelium discoideum*. *Journal of Embryology and Experimental Morphology*, **29**, 647–61.

LODISH, H. F. & ALTON, T. H. (1977). Translational and transcriptional control of protein synthesis during differentiation of *Dictyostelium discoideum*. In *Development and Differentiation in the Cellular Slime Moulds*, eds. P. Cappuccinelli & J. M. Ashworth, pp. 253–72. Elsevier/North Holland, Amsterdam.

LOOMIS, W. F. (1975). *Dictyostelium discoideum – a Developmental System*. Academic Press, New York and London.

McMAHON, D. (1973). A cell contact model for cellular position determination in development. *Proceedings of the National Academy of Sciences, U.S.A.*, **70**, 2396–400.

MAEDA, Y. & MAEDA, M. (1974). Heterogeneity of the cell population of the cellular slime *Dictyostelium discoideum* before aggregation, and its relation to subsequent locations of the cells. *Experimental Cell Research*, **84**, 88–94.

MULLER, V. & HOHL, H. R. (1973). Pattern formation in *Dictostelium discoideum*: temporal and spatial distribution of prespore vacuoles. *Differentiation*, **1**, 267–76.

NEWELL, P. C., LONGLANDS, M. & SUSSMAN, M. (1971). Control of enzyme synthesis by cellular interaction during development of the cellular slime mould *Dictyostelium discoideum*. *Journal of Molecular Biology*, **58**, 541–54.

NEWELL, P. C., FRANKE, J. & SUSSMAN, M. (1972). Regulation of four functionally related enzymes during shifts in the developmental programme of *Dictyostelium discoideum*. *Journal of Molecular Biology*, **63**, 373–82.

PAN, P., BONNER, J. T., WEDNER, H. & PARKER, C. (1974). Immunofluorescence evidence for the distribution of cyclic-AMP in the cells and cell masses of the cellular slime moulds. *Proceedings of the National Academy of Sciences, U.S.A.*, **71**, 1623–5.

RAPER, K. B. (1940). Pseudoplasmodium formation and organisation in *Dictyostelium discoideum*. *Journal of the Elisha Mitchell Science Society*, **56**, 241–82.

RAPER, K. B. & THOM, C. (1941). Interspecific mixtures in the Dictyosteliceae. *American Journal of Botany*, **28**, 69–78.

REITHERMAN, R. W., ROSEN, S. D., FRAZIER, W. A. & BARONDES, S. H. (1975). Cell-surface, species-specific, high-affinity receptors for discoidin: developmental regulation in *Dictyostelium discoideum*. *Proceedings of the National Academy of Sciences, U.S.A.*, **72**, 3541–5.

RICKENBERG, H. V., TIHON, C. & GUZEL, O. (1977). The effect of pulses of 3′:5′ cyclic adenosine monophosphate on enzyme formation in non-aggregated amoebae of *Dictyostelium discoideum*. In *Development and Differentiation in the*

Cellular Slime Moulds, eds. P. Cappuccinelli & J. M. Ashworth, pp. 173–87. Elsevier/North Holland, Amsterdam.

ROSEN, S. D., HAYWOOD, P. L. & BARONDES, S. H. (1976). Inhibition of inter-cellular adhesion in the cellular slime mould by univalent antibody against a cell surface lectin. *Nature, London*, **263**, 425–7.

ROSEN, S. D., KAFKA, J. A., SIMPSON, D. L. & BARONDES, S. H. (1973). Develop-mentally-regulated carbohydrate-binding protein in *Dictyostelium discoideum*. *Proceedings of the National Academy of Sciences, U.S.A.*, **70**, 2554–7.

ROSEN, S. D., SIMPSON, D. L., ROSE, J. E. & BARONDES, S. H. (1974). Carbohy-drate-binding protein from *Polysphondylium pallidum* implicated in inter-cellular adhesion. *Nature, London*, **252**, 149–51.

ROSEN, S. D., REITHERMAN, R. W. & BARONDES, S. H. (1975). Distinct lectin activities from six species of cellular slime moulds. *Experimental Cell Research*, **95**, 159–66.

RUBIN, J. & ROBERTSON, A. (1975). The tip of the *Dictyostelium discoideum* grex as an organiser. *Journal of Embryology and Experimental Morphology*, **33**, 227–41.

SAMPSON, J. (1976). Cell patterning in migrating slugs of *Dictyostelium discoideum*. *Journal of Embryology and Experimental Morphology*, **38**, 663–8.

SAKAI, Y. & TAKEUCHI, I. (1971). Changes in the prespore specific structure during differentiation and cell type conversion of a slime mould cell. *Development, Growth and Differentiation*, **13**, 231–40.

SHAFFER, B. M. (1957a). Aspects of aggregation in cellular slime moulds. I. Orientation and chemotaxis. *American Naturalist*, **91**, 19–35.

SHAFFER, B. M. (1957b). Properties of slime mould amoebae of significance for aggregation. *Quarterly Journal of Microscopical Science*, **98**, 377–92.

SHAFFER, B. M. (1961). The cell founding aggregation centres in the slime mould *Polysphondylium violaceum*. *Journal of Experimental Biology*, **38**, 833–49.

SHAFFER, B. M. (1963). Inhibition by existing aggregations of founder differen-tiation in the cellular slime mould *Polysphondylium violaceum*. *Experimental Cell Research*, **32**, 432–35.

SONNEBORN, D. R., WHITE, G. J. & SUSSMAN, M. (1963). A mutation affecting both rate and pattern of morphogenesis in *Dictyostelium discoideum*. *De-velopmental Biology*, **7**, 79–93.

SPEMANN, H. (1918). Über die Determination der ersten Organlagen des Amphi-bienembryo. I–VI. *Archiv für Entwicklungsmechanik*, **43**, 448–555.

STEINBERG, M. S. (1964). The problem of adhesive selectivity in cellular interac-tion. In *Cellular Membranes in Development*, ed. M. Locke, pp. 321–66. Academic Press, New York and London.

STERNFELD, J. & BONNER, J. T. (1977). Cell differentiation in *Dictyostelium* under submerged conditions. *Proceedings of the National Academy of Sciences, U.S.A.*, **74**, 268–71.

SUSSMAN, M. (1966). Biochemical and genetic methods in the study of cellular slime mould development. *Methods in Cell Physiology*, **2**, 397–410.

SWAN, A. P. & GARROD, D. R. (1975). Cohesive properties of axenically grown cells of the slime mould, *Dictyostelium discoideum*. *Experimental Cell Research*, **93**, 479–84.

SWAN, A. P., GARROD, D. R. & MORRIS, D. (1977). An inhibitor of cell cohesion from axenically grown cells of the slime mould, *Dictyostelium discoideum*. *Journal of Cell Science* **28**, 107–16.

TAKEUCHI, I. (1963). Immunochemical and immunohistochemical studies on the development of the cellular slime mould *Dictyostelium mucoroides*. *Develop-mental Biology*, **8**, 1–26.

TAKEUCHI, I. (1969). Establishment of polar organisation during slime mould development. In *Nucleic Acid Metabolism, Cell Differentiation and Cancer Growth*, eds. E. V. Cowdry & S. Seno, pp. 297–304. Pergamon Press, Oxford.

TAKEUCHI, I., HAYASHI, M. & TASAKA, M. (1977). Cell differentiation and pattern formation in *Dictyostelium*. In *Development and Differentiation of the Cellular Slime Moulds*, eds. P. Cappuccinelli & J. M. Ashworth, pp. 1–16. Elsevier/North Holland, Amsterdam.

TOWN, C. D., GROSS, J. D. & KAY, R. R. (1976). Cell differentiation without morphogenesis in *Dictyostelium discoideum*. *Nature, London*, **262**, 717–8.

TRINKAUS, J. P. (1963). The cellular basis of *Fundulus* epiboly: adhesivity of blastula and gastrula cells in culture. *Developmental Biology*, **7**, 513–32.

TURNER, R. S. (1977). Sponge cell adhesions. In *Specificity of Embryological Interactions*, ed. D. R. Garrod. Chapman & Hall Ltd., London (in press).

TYLER, A. (1946). An auto-antibody concept of cell structure, growth and differntiation. *Growth*, **10**, 7–19.

WEISS, P. (1947). The problem of specificity in growth and development. *Yale Journal of Biology and Medicine*, **19**, 235–78.

WOLPERT, L. (1969). Positional information and the spatial pattern of cellular differentiation. *Journal of Theoretical Biology*, **25**, 1–47.

WOLPERT, L. (1971). Positional Information and pattern formation. *Current Topics in Developmental Biology*, **6**, 183–222.

EXPLANATION OF PLATE

(a) Spherical aggregates formed by shaking *D. discoideum* Ax-2 glucose cells at 140 r.p.m. in 0.017 M phosphate buffer at pH 6.0 and 22 °C for 24 h on a New Brunswick G-86 water bath shaker. (Identical conditions used for other photographs on this plate.)

(b) Polarised aggregates with associated slime sheath form by shaking *D. discoideum* NC-4 cells for 24 h under identical conditions to those used for (a).

(c) Section of spherical aggregate of *D. discoideum* Ax-2 glucose cells after 18 h of shaking. Staining with rabbit anti-*D. mucoroides* spore serum followed by fluorescein-conjugated sheep anti-rabbit serum.

(d) Section of spherical aggregate of *D. discoideum* Ax-2 glucose cells after 24 h of shaking. Staining as in (c).

(e) Phase-contrast picture of *D. discoideum* Ax-2 glucose cells from a single cell suspension shaken under similar condition to those used for (a) but with the shaker speed increased to 260 r.p.m.

(f) Same field of cells as in (e), view under U.V. light after staining as in (c).

SECOND-SET GRAFT REJECTIONS: DO THEY OCCUR IN INVERTEBRATES?

By R. P. DALES

Department of Zoology, Bedford College, University of London, Regent's Park, London NW1 4NS, U.K.

If a small piece of skin is removed from a vertebrate such as a mouse and replaced by a similar piece of skin from another individual, the grafted skin adheres but is sooner or later rejected. Typically, the graft becomes vascularised and its cells proliferate, but after about 10 days there is an abrupt onset of inflammation, the grafted tissue withers away and is sloughed. If then another graft is made from the same donor to the same recipient, the 'second-set' is rejected more quickly, often in about half the time it took to reject the first graft. A 'third party' graft from another donor will be rejected also in a similar manner to the first, but it will not be rejected as quickly as the second-set graft from the original donor. These facts are well known and it is now well established that this rejection of skin grafts is due to an immune response directed against the histocompatibility antigens of the graft tissue. The shorter survival time of the second-set graft is due to the persistence of the immunity acquired from the first-set or possibly from a cellular memory or anamnestic response.

Another characteristic of the vertebrate graft response is that T-lymphocytes transferred from an animal which has received and rejected an allograft confer to the recipient the capacity for a second-set response time when it too receives a graft from the original graft donor. But, as we might expect from the specific nature of graft rejection this reaction is followed by another reaction to the transferred T-lymphocytes, except when both individuals are genetically identical or at least syngeneic as in specially inbred strains of mice.

Rejection is thus primarily cell-mediated but an animal which has rejected an allograft may have serum antibodies for the donor's histocompatibility antigens. Many years ago now, Sir Peter Medawar and his colleagues demonstrated that, in mice, immunological tolerance could be induced by inoculation of recipients with donor cells, usually spleen cells, providing that this was done early in life. Subsequent work has demonstrated that such tolerance is dependent on exposure to the donor's histocompatibility antigens before the recipient's immune system is operating. None the less, adult animals can become tolerant to some antigens

[203]

but explanation of such results appears always to involve the relationship of macrophages and lymphocytes and the phenomenon may well be confined to vertebrates with a well-developed macrophage/B cell/T cell system.

Grafting of tissue from one animal to another is a convenient tool for investigating histocompatibility but is of course a completely unnatural phenomenon in animals. Foreign tissue may be rejected either because it is recognised positively as foreign, or because the animal recognises only its own healthy tissues and has evolved mechanisms to destroy and reject other material. These processes may have arisen together in response to the need to maintain self and resist infection. Grafts are rejected either because of a reaction of the host to tissue which is recognised as not being an integral part of itself – and the basis for this may be various – or because of a mechanism which has been evolved in response to invasion by micro-organisms or multicellular parasites. If the maintenance of the body, vertebrate or invertebrate, primarily depends on histocompatibility, on mechanisms of surveillance and on conformation to pattern, then rejection of an allograft or heterograft (= xenograft) is readily comprehensible, even though reaction to foreign cells may be unspecific or weak. If a graft is regarded as 'foreign' then its rejection could follow for the same reasons which stimulate phagocytes to take up or to encapsulate foreign material. Both systems may be well developed in the most primitive organisms but an enhanced response either to a 'second-set' graft or to a secondary in-fection implies specificity and memory. It is now well established that in vertebrates these abilities are based on immunoglobulins and on a highly developed system of cooperating cells. How far do these phenomena occur in invertebrates?

Work over the last ten years suggests that, in some invertebrates at least, second-set grafts are rejected differently from first-set grafts. I would like to consider the evidence on which such deductions are based and the interpretation of the results of transplanting tissue from one invertebrate to another.

Some of the work on transplantation in invertebrates indicates a high degree of specificity: autografts are successful while allografts are success-ful only in syngeneic strains or genetically closely related individuals or colonies. Thus in coelenterates, genetically identical colonies fuse, while unrelated colonies of the same species do not, although separation or rejection may take some time (Hildemann, Linthicum & Vann, 1975). Actual transplants, whether allografts or xenografts, are rejected. In more advanced invertebrates the same general conclusion may be drawn: allografts and xenografts are rejected and only autografts are successful in the long term. The arthropods appear to be exceptional since both allo-grafts and xenografts between certain species are tolerated and the same appears to be true between some, but not all, species of the nemertine *Lineus* (Langlet & Bierne, 1973).

In transplanting skin of invertebrates we are faced with great diversity of integumentary structure. The arthropods have cuticles which are moulted; the molluscs have rather soft skins often abundantly supplied with mucous cells, the echinoderms have calcareous ossicles. Each presents particular problems for the experimenter. In addition many invertebrates do not live long enough to establish unequivocally the results of first-set and second-set graft responses since these may be very slow. Thus while tissue compatibility may be highly specific, because of technical difficulties or length of life of the experimental animal, very few studies have been made of second-set graft response in cases where a first-set graft has been rejected. As far as I am aware such experiments have been confined to earthworms, sipunculids, echinoderms and nemertines.

Let us look at the work on earthworms first, since this has formed the subject of extensive researches by H. L. Cooper and his colleagues in Los Angeles and P. Chateaureynaud-Duprat and others in Bordeaux. Of the many studies of different aspects of earthworm immunity performed in these laboratories there have been only a few studies of 'second-set' graft responses.

Duprat (1964) was the first to make a second-set graft following a first-set using the earthworm *Eisenia foetida typica* Sav. Duprat found a maximum reaction, defined by amoebocyte attachment, to the second-set graft when the operation was performed 4–8 d after the first-set operation, the amoebocyte response being weaker earlier or later than during that period. Valembois (1963) made implants of small pieces of body wall of another worm *Allolobophora caliginosa* Sav. into *Eisenia foetida*, later grafting body wall to the same recipients. He found that graft destruction proceeded at about twice the rate in implanted worms as compared with controls which had received no implant. In both these types of experiment the 'second-set' grafting performed by Duprat (1964) and the grafting following implantation performed by Valembois (1963) was done a short time after the first-set or implant which may or may not have been rejected or be in process of rejection or destruction. Unfortunately, no data are given. From later histological work, Valembois (1971) concluded that the cells responsible for graft destruction as apposed to wound closure are derived from the splanchnopleure around the gut and that a second-set graft directly stimulates production of cells from that layer. In a later paper Duprat (1967) described her results of second-set grafting using *E. foetida typica* from different geographical regions following different preliminary treatments. First-set grafting between worms from Talence (Bordeaux) as donors and Douai as recipients had shown rapid tissue rejection, from which we could infer strong histoincompatibility. Second-set grafts were said to be more rapidly rejected than second-sets between worms from Lille transferred to worms from Talence which are only feebly histo-incompatible. Duprat performed 170 operations of this type but only conclusions are recorded in that paper. In a second type of experiment, coelomocytes from Bergerac worms were injected into worms from Talence followed 10–15 d later by a body-wall graft from the same donors. Duprat (1967) found an enhanced second-set type reaction in three-quarters of the cases – the remainder showed first-set rejection times. Furthermore, only first-set graft reactions were obtained when the grafts were made 21 d or more after the injection. Unfortunately no data are given. In a third type of experiment, ten Bergerac worms were given grafts from Talence worms. Coelomocytes from the hosts were then transferred 15 d later to other Bergerac worms which received a graft from a Talence worm 2 d after that. All ten of these worms showed second-set type responses. In all these experiments the timing between successive grafts or between cell transfer and subsequent grafting was critical and this has been interpreted by these experimenters as 'short term memory'.

Cooper (1968) exchanged grafts between *Lumbricus terrestris* L. and *Eisenia foetida* and his results in general confirm the results of experiments done at Bordeaux. Forty-six second-set grafts were made after rejection of first-set grafts. Twenty-four showed accelerated rejection times as compared with their individual first-set rejection times, while eighteen second-set grafts survived longer than the corresponding first-sets. In the remaining four worms, second-sets were either still unrejected at the end of the experiment or showed the same rejection time as the first-sets. Cooper analysed the results on the assumption that the second-set rejection times belonged to one of two time-distributions which he described as either accelerated or prolonged. This followed from his interpretation of first-set graft rejections which he described as 'acute' when the graft was rejected in 11 d or less, 'rapid chronic' or 'intermediate chronic' when grafts were rejected in 12–19 or 20–29 d respectively, and prolonged chronic when rejection was not completed in 50 d post-grafting. I will return to the interpretation of these results later, but let us for the moment note the overall result which was that about half the second-set grafts were rejected in an accelerated fashion but that the other half were not. This result suggests that there is no difference in rejection time between first-set grafts and second-set grafts. Nevertheless, in a later paper Cooper (1969) reported an

PLATE I

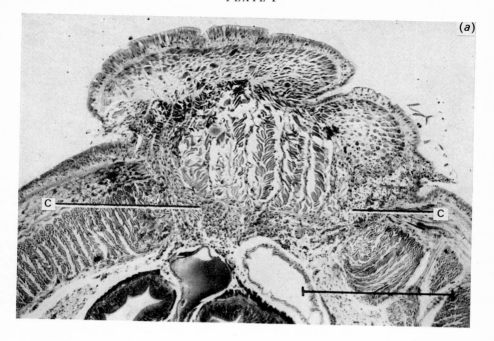

(a)

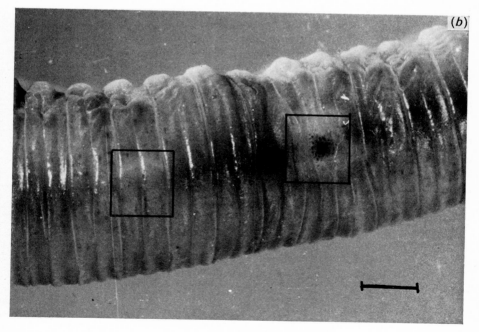

(b)

For explanation see p. 220

(facing p. 206)

overall shorter mean survival time (MST) of second-set grafts made from the same donor *Eisenia* to *Lumbricus* hosts 5 d after the first-set graft had been made.

Note that this technique is different from that reported in the earlier paper (Cooper, 1968) in which second-set grafts were made in the orthodox way after completion of first-set rejection. Comparison of mean survival times of first-sets and second-sets were made 5 d after first-sets were grafted. (Table 1).

Table 1. *Mean survival times of grafts made from* Eisenia *to* Lumbricus
(*from Cooper,* 1969, *Table* 2)

Number grafted	First-set MST (d)	Second-set MST (d)
64	17.8 (range) (7–60)	15.7 (range) (4–32)
92	18.1 (8–72)	15.3 (4–50)

Bailey, Miller & Cooper (1971 also performed the same kind of 'adoptive transfer' experiment as that described by Duprat in 1967, but they injected coelomocytes from *Lumbricus* previously grafted with *Eisenia*. They found that the xenografts from the original donor were rejected in an accelerated fashion. The MST of the two groups was significantly different with samples of 23–39 worms when the data were analysed by the Mann–Whitney U-test. The controls (Bailey *et al.*, 1971, Table 1) showed MSTs of 26.10 and 24.80 d: those injected with coelomocytes from 'non-immunised' worms 24.10 d as compared with MSTs of 15.76 d, 21.40 d and 19.65 d for comparable samples of worms injected with cells from previously grafted or 'immunised' worms. Only one of these experimental groups (with the shortest mean survival time) showed a significantly different MST from the control. Although this group contained 39 worms, a reasonable sample, the mean survival times of the other experimental groups were not statistically different from the controls.

Both Chateaureyneud-Duprat and her colleagues and Cooper and his co-workers believe that these reactions are cell mediated. Hostetter & Cooper (1973) have more recently compared the numbers of coelomocytes associated with different types of graft, both first-set and second-set, autograft, allograft and xenograft. They maintain that there is an increase in the response to second-set grafts which is due to an accelerated rise in cell number of the order of 20–30 % greater at maximum development than that stimulated by a first-set graft.

Second-set rejection time is clearly crucial for the interpretation of the results of such experiments in order to distinguish between simple histo-incompatibility reactions and an immune reaction characterised by both specificity and memory.

My own experiments have been limited so far to xenografts exchanged between *Lumbricus terrestris* and *Eisenia foetida unicolor*, and grafting second-sets in the orthodox manner following first-set rejection. The first experiments were made by transferring *Eisenia* to *Lumbricus* hosts but it is not easy to maintain large *Lumbricus* in a healthy condition for the combined rejection times of first-set and second-set grafts. Another difficulty is in the criteria used to assess rejection. While some experience is clearly essential for achieving comparable judgements of rejection onset and of completion, there remains some latitude in judgement even when done by the same person. Following first-set grafting graft survival was therefore scored at weekly intervals, Relatively few worms survived to reject both grafts, but individual case histories have been gradually built up. Grafting *Lumbricus* to *Eisenia* is more satisfactory, for *Eisenia* is an extremely robust animal which can withstand considerable mutilation and multiple grafting. It is also easier to keep, perhaps because of its surface-living habits and because of its higher lethal temperature which makes it easier to handle in the laboratory. If *Eisenia* is grafted to *Lumbricus* about half the second-set graft rejection times are accelerated and half are prolonged (Table 2): if the grafting is done the other way more are accelerated than prolonged (Table 3). The data are perhaps most easily visualised in the form of a histogram in which the second-set rejection times are expressed as a percentage of the first-set rejection times (Fig. 1). There is clearly no significant difference as far as *Lumbricus* hosts are concerned and while there are many more *Eisenia* hosts rejecting their *Lumbricus* grafts in an accelerated fashion very many do not. A few individual 'case histories' demonstrate some of the problems involved in work of this kind (Table 4). The first worm (no. 3) showed a prolonged rejection time of its second-set graft and an even longer rejection of its third-set. The following worm (no. 15) showed no significant difference in rejection times between its three grafts. Worm no. 37 showed a second-set rejection time only half that of the first-set. Table 4 also reveals another probem with earthworm grafting, that of 'technical' rejection. The Bordeaux school have sutured their grafts with silver wire to the host so that technical loss is, at least initially, less likely. Cooper has simply placed grafts on prepared 'beds' following anaesthesia, and I have followed his technique. Interpretation of whether a graft is rejected soon after operation because of technical failure or because of an immune reaction is critical, especially when these results may be included or excluded

Table 2. *Rejection of second-set and third-set grafts:* Eisenia *to* Lumbricus

Total no. of successfully grafted worms rejecting second-sets following a first-set graft	No. of worms in which second-set graft rejection is *accelerated*	No. of worms in which second-set graft rejection is *prolonged*
23	12	11

	No. of worms in which third-set graft is *accelerated* with respect to second-set	No. of worms in which third-set graft is *prolonged* with respect to second-set
	0	3

Table 3. *Rejection of second-set and third-set grafts:* Lumbricus *to* Eisenia

Total no. of successfully grafted worms rejecting second-sets following a first-set graft	35
No. of worms in which second-set graft rejection is *accelerated*	23
No. of worms in which second-set graft rejection is *prolonged*	12
No. of worms in which third-set graft is *accelerated* with respect to second-set	2
No. of worms in which third-set graft is *prolonged* with respect to second-set	4

from data subjected to statistical analysis. From my own experience the failure of a graft to adhere is different from one which adheres and is subsequently either (1) sloughed or (2) remains attached and is gradually destroyed. I have found that grafts are often rejected within the first seven days and all these I have regarded in the past as 'technical' rejects (Table 5). All those grafts which are attached a week after operation remain to be destroyed. It is possible that some initial 'technical' losses may be due to an immune reaction but a certain number of 'technical failures' always occur and the number seems to be larger in xenografts than in allografts. This fact makes it essential to include all the data and it is particularly unfortunate that some experimenters have not done so in their publications. Autografts always survive once they have adhered so that a measure of real technical failure is perhaps provided by autograft failure. Parry (1975) found about 20 % technical failures with *E. foetida*. With xenografts between *Eisenia* and *Lumbricus* I have found rather higher percentages, often 50 %, as might be expected if loss is at least partly due to a histoincompatibility reaction, but it looks as if the rejection rate may be higher after second-sets or third-sets (Table 6). If grafts are sutured, then technically

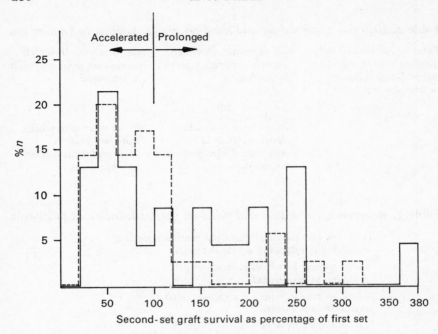

Fig. 1. Rejection times of second-set grafts expressed as percentages of the first-set rejection time. *Eisenia* to *Lumbricus* ($n = 23$) broken line; *Lumbricus* to *Eisenia* ($n = 35$) continuous line.

Table 4. *Examples of case histories:* Lumbricus *to* Eisenia

Worm no.		First-set graft rejection time (d)		Second-set graft rejection time (d)	Third-set graft rejection time (d)
3	T.R. 3 ×	25		50	76
15		48	T.R. 3 ×	35	40
29		73		78	
37		88		40	

(T.R. is technical rejection)

unattached grafts may be scored subsequently as true rejections. If the criterion for technical failure is limited to loss after 24 h (or, say, 1 week), then the results are going to be biassed one way or the other because of the judgement of the experimenter. On the other hand, it may well be that some of the judgements of technical rejection may obscure real immunological reactions. Some 'experienced' worms which have been grafted repeatedly may be described as 'adamant refusers' since it seems impossible to get a graft to adhere. Yet with persistency, or perhaps as a result of loss of memory, a graft may eventually do so. Worm no. 15, for example,

(Table 4) rejected 3 grafts after rejecting its first-set graft, yet accepted a third-set after a similar rejection time. Parry (1975) concluded that these 'hyperacute' rejections could reflect incompatibility between donor and recipient. To test whether such hyperacute or technical rejections were enhanced by previous experience, 3 groups of *Eisenia* were compared (Table 6). The first group consisted of naive worms which had not been grafted previously, the second group consisted of worms each of which had received one or more grafts from the same *Lumbricus* stock; the third group consisted of worms which had received implants 10–12 d previously. Aside from technical rejects there was no significant difference in rejection times of grafts by 'educated', implanted or 'naive' controls.

Table 5. *'Technical' rejection rate*

	First-set grafts rejected by naive worms	Second-set grafts made 3–10 d after first-set rejection	Grafts made 5 d after implantation
Lumbricus to *Eisenia*	$\frac{13}{58} = 22.4\%$	$\frac{33}{51} = 64.7\%$	—
Lumbricus to *Eisenia*	$\frac{21}{40} = 52.5\%$	$\frac{27}{40} = 67.5\%$	$\frac{10}{31} = 32.2\%$
Eisenia to *Lumbricus*	$\frac{68}{151} = 45.0\%$	$\frac{28}{49} = 57.1\%$	—

The fractions (expressed also as percentage of 'N' in each case) indicate the number of instances (numerator) of initial or 'technical' graft failure in each category.

Table 6. *Graft rejection:* Lumbricus *to* Eisenia

	Technical rejects + 48 h	True rejects or graft destruction	
		+ 18 d	+ 30 d
Implants			
Rejected	10	0	8
Unrejected	21	21	13
'Educated'			
Rejected	27	3	5
Unrejected	13	10	8
'Naive' controls			
Rejected	21	3	13
Unrejected	19	16	6

If rejection of second-set grafts was due to recognition of donor tissue previously encountered in a first-set graft made 5–8 d before, then similar tissue when implanted should have the same effect and the grafts should be rejected in an accelerated manner. Therefore, although *Lumbricus* grafts transferred to *Eisenia* a second time show a shorter mean survival

time, and statistical tests confirm that the difference is significant (at the
5 % level), it may be that this difference is not due either to memory or to
specificity in the sense in which these terms have been understood with
respect to skin grafts in mice. Adult mice invariably reject second-set grafts
sooner than first-sets. In *Eisenia* although the mean survival time of second-
sets is less than first-sets, a considerable number of second-set grafts show
prolonged survival. In addition, although few individuals have survived to
destroy third-set grafts, of those that have done so the survival times bear
no obvious relation either to the second or to first-set survival times in the
same individual. The fact that even some second-set (or third-set) grafts
show prolonged survival may mean that we are not dealing here with the
same kind of phenomenon which occurs in mammals. An animal that survives to reject a first-set graft is naturally older, the constituents of the body
fluid, for example, may be different or there may be other reasons of which
we have no clue which cause second-set survival to be less. Perhaps the
meaningful question to ask is not 'is the mean survival time less when
grafted again?' but 'is the survival time always less?'

It will be recalled that Cooper in his original paper (Cooper, 1968) arranged the reaction of first-set grafts into four categories: (1) acute, (2) rapid
chronic, (3) intermediate chronic, and (4) prolonged chronic, according to
rejection onset and completion. Mean survival times were calculated for
each group separately. When onset and survival times of these groups are
considered together they appear to form a single skewed distribution, and
it could be maintained that separation into different categories is unjus-
tified. Unfortunately the data as presented do not enable us to compare the
real mean rejection times of first-sets and second-sets. It will also be
remembered that second-sets were made after rejection of first-sets, and
these facts considered together with my own results, suggests that there is
no evidence for accelerated rejection of second-sets made under these
orthodox conditions. Dr Parry in London (Parry, 1977) has come to much
the same conclusion with regard to the technique (adopted by the Bordeaux
school and later by Cooper, 1969) of grafting second-sets five days or so
after the first, whatever the state of rejection of the first-set.

We cannot argue that there is an appreciable age difference in this case
which might cause a diminution in mean survival time, but this could still
be due to a change in the body fluid constituents or simply to the number
of phagocytes available to remove the graft. The longer survival time shown
by a third-party xenograft from another species as compared with a second-
set graft from the original donor and made at the same time may be irrele-
vant, for each species pair is likely to have a characteristic survival time
depending on the relative incompatibility.

Before abandoning the hypothesis that second-set rejection in these invertebrates is due to the same reaction which occurs in vertebrates, let us see what really happens.

If an earthworm is wounded superficially, the wound is closed by a plug of coelomocytes. During the next 6–18 h the body wall contracts to reduce the size of the wound and amoebocytes attach themselves to damaged cells. After that the epidermis grows together, collagen is laid down beneath and the muscle and other layers are regenerated by growth from the adjacent parent tissue. Beneath the wound, amoebocytes which have taken up damaged cells accumulate as a kind of granuloma or 'brown body'.

If an autograft is placed in an open wound or 'graft bed', the graft becomes attached by the cells forming the plug. The epidermis grows over by the second day and the tissue is gradually removed or reorganised to conform with the pattern prescribed by the host for that region. An allograft appears to be dealt with in the same way, but in practice Parry finds that the graft tissue is invariably removed and replaced by collagen in which the new tissue is gradually organised. Autografts may take 2–3 months to be fully integrated, though the graft appears to the eye to be viable and healthy, allografts may take 4 months to replace. Replacement is by cell migration and division, both on the part of the graft cells and the surrounding tissue as far as autografts are concerned (Parry, 1977), but allograft and xenograft tissue cells do not divide. These results agree with those of Burke (1974) on her studies of simple wounding in *Eisenia*. After all, the autograft is still part of the same body, it is just in the wrong position. The fact that allograft tissue does not divide suggests precise specificity. Apparent acceptance of carefully oriented orthotopic allografts judged by simple visual inspection may be an illusion. Histological examination often reveals a gradual replacement of the graft cells by host cells, although here again we should not fall into generalisations as there may be and probably are, degrees of histocompatibility.

Grafts which are killed by chilling or by other means, as well as xenografts, are removed more quickly than healthy allografts. The simplest explanation of these facts is merely that the different *milieu* presented by the body fluids of other individuals of the same species, or even more by individuals of other species, leads to unhealthy, moribund and eventually dead tissues which are then removed by phagocytes. Completion of removal would be recognised macroscopically as termination of graft destruction, only brown bodies remaining visible through the translucent replacement tissue within the newly laid-down collagen.

If specificity of rejection is individual, as it seems to be in earthworms, there may well be minute biochemical or biophysical differences in the body fluids which present a *milieu* which is different from that in which the graft tissue grew and was orientated. From our knowledge of regeneration, the body pattern of such axially organised animals is very precisely determined, although we know very little about the factors which determine or maintain this pattern. Xenografts are therefore physically different from that of the host even when orthotopic and the most favourable material for experiments are thus allografts, or grafts between subspecies (such as *E. foetida typica* and *E. foetida unicolor*) which is the reason why Parry has concentrated on those varieties.

The lower MST of xenografts may most simply be explained by the fact that the grafts die sooner than allografts, since they are subjected to a markedly different environment.

One of the factors which may be of importance, which both Parry and I have found, is the great individual variation in coelomocyte number. So far I have been unable to convince myself that there is a significant change in coelomocyte number in response to such events as wounding or grafting and I think we need to have much more data before we can draw firm conclusions. Since the graft destruction is brought about by coelomocytes the number initially present may be critical. Further, if there is a real increase in number either by division or by aggregation at or near a graft, then a second-set placed near a first-set may be attacked more rapidly simply because of the locally greater number of cells.

Accelerated rejection of second-set grafts due to a vertebrate-type immune response seems unlikely *a priori*. There is some evidence that some coelomocytes divide in response to wounding (Parry, 1976), although earlier Burke (1974) found evidence for division only amongst the other tissues following wounding. But bearing in mind Valembois' contention that the cells involved in second-set rejection are derived from the splanchnopleure we should perhaps remain open-minded on this point.

Hostetter & Cooper (1973, 1974) have outlined a detailed mechanism of cell interaction and response, but it may be idle to speculate too far before we are satisfied that second-sets are both recognised and rejected sooner than first-sets. Several experimenters also believe that the second-set response is qualitatively different.

Cells which act as memory cells must receive information, yet in general observation would suggest that amoebocytes that come into contact with foreign tissue are eventually lost in brown bodies. It is also by no means clear that the increase in cell number beneath a graft as reported by Hostetter & Cooper is not due simply to active migration or even to passive accumulation. The ^{51}Cr-labelled coelomocytes traced by Lemmi, Cooper & Moore (1974) tends to confirm that cells accumulate at wound areas and that there is no difference between the initial responses to autografts and to xenografts. It should also be noted that worms may reject cells through their dorsal pores when they are irritated. This is not a sensible act by an animal to whom memory cells are important.

Hostetter & Cooper (1973) counted coelomocytes in *Lumbricus* which had received *Eisenia* grafts, both by giving second-sets 5 d after the first-set, and also 24 h after rejection of the first-set. They maintained that cells increase in response to the second-set in both types of experiment. Because of individual variation, however, it may perhaps be wise not to draw conclusions at this stage.

Far less work has been done with echinoderms, though these animals have a special interest for those concerned with the phylogeny of the vertebrate immune response because of their phylogenetic position. The only published works of which I am aware in which first-set and second-set graft responses are compared are those of Hildemann & Dix (1972) working with the holothurian *Cucumaria tricolor* and Karp & Hildemann (1976) working with the starfish *Dermasterias imbricata*.

In *Cucumaria*, Hildemann & Dix (1972) noted that second-set allografts seemed to stimulate a greater reaction and that technical losses were greater. Only three animals survived second-set grafting, but all of them rejected their allografts in less time than the fifteen successful first-sets. Karp & Hildemann (1976) had greater success with *Dermasterias*, but the number of cases was limited, due principally to the very long time required for completion of rejection. Of thirty-five pairs of starfish between which allografts were exchanged, only seventeen rejected the first-sets, and five showed no sign of rejection even after ten months. Of these seventeen successful first-set recipients only seven survived second-set grafting, but of these, five showed a reduced survival time and two showed no signs of rejection at the end of the observations. Four of the rejectors received third-sets, and all of them rejected their grafts in an even shorter time. These are very small samples and while it is premature to draw conclusions, these workers are convinced from their histological examination of the grafts that second-set and third-set responses are different from the first-set.

Second-set transplantation in sipunculids has been looked at only by Triplett, Cushing & Durall (1958) in *Dendrostomum zostericolum*. They implanted allogeneic pieces of tentacle which were encapsulated. They found no evidence of acceleration in response to second-set implants.

More recently, Langlet & Bierne (1975) have reported that survival of second-set grafts made between the incompatible nemertine species *Lineus ruber* (recipient) and *L. sanguineus* (donor) is always less than first-sets.

Here the anterior tip of the head in front of the brain was replaced by a similar piece from the donor species. The foreign tissue is totally replaced by fresh growth in a number of days. The second 'graft' was made 35–45 days after the completion of first-set rejection and replacement, a much longer time incidentally than the period found to be critical in producing an enhanced response in earthworms. These very interesting results thus comply with the criterion that second-set grafts must be rejected sooner. Some of Langlet & Bierne's 'grafts' between species of *Lineus* are of large parts of the body and the tissue interactions in chimaeras may be different from the kind of response induced by a small piece of foreign body wall grafted orthotopically. In earthworms chimaeras between different species survive without obvious signs of rejection for months (Korschelt, 1931).

Much more work clearly remains to be done in clarifying what the events are in graft rejection. In my view there is insufficient evidence to draw the conclusion that second-set grafts are rejected sooner than first-sets because of an immune reaction in any invertebrate. Acceptance of an autograft is due to adhesion and active cooperation of the individuals' coelomocytes, and it is only in autografts that cell division has been seen; allografts and heterografts adhere by infiltration of coelomocytes and by overgrowth of the host epidermis. They are tolerated by variable times which possibly may be merely related to the local number of coelomocytes or to the rate at which the foreign tissue dies.

On the other hand, the specificity of self/non-self recognition in many invertebrates is extremely interesting and we should remember that it is only very recently that it has been appreciated that invertebrates do react at all in this way. Doubts about second-set rejection and alleged 'memory' should not diminish our appreciation of the work that has been done to establish the specificity of these responses. But is there more than non-self recognition and consequent removal of foreign tissue? More interesting, perhaps, than second-set rejection, is the possibility that in the 'hyper-acute' or initial 'technical' rejection of grafts we do have a kind of immune response. It is tempting to speculate that the specificity of self/non-self recognition in these invertebrates is most likely to be based on glycoprotein polymorphism. Such a hypothesis would agree with other observations such specificity of the reactions with different blood-group substances.

We can accept that some, though not all, invertebrates, are able to distinguish non-self in a highly specific manner, so that individuals belonging to the same species but from a different clone or population are recognised as different, so that cells or skin grafts from such individuals are rejected. From what we know of gene mutation, if tissue compatibility depends on recognition of complex surface molecules, whether globular proteins, membrane glycoproteins or glycosyl transferases, it is almost to be expected that widely separated populations will have generated different surface proteins if they have been genetically isolated for long enough. It is then understandable that earthworms, which are local cross-fertilising animals with a short generation time, may well differ markedly in such details from one population to another. There are no reasons why invertebrates should not be as individual as vertebrates so that autografts are accepted and allografts destroyed.

If the elegant immune responses of vertebrates owe their origin to body surveillance with the addition of memory cells and an enhanced response to secondary infection, then there is no reason why a graft which does not conform exactly to the host's body pattern should be rejected differently a second time, when the ability to reject a second challenge from a microorganism is lacking (Cooper, Acton, Weinheimer & Evans, 1969). The cellular and humoral defence of higher vertebrates may well have developed in order to combat the increased vulnerability conferred by high body-temperature, so that response to first-set and second-set grafts are similar to these more natural reactions against infection. Absence of secondarily enhanced responses to natural infection might, therefore, cast doubt on the reality of the apparent differences between first and second grafts in invertebrates. There is no evidence so far that the particular invertebrates we have been discussing do combat a secondary infection more rapidly in the vertebrate manner.

It may be permissible to speculate that invertebrates reject foreign tissue because of surface histocompatibility antigens which are most likely to be glycoproteins, that maintenance of self in a strictly organised body pattern is due to cell–cell interactions which has obvious importance for phagocyte behaviour and that the different degrees of self/non-self recognition could be based on glycoprotein polymorphisms.

The ideas of Roseman (1970) in this regard and the recent suggestions by Parish (1977) that the basis for self/non-self recognition in invertebrates is based on glycosyl transferase polymorphism are particularly interesting. But it is much more difficult to imagine a memory component of such a system since this would depend on cell multiplication with the cells remaining out of contact with the antigenic tissue.

Investigation of immune responses in invertebrates is still in its infancy and we must clearly proceed with caution, resisting the temptation to apply vertebrate interpretations while at the same time remaining alert to the possibility of different solutions having been evolved for similar needs.

Transplantation in invertebrates is difficult and has not, I believe, produced clear-cut answers. It is a difficult procedure because of skin

15

structure, the long graft-survival times, the limited lives of the animals and the difficulty of producing and maintaining syngeneic strains. The follow ing differences remain between vertebrates and invertebrates with regard to second-set graft rejection. In invertebrates studied so far, graft tissues which are rejected (i.e., other than autografts) do not proliferate, vascularisation is doubtful or the tissues are not vascularised in any case, and the second-set grafts are not invariably rejected in an accelerated fashion. It may be that absence of immunoglobulins and a demonstrable dual or multiple cooperative cell system may be fundamental.

Experimental techniques other than transplantation may be more fruitful in attempting to understand the basis of the specificity of self/non-self recognition. Experiments *in vitro* using labelled cells, and studies *in vivo* of the cells themselves and of their reactions to other materials and to cells with or without previous experience of them may be a surer way of putting these transplantation phenomena in their true perspective.

REFERENCES

BAILEY, S., MILLER, B. J. & COOPER, E. L. (1971). Transplantation immunity in Annelids II. Adoptive transfer of the xenograft reaction. *Immunology*, **21**, 81–6.

BURKE, J. M. (1974). Wound healing in *Eisenia foetida* (Oligochaeta). I. Histology and ³H-thymidine radioautography of the epidermis. *Journal of Experimental Zoology*, **188**, 49–64.

COOPER, E. L. (1968). Transplantation immunity in annelids. I. Rejection of xenografts exchanged between *Lumbricus terrestris* and *Eisenia foetida*. *Transplantation*, **6**, 322–37.

COOPER, E. L. (1969). Specific tissue graft rejection in earthworms. *Science*, **166**, 1414–15.

COOPER, E. L., ACTON, R. T., WEINHEIMER, P. F. & EVANS, E. E. (1969). Lack of a bactericidal response in the earthworm *Lumbricus terrestris* after immunization with bacterial antigens. *Journal of Invertebrate Pathology*, **14**, 402–6.

DUPRAT, P. (1964). Mise en evidence de réaction immunitaire dans les homogreffes de paroi du corps chez le lombricien *Eisenia foetida typica*. *Comptes rendus hebdomadaires des séances de l'Académie des Sciences, Paris*, **259**, 4177–9.

DUPRAT, P. (1967). Etude de la prise et du maintien d'un greffon de paroi du corps chez le lombricien *Eisenia foetida typica*. *Annales de l'Institut, Pasteur*, **113**, 867–81.

HILDEMANN, W. H. & DIX, T. G. (1972). Transplantation reactions of tropical Australian echinoderms. *Transplantation*, **15**, 624–33.

HILDEMANN, W. H., LINTHICUM, D. S. & VANN, D. C. (1975). Transplantation and Immunoincompatibility Reactions among Reef-Building Corals. *Immunogenetics*, **2**, 269–84.

HOSTETTER, R. K. & COOPER, E. L. (1972). Coelomocytes as effector cells in earthworms immunity. *Immunological Communications*, **1**, 155–83.

HOSTETTER, R. K. & COOPER, E. L. (1973). Cellular anamnesis in earthworms. *Cellular Immunology*, **9**, 384–92.

HOSTETTER, R. K. & COOPER, E. L. (1974). Earthworm coelomocyte immunity. In *Contemporary Topics in Immunobiology*, vol. 4, *Invertebrate Immunology*, ed. E. L. Cooper, pp. 91–107. Plenum, New York and London.

KARP, R. D. & HILDEMANN, W. H. (1976). Specific allograft reactivity in the sea star *Dermasterias imbricata*. *Transplantation*, 22, 434–9.

KORSCHELT, E. (1931). *Regeneration und Transplantation*, vol. 2 (1), pp. 304–79. Borntrager, Berlin.

LANGLET, C. & BIERNE, J. (1973). Recherches sur l'immunite de greffe chez les Némertiens du genre *Lineus*. Evolution de transplants homospecifiques et hétérospécifiques. *Comptes rendus hebdomadaires des séances de l'Académie des Sciences, Paris*, D, 276, 2485–8.

LANGLET, C. & BIERNE, J. (1975). Recherches sur l'immunite de greffe chez les Némertiens du genre *Lineus*. Rejet accéléré des secondes greffes hétérospéci-fiques incompatibles. *Comptes rendus des séances de l'Académie des Sciences, Paris*, D 281, 595–8.

LEMMI, C. A., COOPER, E. L. & MOORE, T. C. (1974). An approach to studying evolution of cellular immunity. In *Contemporary Topics in Immunobiology*, vol. 4, *Invertebrate Immunology*, ed. E. L. Cooper, pp. 110–19. Plenum, New York and London.

PARISH, C. R. (1977). Simple model for self/non-self discrimination in inverte-brates. *Nature, London*, 267, 711–13.

PARRY, M. J. (1975). Hyperacute graft rejection in *Eisenia foetida typica* and *Eisenia foetida unicolor*. *Experientia*, 31, 117–18.

PARRY, M. J. (1976). Evidence of mitotic division of coelomocytes in the normal, wounded and grafted earthworm *Eisenia foetida*. *Experientia* 32, 449–50.

PARRY, M. J. (1977). 'Body wall grafts in *Eisenia foetida* – an investigation into their survival and the cellular responses involved in the host graft interaction.' Thesis, University of London.

ROSEMAN, S. (1970). The synthesis of complex carbohydrates by multi-glycosyl transferase systems and their potential function in intercellular adhesion. *Chemistry and Physics of Lipids*, 5, 270–97.

TRIPLETT, E. L., CUSHING, J. E. & DURALL, G. L. (1958). Observations on some immune reactions of the sipunculid worm *Dendrostomum zostericolum*. *Ameri-can Naturalist*, 92, 287–93.

VALEMBOIS, P. (1963). Recherches sur la nature de la reaction antigreffe chez le lombricien *Eisenia foetida* Savigny. *Comptes rendus hebdomadaires des séances de l'Académie des Sciences, Paris*, 257, 3489–90.

VALEMBOIS, P. (1971). Origins et function des amoebocytes actifs au cours d'une xenogreffe de paroi du corps chez *Eisenia foetida* Sav. (Lombricien). *Comptes rendus hebdomadaires des séances de l'Académie des Sciences, Paris*, D, 272, 2097–2100.

EXPLANATION OF PLATE

(*a*) Transverse section of a *Lumbricus* graft on an *Eisenia* host two days after grafting. The graft is held in place and walled off by host coelomocytes (C). Scale bar represents 0.5 mm.

(*b*) *Lumbricus* with *Eisenia* grafts. The graft on the left has been completely replaced by host tissue and re-annulated, 110 days after grafting; the second-set graft on the right has been destroyed, agglomerations of host phagocytes being represented by pigment granules seen by transparency through the new host tissue. Annuli have not yet been re-established, 80 days after grafting. Scale bar represents 2.0 mm.

(Photographs Z. Podhorodecki, Bedford College).

MODULATIONS OF THE CELL SURFACE AND THE EFFECTS ON CELLULAR INTERACTIONS

By RICHARD L. HOOVER

Department of Pathology, Harvard Medical School,
25 Shattuck Street, Boston, Massachusetts 02115, U.S.A.

Many cellular functions are controlled by the cell surface. Whether it be by direct involvement of cell surface molecules or by the role of the membrane as a permeability barrier, the composition and stability of the plasma membrane plays an important part. Therefore, any perturbation of this organelle brought about by changes in proteins, lipids and/or carbohydrates or a redistribution of these molecules in the surface can alter particular cell functions. For example, as a cell passes through the cell cycle, morphological changes occur (Crane, Clarke & Thomas, 1977). Concomitant with these are changes in the surface such as the distribution of a major fibroblast protein (Stenman, Wartiovaara & Vaheri, 1977) which may function in adhesion (Yamada, Yamada & Pastan, 1975) and the agglutination of cells with concanavalin A (Smets, 1973). Besides these normal alterations that occur, modulations in the composition can be affected by viral transformation and changes in the environment. In this report, I will present evidence which shows how alterations in the cell surface, brought about through the external environment can effect cellular functions such as adhesion, growth, contact inhibition and cell recognition. Three different systems were used to investigate these cellular interactions: (1) chick embryonic cells; (2) tissue culture cells; and (3) polymorphonuclear leukocytes and endothelium.

EMBRYONIC CELLS

Recent studies have implicated a role for glycosyltransferases in cell-to-cell interactions (Roseman, 1970; Roth, McGuire & Roseman, 1971; McLean & Bosmann, 1975). Accordingly, cells adhere to one another as a result of the interaction between enzymes and substrates on reciprocal cell surfaces. The transfer of carbohydrates from one molecule on one cell to another molecule on a neighbouring cell can account not only for cell-to-cell adhesion but also adhesive specificity, since the particular carbohydrate and enzyme exposed on the surface will determine whether an interaction occurs

[221]

(Roth *et al.*, 1971). Objections have arisen over this hypothesis because of the lack of data demonstrating the existence of this enzyme on the outer cell surface (Deppert, Werchaw & Walter, 1974; Evans, 1974). However, several forms of evidence, in particular, radioautography (Roth *et al.*, 1971; Roth & White, 1972), indicate that these transferases can be ectoenzymes. Recently, Porter & Bernacki (1975) have presented electron micrographs of autoradiographs further indicating the presence of glycosyl transferases on the surface of murine leukemia L-1210 cells.

It might be reasoned from the above theory that a change in the ecto-enzyme activity may result in alteration of adhesive properties, i.e. a more adhesive cell might be expected to show greater transferase activity than a less adhesive cell. Such differences have in fact, been shown to be present between normal and transformed cells which show differences in not only glycosyl transferases activity (Grimes, 1970; Mora *et al.*, 1973) but also adhesive properties (Edwards, Campbell & Williams, 1971; Shields & Pollock, 1974). These first set of experiments examine this relationship between an enzyme (galactosyl transferase) on the cell surface and cell adhesion.

Kirschbaum & Bosmann (1973) have reported that folic acid can increase the activity of collagen:glucosyl and galactosyl transferases 2- to 3-fold in cell fractions from rat kidneys and liver. Based on these findings, I have added folic acid to embryonic chick cells and measured transferase activity and cell adhesiveness.

Neural retinae, hearts, and livers were removed from 7-day chick embryos and dissociated with trypsin (0.25 %) and EDTA (1 mM).

Adhesion measurements were made by two methods – the monolayer collection assay as described by Walther, Ohman & Roseman (1973) and the Couette viscometer according to the method of Curtis (1969).

Table 1 shows the results of adding various concentrations of folic acid to neural retinal cells and measuring the extent of cell attachment to a monolayer of neural retinal cells. Adhesion numbers are based on control values where no folic acid was added. This was done in order to equalize all data because of the daily variability. The data show that as the concentration of folic acid is increased the adhesion of neural retinal cells to homo-typic monolayers is reduced, up to 75 % in 3.5 mM concentration of folic acid. Measurements with the Couette viscometer confirm that the adhesive-ness of the cells is decreased with the addition of folic acid (Table 2). In this instance adhesion is reduced by about 43 % in 0.7 mM folic acid. In conjunction with the adhesion assays, galactosyl transferase activities were measured. Table 3 shows the results of these experiments. The galac-tosyltransferase activity of cells treated with 0.7 mM folic acid is increased compared to controls whether measured as stationary or stirring cultures. Stirring measurements give an indication of the amount of glycosylation occurring on individual cells while stationary measurements monitor glycosylation between cells. In the stationary cultures the increase is 43 % and in the stirring cultures, 87 %. The differences between stationary and stirring conditions in control cultures is small with a slight decrease in activity under suspension conditions. In contrast, the folic acid treated cells show a 21 % increase in the stirring assay as compared to the stationary assay.

Table 1. *Adhesion of neural retinal cells to monolayers of neural retinal cells in the presence of various concentrations of folic acid (Hoover, 1977)*

Adhesion numbers are expressed as a fraction when compared to the control with no added folic acid.

Concentration of folic acid (mM)	Adhesion no.	n
0	1.00	26
0.70	0.72 ± 0.40	26
1.05	0.60 ± 0.20	10
1.40	0.55 ± 0.30	9
2.80	0.27 ± 0.03	3
3.50	0.26 ± 0.02	3

Table 2. *Adhesion of neural retinal cells with and without folic acid using the Couette viscometer method (Hoover, 1977)*

Values are expressed as collision efficiencies

Experimental conditions	Collision efficiencies	n
Control	7.78 ± 1.01	6
Folic acid (0.7 mM)	5.15 ± 0.74	6

Table 3. *Galactosyl transferase activity of neural retinal cells in the presence of folic acid in stirring and stationary cultures (Hoover, 1977)*

Values are expressed as d.p.m. ^{14}C per cell

Experimental conditions	Stationary	Stirring	n
Control	1.78×10^{-5}	1.65×10^{-5}	14
Folic acid (0.7 mM)	2.55×10^{-5}	3.08×10^{-5}	14

Specificity of adhesion was also examined by adding neural retinal cells (treated and untreated with folic acid) to monolayers of heart or liver cells. Table 4 presents the results of these experiments. The number of folic acid treated neural retinal cells adhering to either heart or liver monolayer is less than the number of untreated cells adhering. When these data are compared further with values for adherence of neural retinal cells to neural retinal monolayers, it can be seen that the specificity of adhesion is not changed but rather the number of cells adhering to the monolayer is decreased (30%). In the case of a plastic surface, the percentage of cells adhering remains the same whether treated or untreated.

Table 4. *Adhesion of neural retinal cells to heterotypic surfaces in the presence of folic acid (Hoover, 1977)*

Adhesion values are represented as a fraction of the neural retina on neural retina values in the absence of folic acid

	Control	Folic acid (0.7 mM)	n
Heart monolayer	0.79 ± 0.12	0.63 ± 0.10	10
Liver monolayer	0.66 ± 0.03	0.50 ± 0.01	6
Plastic	0.71 ± 0.12	0.69 ± 0.15	10

These experiments indicate that folic acid can affect galactosyl transferase activity and the adhesion of cells to each other, to plastic and to cell monolayers. It previously has been suggested (Roseman, 1970) that the interaction of glycosyl transferases on the surfaces of neighbouring cells plays a role in adhesion and that alterations would influence the behaviour of cells. The idea that an increase in surface enzyme activity produces a concomitant increase in adhesion is not substantiated by my results because as glycosyl transferase activity increased, adhesion to plastic, monolayers and suspended cells decreased.

Recently a model has been proposed for cis- and trans-glycosylation (Webb & Roth, 1974) which might account for the results observed. According to the model there are two places where glycosylation can occur – between cells (trans) and on a single cell (cis). Since it must be assumed that a cell has both enzyme and substrate on its surface in order for trans-glycosylation to occur, it is not surprising then that glycosylation can occur on the same cell if the two components of the reaction lie next to or near one another. If cells are interacting with each other, cis-glycosylation may not be observed because the membrane sites have reacted with sites on other cells or have been prevented from moving together by other cellular components.

The implication is that folic acid is interfering with this membrane reaction. This is further substantiated by the fact that the structural analogue, amethopterin, which ultimately inhibits the synthesis of purines and pyrimidines does not block the effect of folic acid on adhesion. The explanations for this are speculative. Kirschbaum & Bosmann (1973) have proposed several mechanisms by which folic acid affects transferase activity:

folic acid may be (1) acting as a cofactor; (2) forming an intermediate, possibly a lipid; (3) bringing about conformational changes; (4) acting on a feedback mechanism; or (5) complexing with other factors. In this way it could be affecting the architecture and composition of the cell surface. This could account for the increase in transferase activity because now more of the substrate and enzymes are exposed and in a position to interact. If the lipids are involved, this might result in a change in membrane fluidity in which case the molecules would be freer to interact with molecules not only on their own surface but also on other cell surfaces. Increased membrane fluidity could also decrease the adhesiveness because the amount of cis-glycosylation is probably increased, reducing the number of molecules available for trans-glycosylation and thus adhesion. Roth & White (1972) have postulated that membrane composition may be the cause for the differences in galactosyl transferase activity observed between Balb/c 3T3 and Balb/c 3T12 cells.

It is also important to note that the specificity of adhesion between neural retinal and heart or liver cells was unchanged by the folic acid treatment. The implication is that there are particular molecules on the cell surface accounting for the specificity because the adhesiveness of the neural retinal cells was decreased and yet there was no change in specificity. This does not rule out the possibility that the molecules may be glycosyl transferases because the observed effect was a loss of adhesiveness and not a loss of specificity. But it also should be pointed out that this particular glycosyl transferase may play no role in adhesive specificity since it is assumed that this enzyme is one of many transferases on the cell surface (Roseman, 1970). These experiments represent an example in which a protein on the cell surface, in this case an enzyme, was altered which then effected cell interactions.

TISSUE CULTURE CELLS

In terms of the fluid mosaic model, lipids play a very important role in the structure and function of membranes. Not only do they affect the stability and permeability of the membrane but also the distribution of proteins which in turn can affect other cellular functions. Studies have shown that changes in lipid compositions can modify enzyme activity (Engelhard, Esko, Sturm & Glaser, 1976), morphology (Ginsburg, Salomon, Sreevalsan & Freese, 1973; Hawley & Gordon, 1976), concanavalin A agglutination (Horwitz, Hatten & Burger, 1974), differentiation (Weeks, 1976), and cell adhesion (Curtis, Chandler & Picton, 1975; Hoover, unpublished results). The mechanism by which the lipids are affecting these cellular functions is still unclear. Two possible mechanisms are (1) that there is a direct change in the lipid composition of the cell surface which alters the fluidity of the membrane or (2) that the lipids are acting intracellularly through synthesis of prostaglandins which then alter membrane properties.

In these experiments BHK and PyBHK have been grown in medium in which the type of fatty acids have been defined. Adhesiveness and growth of the cells were monitored as indicators of surface modulations.

The effects of the fatty acids on the adhesion of BHK cells are summarized in Table 5. Since the percentage of control cells adhering ranges between 40–70 %, all data has been standardized based on a value of 1.00 for the control condition. Numbers higher than 1.00 indicate more cells are adhering to the substrate than the control condition and numbers less than 1.00, fewer cells are adhering than the controls. All data have a level of significance of $p \leqslant 0.05$ based on the Student t-test. The data show that as the concentration of linoleic acid is increased, the number of cells adhering decreases. This is true not only for adhesion to monolayers but also to plastic surfaces. Stearic acid (18:0) showed an increase in cell-to-cell adhesion at 10 μg ml^{-1}, a slight decrease at 20 μg ml^{-1} and a 24 % reduction at 30 μg ml^{-1}. The attachment to a plastic substrate followed the same pattern. Fig. 1 presents a representative graph of a time-course study of the adhesion to monolayers. Although the initial rates are similar under the various conditions, the final number of cells attaching is different.

Other fatty acids tested did not produce the same results. Fig. 2 shows the

Table 5. *Effect of fatty acid concentration on adhesion of BHK cells to homotypic monolayers and to a plastic substrate (Hoover, Lynch & Karnovsky, 1977)*

Values expressed as a fraction of the controls; percentages of cells adhering under control conditions are 55.9 ± 1.3 for homotypic monolayers and 43.6 ± 1.0 for plastic

	Homotypic monolayer (n)	Plastic substrate
Control	1.00 (74)	1.00 (13)
Stearic acid (μg ml^{-1})		
10	1.03 ± 0.20 (57)	0.98 ± 0.01 (9)
20	0.92 ± 0.2 (18)	—
30	0.76 ± 0.2 (6)	—
Linoleic acid (μg ml^{-1})		
2.5	0.89 ± 0.10 (4)	0.90 ± 0.07 (11)
5.0	0.62 ± 0.08 (4)	0.57 ± 0.04 (4)
10.0	0.49 ± 0.10 (52)	0.41 ± 0.06 (9)
20.0	0.46 ± 0.09 (13)	—
30.0	0.17 ± 0.09 (6)	—

results of these experiments. The only other fatty acid besides linoleic acid which showed significant changes in adhesion was arachidonic acid (20:4) which reduced the number of cells adhering by 16%. Oleic acid (18:1), palmitic acid (16:0) and stearic acid (18:0) had little effect on reducing the number of cells adhering.

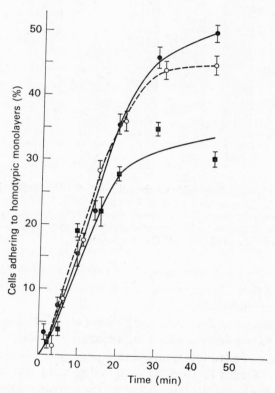

Fig. 1. Adhesion of BHK cells to homotypic monolayers as a function of time. ○, Control; ●, stearic acid (10 μg ml⁻¹); ■, linoleic acid (10 μg ml⁻¹). Reproduced from Hoover *et al.* (1977) by kind permission of the authors and the publishers of *Cell*, MIT Press Journals.

Two other cell types were tested to see how universal the effects of the fatty acids were (Fig. 3). Both Chinese hamster ovary (CHO) and polyoma-transformed baby hamster kidney (Py BHK) cells responded in a manner similar to BHK cells when stearic or linoleic acid was added. In the presence of stearic acid there was no effect but in the presence of linoleic acid, adhesion was decreased by at least 60%.

In the adhesion assay, fatty acids are incubated with the cells for 20 min before being used in the assay. Table 6 shows that 2.2% of the linoleic acid

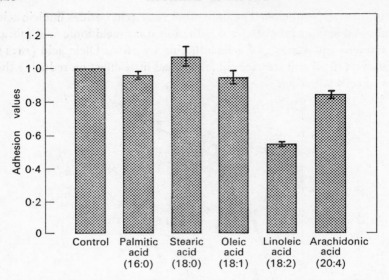

Fig. 2. The effects of various fatty acids on the adhesion of BHK cells to homotypic monolayers at 37 °C. Concentrations for all are 10 μg ml^{-1}. Values are expressed as a fraction of the controls. Percentage of cells adhering in the controls is 80.7 ± 1.6. $n = 6$ for all conditions. Reproduced from Hoover *et al.* (1977) by kind permission of the authors and publishers of *Cell*, MIT Press Journals.

is taken up by the cell and that most of this appears as free fatty acids; however, 20.8 % of that taken up has been incorporated into phospholipids and only 5.5 % into triglycerides.

The fact that changes in fatty acid composition are occurring in the cell when exogenous fatty acids are added is presented in Table 7. Not only does the percentage of 18:2 incorporated into phospholipids increase (7*a*) but also the ratio of 18:2 to 16:0, to 18:0 and to 20:4 increases as the exogenous concentration of linoleic acid (18:2) is increased (7*b*).

Linoleic acid also had effects on the growth of BHK cells. Fig. 4 is a representative growth curve of cells treated with linoleic acid at 10, 20 or 40 μg ml^{-1}. As the concentration of linoleic acid is increased, the maximum cell number decreases. The growth rate remains approximately the same until 40 μm ml^{-1} is reached at which point the rate decreases. In conjunction with these growth differences, morphological differences can be seen – most obvious at the higher concentrations. Plate 1 is a phase-contrast micrograph of normal and linoleic-treated cells at high concentrations of linoleic acid (40 μg ml^{-1}). The controls are typical fibroblasts exhibiting contact inhibition (Plate 1*a*); however, in 40 μg ml^{-1} the cells look transformed in that there is considerable criss-crossing and piling up as if they had lost their contact inhibition properties (Plate 1*b*). At a low concentra-

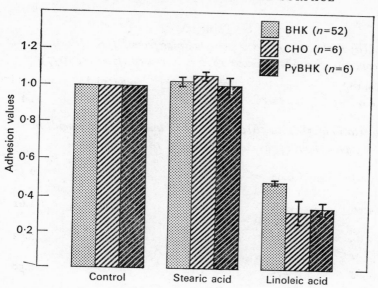

Fig. 3. A comparison of the attachment of BHK, CHO and PyBHK cells to homo-typic monolayers in the presence of stearic acid or linoleic acid (10 μg ml) at 37 °C for 30 min. Percentages of cells adhering in the control conditions for BHK, CHO and PyBHK are 55.9 ± 1.3, 49.2 ± 0.10, respectively. Reproduced from Hoover *et al.* (1977) by kind permission of the authors and publishers of *Cell*, MIT Press Journals.

Table 6. *Incorporation of* [1-^{14}C] *linoleic acid into BHK cells**
(*Hoover* et al., *1977*)

	d.p.m.	%
Phospholipids	60276	20.8
Fatty acids	127600	44.1
Triglycerides	16041	5.5
Cholesterol esters	85475	29.5

* Total incorporation represents 2.2 % of labelled material added.

tion of linoleic acid (10 μg ml^{-1}), scanning electron microscopy revealed no obvious differences in surface morphology between control and treated cells. Because the growth experiments extend for longer periods of time, there is a chance of peroxidation of the fatty acids, in which case by-products might be responsible for the results. The addition of vitamin E (an anti-oxidant) to the medium, however, produced no changes. Indo-methacin, an inhibitor of prostaglandin synthesis, also did not alter the effects on growth when added to the growth medium with the fatty acids.

Since both linoleic and arachidonic acids are precursors to prostaglan-

Table 7.

(a) Percentage of 18:2 phospholipids from BHK cells as concentration of exogenous 18:2 is increased (Hoover, 1977)

Control	10 μg ml^{-1}	20 μg ml^{-1}	30 μg ml^{-1}	40 μg ml^{-1}
9.9	12.5	21.9	22.27	33.4

(b) Ratio of 18:2 to 16:0, 18:0, 18:1 and 20:4 fatty acids in phospholipids as concentration of 18:2 is increased

$\dfrac{18:2}{16:0}$	0.98	0.99	1.32	1.67	2.32
$\dfrac{18:2}{18:0}$	1.06	0.68	0.92	1.47	1.61
$\dfrac{18:2}{18:1}$	0.22	0.38	0.83	2.78	3.03
$\dfrac{18:2}{20:4}$	2.63	0.75	1.12	2.00	2.50

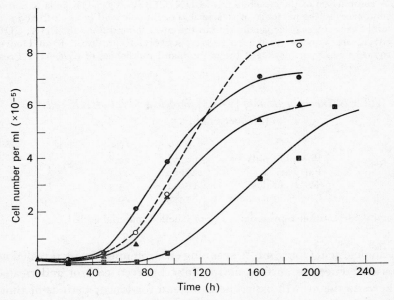

Fig. 4. Typical growth curves for BHK cells grown in various concentrations linoleic acid. Each point represents 2 replicates. O, Control; ●, 10 μg ml^{-1}; ▲, 20 μg ml^{-1}; ■, and 40 μg ml^{-1}. Reproduced from Hoover et al. (1977) by kind permission of the authors and publishers of Cell, MIT Press Journals.

Table 8. *The effects of indomethacin on the adhesion of BHK cells to homotypic monolayers and plastic (Hoover et al., 1977)*

Values are expressed as a fraction of the controls. The percentages of untreated cells adhering to homotypic monolayers and to plastic are 60.5 ± 4.2 and 47.7 ± 2.4, respectively

	Homotypic monolayer (n)	Plastic (n)
Control	1.00 (12)	1.00
Linoleic acid (10 μg ml^{-1})	0.65 ± 0.10 (12)	0.52 ± 0.07 (10)
Linolenic acid (10 μg ml^{-1})	0.54 ± 0.09 (6)	0.74 ± 0.06 (4)
Control + indomethacin	1.13 ± 0.09 (6)	1.16 ± 0.07 (6)
Linoleic acid + indomethacin	0.65 ± 0.11 (12)	0.61 ± 0.06 (10)
Linolenic acid + indomethacin	0.56 ± 0.06 (6)	0.69 ± 0.08 (4)

The concentration of indomethacin was 2 μM; indomethacin was added 3 h before addition of fatty acids and was present during the assay.

Table 9. *The effects of the addition of prostaglandins on the adhesion of BHK cells to homotypic monolayers after 30 min incubation at 37 °C (Hoover et al., 1977)*

Values are expressed as a fraction of the controls (74.9 ± 5.2 %)

Control	1.00 (6)
PGE$_1$	
5 μg ml^{-1}	1.05 ± 0.01 (6)
10 μg ml^{-1}	0.85 ± 0.01 (6)
PGE$_2$	
5 μg ml^{-1}	0.93 ± 0.01 (6)
10 μg ml^{-1}	0.97 ± 0.06 (6)
PGF$_2\alpha$	
5 μg ml^{-1}	1.04 ± 0.01 (6)
10 μg ml^{-1}	1.01 ± 0.06 (6)

dins, three experiments were carried out to see if there was any relationship in this system. Table 8 presents data in which adhesion was measured in the presence of linoleic acid and linoleic acid plus indomethacin. In the control samples, adhesion to both monolayers and plastic increased with the addition of indomethacin; however, in the presence of linoleic acid and indomethacin the adhesion to monolayers was unchanged. Also, linolenic acid (18:3) which is not a precursor to prostaglandins had a similar effect on adhesion to linoleic acid (Table 8). The addition of indomethacin did not affect the adhesion. Addition of exogenous prostaglandins produced only slight changes in percentage of cells adhering to monolayers after 30 min (Table 9). The differences cannot account for the differences seen with the addition of linoleic acid.

Table 10. *Effect of temperature on adhesion of BHK cells to homotypic monolayers in the presence of stearic or linoleic acids (Hoover et al., 1977)*

Concentration of stearic and linoleic acids was 10 μg ml^{-1}; results expressed as fraction of controls at that particular temperature. At 4 and 37 °C, the percentages of cells adhering under control conditions are 34.0 ± 2.5 and 50.8 ± 1.5, respectively

	37 °C	4 °C	
Control	1.00 (6)	1.00 (6)	0.67*
Stearic acid	1.06 ± 0.14 (6)	1.07 ± 0.10 (6)	0.67*
Linoleic acid	0.49 ± 0.06 (6)	0.55 ± 0.12 (6)	0.76*

* These values represent a ratio of the percentages of cells adhering to monolayers after 45 min at 4 and 37 °C.

Table 11. *The effects of linoleic and stearic acids on galactosyltransferase activity in neural retinal cells*

Values are expressed as d.p.m. ^{14}C per cell

Control	Stearic acid 10 μg ml^{-1}	Linoleic acid 10 μg ml^{-1}
9.7 × 10^{-6} (4)	11.1 × 10^{-6} (4)	13.5 × 10^{-6} (4)

The adhesion assays, the apparent loss of contact inhibition, and the negative prostaglandin experiments indicate the fatty acids may be acting at the membrane level. This is substantiated by the results shown in Table 10 in which adhesion assays were carried out at 4 °C. The effects are greater on the control and the stearic acid treated cells than on the linoleic acid treated cells. In all the instances, the number of cells decrease concomitant with a temperature decrease but the decrease in linoleic acid treated cells is less than in the controls, indicating a possible role for membrane fluidity.

Experiments with embryonic tissues support the idea that membrane changes occur when cells are incubated in fatty acids. In Table 11, 7-day embryonic neural retinal cells were incubated in the presence of linoleic or stearic acid and galactosyltransferase activity monitored. Both fatty acids cause an increase in enzyme activity – stearic acid, 14% and linoleic acid, 38%; however, adhesion to plastic was not altered with stearic acid and reduced with linoleic acid by 35% (Table 12). Again it should be pointed out that adhesion of neural retina to heart or liver monolayers was not altered by the lipid additions even though there were apparent alterations in the surface molecules. This may mean that the particular molecules monitored were not involved in adhesive specificity but do play a role in the general adhesiveness of the cell, in this case, involving cis- and trans-glycosylation of molecules on the cell surface.

Table 12. *Effects of fatty acids on adhesion of neural retinal cells to homotypic and heterotypic monolayers and to plastic*

Results are expressed as a proportion of the controls

Experimental conditions	NR on NR	NR on H	NR on L	NR on plastic
Control	1.00 (18)	1.00 (10)	1.00 (10)	1.00 (17)
Stearic acid 10 μg ml^{-1}	0.84 (12)	0.81 (8)	0.77 (8)	0.97 (4)
Linoleic acid 10 μg ml^{-1}	0.76 (18)	0.73 (8)	0.75 (8)	0.65 (4)

H = heart, NR = neural retina, L = liver.

POLYMORPHONUCLEAR LEUKOCYTES AND ENDOTHELIUM

The interaction between polymorphonuclear leukocytes (PMN) and the endothelial cells lining the small blood vessels is one of the initial responses characterizing the acute inflammatory response. Under normal conditions blood-borne PMN flow freely through the vessels or at most adhere only momentarily to the endothelial surface (Atherton & Born, 1972). During acute inflammation the PMN marginate along the vessel wall and subsequently emigrate through the vessel by way of the interendothelial junctions to accumulate at the site of injury. What causes the PMN to adhere at a particular location in the blood vessel and how they become attached is not known. Some possible mechanisms have been investigated, but, *in vivo*, only divalent cations have been implicated (Atherton & Born, 1972; Ryan & Majno, 1977).

Alterations in the surface of either cell could affect this interaction and in the *in vivo* situation have an effect on the inflammatory state. We have investigated further this interaction with cultured endothelial cells and isolated PMN. Using the monolayer collection assay developed by Walther *et al.* (1973) as well as an independent assay utilizing scanning electron microscopy, we have found that there is a specific interaction between PMN and endothelial cells as opposed to many other cell types and that this interaction is influenced by cell surface charge, divalent cations, and chemotactic agents.

Table 13 presents data from experiments in which the number of PMN adhering to cells and to the substrate were counted using a scanning elec-

16

tron microscope. From these numbers the percentage of PMN adhering to the cells was calculated. Over 50% of the calf PMN counted and 36% of the human PMN adhere to the calf endothelium; this value is less than 10% for all other cell types.

Table 13. *Adhesion of calf and human PMN to different cell types; adhesion monitored by direct cell counts in the scanning electron microscope (Hoover, Briggs & Karnovsky, 1978)*

Incubation for 90 min at 37 °C; values are expressed as percentages of PMN adhering to various cell types; number in parenthesis represents the total number of PMN counted; standard deviation in experiments did not exceed 1.5%

Cell type	% calf PMN	% human PMN
Calf endothelium	51.6 (745)	36.2 (1586)
Human endothelium	—	36.0 (261)
Swiss 3T3	9.7 (342)	1.2 (425)
BHK	8.3 (313)	2.7 (485)
PyBHK	5.4 (56)	—
SvPy 3T3	2.5 (157)	0.2 (602)
Embryonic bovine trachea	2.5 (721)	—
B16 F1	2.5 (608)	1.0 (303)
B16 F10	2.5 (632)	1.0 (446)
L929	2.4 (1442)	0 (100)
Rat smooth muscle	1.3 (1185)	4.1 (967)
CHO	—	1.0 (1198)

A majority of the experiments were run with human instead of calf PMN because of the availability of the human material. Table 13 shows that the results seen with human PMN correspond with those from the calf PMN. In general, fewer human PMN adhered to the cells, including the calf endothelium, than did calf PMN; however, the specificity for endothelium remained.

Most of the cells tested were of rat, mouse or hamster origin, raising the possibility that the differences seen in adhesion were due to species differences. To test this possibility, an embryonic bovine trachea culture was tested by adding calf PMN. As the results show in Table 13, the PMN adhered no better to these cells than to those from the other species, implying that the specificity of PMN adhesion is one of cell type and not of species origin. It is interesting to note that only 9.7% of the PMN attached to the surfaces of 3T3 cells which resemble endothelium (Porter, Todaro & Fonte, 1973). There also were no distinct differences between the adhesion of PMN to normal (3T3, BHK, rat smooth muscle and bovine trachea) and transformed (B16 F1, B16 F10, PyBHK, SvPy3T3, and L929) cells.

This specificity is illustrated further in the scanning electron micro-

PLATE I

For explanation see p. 240

PLATE 2

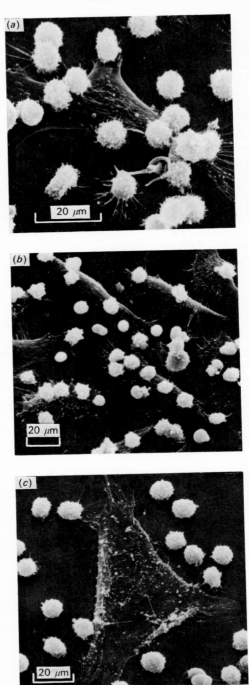

For explanation see p. 240

graphs. Plate 2a shows calf PMN adhering to calf endothelium, roughly half the PMN in the field are attached to the cell. That this specificity is also true for human PMN is shown in Plate 2b; again, a majority of the PMN are found adhering to the endothelial cells. These observations are in sharp contrast to what is found when another cell type is used. Plate 2c shows only one human PMN in contact with a Swiss 3T3 cell, the majority of the PMN apparently preferring the glass as a substrate.

Table 14. *Adhesion of human PMN to different cell types; monitored by the monolayer collection assay (Hoover, et al., 1978)*

Incubation at 37 °C for 45 min; values expressed as percentage of cells adhering; numbers in parentheses represent n

Monolayer	%
Calf endothelium	60.6 ± 0.3 (18)
Plastic	54.4 ± 1.1 (4)
CHO	49.7 ± 0.9 (4)
B16 F1	48.9 ± 0.9 (4)
L929	45.3 ± 0.9 (4)
B16 F10	41.7 ± 0.8 (4)
Swiss 3T3	34.3 ± 0.7 (4)

The specificity of adhesion data was also confirmed using the monolayer collection assay for adhesion (Table 14). As with the data from the SEM experiments, more PMN adhere to the endothelium that to other cell types, although the differences are not as marked as with the SEM technique.

From this set of experiments, we have concluded that the adhesion between PMN and endothelium, *in vitro*, is a specific interaction. Given the opportunity, PMN will adhere significantly better to endothelial cells than to a variety of other cell types or to glass or plastic substrates. This indicates that the endothelial surface provides something in particular for the PMN to adhere to, something which is lacking from the other cell types. Recently, Lackie & de Bono (1977) found that this same specificity occurred between neutrophils and pig aortic endothelium.

By treating either one or both of the cell types, we were able to affect the interaction, as monitored by the monolayer adhesion assay (Table 15). Pre-incubation of the PMN with neuraminidase increased the adhesion of PMN to endothelium by 32%. Pre-incubation of the endothelium produced a similar increase. The data also shows that divalent cations were necessary for the adhesion because the lack of Ca^{2+} and Mg^{2+} in the adhesion assay decreased the number of PMN adhering by 57%. Incubation of the PMN with N-ethylmaleimide which blocks sulphydryl bonds also decreased the adhesion of the PMN to the endothelium. All of these experiments imply a role for charge in this interaction. If, as in this case, the charge is modulated with neuraminidase which reduces negativity (Cook, Heard & Seaman, 1961; Forrester, Ambrose & MacPherson, 1962; Vassar, 1963), the cells can come closer together because the repulsion produced by the negative charges has been reduced. Divalent cations can also function in reducing charge by binding to the negative sites on the surface (Weiss, 1960; Collins, 1966). Likewise, a blocking of sulphydryl bonds which reduced the adhesiveness could be due either to a perturbation of adhesion molecules on the cell surface or to a disturbance of the net surface charge.

Table 15. *Adhesion of human PMN to calf endothelium; monitored by monolayer collection assay in which PMN or endothelium have been treated with chemical agents known to affect adhesion in other systems (Hoover, et al., 1978)*

Values are expressed as a fraction of the controls; numbers in parentheses represent n

Experimental conditions	Adhesion numbers
Control	1.00 (66)
Neuraminidase	
PMN	1.57 ± 0.12 (12)
Endothelium	1.27 ± 0.10 (10)
Ca^{2+}-, Mg^{2+}-free Hank's	0.43 ± 0.16 (24)
N-ethylmaleimide	
PMN	0.73 ± 0.15 (10)
Endothelium	0.39 ± 0.07 (10)

Finally, the monolayer collection assay may be a good model for studying the margination of the PMN during an inflammatory response. We tested the effects of materials which are known to be chemotactic to PMN. Table 16 presents the results of these experiments. All of the agents tested, a low molecular weight bacterial filtrate, a small peptide (N-formyl-methionyl-alanine) and zymosan-activated serum increased the adhesion of PMN to the endothelium. Zymosan treatment of heat-inactivated serum had little effect on the adhesion of PMN to the endothelium.

How these chemotactic agents cause PMN to emigrate from the small vessels at a specific site is not known, but one possible mechanism is suggested by these data. All three of the chemotactic agents produced an increase in the adhesion of the PMN to the endothelial cells. This was true whether the PMN or the endothelium was treated. Characterization of zymosan-activated human serum has shown that the alternate complement pathway is activated, thereby generating, among other components, C5a which has been shown to be chemotactic for PMN (Ward & Newman, 1969; Goldstein, Hoffstein, Gallin & Weissman, 1973). Additionally, Gallin, Durocher & Kaplan (1975) have suggested that C5a decreases the net negative charge on the surface of the PMN. This fits well with the rest of the charge data presented here: in all instances, when cells are treated with an agent that reportedly reduces the surface charge, there is an increase in adhesion. It remains to be tested whether C5a alters the surface charge of endothelial cells, although these data imply that it might. Whether bacterial factors and FMA are acting in a similar manner is however not known.

Table 16. *Adhesion of human PMN to calf endothelium after treatment with chemotactic agents as monitored by the monolayer collection assay (Hoover et al., 1978)*

Values are expressed as a fraction of the controls; numbers in parentheses represent n

Cell treated	Control	FMA	Zymosan-activated serum	Zymosan-activated heat-inactivated serum	Bacterial filtrate
PMN	1.00 (60)	1.27 ±0.02 (24)	1.51 ±0.05 (24)	1.05 ±0.05 (24)	1.18 ±0.04 (12)
Endothelium	1.00 (36)	—	1.49 ±0.09 (24)	—	1.31 ±0.03 (12)

SUMMARY

The results of these experiments have shown that modulations in the cell periphery can affect cellular interactions. The alterations can be introduced through the external environment and can occur in the lipids, proteins and/or carbohydrates. Moreover, an alteration in one of the components can influence the behaviour of the others, e.g. changing the lipid composition of BHK cells causes the glycosyl transferase activity on the surface to change. The modulations affected cell–cell and cell–substrate adhesion, growth and contact inhibition. The specificity of adhesion was unaltered by any of the treatments.

I would like to thank Professors A. S. G. Curtis and Morris J. Karnovsky and Drs Richard T. Briggs and Robert D. Lynch for their support and encouragement. I would also like to thank Sandy McLeod, Jo Ann Buchanan and Russel Campbell for their technical assistance. This work was supported by grants B-RG49099 from the SRC and GM 06235-13 and HL-09125 from the NIH.

238 RICHARD L. HOOVER

REFERENCES

ATHERTON, A. & BORN, G. V. R. (1972). Quantitative investigations of the adhesiveness of circulating polymorphonuclear leucocytes to blood vessel walls. *Journal of Physiology*, **222**, 447–74.

COLLINS, M. (1966). Electrokinetic properties of dissociated chick embryo electro- I. pH–surface charge relationships and the effect of calcium ions. *Journal of Experimental Zoology*, **163**, 23–37.

COOK, G. M., HEARD, D. & SEAMAN, G. V. F. (1961). Sialic acids and the electrokinetic charge of the human erythrocyte. *Nature, London*, **188**, 1011–12.

CRANE, M. ST. J., CLARKE, J. B. & THOMAS, D. B. (1977). Cell cycle dependent changes in morphology. *Experimental Cell Research*, **107**, 89–94.

CURTIS, A. S. G. (1969). The measurement of cell adhesiveness by an absolute method. *Journal of Embryology and Experimental Morphology*, **22**, 305–25.

CURTIS, A. S. G., CHANDLER, C. & PICTON, N. (1975). Cell surface lipids and adhesion. III. The effects on cell adhesion of changes in plasmalemmal lipids. *Journal of Cell Science*, **18**, 375–84.

DEPPERT, W., WERCHAW, H. & WALTER, G. (1974). Differentiation between intracellular cell surface glycosyltransferases: Galactosyltransferase activity in intact cells and in cell homogenate. *Proceedings of the National Academy of Sciences, U.S.A.*, **71** (8), 3068–72.

EDWARDS, J. G., CAMPBELL, J. A. & WILLIAMS, J. (1971). Transformation by polyoma virus affects adhesion of fibroblasts. *Nature, New Biology*, **231**, 147–8.

ENGELHARD, V. H., ESKO, J., STURM, D. & GLASER, M. (1976). Modification of adenylate cyclase activity in LM cells by manipulation of the membrane phospholipid composition *in vivo*. *Proceedings of the National Academy of Sciences, U.S.A.*, **73** (12), 4482–6.

EVANS, W. (1974). Nucleotide pyrophosphatase, a sialoglycoprotein located on the heptatocyte surface. *Nature, London*, **250**, 391–4.

FORRESTER, J. A., AMBROSE, E. I. & MACPHERSON, J. A. (1962). Electrophoretic investigations of a clone of hamster fibroblasts and polyoma transformed cells from the same population. *Nature, London*, **196**, 1068–70.

GALLIN, J. E., DUROCHER, J. & KAPLAN, A. (1975). Interaction of leukocyte chemotactic factors with the cell surface. I. Chemotactic factor-induced changes in human granulocyte surface charge. *Journal of Clinical Investigation*, **55**, 967–74.

GINSBURG, E., SALOMON, D., SREEVALSAN, T. & FREESE, E. (1973). Growth inhibition and morphological changes caused by lipophilic acids in mammalian cells. *Proceedings of the National Academy of Sciences, U.S.A.*, **70** (8), 2457–61.

GOLDSTEIN, I., HOFFSTEIN, S., GALLIN, J. & WEISSMAN, G. (1973). Mechanisms of lysosomal enzyme release from human leukocytes: Microtubule assembly and membrane fusion induced by a component of complement. *Proceedings of the National Academy of Sciences, U.S.A.*, **70** (10), 2916–20.

GRIMES, W. J. (1970). Sialic acid transferases and sialic acid levels in normal and transformed cells. *Biochemistry*, **9** (26), 5083–92.

HAWLEY, H. P. & GORDON, G. B. (1976). The effects of long chain fatty acids on human neutrophil function and structure. *Laboratory Investigation*, **34** (2), 216–22.

HOOVER, R. L. (1977). The effect of folic acid on glycosyltransferase activity and cell adhesion. *Experimental Cell Research*, **106**, 185–9.

HOOVER, R. L., LYNCH, R. & KARNOVSKY, M. J. (1977). Decrease in adhesion of cells cultured in polyunsaturated fatty acids. *Cell*, **12**, 295–300.

MODULATIONS OF THE CELL SURFACE

239

HOOVER, R. L., BRIGGS, R. T. & KARNOVSKY, M. J. (1978). The adhesive interaction between polymorphonuclear leukocytes and endothelial cells *in vitro*.
Cell. **14**, 423–8.

HORWITZ, A. F., HATTEN, M. E. & BURGER, M. M. (1974). Membrane fatty acid
replacements and their effect on growth and lectin-induced agglutinability.
Proceedings of the National Academy of Sciences, U.S.A., **71** (8), 3115–9.

KIRSCHBAUM, B. & BOSMANN, H. B. (1973). Glycoprotein biosynthesis: Folic acid
effects on glycoprotein: glycosyl transferase activities of rat kidney and liver.
Biochemical and Biophysical Research Communications, **50**, 510–16.

LACKIE, J. M. & DE BONO, D. (1977). Interactions of neutrophil granulocytes
(PMNs) and endothelium *in vitro*. *Microvascular Research*, **13**, 107–12.

McLEAN, P. J. & BOSMANN, H. B. (1975). Cell–cell interactions: Enhancement
of glycosyltransferase ectoenzyme systems during *Chlamydomonas* gametic
contact. *Proceedings of the National Academy of Sciences, U.S.A.*, **72** (1),
310–13.

MORA, P., FISHMAN, P., BASSIN, R., BRADY, R. & McFARLAND, V. (1973).
Transformation of Swiss 3T3 cells by murine sarcoma virus is followed by
decrease in a glycolipid glycosyltransferase. *Nature, New Biology, London*, **245**,
226–9.

PORTER, C. W. & BERNACKI, R. S. (1975). Ultrastructural evidence for ectoglycosyl-
transferase systems. *Nature, London*, **256**, 648–50.

PORTER, K., TODARO, G. & FONTE, V. (1973). A scanning electron microscope
study of surface features of viral and spontaneous transformants of mouse
BALB/3T3 cells. *Journal of Cell Biology*, **59**, 633–42.

ROSEMAN, S. (1970). The synthesis of complex carbohydrates by multiglycosyl-
transferase systems and their potential function in intercellular adhesion.
Chemistry and Physics of Lipids, **5**, 270–97.

ROTH, S. & WHITE, D. (1972). Intercellular contact and cell-surface galactosyl
transferase activity. *Proceedings of the National Academy of Sciences, U.S.A.*,
69, 485–9.

ROTH, S., McGUIRE, E. & ROSEMAN, S. (1971). Evidence for cell-surface glycosyltransferases. Their potential role in cell recognition. *Journal of Cell Biology*,
51, 536–47.

RYAN, G. & MAJNO, G. (1977). Acute inflammation. *American Journal of Pathology*,
86 (1), 185–276.

SHIELDS, R. & POLLOCK, K. (1977). The adhesion of BHK and PyBHK cells to the
substratum. *Cell*, **3**, 31–8.

SMETS, L. A. (1973). Agglutination with Con A dependent on cell cycle. *Nature,
New Biology, London*, **245**, 113–5.

STENMAN, S., WARTIOVAARA, J. & VAHERI, A. (1977). Changes in the distribution
of a major fibroblast protein, fibronectin, during mitosis and interphase.
Journal of Cell Biology, **74**, 453–67.

VASSAR, P. S. (1963). The electric charge density of human tumour cell surfaces.
Laboratory Investigation, **12**, 1072–7.

WALTHER, B. T., OHMAN, R. & ROSEMAN, S. (1973). A quantitative assay for intercellular adhesion. *Proceedings of the National Academy of Sciences, U.S.A.*,
70 (8), 1569–73.

WARD, P. A. & NEWMAN, L. J. (1969). A neutrophil chemotactic factor from human
C'5. *Journal of Immunology*, **102**, 93–9.

WEBB, G. & ROTH, S. (1974). Cell contact dependence of surface galactosyltransferase activity as a function of the cell cycle. *Journal of Cell Biology*, **63**,
796–805.

WEEKS, G. (1976). The manipulation of the fatty acid composition of *Dictyostelium discoideum* and its effect on cell differentiation. *Biochimica et Biophysica Acta*, **450**, 21–32.

WEISS, L. (1960). Studies on cellular adhesion in tissue culture—III. Some effects of calcium. *Experimental Cell Research*, **21**, 71–7.

YAMADA, K., YAMADA, S. & PASTAN, I. (1975). The major cell surface glycoprotein of chick embryo fibroblasts is an agglutinin. *Proceedings of the National Academy of Sciences, U.S.A.*, **72**, 3158–62.

EXPLANATION OF PLATES

PLATE 1

Phase micrographs of control (*a*) and linoleic acid-treated (40 μg ml^{-1}) (*b*) cells. Reproduced from Hoover *et al*. (1977) by kind permission of the authors and the publishers of *Cell*, MIT Press Journals.

PLATE 2

(*a*) Scanning electron micrograph of calf PMN and a cultured endothelial cell, Note that approximately half the PMN are attached to the endothelial cell.

(*b*) The interaction of human PMN with calf endothelial cells results in a majority of the PMN attaching to the cell surfaces as opposed to the glass substrate, as can be seen in this scanning electron micrograph.

(*c*) A representative micrograph showing the interaction between human PMN and a Swiss 3T3 cell. An obvious preference for the glass surface as opposed to the 3T3 cell surface is exhibited by the PMN.

(*a*) and (*b*) reproduced from Hoover *et al*. (1978) by kind permission of the authors and the publishers of *Cell*, MIT Press Journals.

GLYCOPROTEINS IN THE RECOGNITION OF SUBSTRATUM BY CULTURED FIBROBLASTS

By D. A. REES, R. A. BADLEY, C. W. LLOYD, D. THOM and C. G. SMITH

Biosciences Division, Unilever Research, Colworth Laboratory, Sharnbrook, Bedford MK44 1LQ, U.K.

INTRODUCTION

The aim of this investigation was to contribute to the understanding, at the molecular level, of events at the cell surface in the attachment and locomotion of fibroblasts on an inert substratum. As the story develops, we hope that it will become clear that this fits into the theme of this volume because the important problem on which we have concentrated is how the cell recognises and responds to areas of the substrate. Earlier work by others, notably Abercrombie and his group, had already shown that the cell was drawn forward by the activity of its thin leading edge which undergoes a cyclical process of membrane ruffling, attachment, cessation of ruffling and contraction (Abercrombie, 1961). Attachment is not over the entire underside, but at discrete foci, each about 1 μm^2 in area, mainly situated near the leading edge of a moving cell (Abercrombie, Heaysman & Pegrum, 1971). The existence of distinct attachment zones is confirmed by scanning electron microscopy (Revel & Wolken, 1973; Revel, Hoch & Ho, 1974), and the structure shows elaborate specialisation when examined with the electron microscope; the inner membrane face carries an electron-dense plaque at which a tract of oriented microfilaments terminates (Abercrombie *et al.*, 1971). In living cells the contacts are conveniently visualised using the interference reflection microscope (Curtis, 1964; Abercrombie & Dunn, 1975) which often shows them lined up along the so-called stress fibres (Izzard & Lochner, 1976) which are now known to be composed of bundles of actin filaments. This assembly of actin filaments terminating in a region which is electron dense with conventional stains, is reminiscent of the well-known Z-line structure of skeletal muscle. Correlation of the interference reflection image with high-voltage stereo electron microscopy shows that focal contacts are always associated with the distal ends of linear bundles of microfilaments and suggests that the formation of microfilament bundles is closely linked to the formation of focal adhesions to the substratum (Heath & Dunn, 1978). In our own laboratory, we have confirmed that whenever

actomyosin bundles are visualised by indirect immunofluorescence they
correspond to stress fibres seen under phase optics and frequently terminate at focal contacts visualised by interference reflection (Plate 1).

From all this we conclude that the formation of the bundle and other
specialised features can be regarded as a cellular response which follows
recognition of a region of substratum which can support focal attachment.
Such an hypothesis would be consistent with observations by scanning
electron microscopy (Witkowski & Brighton, 1971, 1972) and time-lapse
cinematography (Albrecht-Buehler, 1976) which suggest that protrusions
from the cell 'search' for congenial areas of substratum before they attach
and exert tension for cell spreading and locomotion.

Further evidence shows that specialised attachments can only develop
after the prior adsorption of certain serum components onto the surface,
otherwise the characteristic ultrastructure does not develop and adhesion is
passive rather than active (Pegrum & Maroudas, 1975). Likewise the cell
shape, which is characteristic of the given cell line (e.g. epithelial or fibro-
blastic), represents appropriate development of a cytoskeleton constituted
of actomyosin and other fibrillar protein systems, and requires a serum
protein adsorbed on the substratum which need not necessarily be present
in the medium also (Fisher, Puck & Sato, 1958; Lieberman & Ove, 1958;
Lieberman, Lamy & Ove, 1959; Klebe, 1974; Grinnell, 1976a, 1976b).
There is some published evidence (Pearlstein, 1976) that this biological
activity is associated with a form of the fibroblast surface protein which has
recently attracted interest in a number of laboratories because it is found in
much greater abundance on normal interphase cells rather than on their
transformed counterparts. It has been variously named LETS (Large
External Transformation-Sensitive) protein (Hynes, 1973), cell surface
protein (Yamada & Weston, 1974; Yamada, Yamada & Pastan, 1975),
galactoprotein a (Gahmberg & Hakamori, 1973) SF-antigen (Vaheri &
Ruoslahti, 1974), and fibronectin (Keski-Oja, Mosher & Vaheri, 1976). As
well as being retained at or near the cell surface, it is also released into the
medium, perhaps in slightly modified form (Keski-Oja, Mosher & Vaheri,
1977). It is synthesised by a number of other cell types in addition to
fibroblasts, and is antigenically related to a plasma protein which has
been known for many years as cold-insoluble globulin (CIG) (Ruoslahti &
Vaheri, 1975). We have now confirmed that an electrophoretically homogeneous preparation from chick serum is indeed sufficient to facilitate cell
spreading in the absence of other serum components. Further, whole
serum loses its ability to facilitate spreading when passed through an
affinity column prepared from antibody against the purified fraction. (D.
Thom, A. Powell & D. A. Rees, unpublished results). We find that active

protein can be obtained with a molecular weight much lower (140 000) than the 200–250 000 reported by others for preparations with biological activity, suggesting that it can function in variable degrees of polymerisation.

The natural serum component which is necessary for development of the first specialised focal contacts is probably therefore the cold-insoluble globulin, although additional amounts of this, or a closely related, protein appear to be synthesised by normal fibroblasts once they are established in culture and no doubt these play a subsequent part in the interaction with substratum. In general support of this picture is the recent demonstration that this surface protein from normal fibroblasts can not only increase adhesion to the substratum when added to transformed cells which are naturally deficient in it, but evidently does so by enhancing the formation of actin cables (Yamada, Yamada & Pastan, 1976; Willingham, Yamada, Yamada, Pouysségur & Pastan, 1977; Ali, Mautner, Lanza & Hynes, 1977).

The cell surface molecules that interact with adsorbed serum protein in the first stage of adhesion have yet to be characterised. Preliminary indications do however suggest that the carbohydrate parts of membrane glycoproteins could be involved, since mutants which are defective in carbohydrate synthesis, and which appear to insert their proteins into the membrane in underglycosylated form, show dramatically reduced ability to adhere and develop their cytoskeleton unless the defect is bypassed by including precursor carbohydrates in the medium (Pouysségur Willingham & Pastan, 1977; Willingham *et al.*, 1977). Certainly this possibility would be consistent with the theme that we develop later in this paper.

We conclude that the interaction of fibroblast cells with an inert substratum, mediated by adsorbed serum protein, has all the hallmarks of a specific biological recognition coupled to a response of the whole cell. Molecules on the outer membrane face – probably glycoproteins – interact with adsorbed cold-insoluble globulin to trigger the development of specialised structures within the cytoplasm.

FURTHER INVESTIGATION OF THE RECOGNITION EVENT

To probe further the interactions between the cell and substratum-bound serum protein, we developed a method for removal of the main bodies of 16C rat fibroblasts from monolayers in a way that left components of focal adhesions attached to substratum (C. W. Lloyd, R. A. Badley, A. Woods, L. Carruthers, D. A. Rees & D. Thom, unpublished). The cells are grown to confluence and then squirted repeatedly with a stream of buffer. The interference reflection image shows numerous dark grey or black area in patterns indentical to those for intact cells except that the cell bodies themselves are absent (Plate 2). The fine structure was examined by means of scanning electron microscopy and correlated with interference reflection microscopy using a marked coverslip. The similarity in the outline shapes of the images leaves no doubt that they correspond to the same structures (Plate 3a). By either method, the structures remaining on the glass are shown to be either elliptical, about 1 μm long by 0.5 μm wide, or sometimes arranged linearly, as circular patches about 0.75 μm in diameter. In detail (Plate 3b) they are composed of globules varying around 30 nm in diameter and seem to be arranged in clusters with a distinct beaded or perhaps fibrillar boundary composed of more regular particles having similar dimensions.

Fluorescence microscopy using fluorescein-conjugated concanavalin A, or indirect immunofluorescence with an antibody against whole 16C cells (the antibody was shown to be against their outer surface components) indicated that much more cell-derived material remained on the coverslip (Plate 4). Comparison of images of the same area of coverslip by fluorescence and interference reflection microscopy shows that this material surrounds the focal adhesions, and suggests that areas of the underside of the cell have remained on the substratum in addition to the focal adhesions themselves; the focal adhesions stain much less heavily with these reagents. Because of the tenuous nature of this material we believe that it consists of peripheral glycoprotein and/or protein components of membrane regions surrounding the focal adhesions, which do not have a special role in substratum recognition. The process of detachment could then be represented diagramatically, in the light of this evidence and that given in the following paragraph, as shown in Fig. 1.

Fig. 1. Schematic representation of the process of detachment of the main bodies of 16C cells, to leave focal adhesions attached to the glass substratum. Represented here are the cell outline, the nucleus, and actin bundles which terminate at focal adhesions. Close hatching represents materials between the substratum and the membrane at the focal adhesion. Open hatching represents peripheral surface components which surround the focal area.

More evidence about the actual components present in the structures retained on the substrate was obtained when we found that the focal regions, but not the surrounding areas, stained with fluorescein-conjugated ricin, and with antibodies to actin, myosin, α-actinin, and tropomyosin. This could be observed without the use of reagents (e.g. acetone) which are usually necessary to perforate the membrane to achieve similar staining of intact cells, and indeed the staining was not enhanced when such reagents were used. Therefore it is unlikely that these remnants of cytoskeleton are enclosed in membrane vesicles and that the cell, under these conditions, detaches from the focal adhesions by 'pinching them off' in a process resembling exocytosis. No staining at all was observed under any conditions with antibodies to tubulin or bovine serum albumin. Some typical micrographs are shown in Plate 5. We conclude that the specialised structure formed at the focal adhesion in response to recognition of specific serum proteins bound to substrate contains most of the usual components of the actomyosin system in close association with the substrate. This confirms the expectations from observations on whole cells, described in the first section above. Significantly, tubulin appears to be absent from the focal region. This is entirely consistent with our conclusions from a study of the influence of drugs on cell attachment, that cells spread and form focal adhesions in the absence of microtubules (Lloyd, Smith, Woods & Rees, 1977).

The specialised structures also contain binding sites for ricin, namely β-D-galactose end-groups. The adsorbed serum protein itself is known to contain these groups, at least in the form in which it is derived from certain cells (Gahmberg, Kiehn & Hakomori, 1974; Burridge, 1976). However, the stained area has the same size, shape and distribution as the focal adhesions visualised by other methods and if indeed it is serum-derived, the distribution or redistribution must have been controlled by the cell. The possibility that this ricin receptor is antigenically related to the cold-insoluble globulin was tested with anti-human CIG kindly provided by Professor M. W. Mosesson (Plate 5).

Underneath the 16C fibroblasts in monolayer was a deposit of CIG/ LETS protein, similar but less copious than that reported for normal cells by many other groups. (Wartiovaara, Linder, Ruoslahti & Vaheri, 1974; Yamada & Pastan, 1976; Ali et al., 1977; Chen, Maitland, Gallimore & McDougall, 1977). Most of this was removed along with the cell bodies in the preparation of isolated focal adhesions, consistent with earlier reports that most of the LETS protein is associated with the cells rather than the substratum (Yamada et al., 1976; Ali et al., 1977). However, a pattern of cross-reacting spots remained, at least some of which coincided with interference reflection images which we attribute to focal contacts. We conclude that a form of the LETS protein, whether derived from the serum or the cell, is indeed sandwiched between the cell surface and the substratum, as we would expect from its supposed function in adhesion. This form of the protein seems to be distinct from the fibrillar deposit noticed by others.

It seems likely that focal areas differ from surrounding membrane in their composition with respect to other glycoconjugates as well as LETS-type material. This follows because the focal areas stain less heavily, or perhaps not specifically at all, with anti-16C and with con A, and because they not only stain specifically with ricin, but do so to an extent that might be too great to be accounted for in terms of LETS protein alone. We conclude that the external as well as the internal structures are specialised for adhesion and we return to this theme later.

Our conclusions are similar in some respects to those of Culp (1976) who has developed a different method for the preparation of substrate-attached cell components, based on shaking off the cell bodies in EGTA (ethylene glycol *bis* (β-amino ether)-N,N'-tetra-acetic acid). Under these conditions, as we have already shown (Rees, Lloyd & Thom, 1977; see also Rosen & Culp, 1977, and the following section in this paper), cell bodies draw back from the adhesion zones before detaching. The ends of the retraction fibrils so formed seem to be pinched off and remain attached to substratum as membrane-enclosed vesicles which, in contrast to our residual structures, are not permeable to proteins. The preparation is shown by gel electrophresis to contain LETS protein, actin, myosin and several other proteins. In view of the fact that the cytoskeleton is perturbed by EGTA before detachment (see the following section for discussion of this) and that rupture between the cell body and adhesion might occur at a different plane, exact correspondence is not necessarily to be expected between our results and Culp's. An important further conclusion is, however, suggested by his work – that LETS-type material associated with adhesions became labelled metabolically and therefore is at least partly cell-derived rather than serum-derived.

STICK AND GRIP

The picture of the cell's relationship with the substratum as a biological response following recognition rather than passive adhesion like that of a postage stamp to an envelope, has led us to re-examine the mechanisms of detachment of fibroblast cells.

Some evidence already exists that susceptibility to detachment by agents such as trypsin and EGTA can be influenced (strengthened or weakened) by perturbing inner as well as outer structures; for example, detachment is retarded when the intracellular level of cyclic AMP is artificially raised (Johnson & Pastan, 1972) and is accelerated by drugs which destabilise the cytoskeletal elements (Revel *et al.*, 1974; Shields & Pollock, 1974). This evidence together with results from our own further experiments described below and elsewhere, have led us to suggest that, although cell adhesion to substrate must ultimately depend on 'sticking' by physical forces at the outer surface, control is exercised through the 'grip' of the cytoskeleton (Rees et al., 1977). We have distinguished between two levels of control (i) at the actual adhesive site which we call 'focal grip' and (ii) a separate control which augments the adhesion by stiffening and flattening the cell and which we call 'bracing grip' (Rees et al., 1977). Further experiments show that the latter is the first point of action of EGTA, whereas trypsin acts at both levels.

That the primary action of EGTA is on the grip of the cytoskeleton rather than merely unsticking the cells ('dissolving the glue') is readily demonstrated by following the sequence of events by interference reflection microscopy. The focal adhesions remain unchanged (Plate 6) as the cell body rounds up, becomes blebby, and finally leaves the substrate (Rees et al., 1977). Similar conclusions follow from examination of the stages of detachment by scanning electron microscopy, except of course that it is not possible to observe the entire sequence for an individual living cell (Rosen & Culp, 1977).

The mechanism of detachment by trypsin is somewhat more complicated because the first morphological change (Plate 7) is in the image of the focal contacts by interference reflection (Rees et al., 1977). Each area is initially compact and densely black but quickly spreads to become more diffuse and grey. This suggests that focal contacts associated with actin bundles are converted to the so-called close contacts which seem (Heath & Dunn, 1977) to be associated with a more disorganised arrangement of filaments. At this stage the cells remain attached to the substrate and retain any stress fibres that were originally present (Plate 7). Immunofluorescence microscopy shows that all the CIG/LETS antigen that is accessible to antibody, has been removed by this stage. The correlation between this disappearance and the dispersal of the interference reflection image, suggests that the CIG/LETS related protein could be involved in maintaining the focal organisation. Adhesion to substrate can, however, be retained after its removal – presumably through a combination of 'bracing grip' and whatever links remain to the substratum. These could be to a degraded form of the CIG/LETS protein, to a form that is less accessible to enzymes and antibodies, or to some component which is not yet characterised – including bonds that are biologically non-specific. Detachment assays with low levels of trypsin in the cold give further insight into the mechanism of action of trypsin (Fig. 2). Brief treatment under these conditions again leaves the cell flat and attached to substrate, yet it will round up and detach when treated subsequently with trypsin inhibitor and warmed. That the effective action of trypsin is during exposure in the cold rather than with residual enzyme activity in the warm, is shown by the observation that cold incubation must be for a sufficiently long period of time (Fig. 2). Therefore it follows that the tryptic lesion leading to detachment did occur in the cold but could not be expressed until warming allowed metabolic activity and/or membrane fluidity to be re-established to allow the cytoskeleton to re-arrange to loosen the grip. We have shown earlier in this paper that contact with CIG/LETS protein is necessary to establish the flat cellular morphology which is a manifestation of the developed cytoskeleton. These experiments now show that the removal of this protein is coupled to the dispersal of focal adhesions and can occur as a distinct preliminary to cell rounding and detachment. Metabolic activity and/or membrane fluidity is required for the full response – it is not sufficient merely to 'dissolve the glue'. The sequence with trypsin is therefore the converse of that with EGTA, in which rounding precedes any rupture of focal attachments.

We have shown elsewhere (Rees et al., 1977) that the lesions induced by EGTA and trypsin, and which are usually expressed in the release of grip,

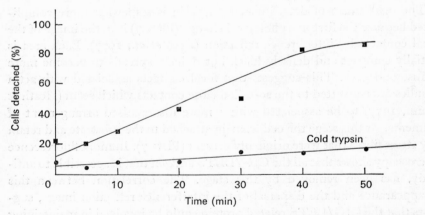

Fig. 2. Detachment of fibroblast cells (16C) by trypsin at 0 °C (●); and by this treatment together with subsequent warming to 37 °C in the presence of soybean trypsin inhibitor (■). Cells were cultured in Leighton tubes and, after tipping off the medium and washing with HBSS, a large volume (10 ml) of pre-chilled trypsin (5 μg ml⁻¹ in HBSS) was added to bring down the temperature quickly and the tube was kept in ice for the time shown on the abscissa, before rocking gently to remove the solution with suspended cells for counting. Soybean trypsin inhibitor (100 μg ml⁻¹ in HBSS) was then added, and after a further 10 min on ice, the tube was placed in an incubator at 37 °C for 20 min and then the rocking and counting was repeated. Counting of cells which remained attached at this stage as well as other techniques were as described by Rees et al. (1977). Each point represents the mean of the results from three tubes.

PLATE I

(a)

10 μm

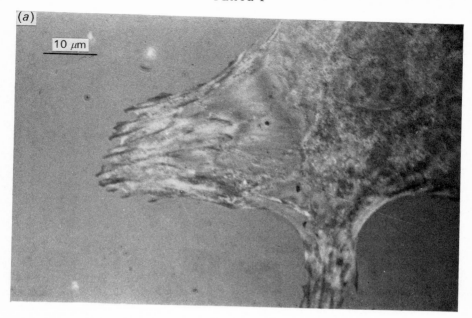

(b)

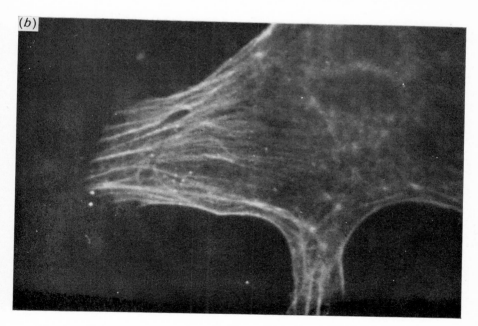

For explanation see p. 258

PLATE 2

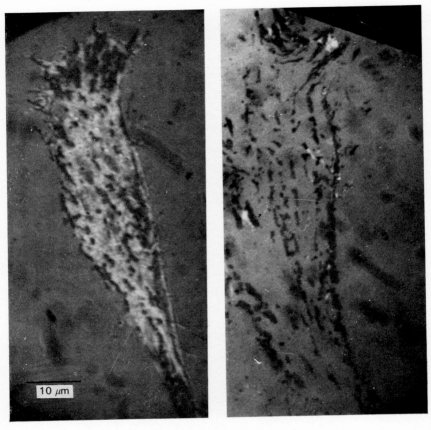

10 μm

For explanation see p. 258

PLATE 3

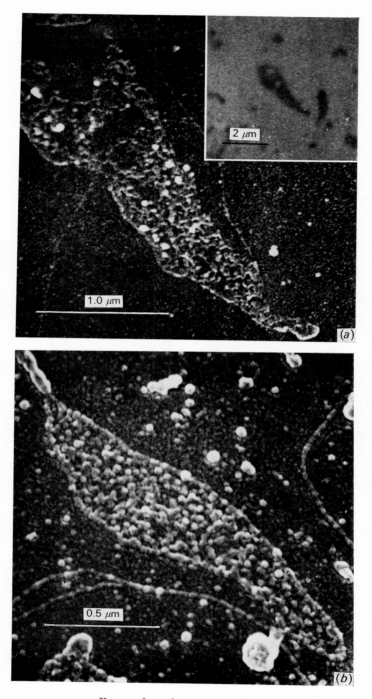

For explanation see p. 258

PLATE 4

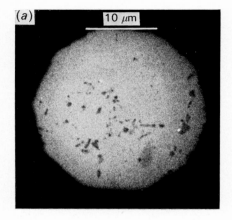

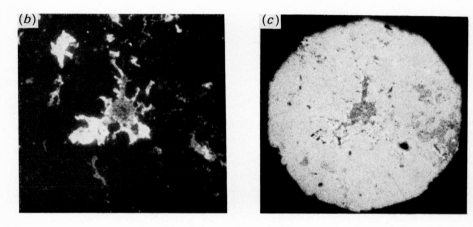

For explanation see p. 259

PLATE 5

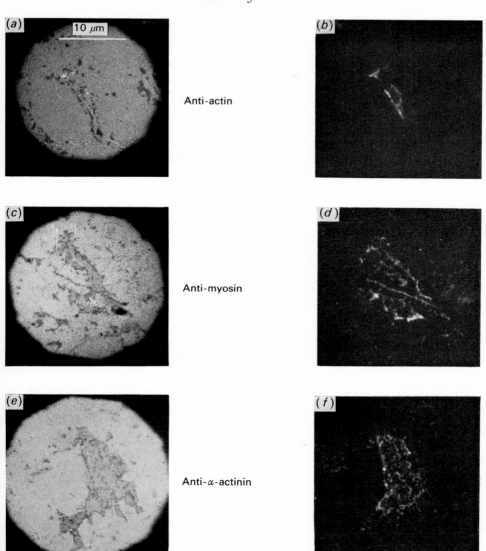

(a) 10 μm

(b) Anti-actin

(c) (d) Anti-myosin

(e) (f) Anti-α-actinin

For explanation see p. 259

PLATE 5

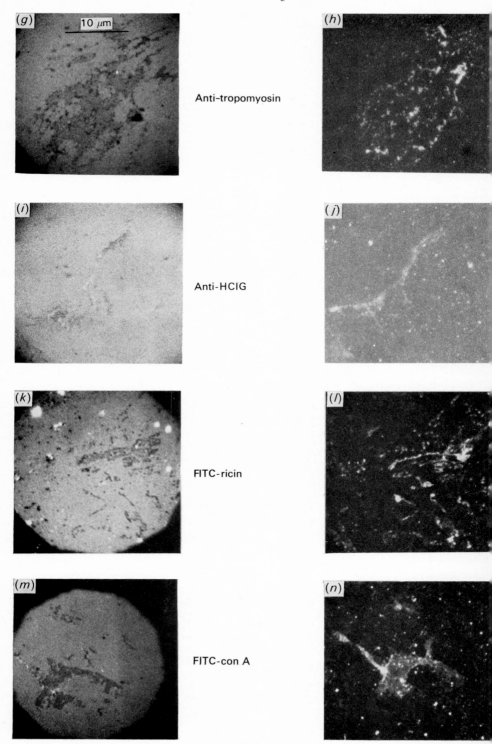

(g) 10 μm

(h) Anti-tropomyosin

(i) (j) Anti-HCIG

(k) (l) FITC-ricin

(m) (n) FITC-con A

For explanation see p. 259

PLATE 6

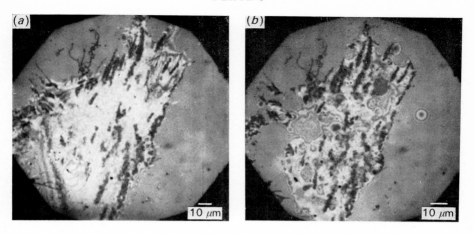

For explanation see p. 259

PLATE 7

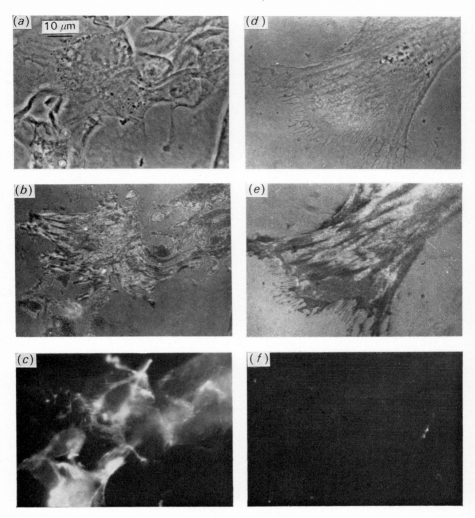

(a) 10 μm

(b)

(c)

(d)

(e)

(f)

For explanation see p. 259

PLATE 8

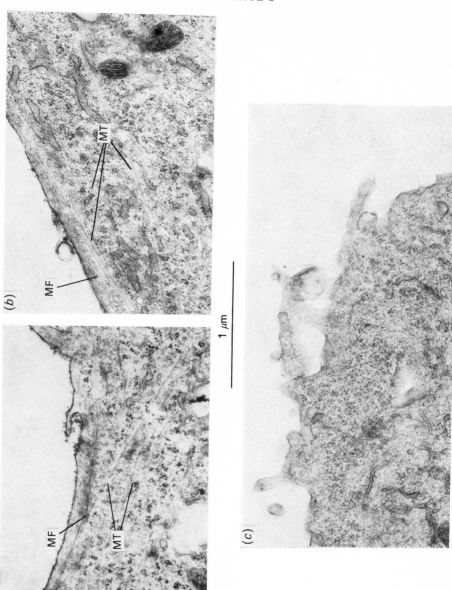

1 μm

For explanation see p. 260

PLATE 9

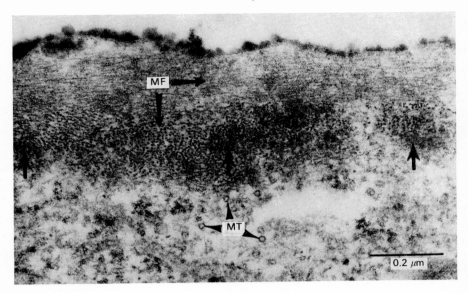

For explanation see p. 260

PLATE 10

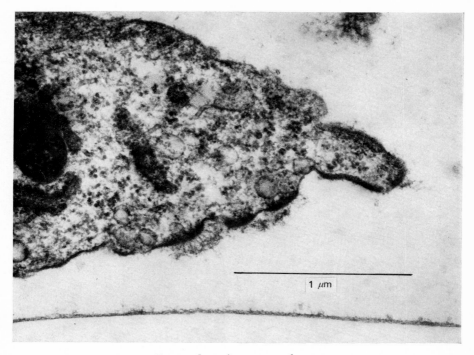

1 μm

For explanation see p. 260

PLATE II

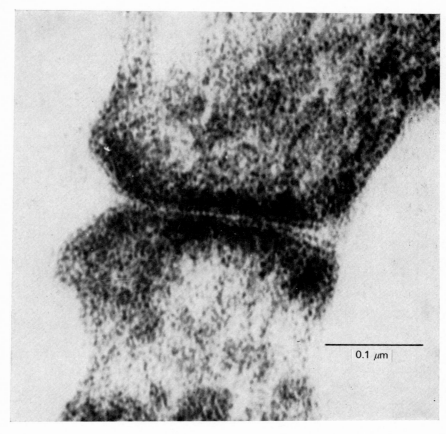

For explanation see p. 260

PLATE 12

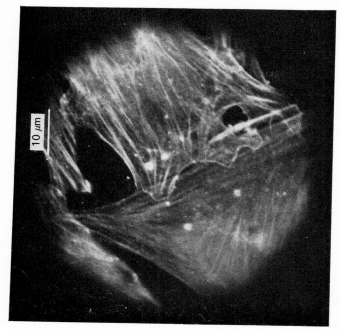

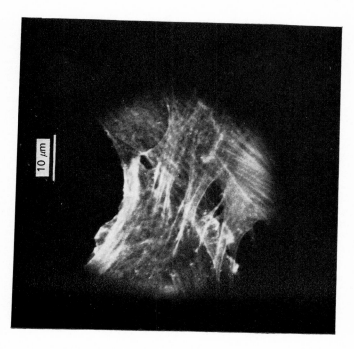

For explanation see p. 260

can be blocked in their expression by pretreatment of the cell with appropriate lectins such as concanavalin A or the *Ricinus* lectin of higher molecular weight: and that the pre-induced lesions are expressed in the absence of detaching agent when the bound lectin is displaced by hapten. We have also given evidence for our interpretation of these observations, that the rearrangement of the cytoskeleton that is necessary for loosening of grip requires lateral movement of glycoprotein receptors in the plane of the membrane (Rees *et al.*, 1977). The cross-linkage by lectin is believed to constrain this movement of receptors and hence, as shown by electron microscopy of thin sections (Plate 8), to block dispersal of the cytoskeleton. Strikingly similar observations have been made for the dispersal of microtubules which results from expression of the transformation gene in tsNY68-infected chick fibroblasts. These cells are changed from the normal to the transformed phenotype by temperature shift, but dispersal of the microtubule network is blocked by pretreatment with concanavalin A before the temperature change (McLain, D'Eustachio & Edelman, 1977).

RAFTING OF SURFACE GLYCOPROTEINS IN THE MECHANISM OF CELLULAR RECOGNITION AND RESPONSE

The simplest type of mechanism that might first come to mind for blockage of expression of the tryptic or EGTA lesion by cross-linkage of cell-surface glycoproteins, is that this linkage 'fixes' the cell surface configuration which in turn some way 'fixes' the cytoskeleton until the cell surface molecules are freed. This cannot be completely true because it is well known that lectins themselves perturb the side-to-side relationships of glycoprotein receptors especially on protease-treated or transformed cells (for a review see Nicolson, 1974) but also to a lesser extent on at least some normal cells (Fernandez & Berlin, 1976; Ash & Singer, 1976). For transformed or protease-treated fibroblasts, the first stage of this rearrangement is a passive clustering into patches followed by a further withdrawal from the cell margin which is known to result from an active contraction because it is inhibited by 2-deoxyglucose or cytochalasin B (Ukena, Borysenko, Karnovsky & Berlin, 1974). This is accompanied by the suppression of ruffling activity of the membrane (Gail, 1973; Brown & Revel, 1976) together with certain shape changes which confirm the impression that the lectin has stimulated a contraction of the cell body. These include the withdrawal of the cell periphery (Brown & Revel, 1976) and a rounding that is sensitive to metabolic inhibitors (Storrie, 1974).

Some light has been shed on the possible origin of these effects by study

17

of the action of concanavalin A on ovarian granulosa cells (Albertini & Anderson, 1977). In their first stages of response these cells resemble fibroblasts in that the clustering of receptors is first passive then active, and that the cell body is stimulated to retract. As shown by transmission electron microscopy, this is accompanied by rearrangement of actin structures from a disorganised meshwork to oriented bundles which then connect the patched receptors (Plate 9). Subsequent changes are complex and lead to, and accompany, endocytosis, but these are not relevant to our discussion here. From this we formulate a working hypothesis that the initial, passive clustering of receptors proceeds to a certain 'trigger point' at which active contraction is induced with a re-organisation of the actomyosin system to bundles which are intimately associated with the inner face of the membrane beneath the clustered receptors.

To further investigate this 'trigger point' we have examined quantitative aspects of lectin binding and correlated this with cell behaviour and membrane properties (D. Thom, D. A. Rees, R. Stafford & D. S. Williams, unpublished). As with some other cell types (Borens, Karsenti & Avremeas, 1976; Gachelin, Buc-Caron, Lis & Sharon, 1976; Schmidt-Ullrich, Wallach & Hendricks, 1976), the binding of concanavalin A to intact cells shows a maximum at low levels in the Scatchard plot (Fig. 3) which could be consistent with positive cooperativity. This would show that the binding of each molecule of lectin facilitates the binding of the next until a point corresponding to the maximum on the plot; after this the binding becomes progressively more difficult as we would expect if fewer sites remain, perhaps of lower affinity. The new observation we have made is that cellular response to certain stimuli is strikingly different on each side of the maximum in the plot, indicating that the cell membrane is in quite different states. Prior to the cooperativity maximum, phagocytosis is active, the cell is readily detached from the substrate with EGTA, and does not show the usual rounding when treated with colchicine; after the maximum, phagocytosis is suppressed, detachment is blocked (Fig. 3) and the cell shows enhanced rounding with colchicine. We suspect that the cooperative phase of binding corresponds to the passive clustering of receptors which is cooperative because the recruitment of each glycoprotein pulls it away from neighbours to facilitate their own binding. When this process has proceeded sufficiently, the actomyosin system is somehow triggered to reorganise, thus altering membrane properties, as we have observed, and influencing the mobility of receptors so that further binding is not cooperative.

This conclusion that the rafting of membrane glycoproteins in a cooperative cross-linking process can trigger an alteration of the cytoskeleton,

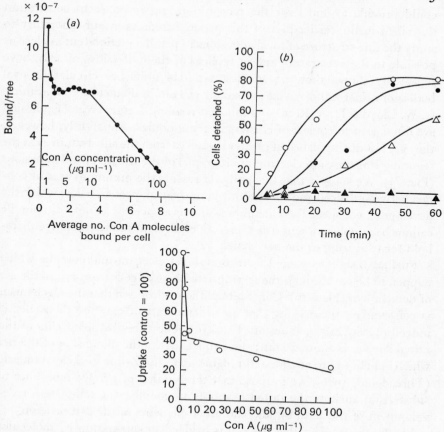

Fig. 3. Correlation of the binding of concanavalin A to rat dermal fibroblasts (16C) with its effects on cell behaviour. Cells were seeded sparsely in 230 ml bottles and grown for 20–36 h. The binding of concanavalin A to the established monolayers was measured using ^{125}I-labelled lectin prepared by lactoperoxidase catalysed iodination. (a) The binding data have been plotted according to Scatchard (1949) and to facilitate the comparison, the concentration of lectin used in the analysis of binding is also given on the abscissa. (b) The rate of detachment of the cells by EGTA (0.05 % in Ca^{2+}- and Mg^{2+}-free HBSS at 37 °C) and the inhibition of this by the binding of concanavalin A were measured by a method closely analogous to that previously described (Rees et al., 1977). ○, Control; ●, con A concentration 2.5 μg ml^{-1}; △, 5 μg ml^{-1}; ▲, 10 μg ml^{-1}. (c) Phagocytosis and its inhibition by concanavalin A were measured using latex spheres (0.8 μm : 2 × 10^{10} per bottle). Cell monolayers were washed with HBSS and incubated with the appropriate concentrations of concanavalin A in HBSS for 20 min at 37 °C. After washing with HBSS, the cells were incubated with the beads for 30 min at 37 °C, thoroughly washed with HBSS and, following detachment with tetracaine (4 mM in Ca^{2+}- and Mg^{2+}-free HBSS), they were collected and rewashed on the centrifuge. The cell pellet was extracted with dioxan (1 ml overnight) and after removal of the cells by centrifugation, the phagocytosis of latex particles was estimated from the extinction coefficient at 259 nm.

could provide a model for the recognition and response to substratum described in the earlier part of this paper. Attempts in our laboratory to study the fine structure of focal adhesions in parallel sections cut as close as possible to the substratum and in replicas of the undersides of cells, have repeatedly shown that the smallest separate units are circular or oval bodies of rather uniform size and about 150 nm in diameter (C. G. Smith, C. W. Lloyd & D. A. Rees, unpublished results). Larger areas of adhesion are often seen to consist of clusters of these bodies. Apparently, however, they are not distinguished at greater distances from the substratum as in the scanning transmission electron micrographs of the isolated focal adhesions (Plate 3). We suggest that this body represents the minimum area of contact between the cell and substratum-bound serum protein below which the development of specialised membrane and its associated actomyosin bundle cannot be supported, and it is therefore equivalent to a cooperative threshold for triggering to the new state.

Further insight is gained from occasional electron micrographs which happen to be cut through the appropriate plane of cells captured in the act of detachment (Plate 10). Consistent with our proposal that the mechanism is cooperative, these show that an ordered raft of external molecules is indeed present at the adhesion (Plate 10). The side-to-side spacing of the rafted species is about 20 nm; since this matches the dimension of the net which is believed to anchor actin filaments to the Z-line in skeletal muscle (Threadgold, 1976), we propose that the binding of surface molecules to substratum anchors the rafted assembly to stabilise the template for development of the actin bundle and other elements of the cytoskeleton.

Finally, we suggest that the cell is bridged to substratum by molecules which are largely flexible. This is likely because the trailing strands in Plate 10 (up to 80 nm in length) could not fit into the space (down to 10 nm) between cell and substratum, without some degree of folding. Although part of the trailing length on the photograph probably represents adsorbed protein which has peeled away from the substratum, the degree of regularity observed in the various dimensions would be difficult to explain if most of it were so derived. Flexible chains would also account for the variable separation distances that are commonly observed between cell and substratum, and they would allow the regularly spaced array of cell surface molecules to be anchored efficiently to serum protein which, being adsorbed to an amorphous, non-biological substratum, must be to some extent irregular in its side-to-side arrangement. With a few special exceptions such as elastin, the only biopolymers that are highly flexible in the biological state are glycoproteins and glycosaminoglycans, and at least one chain of this type is therefore likely to be involved in bridging. Because the spacing of the rafted entities is large in relation to the probable dimensions of the adsorbed cold insoluble globulin to which they are ultimately bound these entitites are probably multivalent – a further property that could readily be provided by glycoproteins or glycosaminoglycans.

Conclusions about the structure which develop in the cellular response to substratum contact are shown in Fig. 4a (a development of this model will be published by Bradley, Smith, Woods & Rees (in prep.)).

In summary, we propose that recognition results from the rafting of

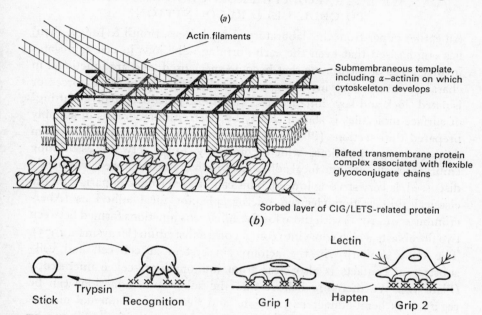

Fig. 4. (a) Minimum model, in schematic terms, for the structure of focal adhesions and their interaction with substratum. An irregular monolayer of CIG/LETS protein adsorbed onto substratum, binds flexible chains of glycoprotein(s) or glycosaminoglycan(s) associated (covalently and/or noncovalently) with a transmembrane protein complex. This is 'rafted' with a regular spacing (20 nm) which facilitates interaction with and stabilisation of the submembranous template for the development of the cytoskeleton, which is probably related in structure to the Z-line net of skeletal muscle and contains α-actinin as an important component.

(b) Outline scheme representing the proposed importance of glycoprotein rafting in cellular recognition and response. The sequence shows a round cell settling on the substratum ('stick'), then sending out filopodia which interact with clusters of CIG protein (shown as X) having appropriate arrangement and dimensions for 'recognition'. This step, which is reversed by trypsin, triggers the development of the organised cytoskeleton (schematically shown by lines which terminate at the adhesions; 'Grip 1'). Modification of the cytoskeleton to a new state ('Grip 2') with change of cell shape, can be caused by rafting surface receptors (rafts are represented by the external branched structures) with an appropriate lectin and reversed by the corresponding hapten.

surface glycoproteins in a cooperative process which, by stabilising the submembranous template, can stimulate actomyosin-driven response when a critical threshold is reached. Whether the rafting is induced in a 'natural' milieu by an endogenous or serum protein, or 'artificially' by lectin, we believe the mechanism could be similar at this general level of description. These cellular responses are summarised schematically in Fig. 4*b*.

CAN WE EXTRAPOLATE FROM CELL–SUBSTRATUM TO CELL–CELL RECOGNITION?

An earlier paper from this laboratory (Lloyd, Rees, Smith & Judge, 1976) has emphasised that even the early-forming adhesions in cell aggregates involve areas of membrane that become specialised, and are junctional in character rather than involving passive obedience to physical forces or isolated 'lock and key' interactions. That these junctions involve 'rafting' of surface molecules is shown clearly by electron micrographs of suitably prepared thin sections (Plate 11). The morphological similarity between adhaerens-type junctions which commonly form between associating embryonic cells, and the focal adhesions to substratum that we have already discussed, is very striking indeed. This can be seen from comparison of the classical thin section electron micrographs for focal adhesions (Abercrombie *et al.*, 1971) with those for the adhaerens junctions formed between two fibroblasts which come into contact on a substratum (Heaysman, 1973). The close similarity in the actomyosin system at cell–cell and cell–substratum contacts is also clear from immunofluorescence microscopy (Plate 12). We therefore believe that the substratum adhesions can be regarded as hemi-adhaerens junctions and that the fundamental mechanisms which we aim to elucidate for substratum recognition, will also be found to apply to important examples of cell–cell recognition.

We thank Professor M. W. Mosesson for the antiserum against human CIG and our colleagues Mrs L. Carruthers, Mrs T. P. Ogden, Miss H. M. Webb, Miss D. S. Williams, Mrs A. Woods, and Messrs C. Allcock, A. J. Powell and R. J. Safford for helpful discussion and for some of the experimental results which we have quoted. We are also grateful to Dr J. Chidlow & Mr B. Goodwin for considerable help in preparing antisera.

REFERENCES

ABERCROMBIE, M. (1961). The basis of the locomotory behaviour of fibroblasts. *Experimental Cell Research* (Suppl.), **8**, 188–98.

ABERCROMBIE, M. & DUNN, G. A. (1975). Adhesions of fibroblasts to substratum during contact inhibition observed by interference reflection microscopy. *Experimental Cell Research*, **92**, 57–62.

ABERCROMBIE, M., HEAYSMAN, J. E. M. & PEGRUM, S. M. (1971). The locomotion of fibroblasts in culture. IV. Electron microscopy of the leading lamella. *Experimental Cell Research*, **67**, 359–67.

ALBERTINI, D. F. & ANDERSON, E. (1977). Microtubule and microfilament re-arrangements during capping of concanavalin A receptors on cultured ovarian granulosa cells. *Journal of Cell Biology*, **73**, 111–27.

ALBRECHT-BUEHLER, G. (1976). Filopodia of spreading 3T3 cells. Do they have a substrate-exploring function? *Journal of Cell Biology*, **69**, 275–86.

ALI, I. U., MAUTNER, V., LANZA, R. & HYNES, R. O. (1977). Restoration of normal morphology, adhesion and cytoskeleton in transformed cells by addition of a transformation-sensitive surface protein. *Cell*, **11**, 115–26.

ASH, J. F. & SINGER, S. J. (1976). Concanavalin A-induced transmembrane linkage of concanavalin A surface receptors to intracellular myosin-containing fila-ments. *Proceedings of the National Academy of Sciences, U.S.A.*, **73**, 4575–9.

BORENS, M., KARSENTI, E. & AVRAMEAS, S. (1976). Cooperative binding of multi-valent ligand to lymphocytes: Cooperative restriction of receptor mobility by concanavalin A. In *Membrane Receptors of Lymphocytes*, (INSERM Sym-posium I), eds. M. Seligman, J. L. Preud-Homme & F. M. Kourilsky, pp. 377–88. North Holland, Amsterdam.

BROWN, S. S. & REVEL, J. P. (1976). Reversibility of cell surface label rearrange-ment. *Journal of Cell Biology*, **68**, 629–41.

BUCKLEY, I. K. & PORTER, K. R. (1967). Cytoplasmic fibrils in living cultured cells. *Protoplasma*, **64**, 349–80.

BURRIDGE, K. (1976). Changes in cellular glycoproteins after transformation: Identification of specific glycoproteins and antigens in sodium dodecyl sul-phate gels. *Proceedings of the National Academy of Sciences, U.S.A.*, **73**, 4457–61.

CHEN, L. B., MAITLAND, N., GALLIMORE, P. H. & McDOUGALL, J. K. (1977). Detection of the large external transformation-sensitive protein on some epithelial cells. *Experimental Cell Research*, **106**, 39–46.

CULP, L. A. (1976). Molecular composition and origin of substrate-attached material from normal and virus-transformed cells. *Journal of Supramolecular Structure*, **5**, 239–55.

CURTIS, A. S. G. (1964). The mechanism of adhesion of cells to glass. A study by interference reflection microscopy. *Journal of Cell Biology*, **20**, 199–215.

FERNANDEZ, S. M. & BERLIN, R. D. (1976). Cell surface distribution of lectin receptors determined by resonance energy transfer. *Nature, London*, **264**, 411–15.

FISHER, H. W., PUCK, T. T. & SATO, G. (1958). Molecular growth requirements of single mammalian cells: The action of fetuin in promoting cell attachment to glass. *Proceedings of the National Academy of Sciences, U.S.A.*, **44**, 4–10.

GACHELIN, G., BUC-CARON, M.-H., LIS, H. & SHARON, N. (1976). Saccharides on teratocarcinoma cell plasma membranes: their investigation with radio-actively labelled lectins. *Biochimica et Biophysica Acta*, **436**, 825–32.

GAHMBERG, C. G. & HAKAMORI, S.-I. (1973). Altered growth behavior of malignant cells associated with changes in externally labeled glycoprotein and glycolipid. *Proceedings of the National Academy of Sciences, U.S.A.*, **70**, 3329–33.

GAHMBERG, C. G., KIEHN, D. & HAKAMORI, S.-I. (1974). Changes in a surface-labelled galactoprotein and in glycolipid concentrations in cells transformed by a temperature-sensitive polyoma virus mutant. *Nature, London*, **248**, 413–15.

GAIL, M. (1973). Time lapse studies on the motility of fibroblasts in tissue culture. In *Locomotion of Tissue Cells*, Ciba Foundation Symposium 14 (new series), pp. 287–302. Elsevier, Amsterdam.

GRINNELL, F. (1976a). The serum dependence of baby hamster kidney cell attachment to a substratum. *Experimental Cell Research*, **97**, 265–74.

GRINNELL, F. (1976b). Cell spreading factor. *Experimental Cell Research*, **102**, 51–62.

HEATH, J. P. & DUNN, G. A. (1978). Cell to substratum contacts of chick fibroblasts and their relation to the microfilament system. A correlated interference reflexion and high voltage electron microscope study. *Journal of Cell Science* **29**, 197–212.

HEAYSMAN, J. (1973). Cell adhesion. In *Locomotion of Tissue Cells*, Ciba Foundation Symposium 14 (new series), pp. 187–94. Elsevier, Amsterdam.

HYNES, R. O. (1973). Alteration of cell-surface proteins by viral transformation and by proteolysis. *Proceedings of the National Academy of Sciences, U.S.A.*, **70**, 3170–74.

IZZARD, C. S. & LOCHNER, L. R. (1976). Cell-to-substrate contacts in living fibroblasts: An interference reflexion study with an evaluation of the technique. *Journal of Cell Science*, **21**, 129–59.

JOHNSON, G. S. & PASTAN, I. (1972). Cyclic AMP increases the adhesion of fibroblasts to substratum. *Nature, New Biology, London*, **236**, 247–9.

KESKI-OJA, J., MOSHER, D. F. & VAHERI, A. (1976). Cross-linking of a major fibroblast surface-associated glycoprotein (fibronectin) catalysed by blood coagulation factor XIII. *Cell*, **9**, 29–35.

KESKI-OJA, J., MOSHER, D. F. & VAHERI, A. (1977). Dimeric character of fibronectin, a major cell surface-associated glycoprotein. *Biochemical and Biophysical Research Communications*, **74**, 699–706.

KLEBE, R. J. (1974). Isolation of a collagen-dependent cell atachment factor. *Nature, London*, **250**, 248–51.

LAZARIDES, E. (1976). Actin, α-actinin and tropomyosin: interaction in the structural organisation of actin filaments in non-muscle cells. *Journal of Cell Biology*, **68**, 202–19.

LIEBERMAN, I. & OVE, P. (1958). A protein growth factor for mammalian cells in culture. *Journal of Biological Chemistry*, **233**, 637–42.

LIEBERMAN, I., LAMY, F. & OVE, P. (1959). Nonidentity of fetuin and protein growth (flattening) factor. *Science*, **129**, 43–4.

LLOYD, C. W., REES, D. A., SMITH, C. G. & JUDGE, F. J. (1976). Mechanisms of cell adhesion: Early-forming junctions between aggregating fibroblasts. *Journal of Cell Science*, **22**, 671–84.

LLOYD, C. W., SMITH, C. G., WOODS, A. & REES, D. A. (1977). Mechanisms of cellular adhesion. II. The interplay between adhesion, the cytoskeleton and morphology in substrate-attached cells. *Experimental Cell Research* **110**, 427–37.

McCLAIN, D. A., D'EUSTACHIO, P. & EDELMAN, G. M. (1977). Role of surface modulating assemblies in growth control of normal and transformed fibroblasts. *Proceedings of the National Academy of Sciences, U.S.A.*, **74**, 666–70.

NICOLSON, G. L. (1974). The interactions of lectins with animal cell surfaces. *International Review of Cytology*, **39**, 89–190.

PEARLSTEIN, E. (1976). Plasma membrane glycoprotein which mediates adhesion of fibroblasts to collagen. *Nature, London*, **262**, 497–500.

PEGRUM, S. M. & MAROUDAS, N. G. (1975). Early events in fibroblast adhesion to glass. *Experimental Cell Research*, **96**, 416–22.

POUYSSÉGUR, J., WILLINGHAM, M. & PASTAN, I. (1977). Role of cell surface carbohydrates and proteins in cell behaviour. Studies on the biochemical reversion of an *N*-acetylglucosamine-deficient fibroblast mutant. *Proceedings of the National Academy of Sciences, U.S.A.*, **74**, 243–7.

REES, D. A., LLOYD, C. W. & THOM, D. (1977). Control of stick and grip in cell adhesion through lateral relationships of membrane glycoproteins. *Nature, London*, **267**, 124–8.

REVEL, J. P., HOCH, P. & HO, D. (1974). Adhesion of culture cells to their substratum. *Experimental Cell Research*, **84**, 207–18.

REVEL, J. P. & WOLKEN, K. (1973). Electron microscope investigations of the underside of cells in culture. *Experimental Cell Research*, **78**, 1–14.

RODRIGUES, J. & DEINHARDT, F. (1960). Preparation of a semipermanent mounting medium for fluorescent antibody studies. *Virology*, **12**, 316–17.

ROSEN, J. J. & CULP, L. A. (1977). Morphology and cellular origins of substrate-attached material from mouse fibroblasts. *Experimental Cell Research*, **107**, 139–149.

RUOSLAHTI, E. & VAHERI, A. (1975). Interaction of soluble fibroblast surface antigen with fibronigen and fibrin. Identity with cold insoluble globulin of human plasma. *The Journal of Experimental Medicine*, **141**, 497–501.

SCATCHARD, G. (1949). The attraction of proteins for small molecules and ions. *Annals of the New York Academy of Sciences*, **51**, 660–72.

SCHMIDT-ULLRICH, R., WALLACH, D. F. H. & HENDRICKS, J. (1976). Interaction of concanavalin A with rabbit thymocyte plasma membranes. Distinction between low-affinity association and positively cooperative binding mediated by a specific glycoprotein. *Biochimica et Biophysica Acta*, **443**, 587–600.

SHIELDS, R. & POLLOCK, K. (1974). The adhesion of BHK and pyBHK cells to the substratum. *Cell*, **3**, 31–8.

STORRIE, B. (1974). Effect of dibutyryl adenosine cyclic 3′,5′-monophosphate and testolactone on concanavalin A binding and cell killing. *Journal of Cell Biology*, **62**, 247–52.

THREADGOLD, L. T. (1976). *The Ultrastructure of the Animal Cell*, 2nd edition, pp. 250–4. Pergamon, Oxford.

UKENA, T. E., BORYSENKO, J. Z., KARNOVSKY, M. J. & BERLIN, R. D. (1974). Effects of colchicine, cytochalasin B, and 2-deoxyglucose on the topographical organisation of surface-bound concanavalin A in normal and transformed fibroblasts. *Journal of Cell Biology*, **61**, 70–82.

VAHERI, A. & RUOSLAHTI, E. (1974). Disappearance of a major cell-type specific surface glycoprotein antigen (SF) after transformation of fibroblasts by Rous sarcoma virus. *International Journal of Cancer*, **13**, 579–86.

WARTIOVAARA, J., LINDER, E., RUOSLAHTI, E. & VAHERI, A. (1974). Distribution of fibroblast surface antigen. Association with fibrillar structures of normal cells and loss upon viral transformation. *Journal of Experimental Medicine*, **140**, 1522–33.

WILLINGHAM, M. C., YAMADA, K. M., YAMADA, S. S., POUYSSÉGUR, J. & PASTAN, I. (1977). Microfilament bundles and cell shape are related to adhesiveness to substratum and are dissociable from growth control in cultured fibroblasts. *Cell*, **10** 375–70.

WITKOWSKI, J. A. & BRIGHTON, W. D. (1971). Stages of spreading of human diploid cells on glass surfaces. *Experimental Cell Research*, **68**, 372–80.

WITKOWSKI, J. A. & BRIGHTON, R. D. (1972). Influence of serum on attachment of tissue cells to glass surfaces. *Experimental Cell Research*, **70**, 41–8.

YAMADA, K. M. & PASTAN, I. (1976). Cell surface protein and neoplastic transformation. *Trends in Biochemical Sciences*, **1**, 222–4.

YAMADA, K. M. & WESTON, J. A. (1974). Isolation of a major cell surface glycoprotein from fibroblasts. *Proceedings of the National Academy of Sciences, U.S.A.*, **71**, 3492–6.

YAMADA, K. M., YAMADA, S. S. & PASTAN, I. (1975). The major cell surface glycoprotein of chick embryo fibroblasts is an agglutinin. *Proceedings of the National Academy of Sciences, U.S.A.*, **72**, 3158–62.

YAMADA, K. M., YAMADA, S. S. & PASTAN, I. (1976). Cell surface protein partially restores morphology, adhesiveness, and contact inhibition of movement to transformed fibroblasts. *Proceedings of the National Academy of Sciences, U.S.A.*, **73**, 1217–21.

EXPLANATION OF PLATES

PLATE 1

Comparison of interference reflection image (*a*), and stress fibre pattern (*b*) for rat embryo cell. Cells were isolated according to Buckley & Porter (1967) and grown as described previously (Rees, Lloyd & Thom, 1977). Sparse cells were fixed for 20 min in 3.5 % formaldehyde in phosphate buffered saline (PBS), cooling from 37 °C to room temperature. Following 30 min incubation in 0.1 M ammonium chloride in PBS the cells were rinsed in PBS and then extracted with acetone (Lazarides, 1976). Stress fibres were visualised by staining with 10 × diluted rabbit anti-actin gamma-globulin fraction (prepared with 50 % ammonium sulphate) for 45 min at 37 °C followed by 3 × 10 min washes with PBS and incubation with 30 × diluted FITC-labelled goat anti-rabbit-gamma-globulin (Nordic Immunologicals, Maidenhead, UK). After rinsing in PBS, cells were mounted in a polyvinyl alcohol–glycerol–PBS mixture (Rodrigues & Deinhardt, 1960). Cells were viewed with a Leitz Ortholux microscope fitted with epi-illumunation.

PLATE 2

Comparison of the interference reflection images of a whole cell with material remaining after removing the cell body. Interference reflection microscopy shows focal adhesion plaques to be more prominent at the leading edge of 16C fibroblasts and that they tend to be aligned. Removing the body of a similar cell (right) by squirting with a jet of buffer clearly leaves the dark adhesive patches remaining on the coverslip.

PLATE 3

(*a*) Correspondence of interference reflection and SEM images of an isolated focal adhesion. To locate particular cells, glass coverslips were thickly coated with gold except for small areas marked with coverslip fragments. Fibroblasts (16C) were grown on, and removed from, the coverslips and after standard fixation the focal adhesions were visualised by interference reflection and photographed along one edge of the uncoated island. The coverslip was then further coated (in this case, with gold) for SEM. The initial thick gold coat presented a contour around the masked island which was easily identified by SEM and when photographically reversed (for easy comparison of the mirror images) the light and electron microscopy images showed exact correspondence.

(b) The fine structure of an isolated focal adhesion. By SEM, focal adhesions appear in circular or ellipsoidal form. Much of the fine detail is obscured by gold coating but this elongated plaque, coated instead with platinum, reveals the structures to consist of seemingly irregular patches of particles of about 30 nm in diameter, surrounded by a more uniform bead chain. These plaques are often seen to occur in lines.

PLATE 4

Comparison of isolated focal adhesions from 16C cells, (a) unlabelled and (b) and (c) labelled with anti-16C antibodies; (a) and (c) show images by interference reflection and (b) shows the image from fluorescence microscopy. The examples in (b) and (c) were labelled with 50 × diluted rabbit antibodies against whole 16C cells. Incubation, washing and labelling with FITC goat anti-rabbit antibodies was as described for Plate 1. Preabsorption with 16C cells eliminated the staining.

PLATE 5

Staining of isolated focal adhesions from 16C cells by a variety of antibodies and fluorescent-labelled lectins. For each label an interference reflection and fluorescence image of the same area is shown. The staining procedure for each antibody was as indicated in Plate 1 except that a 25 × dilution was used. FITC–ConA ($500 \mu g/ml^{-1}$) and FITC–ricin ($100 \mu g/ml^{-1}$) were used as a direct label. In each case, controls consisting of pre-immune or absorbed sera, or in the lectin case the appropriate specific sugar hapten, indicated the specificities of the reactions. The following labels were used: anti-actin (a, b); anti-myosin (c, d); anti-α-actinin (e, f); anti-tropomyosin (g, h); antibody to human cold insoluble globulin (i, j); FITC–ricin (k, l; Miles Laboratories, Slough, UK) and FITC–concanavalin A (m, n; Miles Laboratories). The antibodies against actin, gizzard myosin, α-actinin, tropomyosin and human cold insoluble globulin appear to bind directly to dark. interference reflection images as does FITC–ricin whilst FITC–concanavalin A and anti-16C antibodies (see Fig 4) bind to areas showing white by interference reflection. Details of the antibody preparations will be published elsewhere.

PLATE 6

Photomicrographs of rat dermal fibroblasts (16C line, Colworth strain) showing stages in detachment by ethylene glycol bis (β-aminoethyl ether)-N,N'-tetraacetic acid (EGTA, 0.05 % in Ca^{2+}- and Mg^{2+}-free Hank's balanced salts solution at 37 °C). (a) Before exposure to EGTA; (b) after exposure for 35 min. The size, shape and mutual relationship of the focal adhesions alter little while the cell rounds and peels away from the substrate. Reproduced from Rees et al. (1977).

PLATE 7

Photomicrographs of chick embryo fibroblasts at stages in detachment by trypsin. (a)–(c) Comparison by phase contrast, interference reflection and immunofluorescence microscopy, respectively, of cells before exposure to trypsin; (d)–(f) a similar comparison on another cell exposed to trypsin ($2.5 \mu g ml^{-1}$ in HBSS at 37 °C) for 30 min. Fluorescent labelling was with fluorescein-conjugated goat anti-rabbit IgG antibody (1:30 dilution in PBS) following incubation of the fixed cells with a rabbit antibody ($100 \mu g ml^{-1}$ in PBS) against a cold insoluble globulin preparation purified from chicken serum. Note that the fluorescent staining of the CIG/LETS protein can be completely removed under conditions where the cells remain attached and well spread and their stress fibres are still prominent. Distinct morphological changes are, however, evident by interference reflection microscopy with the dense black focal adhesions spreading to become more diffuse and grey.

PLATE 8

Typical electron micrographs of BHK cells seeded singly and grown on Melinex coverslips for 48 h. After appropriate treatment (see below), cells were fixed, dehydrated and embedded as previously described (Rees *et al.* 1977). Sections were cut parallel to the substrate and stained with lead citrate (*a*) control BHK cell, (*b*) BHK cell after incubation at 37 °C with con A (100 μg ml^{-1} in HBSS) for 20 min followed by trypsin (25 μg ml^{-1} in HBSS) for 30 min, (*c*) BHK cell after incubation at 37 °C with trypsin (25 μg ml^{-1} in HBSS) for 10 min. MT, microtubules; MF, microfilament bundles. Reproduced from Rees *et al.* 1977.

PLATE 9

Transmission electron micrograph illustrating the action of concanavalin A on ovarian granulosa cells. Oriented bundles of actin microfilaments (MF) form concomitantly with the 'patching' of surface receptors for the lectin and are redistributed as the patched receptors move into a central 'cap', MT = microtubules. The concanavalin A has been visualised by labelling with horseradish peroxidase. From Albertini & Anderson, 1977.

PLATE 10

Vertical section through a fibroblast cell (BHK) from a sparse culture grown on Melinex for 20 h. Conditions for fixation, embedding and dehydration were as described by Rees *et al.* (1977).

PLATE 11

The structural nature of an adhaerens junction. When brought together in suspension, 16C fibroblasts rapidly form junctions of the *zonula adhaerens* type (see Lloyd, Rees, Smith & Judge, 1976). This process is so rapid that, even at early times of sampling, microvilli characteristic of the suspended state have disappeared between contacting surfaces where the junctional interface becomes increasingly broader. In order to see within the initial contact we attempted to throw the adhesive process into reverse. Aggregates formed for 2.5 h were treated with 2.5 mM EGTA in tris buffer and when the compact nature of aggregates was lost, after about 75 min, the cells were fixed and processed for electron microscopy as previously described (Lloyd *et al.*, 1976). In this picture, two cells are held together in a focal instead of a broad adhesion. The major point of interest is that as the cells begin to pull apart, ordered structure can be seen within the adhesion. This consists of dumbell shaped entities–approximately 30 nm, in overall length–associated with each membrane. These are aligned in head-to-head register though there is the impression that additional fibrous material is sandwiched between.

PLATE 12

Illustration of involvement of actin cables in cell–cell contacts. Rat embryo cells were treated as in Plate 1. The field of view has been limited to focus attention on the contact areas.

THE POTENTIAL ROLES FOR GANGLIOSIDE GM$_2$ AND GM$_1$ SYNTHETASE IN RETINOTECTAL SPECIFICITY

By MICHAEL PIERCE, RICHARD B. MARCHASE
AND STEPHEN ROTH

Department of Biology, The Johns Hopkins University,
Baltimore, Maryland 21218, U.S.A.

INTRODUCTION

An in-vitro assay has been developed that measures the rate of adhesion of chick neural retinal cells dissected from either dorsal or ventral areas of the retina to dorsal and ventral halves of optic tecta (Barbera, Marchase & Roth, 1973 and Barbera, 1975). Results from this assay show that retinal cells show an adhesive specificity toward the tectal half to which they normally project *in vivo*. Biochemical investigations of the assay have established properties of two kinds of molecules that are involved in the specific adhesion of the retinal cells to the tecta (Marchase, 1977). The structural properties of the ganglioside, GM$_2$, match those of one of the molecules. Assuming GM$_2$ does play a role in retinotectal adhesion, then the properties of the glycosyltransferase, GM$_1$ synthetase, fit those properties of the other molecule implicated by biochemical studies.

The gangliosides are a class of glycolipids that contain sialic acid. The oligosaccharide portion of these molecules is sequentially synthesized by specific glycosyltransferases that utilize the appropriate sugar donors (Kaufman, Basu & Roseman, 1967). Although gangliosides have been found in all vertebrate tissues, those with more complex oligosaccharide side chains are generally more concentrated in neural tissue (Brunngraber, Tettamanti & Berra, 1976). Before he isolated and partly characterized gangliosides from beef brain in 1942, Klenk had shown that patients with Tay–Sachs disease had increased levels of an unidentified class of glycolipid (Klenk, 1939, 1942). More recently, Tay–Sachs disease has been shown to be characterized by the accumulation of one particular ganglioside, whose structure is shown below (Svennerholm, 1962). This molecule is termed GM$_2$ in the Svennerholm notation of gangliosides (Svennerholm 1963).

$$\text{Ceramide} \xleftarrow{(\beta,\,1,\,4)} \text{Glucose} \longleftarrow \text{Galactose} \xleftarrow{(\beta,\,1,\,4)} N\text{-Acetylgalactosamine}$$

$$\uparrow (\alpha\substack{2\\3})$$

$$N\text{-Acetylneuraminic acid}$$

[261]

This paper will summarize the results of the biochemical studies of the in-vitro assay of retinotectal adhesive specificity and describe experiments that suggest roles for GM_2 and GM_1 synthetase in the specific adhesion of neural retinal cells to tectal halves.

First, the rationale for the development of the in-vitro assay as well as a summary of some of the results obtained with the assay will be presented.

Development of chick retinotectal projection

As early as day 5 of development, retinal ganglion axons begin to migrate from the eye. Ultimately, all of the ganglionic cells will send processes from the retina that will cross at the optic chiasma with no mixing and migrate over the surface of the contralateral optic tectum. At about day 2 of development, the axonal tips, which are in place on the surface of the tectum, dive directly down and form synapses inside the tectum (Crossland, Cowan, Roger & Kelly, 1974).

Histological as well as electrophysiological experiments have shown that retinal ganglion axons form a spatially continuous map on the tectum. Cells from a specific area of the retina always project to the same, specific area of the tectum. This projection is inverted along the dorsoventral axis. That is, cells from the dorsal part of the retina project to the ventral part of the tectum, and ventral cells project to dorsal tectum. There is, in addition, a reversal of the projection along the naso-temporal axis (Hamdi & Whitteridge, 1954 and McGill, Powell & Cowan, 1966). The question we wish to focus upon is how the ganglion axonal tips orient on the surface of the tectum.

Chemoaffinity hypothesis of Sperry

R. W. Sperry (1963) has proposed an hypothesis to explain the means by which the axonal tips orient on the tectum based in part upon experiments done earlier by Sperry (1945) and by Matthey (1925) that studied the regeneration of the retinotectal projection in newts. Matthey showed, using behavioural tests, that newts could regain normal vision after their optic nerves had been sectioned. Sperry later repeated these experiments and performed others in which eyes were rotated by 180° and reimplanted after their optic nerves had been sectioned. These animals regained vision, but gave behavioural responses that were systematically inverted 180°. Sperry concluded that, during regeneration, ganglion cells regrew to specific points on the tectum that were independent of the orientation of the eye. He later postulated that the precise orientation of the ganglion cells on the tectum could be explained by affinities of ganglion cells for specific areas of the tectum. These affinities could in turn, be explained by the existence of complementary cytochemical gradients on the surfaces of the axonal tips and tectal surfaces. Accordingly, ganglionic axonal tips would possess a molecular code on their surfaces and the greatest affinity between any axonal tip and the tectum would be that position on the tectum at which the greatest number of molecular complexes could be formed (Sperry, 1963).

In-vitro assay of retinotectal adhesive specificity

We wished to develop an in-vitro assay that could test Sperry's hypothesis and could be used ultimately to characterize the molecules that are important in the formation of the retinotectal projection. A clear prediction of the hypothesis would be that axonal tips and, perhaps, cell bodies from specific areas of the retina should show differential affinities or adhesive preferences for appropriate tectal areas. A number of methods have been devised to measure intercellular adhesion (Marchase, Vosbeck & Roth, 1976). In many of these methods, one cell type is used as a collecting surface for a radioactively labelled cell suspension of another cell type. The amount of radioactivity associated with the collecting cell type as a function of time is then taken as a measure of the rate of adhesion between these cell types.

Barbera et al., 1973, used tectal halves as collecting surfaces for labelled single-cell suspensions of neural retinal cells dissected from dorsal or ventral areas of the retina. The standard experimental protocol (Barbera, 1975) involved bisecting six tecta and pinning the dorsal and ventral halves, in alternating fashion, onto a paraffined dish with cut sides down. Either dorsal or ventral neural retinal fragments were then dissected, labelled with ^{32}P, and dissociated into single cells by treatment with crude trypsin. Dorsal or ventral cell suspensions were added to the pinned tectal halves and the dish was reciprocated at 37 °C. At appropriate times, the tectal halves were removed, washed, and associated radioactivity determined. Throughout this protocol, dorsal and ventral tectal halves receive identical treatment and during any one assay are exposed to the same suspension of either dorsal or ventral cells.

Summarizing the results of many experiments by Barbera, 1975, dorsal retinal cells adhered to ventral tectal halves approximately twice as rapidly as they adhered to dorsal tectal halves. Thus, dorsal cells adhere faster to the tectal half to which they project in vivo. Ventral neural retinal cells adhered to dorsal tectal halves at a rate twice that of their adhesion to ventral tectal halves although this only occurs for collection times greater than 3 h. Therefore, ventral cells also show an adhesive specificity towards their physiologically matching tectal halves.

Since, at early times of collection, both dorsal and ventral cells adhere more rapidly to ventral tectal halves, the adhesion of non-retinal cells to tectal halves was measured to rule out the possibility that ventral tectal halves are, initially, very adhesive in general. Suspensions of cells from liver, cerebrum, and cerebellum showed no preference for either tectal half for at least 6 h.

For long collection times, both dorsal and ventral neural retinal cells show an adhesive preference for their physiologically matching tectal halves. These results are consistent with Sperry's chemo-affinity hypothesis.

The in-vitro assay allowed Barbera, 1975, to vary the developmental age of the tissues and assay combinations of retinal cells and tectal halves that would never come in contact *in vivo*. The adhesion of dorsal cells to tectal halves was measured as a function of developmental age of the retina, and the results are shown in Fig. 1. Each point in this figure represents the

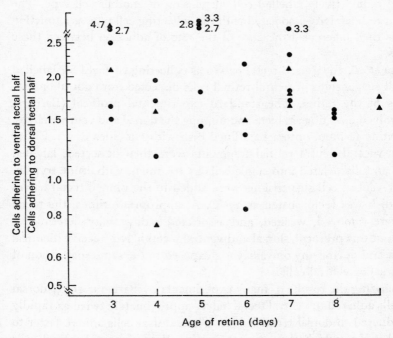

Fig. 1. Ratio of (number of cells adhering to 'matching' ventral half-tecta): (number of cells adhering to 'non-matching' dorsal half-tecta) as a function of the developmental age of the dorsal half-retinas. Each experiment used a 30 min collection time; preincubation times were less than 30 min. Symbols: (●) innervated tectal halves; (▲) non-innervated tectal halves. Reproduced from Barbera, 1975 (by kind permission of Dr Barbera and Academic Press).

average number of cells adhering to six ventral halves divided by the average number of cells adhering to six dorsal halves. The results demonstrate that dorsal retinal cells show an adhesive preference toward their matching 11 day tectal half from as early as day 3.

To rule out the possibility that fragments of retinal axons left on the tectal surface could influence the adhesion of retinal cells to the tectum, tecta from animals that had their eyes removed early in development were used in the assay. Fig. 1 shows that there is no detectable difference in the adhesion of dorsal retina to non-innervated tecta as compared to innervated ones.

Fig. 2 shows the results of the adhesion of ventral neural retinal cells, pre-incubated for 5–6 h, to tectal halves as a function of retinal age. After day 6 of development, the ventral cells show a preference for their matching tectal half, just as dorsal cells do. However, before day 6, ventral cells show a reversed specificity: they adhere preferentially to ventral tectal halves. These results suggest a change in the cell surface properties of

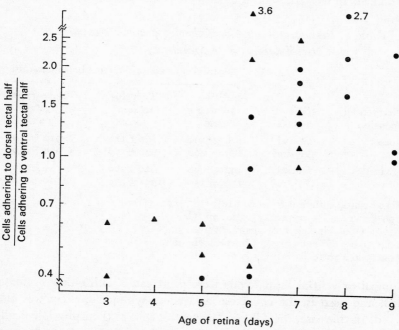

Fig. 2. Ratio of (number of cells adhering to 'matching' dorsal half-tecta) : (number of cells adhering to 'non-matching' central half-tecta) as a function of the ventral half-retinas. Each experiment used a 30 min collection time; preincubation times were 5–6 h. Symbols: (○) innervated tectal halves; (▲) non-innervated tectal halves. Reproduced from Barbera, 1975 (by kind permission of Dr Barbera and Academic Press).

18

ventral cells occurs at around day 6 that allows them to recognize their matching tectal halves. Fig. 2 also shows that ventral retinal cells adhere similarly to innervated versus non-innervated tecta.

Biochemical studies

The strategy for biochemical analysis was to treat either the tectal halves or retinal cells with various, purified, hydrolytic enzymes, and then use these treated components in the collection assay. Since for any given enzymatic treatment four adhesive combinations were measured, a differential effect of the treatment on any of the combinations could be observed. Marchase (1977) treated tectal halves or dorsal and ventral cell suspensions with trypsin or chymotrypsin and used these treated components in the collection assay. Results of a typical experiment are shown in Table 1. When tectal halves were pretreated with either protease, washed, and used in the collection assay all the adhesive rates were decreased slightly. However, an asymmetric effect was evident: the rate of adhesion of dorsal cells to treated, ventral tecta was decreased by a factor of two to three.

Table 1. *Retinotectal adhesion following protease treatment of tectal halves (from Marchase, 1977)*

Labelled retinal suspension	Tectal halves	Retinal cells adhering/tecta half in 30 min		
		PBS-treated tecta	Trypsin-treated[a] tecta	Chymotrypsin-treated[b] tecta
Dorsal[c]	Dorsal	1060 ± 190	890 ± 110	910 ± 40
	Ventral	1850 ± 290	620 ± 70*	670 ± 110*
Ventral[d]	Dorsal	920 ± 90	820 ± 100	740 ± 60
	Ventral	1790 ± 150	1530 ± 270	1410 ± 220

* Significantly different from control, $P < 0.01$.
[a] 0.09 % crystalline trypsin, 15 min, 37 °C.
[b] 0.17 % crystalline chymotrypsin, 15 min, 37 °C.
[c] 7 day, 1.0×10^6 cells ml^{-1}, 4 h preincubation.
[d] 7 day, 1.0×10^6 cells ml^{-1}, 4 h preincubation.

If dorsal or ventral retinal cells were pre-treated with either protease, washed, and used in the collection assay, an asymmetric effect was also observed. In this case, the adhesion of ventral cells to their physiologically matching tectal halves (dorsal) is markedly reduced. These results suggested that protease-sensitive molecules involved in the retina–tecta recognition system were concentrated more on the ventral surfaces of retina and tectum.

A model to explain retinotectal adhesive specificity consistent with these results is shown in Fig. 3. In this model, actually a variant of one of Sperry's models, there are gradients of two complementary molecules on the surfaces of the retina and tectum. One type of molecule is protease-sensitive and more concentrated on the ventral surfaces. The other is relatively protease-insensitive, more dorsally concentrated, and can be bound by the more ventrally localized molecule. Retinal cells would therefore display a specific number of these two molecules and adhere most strongly to the area of the tectum that displayed complementary numbers of these molecules.

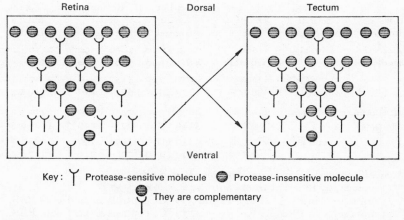

Fig. 3. A model of gradients of complementary molecules in the dorsoventral axis that could explain retinotectal adhesive specificity, and dorsal to ventral affinity. Reproduced from Barbera, 1975 (by kind permission of Dr Barbera and Academic Press).

A number of proteins are known to bind specific carbohydrate residues in a lock-and-key manner. A candidate, then, for the dorsally concentrated molecule was one that contained carbohydrate. These carbohydrates should be sensitive to an appropriate glycosidase. Marchase, 1977, treated tectal halves or retinal cells with various partly purified glycosidases that were then used in the collection assay. Although a number of these glycosidases, including fucosidase, neuraminidase, and β-N-acetylglucosaminidase, slightly decreased rates of adhesion, no asymmetric effects were observed. However, treatment with two enzymes, β-N-acetylgalactosaminidase and β-galactosidase, did show asymmetric effects.

Table 2 shows the results of a typical experiment in which retinal cells or tectal halves were pretreated with β-N-acetylgalactosaminidase. When retinal cells were treated, the adhesion of dorsal retina to ventral tectum was greatly reduced. When tectal halves were treated, the adhesion of ventral retina to dorsal tectum was reduced more than the other adhesive rates. Table 3 shows a similar experiment with β-galactosidase. Treatment with this enzyme slightly stimulated all adhesive rates but greatly stimulated only the binding of ventral retina to ventral tectum when either retinal cells or tectal halves were treated.

Table 2. *Retinotectal adhesion following treatment with bacterial*
β-N-acetylhexosaminidase (reproduced from Marchase, 1977)

			Retinal cells adhering/tectal half in 1 h		
Enzyme source	Labelled retinal suspension	Tectal half	Neither tissue treated	Retinal cells treated	Tectal halves treated
Diplococcus pneumoniae[a]	Dorsal[c]	Dorsal	2430 ± 210	1860 ± 90*	2070 ± 260
		Ventral	4560 ± 620	1990 ± 140**	4240 ± 390
	Ventral[d]	Dorsal	6170 ± 390	5850 ± 610	2820 ± 140**
		Ventral	3990 ± 470	3740 ± 210	2810 ± 310*
Clostridium perfringens[b]	Dorsal[e]	Dorsal	2860 ± 350	2450 ± 210	2230 ± 270
		Ventral	5320 ± 390	2730 ± 240**	4820 ± 300
	Ventral[f]	Dorsal	4250 ± 610	3970 ± 390	1960 ± 210**
		Ventral	2290 ± 100	2060 ± 240	1720 ± 260*

* Significantly different from controls, $P < 0.05$.
** Significantly different from controls, $P < 0.01$.
[a] 25 units ml^{-1} in PBS pH 6.0, 15 min, 37 °C; controls treated with PBS.
[b] 25 units ml^{-1} in PBS pH 6.5, 15 min, 37 °C; controls treated with PBS.
[c] 8 day, 1.0×10^6 cells ml^{-1}, 0.09 c.p.m. cell^{-1}, 4 h preincubation.
[d] 7 day, 1.0×10^6 cells ml^{-1}, 0.24 c.p.m. cell^{-1}, 4 h preincubation.
[e] 7 day, 0.9×10^6 cells ml^{-1}, 0.23 c.p.m. cell^{-1}, 4 h preincubation.
[f] 7 day, 1.0×10^6 cells ml^{-1}, 0.19 c.p.m. cell^{-1}, 4 h preincubation.

The model presented in Fig. 3 can account for these results. A molecule relatively insensitive to protease treatment and containing a terminal β-N-acetylgalactosamine residue would be more concentrated in the dorsal aspects of retina and tectum. In the ventral aspects, this carbohydrate residue could be penultimate to a terminal galactose. The protein molecule localized ventrally would bind to the terminal β-N-acetylgalactosamine residue located dorsally, but not the same molecule terminated by a galactose in the ventral aspects.

The criteria for the dorsally concentrated molecule are met by the ganglioside GM_2. This molecule is protease-insensitive and has a terminal β-N-acetylgalactosamine residue. In addition, the enzyme, GM_1 synthetase, can add a molecule of galactose from the donor, UDP-galactose, to the 3' hydroxyl of the N-acetylgalactosamine residue through β linkage and thereby synthesize the ganglioside GM_1.

Table 3. *Retinotectal adhesion following treatment with bacterial β-galactosidase* (*from Marchase, 1977*)

Enzyme source	Labelled retinal suspension	Tectal half	Retinal cells adhering/tectal half in 1 h		
			Neither tissue treated	Retinal cells treated	Tectal halves treated
Diplococcus pneumoniae[a]	Dorsal[c]	Dorsal	2850 ± 360	3130 ± 140	3070 ± 290
		Ventral	4640 ± 210	4820 ± 260	4780 ± 410
	Ventral[d]	Dorsal	5190 ± 430	5970 ± 360	5860 ± 610
		Ventral	2860 ± 360	4640 ± 190*	5130 ± 470*
Clostridium perfringens[b]	Dorsal[e]	Dorsal	3620 ± 260	3970 ± 380	3920 ± 210
		Ventral	6170 ± 390	6820 ± 850	6350 ± 470
	Ventral[f]	Dorsal	4890 ± 410	5130 ± 290	5470 ± 410
		Ventral	2720 ± 290	4420 ± 350*	4430 ± 280*

* Significantly different from controls, $P < 0.01$.
[a] 25 units ml^{-1} in PBS pH 6.5, 15 min, 37 °C; controls treated with PBS.
[b] 25 units ml^{-1} in PBS pH 6.5, 15 min, 37 °C; controls treated with PBS.
[c] 8 day, 1.0×10^6 cells ml^{-1}, 0.23 c.p.m. cell^{-1}, 3 h preincubation.
[d] 8 day, 0.9×10^6 cells ml^{-1}, 0.19 c.p.m. cell^{-1}, 3 h preincubation.
[e] 7 day, 1.0×10^6 cells ml^{-1}, 0.29 c.p.m. cell^{-1}, 4 h preincubation.
[f] 7 day, 1.1×10^6 cells ml^{-1}, 0.21 c.p.m. cell^{-1}, 4 h preincubation.

To determine if GM_2 could be recognized by molecules present on the surface of the tectum, GM_2 was incorporated into labelled lecithin vesicles by co-sonicating GM_2 and labelled lecithin. These vesicles were then used in the collection assay in place of the labelled suspension of retinal cells (Marchase, 1977). Results of this experiment are shown in Table 4. Compared to vesicles with no glycolipid incorporated or vesicles with glycolipids other than GM_2 incorporated in them, the GM_2 vesicles were bound at a greater rate to ventral tectal halves than to dorsal ones. Preliminary experiments using GM_2 labelled with a galactose oxidase/NaB[^{3}H]$_4$ treatment and used in the assay in place of the retinal cells show that labelled GM_2 associated preferentially with the ventral tectal halves. These results are consistent with the hypothesis that a molecule concentrated more on ventral tectal surfaces can recognize and bind GM_2 or a carbohydrate sequence similar to the one present in GM_2. Attempts are in progress to generate an antiserum against GM_2 that can be labelled and used to determine directly if a gradient of GM_2 does exist on the surfaces of the retina and tectum.

GM_1 synthetase, the enzyme that recognizes GM_2 and adds galactose to the terminal N-acetylgalactosamine is a glycosyltransferase. Enzymes of this class have been localized on the surfaces of many cell types and have

Table 4. *The adhesion of lecithin vesicles to tectal halves in 1 h (from Marchase, 1977)*

Vesicle composition[a]	c.p.m. [^{3}H]lecithin adhering/tectal half	
	to dorsal half-tecta	to ventral half-tecta
Lecithin	440 ± 30	450 ± 20
Lecithin + 2 % GM$_2$	$680 \pm 60*$	$780 \pm 30*\dagger$
Lecithin + 2 % GM$_3$	410 ± 30	430 ± 20
Lecithin (GM$_2$ added after sonication)	440 ± 30	450 ± 30
Lecithin	560 ± 30	590 ± 40
Lecithin + 2 % GM$_2$	630 ± 20	$860 \pm 60\dagger*$
Lecithin + 2 % cer-glucose	520 ± 10	550 ± 20
Lecithin + 2 % GD$_{1b}$	520 ± 40	540 ± 30
Lecithin	420 ± 20	430 ± 40
Lecithin + 2 % GM$_2$	$580 \pm 20*$	$670 \pm 30*\dagger$
Lecithin + 2 % cer-lactose	400 ± 20	390 ± 30
Lecithin + 2 % GM$_1$	400 ± 10	420 ± 20

* Significantly greater than the value with lecithin alone, $P < 0.05$.
† Significantly greater than the corresponding dorsal value, $P < 0.05$.
[a] 200000 c.p.m. and 4 mg lecithin in 4 ml PBS per assay plate.

been postulated to play direct roles in morphogenesis (Shur, 1977), sperm–egg interactions (Durr, Shur & Roth, 1977), and other systems of intercellular recognition (Shur & Roth, 1975). Thus, GM$_1$ synthetase is a logical candidate for the ventrally localized protein. Accordingly, this enzyme should show a higher specific activity on ventral cells than on dorsal ones. As a first attempt to test this prediction, GM$_1$ synthetase specific activity was measured in sonicates of dorsal and ventral neural retinal cells in the presence of detergent, under optimal conditions, and as a function of retinal age (Marchase, 1977).

The results in Fig. 4 show that after day 6 of development, the specific activity of GM$_1$ synthetase increases and is greater ventrally than dorsally. Another galactosyltransferase, N-acetylglucosamine:UDP-galactosyltransferase was assayed concurrently and the results show equal specific activities of this enzyme in dorsal and ventral sonicates. Interestingly, the point at which the increase and asymmetry of GM$_1$ synthetase specific activity appears, day 6, is the same point at which ventral retinal cells begin to show an adhesive specificity for their physiologically matching tectal halves in the collection assay (Fig. 2). Preliminary experiments show that GM$_1$ synthetase is localized on the surfaces of 9-day neural retinal cells. Experiments are in progress to determine if there is indeed an unequal distribution of the enzyme on dorsal and ventral cell surfaces.

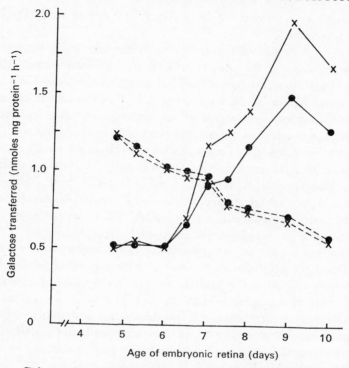

Fig. 4. Galactosyltransferase activity in sonicates of dorsal (●) and ventral (X) halves of neural retina as a function of embryonic age. ———, GM_1 synthetase activity; — —, N-acetylglucosamine:UDP-galactosyltransferase activity. Reproduced from Marchase, 1977 (by kind permission of Dr Marchase and Rockefeller University Press).

DISCUSSION

Recently, McClay, Gooding & Fransen (1977) have studied the adhesion of dorsal and ventral neural retinal cells to dorsal and ventral aggregates of tectal cells. These studies showed that when the rates of collection of 10-day dorsal retinal cells to dorsal and ventral tectal aggregates were measured, the dorsal cells preferentially adhered to ventral aggregates. The suspension of retinal cells was prepared by trypsin dissociation and their preferential adhesion to ventral aggregates was affected little by first allowing the retinal cells to recover for 5 h. Ten-day ventral retinal cells, in contrast, showed no specificity for dorsal and ventral tectal aggregates until they were allowed to recover for 5 h. After recovery, the ventral cells adhered preferentially to dorsal tectal aggregates.

These results are thus similar to those obtained by Barbera, 1975, using tectal halves instead of tectal aggregates and are consistent with the model of retinotectal adhesive specificity shown in Fig. 4. Both these studies suggest that an in-vitro assay can be successfully used to study in-vivo intercellular recognition.

Although the projection of the retina upon the tectum is extremely complex, the anatomy of the projection is fairly well understood. This complexity was used to advantage in the biochemical studies of the in-vitro assay using tectal halves by Marchase, 1977, in that the effects of each enzyme perturbation could be measured on four separate adhesive rates. Thus an asymmetric effect of an enzyme treatment could be noted. The biochemical studies suggest that a specific oligosaccharide sequence plays a role in intercellular recognition. Tay–Sachs ganglioside, GM_2, has such a sequence. Experiments with GM_2:lecithin vesicles and [^{3}H]GM_2 suggest that GM_2 can be recognized and bound by molecules on the surfaces of tecta and more concentrated ventrally. These results suggest a function for a particular ganglioside in intercellular recognition.

No function was known for any ganglioside until recently a number of bacterial toxins have been shown to interact with a number of gangliosides. Best characterized is the binding of ganglioside GM_1 by cholera toxin. Since then, five trophic hormones, thyroid stimulating hormone, lutenizing hormone, human chorionic gonadotropin, and follicle stimulating hormone, have been shown to interact with particular gangliosides and display some amino-acid sequence analogies among themselves and cholera toxin (reviewed in Fishman & Brady, 1976).

The structure of glycolipids in general would make them logical candidates for molecules involved in cell–cell recognition. The fatty-acid portion of these molecules has been shown to be inserted in the lipid bilayer of membranes while the oligosaccharide portion is oriented outside of the bilayer (Clowes, Cherry & Chapman, 1972). If a glycolipid is associated with the cell surface, then its oligosaccharides would naturally be available for interaction with other cell surfaces or circulating molecules such as the trophic hormones.

GM_1 synthetase fulfils the criteria for the ventrally localized molecule. This enzyme is present on neural retinal cell surfaces and, in cell sonicates, shows a higher activity ventrally after day 6 of development. These results provide some indirect evidence for the proposed model for retinotectal adhesive specificity.

Confirmation of roles for GM_2 and GM_1 synthetase, or any specific molecules, in the in-vivo retinotectal recognition will only occur when the recognition can be perturbed by the in-vivo addition of either the molecules themselves, their inhibitors, or antisera directed against them.

REFERENCES

BARBERA, A. J. (1975). Adhesive recognition between developing retinal cells and the optic tecta of the chick embryo. *Developmental Biology*, **46**, 167–91.

BARBERA, A. J., MARCHASE, R. B., & ROTH, S. (1973). Adhesive recognition and retinotectal specificity. *Proceedings of the National Academy of Sciences, U.S.A.*, **70**, 2482–6.

BRUNNGRABER, E. G., TETTAMANTI, G. & BERRA, B. (1976). Extraction and analysis of materials containing lipid-bound sialic acid. In *Glycolipid Methodology*, ed. L. A. Witting, pp. 159–86. The American Oil Chemists Society, Champaign, Illinois.

CLOWES, A. W., CHERRY, R. I. & CHAPMAN, D. (1972). Physical effects of tetanus toxin on model membranes containing gangliosides. *Journal of Cell Biology*, **67**, 49–57.

CROSSLAND, W. J., COWAN, W. M., ROGERS, L. A. & KELLEY, J. (1974). The specification of the retino-tectal projection in the chick. *Journal of Comparative Neurology*, **155**, 127–64.

DURR, R., SHUR, B. & ROTH, S. (1977). Surface sialyltransferases on mouse eggs and sperm. *Nature, London*, **265**, 547–8.

FISHMAN, P. H. & BRADY, R. O. (1976). Biosynthesis and function of gangliosides. *Science*, **194**, 906–15.

HAMDI, F. A. & WHITTERIDGE, D. (1954). The representation of the retina on the optic lobe of the pigeon. *Quarterly Journal of Experimental Physiology*, **39**, 111–19.

KAUFMAN, B., BASU, S. & ROSEMAN, S. (1967). Studies on the biosynthesis of gangliosides. In *Inborn Disorders of Sphingolipid Metabolism*, eds. B. W. Volk & S. M. Ironson, pp. 193–213. Pergamon Press, New York.

KLENK, E. (1939). Beitrage zur Chemi dur lipoidosen, Niemann-Picksche Krankheit and amaurotische Idiotie. *Zeitschrift für Physiologische Chemie*, **262**, 128–35.

KLENK, E. (1942). Uber die Ganglioside, eine neue Gruppe non zuckerhaltigen gehirnlipoiden. *Zeitschrift für Physiologische Chemie*, **273**, 76–85.

McCLAY, D. R., GOODING, L. R. & FRANSEN, M. E. (1977). A requirement for trypsin-sensitive cell-surface components for cell–cell interactions of embryonic neural retina cells. *Journal of Cell Biology*, **75**, 56–66.

McGILL, J. I., POWELL, T. P. S. & COWAN, M. W. (1966). The retinal representation upon the optic tectum and isthmo-optic nucleus in the pigeon. *Journal of Anatomy*, **100**, 5–33.

MARCHASE, R. B. (1977). Biochemical investigations of reitnotectal adhesive specificity. *Journal of Cell Biology*, **75**, 237–57.

MARCHASE, R. B., VOSBECK, K. & ROTH, S. (1976). Intercellular adhesive specificity. *Biochimica et Biophysica Acta*, **457**, 385–416.

MATTHEY R. (1925). Recuperation de la vue apres resection des nerfs optiques chez le triton. *Comptes Rendus des Séances de la Société de Biologie*, **93**, 904–6.

SHUR, B. (1977). Temporally and spatially specific cell surface glycosyltransferase activities in gastrulating chick embryos. *Developmental Biology*, **58**, 23–39.

SHUR, B. & ROTH, S. (1975). Cell surface glycosyltransferases. *Biochimica et Biophysica Acta*, **415**, 473–512.

SPERRY, R. W. (1945). Restoration of vision after crossing of optic nerves and after contralateral transposition of the eye. *Journal of Neurophysiology*, **8**, 15–28.

SPERRY, R. W. (1963). Chemoaffinity in the orderly growth of nerve fiber patterns and connections. *Proceedings of the National Academy of Sciences, U.S.A.*, **50**, 703–10.

Svennerholm, L. (1962). The chemical structure of normal human brain and Tai–Sachs gangliosides, *Biochemical and Biophysical Research Communications*, 9, 436–442.

Svennerholm, L. (1963). Chromatographic separation of human brain gangliosides. *Journal of Neurochemistry*, 10, 613–23.

These studies have been sponsored by a research grant and career development award to S.R. from the National Institute of Child Health and Human Development. This is publication 955 from the McCullum–Pratt Institute.

MORPHOGENETICS AS AN ALTERNATIVE TO CHEMOSPECIFICITY IN THE FORMATION OF NERVE CONNECTIONS

A REVIEW OF LITERATURE, BEFORE 1978, CONCERNING THE CONTROL OF GROWTH OF REGENERATING OPTIC NERVE FIBRES TO SPECIFIC LOCATIONS IN THE OPTIC TECTUM AND A NEW INTERPRETATION BASED ON CONTACT GUIDANCE

BY T. J. HORDER AND K. A. C. MARTIN

Department of Human Anatomy, University of Oxford,
South Parks Road, Oxford OX1 3QX, U.K.

INTRODUCTION

Among various attempts that have been made to elucidate the nature of mechanisms controlling the formation of nerve connections those concerned with the projection of the retina to the midbrain optic tectum occupy a unique and exposed position. The problem that arises is that of explaining how each member of a large population of incoming optic nerve fibres forms terminal connections at separate and, in each case, highly characteristic sites in the tectum. Since 1942 when R. W. Sperry first drew attention to this problem, the retinotectal system has remained the most intensively studied and most adequately understood instance of discriminative interactions between cells seemingly explicable only in terms of specific cell recognition mechanisms. Our views concerning this system, which is perhaps unmatched in its complexity and precision elsewhere in biology, must strongly influence our attitudes to the potentialities of cell recognition in other developmental situations.

One of the unusual features of this field has been the stranglehold of a concept which was formulated on limited evidence in the very first papers to consider the problem. The concept was, as is still the case, based purely on inference. Despite repeated disproofs of the validity of the concept in its original and later modified forms, no real alternative has been proposed. Because of the dominance of the concept of chemospecificity it is unavoidable that this review will appear to be a critique of it. However, this

There are two systems of references in this paper: a name–date system, as in the rest of the volume; and a numerical system.

The numerical system is used to cite *groups* of references. Throughout the text there are bold numbers, e.g. (**22**), these refer the reader to those references identified by this number in the right-hand margin of the reference list. A key to the groups of references is given on pp. 331–2.

is not our primary purpose, which is to consider the question of what an alternative model would be like.

Sperry argued that growing optic fibres are actively guided to their specific termination sites since they reach them under conditions in which they had been given the opportunity to innervate quite different sites. He proposed that termination sites must be distinctive by virtue of unique, presumably chemical, differentiation which he termed '*chemospecificity*'. We will use this term rather than the alternative, less committed, '*neural specificity*' because it avoids possible confusion in the use of the word 'specificity' which can refer to exclusivity of connections (in this sense termed '*locus specificity*' by Hunt & Jacobson, 1974*a*) as well as to the supposed underlying guidance cues. Clearly the greatest difficulty with the chemospecificity hypothesis is the great demands that it makes on the ability of otherwise homogeneous tissues, such as the retina or the tectum, to generate high degrees of regional differentiation in, moreover, exactly matching fashion, despite their independent embryonic development. Does such a model not exceed the limits on possible differentiated cell states and on the machinery in cells enabling them to recognize such states set by the very restricted number of genetic instructions available to supervise embryonic development?

Biologists may have allowed themselves to believe that they are disqualified from passing judgement on this field partly, one suspects, due to their impression that it hinges on 'difficult' neurophysiological techniques. This impression we hasten to dispel in the first section of this review. The interpretation and reliability of electrophysiological methods for mapping the arrangement of optic fibres in the tectum is not an issue since the method is simple and well founded. This is a subject in which speculation and model building have often appeared to stand in for hard evidence. During the slow advance of research, assumptions that were made in formulating early concepts have been forgotten, with very serious consequences for later interpretations of evidence. The second part of the review is devoted to a consideration of the adequacy of the evidence for and against chemospecificity. In tracing, in roughly historical order, the development of modifications to the original concept, an attempt is made to show how interpretations were influenced by the assumptions made. A primary aim is to assist biologists outside this immediate field to catch up with arguments based on several strata of earlier assumptions no longer explicitly stated.

In a situation where the techniques used are unable to validate a hypothesis directly, a crucial safeguard against over-interpretation is to take into consideration *all* the data available. This in particular is true in the present field where, without exception, single results, which were often so unexpected that workers have gone to great interpretative lengths to dispose of them, have always been confirmed eventually and frequently served to trigger the next model. This is one justification for a new review of the subject. For the first time a complete coverage of the literature is being attempted. All references, including reviews and abstracts but not theses, available to us at the end of 1977 are listed. In the text each class of result which features in more than two or three papers is referred to collectively by a number with which the relevant individual papers are classified in the reference list. Thus all previous review articles specifically concerned with this subject, omitting articles and books of more general coverage, are given the number (1) in the right-hand margin of the references. Theoretical papers are designated (2).

This review is solely concerned with the question of the factors which control the growth of optic axons to particular places in the optic tectum: only references strictly concerned with this question are automatically included, along with references taken from related areas which are not themselves covered exhaustively. The topics for which we have chosen to give a complete list of references are detailed on pp. 331–2 in the key to the numerical classification. The review will be primarily concerned with evidence from the study of *regeneration* of optic fibres in lower vertebrates because this is the major source of our information. For the same reason, unless otherwise stated, the evidence will be based on experiments on goldfish. Data from all species including mammals will for the most part be lumped together since, gratifyingly, they usually complement each other. For reasons that will be discussed later the evidence from embryonic development of the visual system and biochemical or cell biological studies of differentiation within the retinotectal system are of uncertain relevance to the central issue of this review. No attempt has been made to cover these fundamental fields definitively. However, at the end of the review we will take up the extremely important question of the relevance of regeneration studies to normal embryonic development, which it must be our ultimate aim to understand.

In Part 3 we will introduce an alternative to the chemospecificity hypothesis. Regardless of whether this approach proves of any lasting merit, it serves to demonstrate in concrete terms what assumptions underlie other present-day interpretations and to pinpoint the criteria which must be met by any future hypotheses. The approach presented here is inspired by the conviction that biological mechanisms are more likely to be simple than complex. It has the positive virtues that it does not go beyond the available evidence and can be very easily disconfirmed.

PART 1. TECHNIQUES

A brief review of techniques used to demonstrate the pattern of termination of optic fibres in the tectum: their consistency, adequacy and simplicity

Electrophysiological mapping of the retinotectal projection has long been the standard and preferred method for assessing the pattern in which the axons from particular retinal ganglion cells arrive in the optic tectum. Electrodes inserted in the superficial layers at a chosen site on the tectum record nerve impulses evoked by visual stimuli in single, restricted locations in the visual field. The characteristics of electrodes do not need to be closely defined; platinum-tipped, glass-coated or resin-insulated, Woods metal or tungsten electrodes have all been used, but fluid-filled glass microelectrodes appear to select different, post-synaptic, responses. Early concern as to whether these responses originate from the pre-synaptic terminal arborizations of optic axons or post-synaptically from the tectal cells supplied by the fibres, may explain occasional lingering suspicions regarding the adequacy of the method for mapping the projection of the optic nerve. The early uncertainties, though never justified given considerable circumstantial evidence supporting their pre-synaptic origin (Maturana, Lettvin, McCulloch & Pitts, 1960; Gaze 1970; Gaze & Sharma, 1970), have since been further resolved (Chung, Bliss & Keating, 1974; George & Marks, 1974). No single known criterion is utterly decisive but many criteria agree while none disagree. Particularly persuasive is the evidence that the electrical responses originate from structures whose shape and depth distribution correspond with the known anatomy of classes of optic fibre terminal arborizations. Even if responses were post-synaptic, their evocation by stimulation of small regions of the retina would best be explained as the direct reflection of the particular selection of fibres supplying the cell. Such selectivity of response in the face of non-patterned distributions of optic fibres would itself require elaborate explanation in terms of highly selective patterns of inhibitory connections.

Anatomical studies in normal animals of various species directly confirm that the electrophysiological maps correspond to the arrangement of fibre terminals; compare Clarke & Whitteridge (1976) with McGill, Powell & Cowan (1966) for birds, Gaze & Jacobson (1963) with Lázár (1971) for amphibians, and Jacobson & Gaze (1964) with Attardi & Sperry (1963) for fish. Even in animals with experimentally modified projections, anatomical and electrophysiological assessments of fibre distributions agree quite closely (Cook & Horder, 1977). Compared to electrophysiological methods for mapping the retinotectal projection, anatomical methods suffer from the disadvantage that the animal must be lesioned and killed and that only a single gross fibre subpopulation can generally be mapped in a single animal. These methods involve tracing of degeneration after selected retinal lesions or tracing the distribution of surviving axons by conventional silver stains and, more recently, by transport of radioactively labelled amino-acid precursor (e.g. proline) or enzymes (e.g. horseradish peroxidase). Mapping deficiencies that result from section of selected fibre bundles prove that the visible fan-shaped array of fascicles on the normal goldfish tectum (Sperry & Hibbard, 1968; Horder, 1974b; Cook & Horder 1977) is an accurate indication of the orderly arrangement in which optic fibres enter and cross the tectum (Scalia & Fite, 1974).

Electrophysiological mapping of the positions in the visual field from which a regular sequence of recording positions can be excited therefore gives a rapid and realistic summary of where the axons of ganglion cells so stimulated terminate in the tectum. In normal animals fibres terminate in a highly orderly and predictable arrangement which corresponds with the mirror-image inversion (Bunt & Horder, 1977) of the spatial arrangement of their cells of origin across the retina, an arrangement which will be referred to as 'retinotopic'. Strictly speaking maps obtained in this way are 'visuotectal' but, allowing for the reversal of the image falling on the retina due to the lens, this corresponds directly to the positions of ganglion cells mapped, and in this review all maps will be referred to as 'retino-tectal'. Depth of termination in the tectum is not considered.

Conventional maps do not include any estimate for errors of measurement. The centres of the receptive fields of recorded responses are plotted. An ideally unambiguous presentation of a retinotectal projection would be one which showed only those populations of ganglion cells which exclusively and uniquely innervate the tectal points under consideration. Given the spread of terminal arborizations of optic fibres, intermediate recording points would show overlapping fields. The spacing of electrodes would have to be just greater than the extent of the average arborization to ensure that exclusive ganglion cell populations are mapped and then only if their receptive fields can be separated without overlap. In practice multiple optic fibres are recorded from simultaneously at each tectal point. If these were not perfectly consistent in their retinal origins spacing of recording points would have to be further increased. However the practice of mapping the centres of multiunit receptive fields allows further, more detailed information to be extracted. The central point of a multiunit receptive field does not correspond to any real anatomical feature of organization but it does, statistically, sum up a property intrinsic to the whole population of units. Although it would be a hazardous task to use these data to infer the amount of 'information' inherent in the degree of non-randomness of the projection, the method does reveal real retinotopic organization. So long as sequentially recorded points, mapped 'blind', show consistent shifts in receptive field centres, real order must exist. In such a situation it is relevant that the reliability of plotting of each individual point is being constantly verified by its consistency with later mapped points, where these all fall into a coherent pattern.

Deficiencies in the optics of the eye or the fact that mapping is done not in water, but in air, could only blur distinctions between receptive fields. The features of such maps that we are usually interested in, particularly retinotopic orderliness, linearity of map and its orientation on the tectum, cannot be affected by known optic errors. When highly irregular maps are obtained this could reflect real patterns of misconnection by fibres or experimental errors which could only be eliminated by carefully repeated mapping. However, the claim that a projection is completely random can be verified by measurement of multiunit receptive fields; these should be greatly enlarged if fibres from a random scatter of retinal areas are terminating near each recording point. Possible enlargement of fields due to optical distortions can be eliminated by checking the normality of the fields of individual fibres. Limitations of the technique do however arise from the fact that the method does not directly trace the pattern of projection of chosen ganglion cell populations but, since it involves sampling of recording positions at the tectum, can only survey ganglion cells which happened to form terminations at recording sites. The sampling procedure leaves open the possibility that fibre populations compressed into small tectal areas will be missed, with the result that in the projection map the relevant retinal area will appear not to be represented in the map. The absence of a part of the field can never be taken to mean that it is not represented in the tectal map. The method makes it very difficult to locate or follow any aberrant projection from a given retinal area. It is also difficult with this method to establish whether a single retinal area simultaneously projects to several tectal regions unless each projection forms part of separate internally consistent retinotopic mappings.

PART 2. EVIDENCE FOR AND AGAINST CHEMOSPECIFICITY AND ITS VARIOUS FORMULATIONS

NO SPECIFICITY WITHOUT SELECTIVITY: A CLASSIFICATION OF EXPERIMENTAL DESIGNS FOR DEMONSTRATING SELECTIVE FIBRE GUIDANCE

Sperry's arguments for the existence of chemospecific guidance cues rest entirely on the availability of experimental techniques which provide particular optic fibres with potentially equal opportunities to reach and innervate alternative termination sites. Only in so far as the fibres demonstrate selection between the alternatives must chemospecific distinctions be invoked as the basis for the choice. The following approaches have been taken in demonstrating the choices open to fibres.

(i) *If optic fibres are randomly arranged in the normal optic nerve or become so during growth across the site of nerve transection, then during regeneration fibres will have random access to all possible terminal sites*

Following the then contemporary evidence of Herrick (1941 *a, b, c*, 1942) which failed to show that optic fibres from various retinal origins are topographically arranged in the optic nerve of urodeles and fish, and similar work by Lashley (1934) in the rat, Sperry (1943 *a*) could argue that if the nerve was cut, fibres regenerating centrally would have to be actively guided in order to reach orderly specific termination sites. However this particular argument could not be invoked for species in which the evidence (3) then (Dean & Usher, 1896; Ströer, 1939, 1940 *a, b*) and more recently (Polyak, 1957; Horder, 1974 *a*; Roth, 1974; Bunt & Horder, 1977) showed that fibres maintain a highly retinotopic arrangement throughout the optic pathway. An entirely independent approach can still be applied to all species. Based on the appearance of regenerated fibres as seen in silver-stained sections of the site of the original nerve transection, of which a few pictures have been published (4), it would appear that scarring has resulted in random disarrangement of centrally growing fibres. Regenerating fibres in general were thought to grow haphazardly (Sperry, 1945 *a*; Weiss, 1955).

This evidence is of crucial significance because it forms the basis for the following argument. Individual regenerating fibres, being initially channelled towards any of a large number of possible termination sites, must be individually guided to specific appropriate sites. Chemospecificity must be a property of individual fibres and must be capable of allowing them to make highly accurate choices between many alternatives.

In so far as retinotectal projections are regenerated with organizations remarkably close to that of normal maps (5) this is strong evidence for high levels of chemospecific differentiation.

(ii) *Active selection by fibres of their appropriate termination sites can be tested by rotation or displacement of the incoming array of fibres with respect to their targets which will give regenerating fibres most ready access to inappropriate central tissue*

(a) *Optic nerve and eye rotation*

In Sperry's original studies (Sperry, 1942, 1943 a, b) the assessment of the nature of regenerated optic nerve connections relied on the technique of eye rotation (6) and observation of the behavioural consequences. His finding that the visual strike responses were systematically reversed was consistent with the interpretation that the original anatomical connections had been reformed and the maladaptive behavioural results incidentally made it possible to exclude learning as a determinant of visual localization behaviour (Sperry, 1943 b). Behavioural criteria (Székely, 1954 a, b, 1957; Sperry, 1945 b, 1948 a; Hibbard, 1967; Misantone & Stelzner, 1974 b) are poorly suited for assessing regionally specific tectal connections: they depend on motor connections and other visual centres. For this reason results employing this assay will not be considered further in detail.

As we now know from electrophysiological mapping of projections from rotated eyes (Jacobson, 1968 b) this inference by Sperry was entirely justified. Moreover, this technique also introduced a test for the existence of mechanisms for the selective growth of fibres which is quite independent of the question of random fibre access. Rotation of the nerve stump attached to the eye effectively brings regenerating fibres into proximity with the central channels previously occupied by fibres travelling to quite different parts of the tectum and yet they connect correctly.

Active re-orientation of the fibres must be invoked to account for the fact that fibres did not simply follow these nearest but inappropriate channels which would have led them into a rotated pattern of termination in the tectum. Clearly it is essential, if this reasoning is to be applicable, to ensure that the peripheral and central nerve stumps have been, and remain, relatively rotated. Although this does not necessarily obtain after eye rotation *in situ*, and although it has never been explicitly checked, it seems likely that this must be true in experiments involving eye removal and re-implantation (Stone, 1944; Sperry, 1945 b; Székely, 1954 b; Jacobson, 1968 b). Rotation of parts of the eye effectively achieves the same thing (Székely, 1954 a; Burgen & Grafstein, 1962; Gaze, Jacobson & Székely 1963; Hunt & Jacobson, 1974 a; Levine & Cronly-Dillon, 1974).

An extension of these approaches is the surgical deflection of selected bundles of optic fibres towards chosen tectal regions (Arora, 1963; De Long & Coulombre, 1967; (7)) a technique most easily done by separating out fibre fascicles as they cross the tectum itself.

(b) Innervation of the tectum by the ipsilateral eye

The same logic applies to experiments involving the bilateral exchange of eyes (Stone, 1944; Sperry, 1945b, (8)) or deflection of the optic tract into the tectum on the same side as the innervating eye (Sperry, 1945b, (9)). Tract deflection can be achieved at the optic chiasma (Sperry, 1945b, (10)) – 'uncrossing the chiasma' – or by deflection of fibres across the midline at points along the optic tract or at the level of the tectum (Arora, 1966, (7)). Since the projections to the two tecta are identical in every respect but for their symmetrical, and therefore non-congruent, arrangement with respect to the midline of the animal, it is hardly surprising that, so far as can be judged from the available evidence, fibres from the ipsilateral eye establish a projection as readily as the normal contralateral eye, and according to the same rules: namely nasal retina projecting to caudal tectum, ventral retina to medial tectum and so on. While the organization of the projections to the two tecta are homologous rostro-caudally their mirror-image disposition across the midline means that fibres deflected ipsilaterally arrive with inappropriate organization in the medio-lateral axis. Therefore, in so far as they successfully establish a projection which conforms to the normal organization of the host tectum, fibres must have been actively re-arranged with respect to positions in this axis and this re-arrangement has occurred independently of order in the other axis.

(c) Tectal graft rotation

The same effects can be achieved if, instead of rotating the array of incoming optic fibres, the target tissues themselves are rotated (11, 12). For example, the tectum itself has been rotated in situ either in part or as a whole (Crelin, 1952; Wiemer, 1955; Sharma, 1969; Sharma & Gaze, 1971; (11)). One axis can be rotated independently of the other by exchange of tissue between the two tecta (Jacobson & Levine, 1975a, b; Sharma, 1975a) or by inversion of the tissue (Yoon, 1975d).

It is important to note that tests involving eye, nerve, tract or tectum rotation entail rotation of tectal tissue with respect to the fibre array as a whole. Unless the operative procedures can be shown independently to cause random scrambling of regenerating fibres then the only form of chemospecific guidance that they demonstrate is that which can control *orientation* of the array. The retinotopic arrangement of the members of the fibre population may have been a property already intrinsic to the fibre array and may have been retained during its re-orientation.

(d) Tectal graft translocation

Graft translocation (**12**), in which a graft taken from one position in the tectum is implanted in a non-corresponding position in the same tectum or the tectum of the other side (Jacobson & Levine, 1975a, b; Sharma, 1975a; Hope, Hammond & Gaze, 1976; Martin, 1978) tests a further aspect of tectal differentiation, one which may itself be the basis for the ability of the tectum to control orientation of fibre populations. It potentially tests the accuracy with which portions of tectal tissue can select their appropriate fibres, irrespective of fibres in the remaining population. Where half-tecta are involved 'double rostral' or 'double caudal' 'compound tecta' are formed (Jacobson & Levine, 1975b; Sharma, 1975a).

(iii) *Depriving certain tectal areas of their fibre supplies may give other fibres more ready access to alternative termination sites*

In 1963 Attardi & Sperry published the paradigm of a class of what have since come to be known as '*size disparity*' experiments. After ablation of parts of the retina (**13**) and in spite of the consequential emptying of various parts of the tectum of fibres, remaining regenerating fibres were shown to be accurately confined to their normal tectal areas. Arrival of partial sets of regenerating fibres can also be achieved by preventing arrival of the remainder by incomplete optic nerve section (Jacobson & Gaze, 1965). That fibres had actively avoided terminating in inappropriate tectal areas can be asserted, particularly with respect to fibres which had grown through unoccupied territory in order to reach their own termination sites. A comparable approach was introduced by Székely (1954a) and Gaze, Jacobson & Székely (1963) involving the formation in *Xenopus* embryos of 'compound eyes' (**14**) composed of two nasal or temporal half-retinae. Although of considerable influence on thinking, this approach has not proved of analytical significance because of numerous uncontrolled variables affecting interpretation of results.

The complementary technique of partial tectal ablation (**15**) was introduced by Jacobson & Gaze (1965), again with results which in no way departed from expectation – original connections were restored during regeneration and inappropriate fibres were not represented.

FIRST EVIDENCE INCONSISTENT WITH
CHEMOSPECIFICITY: PLASTICITY

Although the possibility that fibres could form connections with inappropriate tectal areas had been raised earlier in the context of experiments
with compound eyes, the first clear evidence came from goldfish in which
caudal tectum had been ablated (Gaze & Sharma, 1968, 1970). It was
discovered that fibres initially deprived of their own tectal areas came to be
represented on the surviving tectal fragment with the result that a 'compressed' complete retinotopic projection was formed (**16, 17**). If this
technique is combined with partial retinal ablation it is possible to obtain
complete *transposition* (**19**) of the projection from half the retina into
non-corresponding tectum (Roth, 1967, 1972; Horder, 1971*a*; Yoon,
1972*c, f*). A further aspect of the readiness with which fibres can innervate
sites other than those to which they are normally confined was revealed in
the demonstration (Horder, 1971*a*) of *'expansion'* (**20**). Here the projection from a portion of the retinal array comes to occupy the whole tectal
surface in retinotopic order. Taking together the results of a growing
variety of experimental manipulations of this kind, it can be concluded
that any given fibre is capable of forming terminations at virtually any site
in the tectum.

It is worth emphasizing that there can be no doubt that these phenomena
involve true formation of fibre connection in entirely new anatomical
regions of the tectum. It cannot for example be argued that the halved
tectum has replaced the missing parts by regeneration with the result that a
fibre's original termination site has been restored. Although tectal regeneration can occur in goldfish (Kirsche & Kirsche, 1961) this only involves its marginal portions. It is likely (Stevenson, 1977) that after caudal
tectal removal, augmented mitosis is quite insufficient to replace the missing tissue. In any case the lesioned tectum visibly retains a constant form
and size. Compression can occur without tectal lesions (Freeman, 1977).
After partial retinal lesions regeneration, although it occurs (Schmidt,
Cicerone & Easter, 1978) can be excluded if parts of the visual field
are still not represented in the recorded maps. What is more, expansion
can be demonstrated without involving retinal lesions by supplying a tectum with restricted portions of the optic nerve supply from an intact eye
(Feldman, Keating & Gaze, 1975; Meyer, 1976*a*, Bunt, Horder & Martin,
1978).

CAN THE CHEMOSPECIFICITY MODEL BE MODIFIED TO
ACCOMMODATE INSTANCES IN WHICH FIBRES MAKE
CONNECTIONS AT SITES OTHER THAN THEIR NORMAL
TERMINATION SITES?

Since 1970 the pre-eminent concern in the field has been the need to
accommodate the mounting evidence that fibres are capable, in effect, of
forming terminations at any site in the tectum. On the face of it this is an
outright disproof of the chemospecificity hypothesis, but great efforts have
been made to modify the hypothesis so that it can incorporate the new
evidence. The problem was encapsulated at an early stage in the term
'*systems matching*' (Gaze & Keating, 1972) which, without implication as
to mechanisms, sums up the property whereby the available fibre array
tends to fill the available expanse of tectum. A more cautious term for the
property was '*contextuality*' (Hunt & Jacobson, 1974a) which left room for
the possibility that the system would behave differently according to
conditions.

A number of models have been proposed which attempt, by reference to
the form or modifiability of chemospecificity cues, to marry evidence and
theory. The concept of *regulation*, derived from studies of embryonic
systems, proposes that ablation of part of a tissue initiates a process of
reorganization of its pattern of differentiation resulting in re-instatement
of a complete pattern in remaining tissue. If this happens after removal of
half the tectum then, by the normal rules of connection formation, a complete projection will be the result (Sperry, 1965; Meyer & Sperry, 1974;
Yoon, 1976b). As an alternative to this model, which preserves intact the
'hard-line' chemospecificity view that termination at exclusive sites is the
direct reflection of exclusive affinity cues at those sites, it has been proposed
that specificity cues may exist in a form which allows for both normal
retinotectal mapping and for flexibility. Such models commonly take as a
starting point the idea that tectal cues exist in a *graded* form which would
automatically allow fibres to have degrees of affinity for sites other than
their own (Gaze, Jacobson & Székely, 1963; Gaze, 1970). Such a view has
an added attraction. The idea of graded tectal differences provides for
considerable simplification of the information required to be programmed
into the tectum. It happens that the concept of a gradient has also been
used as a convenient way to account for regulation. A source of a diffusible
chemical gradient could, if it were constantly renewable, restore a complete range of values across a tissue after part had been ablated. But a
graded form of cue could account for flexibility, while retaining fixed
values within the tissue, if, for example, fibres obey the general rule of

seeking to distribute themselves between available minima and maxima of the property concerned. An extreme version of this class of models is the 'arrow' model (Hope et al., 1976) which simplifies the tectal cues to a minimum such that they effectively only provide information about orientation.

However from their inception these modifications of the chemospecificity model could not claim to offer a complete solution to the problem. Given that the projection could display systems-matching properties so readily, for example after partial tectal ablations, situations such as those described by Attardi & Sperry (1963), where fibres were confined to their appropriate territories despite the opportunity and capability to form an expanded projection, were paradoxical. Something of a break-through was achieved as it became clear (Gaze & Sharma, 1970; Sharma, 1972b, c; Yoon, 1972a, e; Cook & Horder, 1974) that compression is the result of secondary gradual translocation of terminals, following return of fibres to their normal termination sites as the initial outcome of regeneration (21). That fibre terminals are capable of extensive, progressive translocation within the tectum (22) is a new and important generalization. Since compression also occurs among quite normal, intact optic terminals (17) this must now be recognized as a real potentiality of all optic terminals and one which is not incompatible with simultaneous, normal, functional synaptic contact with post-synaptic cells (Scott 1975, 1977; Arora & Grinnell, 1976; Scott & Meyer, 1976).

On the basis of this information it was now possible to explain Attardi & Sperry's findings as the result of early innervation by fibres of their appropriate termination sites prior to expected later expansion. However none of the models described above automatically predict such a sequence of events or explain why fibres should be confined to their appropriate places initially. The regulation model can be suitably adjusted by supposing that the regulation process is delayed in time. Yoon (1975a, 1976b) has extended this argument to the point of arguing that the progressive transposition of fibres across the tectum is a direct reflection of, and can be used as an assay for, the progress of regulation. From the point of view of adding support to the evidence for regulation this is a circular argument which does nothing to substantiate the model. The idea that regulation occurs autonomously as the result of changes following halving of the tectum is also disproved (Cook & Horder, 1974, 1977; Cook, 1978) by the finding that even after compression, when regulation would have been expected to have been complete, newly regenerating fibres again innervate their normal uncompressed sites in the first instance. The process is not autonomous because fibres on rostral tectum do not move in the absence of fibres normally destined for the ablated caudal tectum (Meyer 1975; Cook & Horder, 1977). Recently Meyer & Sperry (1976) have proposed what is essentially a hybrid model which accounts for compression into a rostral half-tectum as follows. On the basis of slight gradations in chemospecificity cues in the tectum, fibres which would normally project to a region just caudal to the lesion are able to form terminations at neighbouring sites on the caudal edge of the half-tectum. These fibres then cause a change in the tectal chemospecificity of their new location towards their own, more caudal, chemospecificity (a process of 'modulation') with the result that further increasingly caudal fibres can form terminations there and so on in domino fashion. Although this model contains elements of a gradient model it is ultimately equivalent to regulation models in general and it is disconfirmed by the same evidence.

It is evident that various models can be made to fit the data concerning the initial return of fibres to their normal tectal locations and their subsequent translocations, but only by the arbitrary addition of new provisions. Regulation models have to make arbitrary and complex assumptions concerning the dynamics of the regulation process. Gradient models cannot, without additional provisions, explain why fibres initially regenerate to their normal locations or why, having done so, they are able to revise their first preferences: as Prestige & Willshaw (1975) have emphasized, an additional parameter of competition between fibres for space must be invoked. The most serious problem with any theory which involves simplification of the guidance cues provided by the tectum in order to explain plasticity of connections – and this is particularly true of the arrow model – is that these cues can no longer explain the establishment of detailed retinotopic rank-order of fibres.

In the sections which follow we review further evidence concerning projections in which equivalent optic fibres (that is fibres originating from comparable retinal regions in the same eye or both eyes) project to two different places in the same tectum (24) or in which two different regions of the retina project simultaneously to the same region of the tectum (25). This evidence disproves at a stroke any model which refers fibre ordering to a single integrated system of ordering cues which is essentially what all chemospecificity models, by merely modifying Sperry's model, attempt to do.

RE-EXAMINATION OF AN EARLY ASSUMPTION CRUCIAL FOR CHEMOSPECIFICITY

Do regenerating optic fibres initially arrive at a random assortment of tectal locations?

Sperry's original solution to the problem of retinotopic ordering was simple and direct. The final orderliness which is reached in the organization of fibre terminals was immediately explicable because it was the result of cues distinguishing the termination sites themselves. A direct choice between termination sites themselves is the most reasonable way to explain selection if one starts out with a situation, as Sperry assumed one must, where regenerated fibres were initially randomly distributed among various potential termination sites. All subsequent modifications to the chemospecificity model, perhaps through unconscious carry-over of the original assumption of initial random regeneration, refer fibre ordering largely to events within the tectum. This may also be due in part to the methods used, which concentrate attention on fibre terminals and on the end result of regeneration while disregarding the means whereby fibres reach their terminal locations. The evidence for this assumption must now be re-examined. Certainly the fact that fibres from any given retinal region can come to terminate anywhere in the tectum cannot be taken to show initial access to this range of possible termination sites because, as we have argued, this could be the result of secondary translocations.

The first species in which the time course of events in the regeneration of the retinotectal map was examined was the frog (Jacobson, 1961a, b; Gaze & Jacobson, 1963). It was found that the earliest regenerated projections were either normal or of a form termed 'Pattern 1', which typically consisted of a representation of only a restricted part of the retina spread, in disorderly array, over abnormally wide tectal areas. That this projection involves extensive overlapping of fibres is confirmed by the enlargement of multiunit receptive fields (Gaze & Jacobson, 1963; Jacobson, 1966) which could not be attributed to any changes in the fields of individual optic fibres (Jacobson, 1966). This has since been confirmed in *Rana* and *Xenopus* (Gaze & Keating, 1970; Beazley, 1975a, b; 1977) and has been taken to mean that fibres initially extend haphazardly and widely throughout the tectum.

However, with regard to goldfish many reports (21, 22) testify to the fact that fibres are in retinotopic array from the earliest possible mapping stages during regeneration.

The limited degree to which fibres deviate from direct lines of growth towards their targets is further documented by anatomical observations of regenerating fibres in their approaches to the tectum (Attardi & Sperry, 1963; Roth, 1972; Horder, 1974a) and within the tectum (Attardi & Sperry, 1963; Westerman, 1965).

The significance of Pattern 2 in frogs, which is never seen in goldfish (Horder, 1971b), will remain in doubt until the following possible sources of confusion in interpretation have been resolved. In the one reported case in which an animal initially showing this projection was mapped again later (Gaze & Jacobson, 1963) many of the features of the earlier map remained even though added to by a superimposed anomalous commissural projection ('Pattern 4'). It thus remains a possibility that some errors are of a permanent nature, a conclusion which is reinforced by the fact that some abnormalities, referred to as 'Pattern 2' and consisting of fibre projections randomly displaced along the rostro-caudal axis of the tectum, remain at the latest mapping times (Gaze & Jacobson, 1963; Gaze & Keating, 1970; Beazley, 1975a). Ingle & Dudek (1977) and Udin (1977) also report highly disorderly projections in halved tecta in *Rana*. Beazley (1975a; 1977) has suggested that low density of fibres, which would apply during the early stages of regeneration, may promote extensive exploratory growth of fibres. It remains uncertain whether the enlarged multiunit receptive fields seen in Pattern 1 are due to enlarged terminal arborizations of fibres which had initially regenerated accurately to their normal termination sites (during embryonic development terminals are large and later contract (Lázár, 1973)) or to relatively imperfect selection of initial sites. In any event these irregularities may later be corrected by the gradual insertion of the full complement of arriving fibres and improved orderliness imposed as a result of whatever orderliness is intrinsic to the fibre population as a whole. The frog may only differ in degree from the goldfish, in which limited multiunit field enlargement is also seen initially (Horder, 1971b) as it is in other amphibians (Cronly-Dillon, 1968): in all cases fibres may arrive at the tectum with some retinotopic order established.

Implications for projection formation of the restricted patterns of access to tectal locations of normally regenerating fibres

The importance of the evidence that fibres arrive in the tectum in an arrangement which allows them to grow directly towards their normal termination sites lies in the fact that it calls into question the most direct evidence for chemospecificity in the tectum. If, simply as a result of a lesion of the optic nerve or tract, fibres are not given initial access to a variety of possible termination sites, then the fact that they reach one particular site cannot be taken as evidence that active selection between sites was involved. However, the evidence for chemospecificity does not rest entirely on this argument but also on a variety of rotation and exchange techniques as discussed in Part 2. We must now consider how these other arguments for chemospecificity are affected if fibres approach the tectum in orderly array.

Many aspects of the results of Attardi & Sperry (1963) now take on a new complexion. If, as Attardi & Sperry show, fibres approach the tectum in an orderly array and reach their targets by the most direct routes then few alternative termination sites will have been open to them. The restriction of fibres to the appropriate mediolateral region of the tectum can be attributed to their confinement to one of the fascicles of entering fibres (Sperry & Hibbard, 1968; Cook & Horder, 1977). There is no reason for fibres to spread beyond their normal terminal regions in a caudal direction; electrophysiological repetition of Attardi & Sperry's experiment (Horder, 1971 a) shows that in the short term they do not. Restrictions of regenerated fibres to their appropriate tectal areas reported in similar electrophysiological studies by Westerman (1965) and Schmidt (1978) may be explained in the same way. Only where retinal lesions have removed fibres occupying positions at intermediate points along the paths of surviving fibres might misconnections be expected during the actual course of initial regeneration. Active selection can only be invoked in cases where fibres had grown through and left vacant sites in emptied tectal areas on their way to their own sites. It is not clear whether previous experiments employing electrophysiological methods specifically and adequately examined this aspect, but Bunt, Horder & Martin (1978) did so and could detect no such vacant sites at the earliest times during regeneration at which responses could be recorded.

This new finding is not necessarily incompatible with previous evidence (Attardi & Sperry, 1963) if it is assumed that electrophysiological methods are capable of detecting sparse innervation which could easily have been overlooked anatomically. Springer, Heacock, Schmidt & Agranoff (1977) found that normal electrophysiological maps could be mediated by $c.$ 5 % of the normal optic fibre population. Only low innervation densities would be expected in the experiment of Bunt et al. (1978) since, as their maps show, fibres from central retina effectively become spread out to occupy non-corresponding peripheral tectum and therefore cover a larger area than usual. It is interesting that at all stages during regeneration in studies using proline labelling of projections from hemiretinae (Meyer, 1976a; Meyer & Sperry, 1976) and using Nauta and Bodian techniques (Roth, 1972), experimenters have found sparse innervation outside the tectal areas strictly appropriate to the fibres, particularly in rostral tectum after regeneration of fibres remaining after removal of the temporal hemiretina.

These particular misconnections, which were also seen by Attardi & Sperry (1963) who noted 'a few' stray fibres in rostral tectum, may reflect the fact that removing a temporal hemiretina creates vacant sites in rostral tectum *some* of which may be open to surviving fibres as they travel towards caudal tectum, although to an extent which will be limited by the fact that fibres enter the tectum at the media lrather than rostral edge. Such considerations may explain why expansion of projections from nasal hemiretinae have been detected earlier after regeneration than for temporal hemiretinae (Horder, 1971 a; Meyer, 1976 a; Meyer & Sperry, 1976; Cronly-Dillon, personal communication) though not by Yoon (1972 c, f) or Schmidt, Cicerone & Easter (1974). Meyer (1976 a) suggests that expansion of the projection from a temporal hemiretina is never complete anatomically.

FURTHER EVIDENCE FAILING TO SHOW DETAILED CHEMOSPECIFICITY IN THE TECTUM

The most direct tests of the capacity of tectum to influence fibre growth selectively, once constraints due to surrounding fibres and to restricted access have been removed, fail to reveal chemospecificity

The following experimental approach has been used in an attempt to ensure access of regenerating fibres to a variety of foreign tectal sites so that their preferences can be systematically explored. Fascicles of fibres are dissected free from one tectum and inserted at various locations into the second tectum which is otherwise intact except that, to eliminate the possible influence of other optic fibres, the normal optic innervation of the host tectum is removed. Fibres invariably first innervated the nearest tectum regardless of where they had been implanted (Bunt *et al.*, 1978), thus confirming that they occupy tectal sites on a 'first come, first served' basis. This technique, which in principle would be expected to leave individual fibres free to express any intrinsic preferences that they may have for any of the complete range of tectal sites open to them, failed to demonstrate any such preferences with regard to the location at which terminals were formed: responses could be obtained in extreme rostral tectum from fibres which would normally innervate caudal tissue and in extreme medial tectum from fibres normally destined for central tectum and so on. Maps contained many errors in detail, and a bias in orientation due to nasal retinal fibres terminating too far rostrally in lateral tectum, but even if fibres enter caudally temporal, retinal fibres occupy more rostromedial tectum. Meyer & Sperry (1976) report that transplanted fascicles do show a tendency to regenerate to an appropriate location but this only applied when the host tectum was itself innervated and Meyer (1976 a, b indicates that the accuracy may have been no greater than that fibres normally innervating the caudal half of the tectum invade this half, rather than rostral tectum, after being implanted rostrally. This tendency, also reported by Arora (1966) and Cronly-Dillon & Glaizner (1974), may reflect a non-specific caudalwards inclination of growing fibres: Meyer (1976 a) discusses the possibility that transplanted fibres are simply occupying regions within the resident projection whose fibres had been cut during fascicle implantation. When the tectum was uninnervated the projection was so expanded, according to the description of Meyer (1976 a), that it was impossible to detect any initial discrimination that fibres may have shown early in regeneration.

In the light of these findings it becomes possible to explain the following observations. Bunt et al., (1978), like Cunningham & Speas (1975), noted on a number of occasions that the organization of transplanted fibres mediolaterally was the reverse of the normal in that tectum. This presumably reflects the arrangement of fibres within the transplanted fascicles which, since they come from the opposite tectum, is reversed in this dimension. Imposition of the organization intrinsic to the fibre array, without regard to the pattern of terminations normally appropriate to the tectum, is presumably also the explanation for cases (26) in which normal projections have been seen after graft rotation or exchanges. Martin (1978) finds that after exchange of grafts between the two tecta, the mediolateral polarity of their innervation was inappropriate and such as could be explained in terms of the arrangement in which fibres reach them. Martin (1978) also finds that grafts taken from medial and lateral parts of dorsal tectum, when exchanged, again become inappropriately innervated by fibres expected to reach them according to their new positions in the fan of incoming fibres, unlike Yoon (1977 d) whose grafts, however, were unrotated.

Some tests indicate a degree of regional tectal differentiation but only require that it be of an order insufficient to account for retinotopic ordering of a projection

Complete retinotopic occupation of a rostral half-tectum entirely by non-corresponding fibres (Horder, 1971 a; Yoon, 1972f) would appear to disprove the existence of any differences between rostral and caudal tectum but here electrophysiological results may be deceptive. Roth (1967) and Meyer (1975, 1976 a) showed anatomically that extreme rostral tectum remains relatively sparsely innervated while most of the transposed fibres are concentrated towards the caudal part of the rostral fragment. Relatively incomplete innervation of the most rostral parts of the tectum in this situation could easily be missed during electrophysiological mapping because of the difficulty in reaching extreme rostral tectum with an electrode. Meyer (1977) reports that even in compression of a whole projection into a rostral half-tectum the projection can be shown to be non-linear, implying that differences may exist in rostro-caudal parts of the tectum which prevent them being equally readily innervated by any fibre.

There is direct evidence from another source which appears to strongly support a case for tectal differentiation. In contrast to the findings (Martin, 1978) when tectal grafts are exchanged in the mediolateral dimension, Hope *et al.* (1976), following earlier evidence from Jacobson & Levine (1975 *a*, *b*) and Sharma (1975 *a*), reported that small rostro-caudally exchanged grafts became innervated by appropriate fibres. This is surprising given the way in which fibres enter the tectum in orderly array. It is particularly difficult to see, for rostral grafts transplanted into caudal tectum, how the appropriate fibres were impelled or able to grow towards the caudal tectum which they would never normally reach. Martin (1978) has proposed that this is due to the influence of the rostrally placed graft from caudal tectum which in some way deflects these fibres caudally: in the absence of any surgical intervention in rostral tectum the caudally placed graft is innervated solely by fibres native to the site of the graft and therefore quite inappropriate. No such deflection has been detected simply as the result of implanting a graft (Gaze, Jacobson & Sharma, 1966; Levine & Jacobson, 1974) so it is likely that deflection is the result of 'rejection' of inappropriate fibres by the rostrally placed graft perhaps on the basis of its preferential occupation by appropriate fibres. Even in the original experiment involving rostro-caudal graft exchange the set of fibres innervating the caudally placed graft does not accurately correspond to its normal supply. It includes occasional, misplaced fibres appropriate to the surrounding tectum. Indeed the most appropriate fibres are commonly not represented anywhere in the recorded map. Thus innervation may be based not so much on chemospecificity of the graft but on the choice of fibres which happen to gain access to it.

In the case of caudal tectum placed rostrally the graft could presumably be reached by both local and appropriate fibres and its innervation by the latter appears to demonstrate a degree of real selectivity as also do the results, involving larger grafts, of Jacobson & Levine (1975 *b*) and the early stages of Sharma's (1975 *a*) results. However, the innervation of rostrally placed grafts (Hope *et al.*, 1976; Martin, 1978) was also not precisely appropriate. Projections to these grafts tended to include fibres that would normally innervate larger areas of the caudal tectum than the graft originally occupied and in general the results were variable between individual cases. Because grafts were in all cases taken from rostral and caudal halves of the tectum none of the results test for levels of tectal differentiation greater than the single distinction between the two halves. This, together with the lack of precision of the innervation of the grafts, justifies the conclusion that whatever chemospecific selectivity is operating is only of a low order of discrimination.

With respect to possible large-scale differentiation in the mediolateral dimension of the tectum the following is relevant. Arora & Sperry (1962) exchanged the medial and lateral divisions of the optic tract surgically at the rostral pole of the tectum and found that emerging regenerating fibres immediately grew back towards their appropriate tectal region rather than towards the foreign half of the tectum to which they had been deflected. Evidence of a similar selection between dorsal and ventrolateral halves of the tectum came from surgical deflection of fibres from the lateral tract division into dorsal tectum (Arora, 1963, 1966; Sperry, 1965). Horder (1974b) found that fibres which regenerate into the inappropriate tract division following optic nerve section can double back across the foreign half of the tectum to reach relatively normal termination sites. This evidence (30) for chemospecificity at the level of dorsal and ventrolateral halves of the tectum is not necessarily incompatible with previously reviewed evidence against chemospecificity on the smaller scale of mediolateral regions *within* the dorsal half. The accuracy with which fibres selected their appropriate termination sites in the studies of Arora (1963, 1966) and Arora & Sperry (1962) are unknown. The accuracy of termination of fibres in Horder's (1974b) study was in most cases difficult to assess, because of the few fibres involved. Most of the fibres that could be localized came from a region of the retina whose fibres travel through either of the two tracts and overlap in the tectum in the normal animal. One feature of the results of Bunt et al. (1978) may also reflect a degree of tectal differentiation. It was found that fibres from central retina formed an expanded projection occupying the entire peripheral to central extent of dorsal tectum. A bias towards the innervation of appropriate, central tectum may have to be invoked to explain this because without such a bias a simple transposition into vacant, peripheral tectal areas might have been anticipated. Such a bias might explain why Attardi & Sperry (1963) saw a concentration of fibres in central tectum.

Put briefly, the available evidence demonstrates that, under conditions in which fibres have been directed into inappropriate tectal regions deliberately and with certainty, the tectum can do little to redirect them to appropriate termination sites. A degree of tectal differentiation can be detected but in no case does it control the growth of a fibre with sufficient accuracy to explain the orderliness with which that fibre terminates in the tectum during normal regeneration. If higher degrees of differentiation, so far undetected by unambiguous tests, are to be of sufficent potency to explain why such a fibre normally innervates an exclusive and precisely-defined tectal location, then it is fair to ask why their influence should have been ignored in the experiments discussed above.

DETAILED RETINOTOPIC ORDERING IS A PROPERTY INTRINSIC TO FIBRES THEMSELVES INDEPENDENT OF THE TECTUM

In view of the lack of evidence for high levels of tectal chemospecificity and the likelihood that detailed fibre organization is a reflection of the pattern in which fibres arrive at the tectum, it seems certain that the tectum can only contribute to the most general aspects of fibre distribution. One is therefore led to conclude that detailed retinotopic ordering of fibres in the tectum is a property intrinsic to the fibre population itself. In such terms it is then easy to explain the freedom with which fibres can innervate foreign tectum and the maintenance of full retinotopic ordering during translocations such as occur during compression and expansion. This is essentially the position taken in the model of Cook & Horder (1974, 1977), an explicitly multifactorial approach in which an attempt is made to pinpoint the contribution of various factors by identifying them in concrete terms of components of the sequence of events in regeneration. Having proposed that interactions directly between fibres themselves are responsible for the maintenance of retinotopic order in the tectal projection, Cook & Horder (1977) explained the initial return of fibres to their correct terminations by *pathway guidance*. If chemospecific cues exist in the tissue of the pathways leading into the tectum, fibres could be gradually channelled towards the appropriate tectal region, even after rotation of the regenerating fibre array. By supposing that fibre–fibre interactions operate during this initial setting up of the fibre array it became possible to explain the high degree of orderliness existing among fibres by the time they arrive at the tectum, while it would then only be necessary to hypothesize extrinsic pathway guidance cues of limited number and precision, since the details of organization could be left to the fibres themselves. Only in terms of autonomy in the ordering of fibres without recourse to cues in the target tissue can one begin to explain situations of the kind reported by Yoon (1977a) demonstrating retinotopic order in an optic projection to cerebellar tissue implanted into the tectum.

Emphasis on the dominant role played in determining fibre distributions by early events in the pathways prior to fibres' arrival in the tectum also makes it possible to account for some of the more complex of the projections on record in the literature. As the result of 180° rotation of fragments of the neural tube in *Xenopus* embryos Chung & Cooke (1975) and Cooke (1977) were able to generate a tectal structure which, on the basis of its anatomical features and those of surrounding nuclei, contained at its caudal pole a second, rostro-caudally rotated, rostral tectal component. The formation of two overlapping and mutually reversed projections to this compound tectum from the contralateral eye could be explained if two sets of optic fibres had arrived at both 'rostral' tectal poles, each had established a map in the normal way with respect to its own point of entry and each had later expanded to cover the remaining tectum. Although the paths of fibres were not deliberately controlled and are unknown in this experiment, the above interpretation serves to show how a simple and reasonable extension of a model based on pathway guidance can account for a most dramatic finding. Similar conclusions can be drawn concerning the results of Sharma (1975 a) in goldfish in which double-rostral or double-caudal compound tecta had been created. After each half had initially become innervated by appropriate fibres – a result obtained also by Jacobson & Levine (1975 b) – the transplanted half-tectum became secondarily innervated by foreign fibres laid down in retinotopic conformity with the projection to the other half-tectum. As a result the transplant became innervated by inappropriate fibres in the reverse of the appropriate polarity. Such a situation has also been observed in transposed projections following removal of caudal tectum in goldfish (Yoon, 1972f; Horder & Martin, 1977a). Just as we have already seen that the polarity of innervation of a graft can be the complete reverse of the normal under conditions in which the pattern of arrival of fibres are likely to dictate the polarity of innervation (26) the above results may be interpreted as being due to the pattern of arrival of fibres from the caudal direction.

However, powerful as the concept of fibre–fibre interactions is in explaining the establishment of the retinotopic order which prevails among fibres in all above described results, it is open to various objections. Although the approach of Cook & Horder (1977), in common with previous models, reduces the amount of information required of guidance cues provided by central tissue, it does not overcome the problem of requiring an implausible abundance of states of cellular differentiation, together with the means for recognizing them, inherent in all versions of the chemospecificity hypothesis. In order for fibre–fibre interactions to generate retinotopic order, fibres must themselves still be individually distinguishable on the basis of features distinguishing the positions of their retinal ganglion cells of origin. This must also be true, though not explicitly discussed, in the

20

arrow model (Hope *et al.*, 1976) if it is to explain rank-ordering of fibres. On the gradient model fibres would still be required to make fine distinctions between levels of concentration of the graded factor. The possibility has been raised (Chung, Gaze & Stirling, 1973; Chung, 1974; Willshaw & v.d. Marlsburg, 1976) that optic fibres could be organized, without chemospecificity, on the basis of the relatedness of nerve impulse patterns in neighbouring retinal ganglion cells, a mechanism which appears to apply in other visual pathways (Gaze, Keating, Székely & Beazley, 1970; Horder & Martin, 1977*b*; Keating. 1977). But what evidence there is (Chung *et al.*, 1973; Cook & Horder, 1977) does not offer any support for this possibility. Although Cook & Horder's model requires regional differentiation of retinal ganglion cells it is argued that a combinatorial system of multiple cue properties could considerably reduce the amount of differentiation that would have to be generated during embryonic development of the retina to account for retinotopic sorting of fibres. Hunt (1976*c*, 1977*d*) has argued in a similar way.

Evidence that fibre self-ordering is not due to interactions based on chemospecificity

Mutual exclusion of fibres arriving by different routes but of comparable retinal origins

The claim that optic fibres are able to undergo active chemospecific sorting with respect to one another cannot readily account for evidence of the following kind. Schneider (1971) using silver stains and Levine & Jacobson (1975), Meyer (1976*a*), Meyer & Sperry (1976) and Schmidt (1978) using autoradiography report that fibres from the two eyes innervating the same tectum tend to become segregated into discrete and mutually exclusive patches. These patches may be difficult to detect electrophysiologically because a few overlapping fibres, giving only diffuse autoradiographic labelling, may be detectable electrically. The autoradiographic method may give an exaggerated impression of exclusive areas of tectal occupation by fibres because some patches may represent bundles of fibres in transit. The absence of representation of certain retinal areas in some projection maps (e.g. temporal retina of the ipsilateral eye in the report by Cronly-Dillon & Glaizner, 1974) raises the possibility that some populations of fibres may be bunched together and therefore easily missed in the sampling procedure used in electrophysiological mapping. These considerations may help to explain the variety of published results including cases in which no patch formation was detected (Gaze & Sharma, 1970; Straznicky *et al.*, 1971*a*; Sharma, 1972*a*, 1973*b*; Horder & Martin, 1977*a*; (**14, 18,**

23)), even though two sets of fibres were projecting to the same tectal area. A further explanation, for which there is some evidence (Schmidt, 1978), might be that patch formation increases in time, starting from a condition of complete overlap of projections immediately following regeneration. Phenomena such as these are quite contrary to what would be predicted by the fibre–fibre interaction model. If, in order to generate an orderly and coherent map, fibres were able to move into closer proximity to fibres of related retinal origins, then it would be expected that fibres from the two eyes would tend towards exact overlap. Long-term adjustment of projections is possible on a large scale (Sharma, 1975a) so that there is no apparent reason why patches should not tend to be eliminated.

Patch formation can most simply be explained as a manifestation of a tendency for fibres of the same origin, that is in similar regions of one eye, to remain in tight conjunction with each other. In doing this fibres originating in one eye will compete with, and effectively displace, fibres from the other. There is a wide selection of reported projection maps consistent with this idea (24). Cronly-Dillon & Glaizner (1974) reported that, following deflection of the tract of one side onto the surface of the tectum of the other side, the resident projection to the host tectum was displaced into rostral tectum. The inclination of transplanted fibres to travel in a caudal direction, which we have noted in our own studies with transplanted optic fascicles and which is apparent in the extension of fibres across transverse gaps in the tectum (Sharma, 1972c; Murray & Sharma, 1977) and along the line of rostro-caudal tectal lesions (Martin, 1978), may be explained as a result of impetus to follow a line of growth continuous with the direction in which a fibre is already pointing or as a result of the alignment of the pre-existing fan of fibres bundles or their debris. Wiemer (1955) and Bunt (unpublished observations) have shown that there is no intrinsic prohibition on the growth of fibres in the reverse direction across the tectum. Subsequent work (Levine & Jacobson, 1975; Meyer, 1976a, b Schneider, Singer, Finlay & Wilson, 1975) has shown that a wide variety of end results can follow what were intended to be consistent forms of deflection of the two projections into one tectum. Cronly-Dillon (personal communication), in a recent repetition of his earlier experiments, has obtained variable patterns of fibre distribution including cases in which the transplanted projection occupies rostral tectum with the resident projection displaced caudally and cases in which the two projections were completely overlaid. Such a variety of outcomes could only be accounted for on any chemospecificity model with the greatest difficulty, whereas it is simple to interpret them as a reflection of patterns of growth and access to synaptic sites of transplanted fibres in an already innervated tectum. It is far from clear what determines whether a fibre will superimpose on resident fibres or will displace them, presumably as a result of competition for tectal space. The phenomenon of expansion suggests that fibres may have an intrinsic tendency to repel one another's terminals. A similar choice is demonstrated in the response of fibres to removal of caudal tectum: fibres may compete, with the result that compression occurs, or fibres deprived of tectum by the lesion may superimpose on fibres innervating rostral tectum (Gaze & Sharma, 1970, (**18**)). Horder & Martin (1977a) discuss the possibility that the numbers of competing fibres and the relative stability of resident fibres decide the issue.

Conclusion

The available evidence offers no support for the idea that fibres owe their relative retinotopic ordering to direct selective interactions based on chemospecific differentiation. In view of the lack of evidence that individual growing fibres can be influenced in a selective manner by any surrounding structures capable of giving them retinotopic order, including tectal tissue as well as other optic fibres, our only remaining course is to reconsider the arguments which have led us to assume that fibres need to be guided chemospecifically.

At the beginning of Part 2 of this review we described a variety of techniques which have been employed to deflect fibres towards inappropriate regions of the tectum: if fibres nonetheless terminate retinotopically then selective fibre growth has to be invoked. However, the problem of explaining detailed, accurate retinotopic mapping of fibres hinges on properties intrinsic to the fibres themselves: the tectum itself cannot be responsible. We have seen that, even after regeneration after optic nerve section, fibres enter the tectum in retinotopic arrangement: this organization could be retained passively by fibres when they are displaced on the tectum, since procedures used to deflect fibres into foreign tectum do not in themselves disturb fibre relationships. Only if it is claimed that, in regenerating across any scar formed at the site of a lesion in the optic pathway, fibres become randomly disorganized, can it be argued that active fibre guidance is required to explain this level of fibre ordering.

In searching for an alternative to chemospecificity models it is clear that we must re-examine the evidence that underlies the crucial argument that regenerating fibres grow at random. Since fibres arrive at the tectum already in retinotopic array, it is also clear that we must concentrate on the actual nature of events at the site of a lesion earlier in the pathway.

PART 3. TOWARDS AN ALTERNATIVE TO CHEMO-SPECIFICITY

THE ARRANGEMENT OF OPTIC FIBRES MAY NOT BE RANDOMLY DISORGANIZED IN THE REGENERATING OPTIC PATHWAY

As the result of work in our laboratory and elsewhere it is now certain that retinotopic organization is rigidly adhered to among optic fibres throughout their transit of the optic pathway in the following animals (3): fish (Ströer, 1940a; Anders & Hibbard, 1974; Horder, 1974a; Roth, 1974; Bunt & Horder, 1977), reptiles (Ströer, 1940a, Armstrong, 1951), birds (Ströer, 1940a; Bunt, unpublished data) and mammals (Polyak, 1957; Hoyt, 1962). Using the same method, that is tracing fibre degeneration resulting from focal retinal lesions light- and electron-microscopically, we, like Maturana *et al.* (1960), have failed to demonstrate such organization in frogs: we have examined *Rana temporaria* and *Xenopus laevis*. Rather than suppose that these two species are isolated exceptions in a general evolutionary sequence – even in other amphibians such as *Triturus vulgaris, alpestris* and *cristatus* Bunt (unpublished observations) has found perfect retinotopic organization – we suspect that this result can be explained by the technique used. In these species it may be difficult to achieve a small and restricted patch of degeneration in the retina. We have also noted that the degenerating debris of myelinated fibres becomes aligned radially in the optic nerve of *Rana*, implying that glial cells can move debris across the nerve. Such movement would prevent the tracing of the original path of the fibres.

The original evidence for random disorganization of regenerating fibres was based on the appearance of the scar region of the nerve as seen in silver-stained sections. In the published pictures (4) one is immediately struck by instances in which fibre bundles can be seen to be travelling in directions oblique to the line of the nerve as a whole. The significance of such bundles depends on where they get to. Reier & Webster (1974) and Beazley (1977) describe in *Xenopus* the escape of regenerating fibre fascicles into connective tissue surrounding the optic nerve. Once trapped outside the nerve stump it is likely that the fibres end blindly and extremely unlikely that they could rejoin the optic nerve by penetrating its sheath further centrally. A similar phenomenon may explain the escape of fibres after intracranial section, but not crush, of the nerve in *Rana*, which results in their anomalous direct innervation of the overlying ipsilateral brain (Gaze & Jacobson, 1963; Gaze & Keating 1970). Highly oblique fascicles in some published pictures could represent such fibres or even fibres entering the optic nerve from immediately neighbouring central nervous tissue.

But much more significant than stray, oblique fibres is the fact that in a number of cases fibres, and apparently the majority of them, can be seen to pass in large bundles straight across the lesion. The maintenance of parallel alignment by fibres crossing a lesion is described in newts by Stensaas & Feringa (1976). Clearly swelling of connective tissue partitions at the site of the lesion can cause considerable displacement of neighbouring fibre fascicles from parallel alignment without entailing any actual crossing or intermixing. In histological sections this would give an impression consistent with fibres generally travelling in random array. But it is important to realize that this cannot be established from simple observation of sections: this does not allow the tracing of the path of any one fibre for any useful distance. Two recent studies in regenerated goldfish optic nerves have, however, set out to do this. Horder (1974a) has shown that the majority of fibres do not get diverted from their normal parallel lines of growth as they traverse the scar region while Roth (1972) reports that there is 'some' scrambling of regenerated fibres at the lesion and that fibres are in orderly retinotopic array in central tracts shortly beyond it. Maturana (1958) shows a picture of a sectioned optic nerve stump of *Bufo* at an early stage of outgrowth of regenerating fibres in which growing fibres appear to be maintaining a generally parallel arrangement. Ortin & Arcaute (1913) report the case of a rabbit's optic nerve in which an isolated fascicle of regrowing fibres can be seen to span a lesion in a roughly straight line to enter part of a nerve stump directly opposite its channel of origin. It is interesting to detect a note of caution in Attardi & Sperry's (1963) description of the organization of fibres in the scar region in teleosts.

The conclusion that there is not necessarily any major random disorganization of optic fibres regenerating across a region of interruption of the optic pathway should not have been entirely unexpected. After all surgical interruption of the nerve does not in itself introduce any changes in the positions or orientations of the cut fibres. It is also likely that there are powerful forces operating to keep regenerating fibres in parallel alignment. A prominent feature of regenerating optic fibres is their grouping into fascicles so that individual fibres are in immediate contact with other fibres (Attardi & Sperry, 1963; Grafstein, 1971; Horder, 1971c; Reier & Webster, 1974; Murray, 1976). Fasciculation is a feature common to growing nerve fibres generally (Harrison, 1910; Ramón y Cajal, 1928, p. 369; 1960; Speidel, 1933; Weiss, 1955) including fibres in culture (Nakai, 1960; Grainger, James & Tresman, 1968; Agranoff, Field & Gaze, 1976). It can most easily be explained as an instance of the general phenomenon of *contact guidance* (Harrison, 1910; Weiss, 1955), the following by fibres of the planes of their mechanical substrate. Late-arriving fibres will be affected mechanically by first-formed fibres ('pioneer fibres'), whose elongated structure is likely to be a potent orientating influence like filaments of fibrin (Weiss, 1934) or grooves (Weiss, 1945).

The importance of these considerations lies, of course, in the fact that if growing fibres can maintain parallel alignment, then the retinotopic arrangement with which they leave the retina will be directly carried through to their arrival at the tectum. In the following sections we review evidence concerning the mechanisms of growth of optic nerve fibres which reinforces this proposal based on contact guidance and shows that purely morphogenetic considerations can satisfactorily account for the way that fibres arrive at specific places in the brain.

A REVIEW OF ASPECTS OF OPTIC FIBRE GROWTH WHICH DEMONSTRATE THAT FIBRES ARE SUBJECT TO CONTACT GUIDANCE

The growth cone as an indicator of the prevailing mechanical conditions

A growing axon advances solely by lengthening at its tip. So the future path of the nerve fibre is the product of the behaviour of the terminal growth cone and its predecessors. The vivid descriptions of Ramon y Cajal (1928, 1960) testify to the way in which the shapes of growth cones vary. Ramón y Cajal ascribes this to their accommodation to the shape of the intercellular spaces in which they lie. The swollen amoeboid or bulbous appearance commonly ascribed to growing nerve tips may be a feature typical of contexts in which barriers to forward growth are considerable and growth is therefore slow (Ramón y Cajal, 1928, p. 365; Speidel, 1933) or absent (Ramón y Cajal 1928, p. 212, 'arrested' or 'gigantic' clubs). Immobile giant expansions are seen in cut optic nerves and may be due to accumulated or leaking axoplasm (Maturana, 1958; Grafstein, 1971). Under conditions where a fibre is in a regularly orientated mechanical context, such as on entering a nerve stump or joining already formed fibres, growth is fast (Ramón y Cajal, 1928, p. 234) and associated with an elongated, tapering growth cone which may be very difficult to see (Ramón y Cajal, 1928, p. 234, p. 365; Holmes & Young, 1942; Weddell, 1942) and structurally simple without filopodia (Harrison, 1901). Irregular and enlarged cones would be expected among isolated fibres at the earliest stages of development while later-formed fibres, following by fasciculation, may be much simpler (Lopresti, Macagno & Levinthal, 1973). Hence the great variety of forms of growth cones reported (Hinds & Hinds, 1974), although tissue culture conditions and incipient formation of terminal synaptic arborizations may account for some of the most extensive cones described.

Growth cones cannot be seen in the developing optic tract or tectum (Goldberg, 1974) or in regenerating optic nerve (Grafstein, 1971) despite the fact that optic fibres appear to differ in no way from any other fibres in their formation of large amoeboid growth cones in tissue culture (Agranoff, Field & Gaze, 1976; Landreth & Agranoff, 1976). The fact that the only place at which large, branched growth cones of optic fibres are regularly seen is the inner retinal lamina (Ramón y Cajal, 1960; Goldberg & Coulombre, 1972; Hind & Hinds, 1974) is easily explained since this is where fibres meet a barrier perpendicularly. In electron-microscopic observations of regenerating goldfish optic nerves (Horder, 1971c), the only specialized structures which can be identified as growth cones are slight enlargements of newly formed axonal sprouts, lacking filopodia and containing branching or vesicular smooth endoplasmic reticulum, like that described by Bray & Bunge (1973). Structures identified as growth cones in the embryonic visual system (Berthoud, 1943; Ramón y Cajal, 1960; Goldberg & Coulombre, 1972; Hinds & Hinds, 1974) are consistent with this description, as is Murray's (1976) account of growth cones in the tectum. The absence of large amoeboid growth cones in the regenerating optic nerve is consistent with the inference that fibres are freely and rapidly advancing through an aligned environment.

Axonal branch formation

Perhaps the most direct manifestation of the growth cone's responsiveness to mechanical conditions occurs when the fibre meets an impenetrable obstacle. Harrison (1910), Ramón y Cajal (1928, 1960, p. 100) and Speidel (1933) describe the resultant diversion of the growth cone and the frequent resulting splitting of the cone into two parts which pass in opposite directions around the obstacle, leading to the formation of two independently growing fibre branches.

During embryonic development branching is certainly not a common feature of fibres growing through major tracts (Ramón y Cajal, 1960). It is rarely seen among optic fibres traversing the optic retinr or tectum (Goldberg, 1974; Rager, 1976). In *Xenopus* (Wilson, 1971) the numbers of ganglion cells and axons in the optic nerve correspond closely at all stages, which implies that branching is not a major feature of axonal growth in normal development. However, late in the development of the avian optic nerve there is a 50% reduction in the number of axons (Rager & Rager, 1976) which may be accounted for by death of ganglion cells rather than being due to loss of one of two initially formed branches from each cell. Some branching of optic fibres is visible during early outgrowth of avian optic fibres (Berthoud, 1943). During regeneration of the optic nerve in goldfish, it is estimated (Horder, unpublished observations) that the average optic fibre produces up to five branches at the most: judging from the average diameter of regenerating fibres and the cross sectional area of the central nerve stump occupied by fibres, this is the maximum number that could be accommodated. In regenerating peripheral nerves between three (Shawe, 1955) and fifty (Ranson, 1912) branches are formed by the average fibre. It is likely that the branching of regenerating fibres is a direct or indirect response to the trauma of axonal section. As can be seen in the regenerating optic nerve (Horder, 1971c), most branch formation occurs at some distance back from the point of interruption of the axon (Ramón y Cajal, 1928) – new growth cone formation can occur at all points along an already formed axon (Ramón y Cajal, 1928, p. 364; Speidel, 1932) – and therefore cannot be ascribed to any direct mechanical effect of the lesion. Fibres are also known to form collaterals in response to the degeneration and removal of fibres innervating neighbouring synaptic sites (Goodman & Horel, 1966; Cunningham, 1972; Goodman, Bogdasarian & Horel, 1973; Schneider, 1973).

After regeneration in peripheral nerves a gradual reduction in the number of branches is seen (Weddell, 1942; Shawe, 1955). The same must be true in the regenerating optic nerve, although an excess of unmyelinated nerve fibres survives for a considerable period (Horder, 1971c). In a goldfish (3.6 cm nose to tail-fin base in length) it has been estimated (Horder, 1971c) that between 50 and 90% of the number of myelinated fibres normally travelling through the nerve (80000) survive and remyelinate. The normal nerve is effectively totally myelinated because the remaining 10000 unmyelinated fibres are confined to a single, ventrally lying bundle. On the grounds that fibres only mature if they have successfully formed terminal connections (Aitken, Sharman & Young, 1947; Maturana, 1958) and that only previously myelinated fibres tend to remyelinate after regeneration (Speidel, 1932, 1933), this estimate may give us a direct measure of the proportion of ganglion cells which successfully re-establish central connections. It is likely (Grafstein, 1971; Horder, 1971c) that the largest myelinated fibres of the normal nerve fail to regenerate.

It might be thought that the filopodia of growth cones and the branches of stem axons could provide the means whereby single fibres could gain access to a wide variety of possible growth paths, to be followed, when they have selected the most favourable alternative, by the withdrawal or degeneration of the remaining processes. However it is doubtful that all initially formed fibre branches succeed in traversing the lesion and, typically branching does not occur once a fibre has successfully crossed the scar (Ramón y Cajal, 1928; Shawe, 1955), although Murray (1976) describes some branching within the tectum. For the same reasons that fasciculation tends to hold neighbouring fibres together as they regenerate, there is no reason to suppose that the separate branches formed from a single stem axon should be able to enter widely different growth channels. So, although branches of an individual fibre do die back on some selective basis (Speidel, 1932; Weddell, 1942), the processes formed by a single regenerating fibre do not necessarily play an extensive exploratory role.

As in the developing nervous system in general (Ramón y Cajal, 1960), the occasional fibres can be seen in the early embryo to be growing in highly aberrant courses in the retina (Ramón y Cajal, 1960), at the entrance to the optic stalk (Tello, 1923; Berthoud, 1943), near the chiasma (Berthoud, 1943) and at the advancing front of fibres invading the tectum (Goldberg, 1974). Equally bizarre patterns of growth can be seen in adult mammalian ganglion cells caused to undertake new axonal growth within the retina by the trauma of optic nerve section (Tello, 1907; Ortin & Arcaute, 1913). Goldberg (1977a) describes randomly orientated growth of optic fibres travelling outside the optic fibre layer of the retina. Aberrant fibres or branches die off during later embryonic development (Ramón y Cajal, 1960) presumably as a result of failure to reach, or to form connection with, central nuclei (Ramón y Cajal, 1928, p. 367; van Buran, 1963; Hughes & La Velle, 1975). In frogs there is little death of ganglion cells, once they have differentiated (Glücksmann, 1940) and before metamorphosis (Hunt, 1976c), so it is likely that few cells make such aberrant connections in the process of normal establishment of the retinotectal projection.

Like branching, the formation of aberrant fibres should perhaps not be regarded primarily as a device for sampling a variety of possible pathways, but as a direct and inevitable reflection of contact guidance. Both phenomena occur most commonly at sites along the optic pathway, such as the entrance to the optic stalk and at the optic chiasma (Tello, 1923; Berthoud, 1943), where the cellular environment met by pioneer fibres is least regularly arranged. Only later-formed fibres will benefit from the orientation given to the tissue by these earliest fibres and so follow straighter and more consistent paths. Contact guidance accommodates, within the same explanatory framework, both aberrant fibre growth and the highly orderly growth that is eventually achieved.

*Further evidence that optic fibres are influenced at various points
along their growth paths by the mechanics of the substrate*

Fibre guidance in the retina

Accounts by Ramón y Cajal (1960), Hinds & Hinds (1974) and Goldberg & Coulombre (1972) of the normal development of the retina indicate that in general fibres follow direct paths in radial fascicles to the entrance of the optic stalk. Highly abnormal patterns of growth of retinal fascicles have been produced experimentally (Rogers, 1952; Szentágothai & Székely, 1956; Goldberg, 1976b, 1977a). Goldberg (1977a) demonstrated that bundles follow new paths such as would be predicted simply in terms of the imposed change in the morphology of the retinal sheet. This is strong evidence that the normal pattern of fascicles is the result of contact guidance. Even the normal choice of a centripetal direction of growth by fibres entering the optic fibre layer is not in any sense obligatory. Szentágothai & Székely (1956), Ramón y Cajal (1960) and Goldberg (1977a) describe instances in which fibres grow against their normal centripetal direction, either by turning centrifugally on meeting the inner lamina or by overshooting the region of the optic nerve head and joining fascicles beyond. The normal centripetal inclination of fibres may be dictated by the mechanical conditions resulting from the asymmetrical disposition of the ciliary margin where new ganglion cells are formed.

How do optic nerve fibres reach the brain?

Given the straightforward manner of growth of most axons during embryonic development it seems certain that the retinotopic arrangement of fibres found in the adult optic nerve reflects the arrangement taken up by fibres during their original formation. Available descriptions of embryonic optic nerves (Tello, 1923; Ströer, 1940a; Berthoud, 1943) suggest the general parallel arrangement of fibres. During their traverse of the optic pathway fibres undergo a systematic rearrangement (Bunt & Horder, 1977) which can be explained as the result of the tissue planes of the choroid fissure along which fibres are growing. The tendency of fibres of the optic nerve to follow blood vessels, muscles or nerve trunks (Reier & Webster, 1974; Beazley, 1977) or the plane of the meninges (Hibbard, 1967) suggest that purely mechanical guidance, such as is provided by the optic stalk, together with an intrinsic potential for linear outgrowth (Harrison, 1910; Szentágothai & Székely, 1956; Agranoff et al., 1976), is involved in the guidance of optic fibres to the region of the optic chiasma.

The decision whether to remain uncrossed at the optic chiasma

Despite the decussation of the major part of the optic projection in lower vertebrates it should be emphasized that in all classes of vertebrates an uncrossed component has now been described; in reptiles (Armstrong, 1951), in fish (Springer & Landreth, 1977), amphibia (Currie & Cowan, 1974; Kicliter, 1974; Stelzner, 1976), in birds (Berthoud, 1943; Ferreira-Berrutti, 1951; Knowlton, 1964) and in mammals (Polyak, 1957; Sanderson, Guillery & Shackelford, 1974).

Crossing the midline towards the tectum of the opposite side must be considered mechanically the easiest option for fibres given their direction of approach – the lateral position of the eyes in all early vertebrate embryos causes fibres to approach the chiasmatic region perpendicularly to the neuraxis – and given the wealth of fibres already forming commissural fascicles in the ventral midline of the neural tube at the time (Tello, 1923; Ramón y Cajal, 1960). Oppenheimer (1950), Szentágothai & Székely (1956) and Schmatolla (1974) all have direct evidence that the side chosen by optic fibres in the embryo is influenced by the angle of the fibres' approach. In the circumstances it is perhaps not surprising that bilateral exchange of eyes (Székely, 1954*b*; Beazley, 1975*a*) did not reveal any preference to connect with the tectum normally appropriate to the eye. Such a test does not exclude 'laterality specificity' of the eye as a determinant of decussation since it may not offer fibres any alternative access. But other evidence is against it, particularly the readiness with which prospective decussating fibres can erroneously innervate ipsilateral pathways (**27, 28, 29**) and the equal ease with which the optic nerve will innervate the tecta of either side (**8, 9, 29**). Many studies describe optic fibres which have 'spontaneously' found their way into commissural tracts and through them crossed to the opposite tectum, including intertectal commissures (Schneider & Nauta, 1969; Sharma, 1973*b*; Cronly-Dillon & Glaizner 1974; **29**), posterior commissure (R. L. Roth, 1967; Levine & Jacobson, 1975), anterior commissure (Sharma, 1973*b*) and pretectal commissures (Rogers *et al.*, 1977). The tendency for optic fibres to cross the midline, immediately after arriving perpendicularly through the oculomotor nerve root presumably reflects the operation of similar forces (Hibbard, 1967). It occurs at other levels (Szentágothai & Székely, 1956; Schmatolla, 1974) and is similarly shown by other fibres of quite different types of neurones (Hibbard, 1965).

The conditions which control the proportion of fibres selecting the ipsilateral route at the chiasma are complex and finely balanced (Lund, 1975) which suggests that the explanation for its genetic control (Sanderson *et al.*, 1974) will not be resolved easily. Removal of one eye in the embryo increases the ipsilateral component of the remaining projection (**27**) although this was not detected in tectal recording by Gaze, Keating, Székely & Beazley (1970), Hirsch & Jacobson (1973) or Beazley (1975 a). This could be the result of non-occupancy of one optic tract facilitating ipsilateral collateralization. The evidence from regeneration throws some light on this possibility. After removal of one eye regenerating fibres from the other eye form anomalous ipsilateral projections. Against the possibility that this is a direct response to degeneration debris in the empty tract, it does not happen in a non-regenerating projection (**27**). Further evidence that non-occupancy is involved comes from the finding that anomalous projections do not form during simultaneous regeneration of both nerves when debris will be present (Springer *et al.*, 1977). The increased probability of anomalous ipsilateral projection following bilateral eye exchange (Beazley, 1975 a) may be the result of the change, at critical times, in the arrangement of fibres derived from specific regions of the retina as they arrive within the chiasmatic region. This is a change that may well follow from a procedure which is effectively equivalent to rotation of an eye in one axis. The effect is not due to the surgery of eye removal and reimplantation (Beazley, 1975 a). Jacobson & Hirsch's (1973) demonstration that it follows eye rotation alone, without bilateral exchange, supsupports this interpretation. When they rotated the eye without optic nerve section the phenomenon did not happen, presumably because fibres then arrived in normal array at the chiasma.

The relevance of timing on the selection of fibres projecting ipsilaterally is demonstrated by So *et al.* (1977) in mammals and in frogs by Currie & Cowan (1974) who showed that the normal ipsilateral projection is composed of fibres formed later than those forming the crossed projection. The ipsilateral fibres only innervate diencephalon (Currie & Cowan, 1974) and Tay & Straznicky (1977) have shown that the diencephalic nuclei are only innervated by part of the optic projection. Murray (1977) shows that diencephalic innervation during regeneration in goldfish may entail selection based on timing.

Once they have entered the neuraxis how do optic fibres find the tectum?

The optic nerve has been caused to enter the central nervous system by way of the roots of cranial nerves V and VIII (Hibbard, 1967; Constantine-Paton & Caprinica, 1975, 1976) or caudally (Katz & Lasek, 1977; Piatt, 949). Fibres failed to show evidence for guidance towards the tectum: they either terminated locally (Paton & Capranica, 974) or underwent major and inappropriate diversion joining, for example, tracts which descend the spinal cord caudally. By contrast fibres entering the neuraxis through the root of cranial nerve III (Gaze, 1959a, 1970; Hibbard, 1967; Beazley, 1977), and through the dorsal diencephalon (Hibbard, 1959; Sharma, 1972a; Hibbard & Ornberg, 1976) reach the tectum. The difference between these two sets of results may lie in the shorter distances that fibres have to travel in cases where they successfully reach the tectum.

The distances involved at the time of fibre growth are small because operations were performed in embryos or tadpoles. The apparent purposefulness with which fibres head for the tectum may give a misleading impression of how fibres actually achieved the result at the time. The directness of fibre bundles could be partly due to stretching during later growth of the brain. The initially forward inclination of fibres away from the tectum described by Hibbard & Ornberg (1976) could be due to later differential growth of brain parts. Moreover, the fibres reaching the tectum could have been those which happened to reach it in the course of initially haphazard growth through the relatively short distances involved. Subsequently, unsuccessful fibres may atrophy and later-added fibres may be guided through fasciculation by successful fibres which have thrived and matured. Consistent with the view that initial growth was haphazard is the evidence that – despite the single and fixed point of arrival of optic fibre dictated by the exit of the oculomotor nerve – within Hibbard's series some surviving fibres could be seen travelling in various directions in the brain. Furthermore, in Gaze's case fibres reached the tectum by an altogether different route from those in Hibbard's series. Further evidence for the indiscriminateness with which optic fibres will sample a variety of pathways comes from the following findings. After deflection of the optic nerve into the optic tract of the same side, fibres arrive at a number of sites to which there would never normally be any direct optic projection, including telencephalon, tegmentum, cerebellum, medulla, nucleus isthmi and reticular formation (Sharma, 1973b; Levine & Jacobson, 1975; Yager & Sharma, 1976; Springer et al., 1977; Schmidt 1978).

Instances such as the growth of optic fibres away from the tectum into the spinal cord (Constantine-Paton & Capranica, 1975, 1976) make no sense in terms of any model which seeks to explain fibres' arrival at the tectum by means of a system of guidance cues coordinated throughout the brain. But such a result can be readily explained by contact guidance since tracts passing to and from the spinal cord would be expected to provide a highly orientated environment to which optic fibres would respond by fasciculation. The potency of this influence is indicated by its indiscriminate effectiveness on fibres of very different types; axons of Mauthner neurones show very similar behaviour (Swisher & Hibbard,

1967). The Mauthner axon also demonstrates the irrelevance of rostro-caudal polarity in the influence of contact guidance (Hibbard, 1965). Further proof of the lack of chemospecificity of brain regions comes from the fact that optic fibres can successfully terminate in virtually any region: the list (31) of regions not normally receiving direct optic projections now includes auditory and vestibular nuclei (Paton & Capranica, 1974), non-optic pretectal and thalamic nuclei (Schneider & Nauta, 1969; Cunningham, 1972; Sharma, 1973b; Kicliter, Misantone & Stelzner, 1974; Baisinger, Lund & Miller, 1977; Rogers et al., 1977), the medial geniculate (Schneider, 1973), deep tectal layers and tegmentum (Schneider & Nauta, 1969; Schneider, 1974; Levine & Jacobson, 1975; Miller & Lund, 1975; Crain & Owens, 1976), various cranial ganglia and the medulla (May & Detwiler, 1925), the forebrain (Wiemer, 1955), efferent pathways running from the tectum (R. L. Roth, 1967), the pharyngeal epithelium (Ferreira-Berrutti, 1951), in addition to cases already referred to. Although not established, it can be argued that in all these cases optic fibres had formed some kind of synaptic connection because failure to do so is known to lead to fibre atrophy. Synaptic connections can be formed between highly inappropriate cells (McLachlan, 1974; Puro, De Mello & Nirenberg, 1977).

In practice it seems likely that arrival of optic fibres at the embryonic tectum is virtually unavoidable. At the time of arrival of the first fibres the anlage of the tectum is immediately ahead of them as they lie crossing the chiasma and a very small distance away; most nuclei which later intervene between chiasma and tectum are not yet formed.

Is there fibre-specific pathway guidance during entry to and transit of the tectum?

The claim that fibres can actively select between the pathways approaching the tectum rests on evidence that regenerating fibres segregate, according to retinal origin, into either the medial or lateral divisions of the optic tract (Sperry 1955; Attardi & Sperry, 1960, 1963; Jacobson, 1961a, 1966; Horder, 1974b). However, in none of these experiments was rotation of nerve stumps involved. They all involved regeneration alone. Given that fibres may have maintained a segregated arrangement throughout regeneration towards the tract division, this is inconclusive as evidence for active guidance. Similarly segregation by fibres from double-ventral compound eyes (Cook & Horder, 1977) is not necessarily due to selective events at the division: it could be dictated by the paths of arrival of fibres as determined, for example, by the pattern of initial fibre entry into the optic nerve. The evidence that *some* fibres take the wrong route, in that segregation is not as clear-cut as in the normal animal (Sperry, 1955; Horder, 1974*a*, *b*; Udin, 1978; (30)), indicates, if anything, against active guidance. The only attempt to observe the paths of fibres after eye rotation was that of Sperry (1945*b*) who saw no evidence for systematic reorientation of the fibres at the lesion or within the tract.

While evidence in favour of active selection of pathways by fibres is simply not available, there is one reason for thinking that it need not happen in the tract but could be delayed until arrival at the tectum. It is clear that re-sorting of the fibres of the medial and lateral divisions of the tract can occur at the tectal level. Arora & Sperry (1962) transposed the two divisions and observed the regrowth of their fibres towards their matching tectal areas within the tissue at the very rostral pole of the tectum, thus suggesting a basis for the ability of fibres to return to their matching medial or lateral tectal regions after deflection (Arora, 1963, 1966; Horder, 1974*b*). As discussed in earlier sections it is unlikely that fibres are guided in this way with any greater degree of accuracy than to within the appropriate quadrant of the tectum.

Against suggestions of chemospecific guidance of fibre growth across the tectum must be set some evidence which can only be explained by susceptibility of tectal fibre bundles to purely mechanical influences. Sperry & Hibbard (1968) describe how optic fibres travel along the line of a barrier placed across their paths in the tectum, an 'adherence to obstacles' as discussed by Ramón y Cajal (1928, p. 369). The surgery of inserting the barrier does not in itself deflect fibres so that the aberrant paths of the fibres must be the result of their impetus to advance even when this means growing away from their normal targets, and their selection of mechanically the most accessible available path. Sharma & Gaze (1971) describe the deflection of fibres around grafts. We have frequently observed the formation of massive regenerated fascicles along cut edges of the tectum. The remarkable arrival in caudal tectum of fibres normally confined to rostral tectum (Martin, 1978) can be explained by deflection by grafts or by contact following of the cut lateral edge of the tectum. We, like Arora (1966), have also noted a tendency for implanted fibre fascicles to follow caudally the line of the fan of fascicles of the normal tectum which could again be the result of contact guidance.

Summing up the case for the establishment of a retinotopic projection through contact guidance

A wide variety of evidence demonstrates that optic nerve fibres are no different from any others in their mode of growth: the constant theme that growing nerve fibres in general are dependent upon and guided by their mechanical substrate (Harrison, 1910, 1935; Ramón y Cajal, 1928, 1960; Speidel, 1933; Weiss, 1955) applies to optic fibres also.

In previous sections of this review we reached the conclusion that the case for the chemospecific control of detailed retinotopic ordering of optic fibres in the tectum rested squarely on the assumption that regenerating fibres were caused to grow centrally in random array due to surgical interruption of the optic pathway. This assumption must now surely be revised. There is no reason that transection of the optic nerve should deflect fibres and there is every reason, given contact guidance, to expect fibres to retain the parallel alignment which will account for their arrival in the tectum with high degrees of retinotopic order.

Pictures of the sites of initial regeneration elsewhere in the central nervous system, in cranial nerves and even in peripheral nerves (32), leave entirely open the possibility that the majority of fibres grow across a scar without crossing their neighbours. Unless otherwise deflected, fibres often demonstrate an intrinsic predisposition to grow out in straight lines (Ramón y Cajal, 1928, p. 363), as can be seen in optic fibres in culture (Agranoff, Field & Gaze, 1976; Landreth & Agranoff, 1976; (33)). Since cut fibres die back some distance into the retinal optic nerve stump before regenerating (Horder, 1971c) they will necessarily start their growth in regular parallel array. The alignment of the vacant growth channels in the central nerve stump may in itself favour the entrance of regenerating fibres travelling straight across the lesion. They way in which fibre fascicles are parcelled together by a tight sheath of astrocytes (Horder, 1971c; Murray, 1976) does not automatically mean that fibres have been guided by astrocytes, whose own dispositions would then require explanation. It is likely that these cells become secondarily arranged around first-formed fibres, like sheath cells in general (Speidel, 1932), since at the earliest stages of regeneration fibres are not invested in this way (Horder, 1971c; Reier, 1978).

Any disorganization in the paths of isolated, initially formed fibres will tend to be secondarily swept away by the intercalation of large parallel fibre bundles, an effect illustrated by Goldberg (1974) in the embryonic tectum. The paths of the earliest optic fibres to invade the tectum are detectably irregular but these subsequently disappear, and are presumably straightened out by consolidation with later-formed fibres into bundles with highly regular alignment. The presence of some strikingly irregular fascicles in sections of a scar region of regenerated nerves cannot be taken to invalidate what has been concluded about the orderliness of the remaining majority of fibres. Indeed these aberrant fibres may be accounted for in terms of contact guidance as fibres which happen to have followed struc-

21

tures orientated in the line of the transection such as ingrowing blood vessels. Fibres often lie at right angles to their original line of growth (Dunn, 1917, Piatt, 1955). Aberrant fibres, particularly in their extreme forms, are unlikely to contribute to the central optic projection because they are presumably travelling out of the optic nerve sheath and usually end blindly. Finally it may be mentioned that certain degrees of scrambling of fibres within the scar region may be compatible with an orderly projection as recorded electrophysiologically, because this technique tends to disregard inconsistencies on the part of minorities among the fibres recorded. The considerable permanent enlargement of multiunit receptive fields observed after regeneration in some conditions (30°, Yoon, 1975d; 45°, Bunt, unpublished result, following intertectal transplantation of optic fascicles) suggests the contribution of such aberrant fibres.

It might be argued that much of the evidence for contact guidance is unreasonably weighted against revealing chemospecificity, which may be subtle, because it involves pathological conditions resulting from gross surgical disturbances or because it involves massive deflections of large numbers of fibres. For the most direct test of the two rival theories the most critical evidence would be the possibility of obtaining projection maps which contain little or no retinotopic ordering at all. The chemospecificity model, although tolerant of a temporary initial random regeneration of fibres, would predict gradual sorting out. Indeed it was designed precisely to account for such an effect. According to the contact guidance model, on the other hand, it would be expected that if any procedure did succeed in producing random access of fibres to sites in the tectum, then a permanently random map will be seen. Following a variety of as yet poorly defined procedures including repeated rotation of the eye or ablation of the entrance to the optic stalk in early *Xenopus* embryos, maps have been recorded which are entirely random (Hunt, 1975a, 1976a, b, 1978b; Hunt & Jacobson, 1974b; Hunt & Piatt, 1974, 1978) or highly disorganized in part (Hunt & Jacobson, 1974b; Berman & Hunt, 1975; Feldman & Gaze, 1975; Hunt, 1975a; Hunt & Frank, 1975). Although one cannot exclude the possibility that fibres are still being guided to tectal locations on the basis of individual chemospecificities and that the random pattern is due to randomization of the developmental assignment of retinal specificities as a common result of these two quite different experimental procedures, a very much simpler explanation is that fibres were caused to enter the optic stalk in random array. Both procedures have in common their mechanical interference with the optic stalk. As a result of what were gross mechanical disruptions of the developing eyecup, including rotation, Goldberg (1976b, 1977a) obtained some highly disorganized patterns of fibre growth across the retina. Beazley (1975 a) also reports some highly irregular projections which are the result of the application of different procedures again in *Xenopus* embryos. Although similar to Patterns 1 and 2 (Gaze & Jacobson, 1963) these are the result of events in embryonic development rather than regeneration and are unmodified over long periods of time. Projections which are generally, or in part, highly irregular or random have been reported frequently (34). It is a general rule that projections mapped after regeneration contain some irregularities when compared to the projections of normal animals, which remain uncorrected long-term. The best measure of this is the enlargement of multiunit receptive fields which are not corrected long after regeneration (Cronly-Dillon, 1968; Horder, 1971 b).

In conclusion we arrive at a model of retinotectal organization which meets all the requirements laid out in Part 2 of this review including the property of the maintenance of retinotopic order within a fibre population independent of tectal location. In place of the idea of chemospecific sorting between fibres themselves (Cook & Horder, 1977) we have simply substituted the idea that fibre ordering is passively maintained as a result of contact guidance forces which prevent fibres from changing their relative positions during passage from the retina.

SOME REMAINING UNRESOLVED ISSUES AND SOME SPECULATIONS

A final test of validity to which any model should be subjected is to identify the sort of evidence which might in principle disprove it and to survey any as yet unexplained evidence with a view to assessing whether this evidence may eventually prove critical in this way. In the sections that remain we discuss some of these issues, starting with the important question of whether the mechanisms we have discovered in regeneration can be applied to normal embryonic development of the retinotectal projection.

How good a model for normal embryonic development of the retinotectal projection is regeneration? The question of modulation

The formation of the retinotectal projection during normal embryogenesis can be readily accounted for in terms of contact guidance (Bunt & Horder, 1977). The orderly arrangement of fibres seen throughout the optic pathway in many adult vertebrates indicates that during development each new fibre may simply have grown by following the fibres of its neighbours in the retina and this is reinforced by what we have already said about the lack of alternative avenues of growth pursued by developing nerve fibres. What sense can then be made of evidence from regeneration studies of the adult brain which indicate chemospecific differentiation of widely separated tectal regions? What purpose can be served by such differentiation if fibres arrive at the tectum in organized array, particularly since any given developing fibre would not then have access to the grossly inappropriate tectal areas which are capable of influencing its growth in regeneration? The significance of tectal differentiation as revealed in the adult is made even more obscure by the claim (Gaze, Chung & Keating, 1972; Gaze, Keating & Chung, 1974) that, during embryogenesis in *Xenopus*, fibres are continually adjusting their positions on the tectum. This raises the problem of how the adult levels of cues are able to be ignored.

Although transposition in adult brains shows this to be plausible, direct tests of this claim in *Xenopus* leave the matter unresolved (Scott & Lázár, 1976; Jacobson, 1977) and, according to the evidence of De Long & Coulombre (1965), Crossland, Cowan, Rogers & Kelly (1974) and Rager (1976), no transposition is seen during embryogenesis in chicks.

One solution to this paradox has come with the suggestion that chemospecificity cues revealed during regeneration are attributable to the fact that, unlike embryonic fibres, regenerating fibres grow through tissue that may have been modified by its earlier exposure to an optic projection. In effect the tectum may become modulated by fibres laid down during embryogenesis, regardless of how embryogenesis itself is explained (Marlsburg & Willshaw, 1977; Gaze, 1978; Schmidt, 1978). According to the intriguing findings of Schmidt (1978) fibres from a normal eye regenerating into an ipsilateral tectum which was already supplied by an expanded projection from a temporal half-retina, formed a similar, expanded projection of fibres from temporal retina while fibres from nasal retina were not represented anywhere in the recorded map. After a resident expanded projection to the host tectum had been removed the regenerating fibres of the normal eye formed a double projection consisting of a normal map together with the expanded map from temporal retina. The author concludes that the guidance cues of the tectum itself have been modified by cues provided by the previously innervating expanded array of optic fibres. These findings contrast with those of Cook & Horder (1974), where there was no evidence that fibres were influenced to regenerate immediately into a compressed arrangement, as opposed to growing initially to their normal termination sites when regenerated into a tectum previously occupied by a compressed projection. It may be that compression and expansion are not directly comparable. Expansion of a projection normally confined to rostral tectum must involve the laying down of new pathways into caudal tectum. The immediate re-establishment of a caudally expanded projection on second regeneration may therefore be explained by the debris of this modified pattern of fibre pathways facilitating this alternative pattern of fibre distribution during re-innervation. We have already referred to the inclination of fibres to grow in a caudal direction.

Schmidt (1978) further reports that if a tectum has been allowed to remain denervated for a considerable period then, in contrast to the normal situation (Schmidt et al., 1978), the regenerating projection from a temporal half-retina forms an immediately expanded map, even though the tectum had previously been innervated by a complete retina. Such a finding may indicate the loss of tectal cues marking normal termination sites which are due, during immediate regeneration, to the degenerating debris of the original fibre projection, but other explanations are available. A number of variables have been shown to affect the occurrence, rate and form of translocation of terminals in the tectum, including lighting conditions (35) and optic fibre maturity, in the sense of whether fibres are freshly re-generated or long-established (Gaze & Sharma, 1970; Horder & Martin, 1977a). Further variables affect fibres in a way that might have a bearing on all aspects of fibre movement: the rate of fibre regeneration is affected by the number of times the fibres have been cut (Agranoff et al., 1976; McQuarrie, Grafstein & Gershon, 1977), by temperature (Springer & Agranoff, 1977) and possibly the distance of the lesion along the axon (Lubińska, 1964) and the axonal type (Ramón y Cajal, 1928; Speidel, 1932). Optic nerve growth is even affected by concentration of nerve growth factor (Turner & Glaze, 1977).

It might also be anticipated that the denervated state of the tectum could also affect the mobility of fibre terminals: Springer et al. (1977) and Schmidt (1978) show that such a state affects the number of fibres successfully in-vading the tectum. Thus it may be that long-term denervation has its effect of producing expanded projections soon after regeneration, by accelerating the rate of formation of the expanded projection.

In circumstances in which the translocation of fibre terminals is known to be subject to many extraneous variables it is important not to use the phenomenon as a direct assay of guidance mechanisms. Following section of the optic tract Yoon (1976b) has compared the time of appearance of compressed maps in a freshly created rostral half-tectum and in a half-tectum which had already received a compressed projection, following earlier halving and regeneration of the optic nerve. In the latter case com-pressed maps were formed at times when still uncompressed initial re-generated projections could be recorded in the former case. Here the dif-ferent histories of the optic fibres and the tectal fragments under compari-son may be expected to affect the relative rates of achieving a compressed arrangement. Romeskie & Sharma (1977) and Sharma & Romeskie (1977b) have shown short-term compressed projections in tecta which had been denervated long-term. The claim in all the above cases that fibres have reached their transposed arrangements immediately during regeneration without initially innervating their normal termination sites, can only be justified by careful survey of the earliest stages of regeneration and even then it remains possible that early transitory fibres are unrecordable by available methods. It is therefore unfortunate that in some reports this claim is based on the examination of very small numbers of animals. In all studies of this kind it is essential to make a distinction between what controls retinotopic ordering on the one hand and what controls the actual location or transposition of the array on the other. In all transpositions, including compression and expansion, the rank order may be determined prior to arrival at the tectum and may be maintained without the need for active chemospecific interactions during all stages of transposition since, as far as we can say, none of the surgery involved in initiating transpositions randomly disorganizes fibres. Therefore no procedures which affect the dynamics of transpositions necessarily test variables affecting forces which generate retinotopic order and these procedures may all have their effects very simply by way of affecting the threshold which determines whether or not potentialities for reorganization intrinsic to the projection as a whole are released.

The concept of modulation creates as many puzzles as it solves. What determines fibres' choices of new termination sites if, after long-term denervation, their normal sites lose their markers? The double projections described by Schmidt (1978) require that normal markers are still present and therefore make it impossible to explain previous expansion by changes in markers. Such a concept leaves entirely unexplained why and how fibres acquire the chemospecific distinctions that they later deposit in the tectum. It is hard to see why such elaborate mechanisms should be provided for an eventuality, namely section of the optic nerve leading to regeneration, which can hardly be of major selective advantage in evolution. Regeneration should surely be regarded less as a repair mechanism than as a re-expression of potentialities which had to exist in embryogenesis. Recent findings (Hunt, 1976c; Romeskie & Sharma, 1977; Straznicky, 1978) argue against modulation and in favour of the establishment of tectal differentiation during embryonic development independent of the fibre supply: rotated tectal grafts caused rotation of their projections despite long-term denervation and previous innervation by a random projection.

Contrary to the assumption implicit in modulation, there is evidence that during embryogenesis optic fibres are subject to active guidance by central tissue. Eyes rotated before the outgrowth of optic fibres (Jacobson, 1968 b; the transplanted parts of compound eyes (14)) form normally orientated projections from their time of first arrival (Feldman & Gaze, 1975 a). Therefore at the time fibres first reach them, central tissues must have regional cues capable of adjusting the orientation of the projection. Acquisition of cues by the tectum independent of the arrival of a retinal projection is also proved by the establishment of an appropriate projection – despite the inversion of one axis of arriving fibres that results from lateral transplantation of the optic tract – in a mature tectum which had never experienced an original embryonic optic innervation (Feldman, Gaze & Keating, 1971). The similarity of the results of guidance during regeneration and embryonic development suggest an identity of mechanisms and the simplest explanation would be that the same tectal cue systems are used. On this view then, the establishment of the cues demonstrated in regeneration studies is simply the result of the laying down of differentiation during embryogenesis. Perhaps the contrast between regeneration and development has been exaggerated: certainly they overlap in the sense that development of new retinotectal connections continues in adult stages (Jacobson, 1976 a; Johns, 1977) and must therefore be subject to forces which apply to regeneration under the same conditions. Gaze & Grant (1978) show that regenerating fibres are influenced by morphogenetic forces in the same way as normally developing fibres.

The nature of tectal differentiation – does the gross tectal differentiation that is revealed by regeneration studies justify the term chemospecificity?

Our review of the evidence from regeneration demonstrated that although the adult tectum is not differentiated to a degree sufficient to account for detailed retinotopic ordering of fibres, it is differentiated on a gross scale. Here we want to speculate briefly on the possibility that purely anatomical differences between tectal regions could be responsible for the distinctive fibre growth which demonstrates this differentiation. The histology of the tectum is sufficiently complex and multivariate in its components that the tissue may be able to contain regional differences with a selective effect on fibres. Even variations in tectal thickness or density could have such an effect provided that fibres themselves differ in their growth properties in a way which is differentially affected by the tectal differences: the size, growth rates or degrees of fasciculation of fibres might be sufficient.

One particularly appealing basis for tectal differentiation is what could be loosely called 'relative age'. The evidence of Gaze et al. (1972) can be taken to show that, during larval development of the retinotectal projection, the distribution of optic fibres is much affected by the state of maturity of the tectal tissue. The tectum is laid down progressively in a roughly rostro-caudal sequence by a caudal rim of mitotic cells (Straznicky & Gaze, 1972). The failure of projections to invade caudal tectum during larval stages, confirmed anatomically by Scott (1974) may be due to the immaturity of caudal tissue. Relative immaturity may explain the relative expansion of the projection in neighbouring regions, a phenomenon which may correlate with Lázár's (1973) demonstration that terminal arborizations are larger at sites of new terminal formation. Concentration on this transitional region of the developing projection, which may be unrepresentative as regards fibre terminal modification, could possibly account for the conflicting reports of Scott & Lázár (1976) and Jacobson (1977) concerning anatomical relations between newly arrived optic fibres and tectal cells. The results of Crossland, Cowan & Rogers (1975) can be similarly interpreted. They show that, although fibres cover all rostral tectum, terminal arborizations are first established in central regions of the tectum where differentiation of tectal cells first begins. On the grounds that early tectal development is automonous (Cowan, 1971) it seems likely that these features can be explained as an example of tectal maturation controlling optic fibre behaviour rather than the reverse. Incidentally this mechanism may explain why central retinal fibres selectively innervate their appropriate tectal areas in the avian embryo leaving vacant sites along their paths (De Long & Coulombre, 1965; Crossland et al., 1974) although this result was not confirmed by Goldberg (1974).

A correlation between generations of developing diencephalic cells and the paths in the optic tract followed by developing optic fibres (Straznicky & Gaze, 1971) described by Gaze & Grant (1978) strongly suggests that successively laid down cells control the ultimate tectal distributions of fibres according to when they are formed. Another such correlation may explain the distribution in depth in the tectum of specific functional classes of optic fibre (Maturana et al., 1960; Jacobson & Gaze, 1964; Llinás & Precht, 1976). In frogs deeper-terminating fibres originate in earlier-formed types of ganglion cell (Pomeranz, 1972; Chung, Stirling & Gaze, 1975). On the reasonable supposition (Cowan, 1971) that outer layers of the tectum are laid down in an inside-to-outside sequence, like the region traversing the optic tract (Gaze & Grant, 1978), the distribution of fibres in depth may be due to first arriving fibres innervating earlier-formed tectal layers and indeed Potter (1976) noted that deeper fibres are the first to arrive. The correlation seen in the adult between axonal diameter, together with associated ganglion cell size, and fibre depth may be explained by saying that earlier-formed neurones tend to reach larger sizes at maturity rather than by supposing that axons differ in size as they grow into the tectum. Similar correlations between axonal diameter or ganglion cell type and specific patterns of projection in the central nervous system (Fukuda & Stone, 1974; Levick, 1975; Kelly & Gilbert, 1975; Karten *et al.* 1977) may all indicate that the selection of different pathways during embryogenesis was the result of different times of fibre outgrowth. The axonal diameters at the time may all have been identical. In the adult brain, following partial tectal ablation without interruption of fibres innervating remaining tectum (**17**), fibres dispossessed of their targets sometimes form a duplicate transposed projection instead of undergoing compression (Gaze & Sharma, 1970; Horder & Martin, 1977). This might conceivably be due to a tendency for small regenerating fibres from ablated parts of the tectum to grow in a separate, more superficial tectal stratum than the larger intact fibres, in a way that parallels events in embryonic development.

As a basis for the differentiation of tectal regions, relative age has the advantage that it would explain how differentiation of the adult tectum can arise out of, and be similar in arrangement to, known anatomical variables in embryogenesis. The sequence of embryological events by which the tectum is formed (Straznicky & Gaze, 1972) remains reflected in the adult tectum as a rostro-lateral to caudomedial gradient in thickness. Correlations of the kind discussed above have been taken as evidence for the '*timing*' hypothesis of fibre guidance (Gaze & Hope, 1977; Gaze, 1978) but it would be misleading to imply that time itself is the operative parameter affecting fibre behaviour. Since development of structures such as the tract and tectum is progressive the conditions met by succeeding generations of arriving optic fibres will change in step with the passage of time but these varying conditions may only consist of changing spatial arrangements of cells of varying maturity. Therefore fibres may behave differently in correlation with stages of development but by responding to equivalent morphogenetic influences resulting from relative age throughout. Gaze & Grant's (1978) demonstration that during regeneration fibres do not recover their original layered arrangement in the optic tract shows that this arrangement is not controlled by chemospecific cues. The change in the pathways occupied by regenerating fibres does not prevent them from forming a retinotopic projection at the tectum. This indicates that fibres can maintain their retinotopic relationships autonomously and that the sequence of their arrival during embryogenesis and also their normal pattern of layering in the tract are incidental to retinotopic ordering of fibres.

Direct attempts, using the methods of cell biology and biochemistry, to isolate tectal factors which may selectively affect the adhesion of different portions of the retina (Barbera, Marchase & Roth, 1973; Barbera, 1975; Balsamo, McDonough & Lilien, 1976; Roth & Marchase, 1976) have characteristically involved comparisons of the properties of relatively gross tissue fractions, such as halves or quadrants of the retina or tectum. Given that the adhesive properties of the retina (Moscona, 1974; Balsamo *et al.* 1976; Gottlieb, Rock & Glaser, 1976; Rutishauser, Thiery, Brackenbury, Sela & Edelman, 1976) and possibly of the tectum (Balsamo *et al.* 1976; Irwin, Chen & Barraco, 1976; Rutishauser *et al.* 1976) may change with developmental age, it is difficult to exclude the possibility that regional differences in adhesivity between retina and tectum are not related simply to differences in maturity. This might explain the otherwise puzzling finding (Pierce, this volume) that cells of the pigment epithelium of the eye show the same selective adhesivity for tectal fragments as neural retina from the same region of the eye: the two cell types are laid down in the same spatially varying sequence during the embryonic development of the eye. It is tempting indeed to equate such differences in cell adhesivity with selective events in the normal formation of a retinotectal projection based on chemospecificity, but the following point should be borne in mind. When comparisons are made between relatively gross tissue fractions purely incidental differences with no bearing on normal ordering of optic fibres, such as differences in cellular composition across the tectum, could give rise to adhesive differences between pooled fractions.

The role of gross differentiation of tectal regions – 'polarity'

A number of studies (Yoon, 1973*b*; Levine & Jacobson, 1974; Martin, unpublished results) show that quite small components of the tectum, when rotated, can guide optic fibres into a projection with a correspondingly rotated orientation: in goldfish fragments may be as small as one sixth of the area of the dorsal half of the tectum. Given that this requires regional differentiation within these tectal fragments, these results might be taken as evidence for high degrees of tectal differentiation. Some insight into the significance of this evidence comes from the following source. By rotating a tectal graft *in situ* within a rostral half-tectum receiving either a compressed whole projection (Yoon, 1977*b*) or a transposed projection from completely non-corresponding, nasal retina (Martin, unpublished results), it has been shown that the orientation of the projections can be actively matched to the new orientation of the grafts even though the grafts in this case are of tissue that the re-orientated fibres would not normally innervate. We can conclude that, whatever property it is which allows tectal grafts to control the orientation of their innervation, it is a generalized property, held by all parts of the tectum and not based on absolute regional differences. A single, continuous characteristic such as relative age would provide this property. It would also help to explain why, when in the course of extreme experimental manipulations fibres have been directed into highly inappropriate tectal regions which they would never meet during normal embryonic development, fibres can detect differences in the tissue and assess how to reach preferred tectal regions on the basis of them. The main point that we wish to make here is that the control of the orientation of a population of fibres does not necessarily require us to revise our previous conclusion that tectal differentiation is of a low order. The control of orientation is a quite different measure of tectal properties from already discussed measures of the ability of fibres to discriminate absolutely between parts of the tectum. This distinction is illustrated by the finding (Romeskie & Sharma, 1977; Sharma & Romeskie, 1977 a, b) that while long-term denervation of goldfish tectum promotes immediate compression of a regenerating projection, rotated grafts retain their ability to rotate incoming fibres.

Another aspect of the control of orientation within populations of fibres is not immediately explicable in terms of the pattern in which fibres arrive at the tectum. In all cases in which a projection from one eye is caused to innervate the ipsilateral tectum (9), and after graft inversion (Yoon, 1975 d) or exchange between tecta (Jacobson & Levine, 1975 a, b; Sharma, 1975 a), the projection becomes re-orientated in its organization with respect to the mediolateral dimension of the tectum independently of its organization in the rostro-caudal dimension. In order to achieve this form of re-organization, which can occur within the tectum itself, fibres cannot retain their usual relationships with their neighbours. How they do this is unknown, but it does not necessarily entail the use of rectangularly arranged, independent axial cues. Although this was implied in the above description of the reorganization, such a description is an arbitrary one and the fibres may achieve the same end-result by using quite different principles of assortment.

Although much remains to be learned about these matters, it can be argued that none of the evidence concerning the control of orientation of projections necessarily requires tectal differentiation of a form and at a level of detail sufficient to account for the normal retinotopicity of a projection. In the face of the strong evidence against the existence of absolute distinctions between tectal locations sufficient to control fibre behaviour, one should perhaps for the moment regard the control of orientation as reflecting different properties of projecting fibres. Since orientation is a property shared within a *population* of fibres, competitive forces between fibres may help the population to make use of tectal distinctions in a way that fibres in isolation cannot. There is strong evidence that none of these aspects of the control of orientation depends on hitherto unsuspected, absolute tectal distinctions. Under some conditions fibres from ablated caudal tectum form projections on rostral tectum with rostro-caudally reversed orientation, independent of normal mediolateral organization (Yoon 1972 f; Horder & Martin, 1977 a). It seems likely that the re-organization reflects interactions between fibres themselves (Horder & Martin, 1977 a) it cannot be controlled by the tectum because the tectum is a foreign part and also has normal orientation.

The nature of retinal differentiation – the interpretation of compound eyes

If future evidence eliminates the morphogenetic explanation for the retinotopic organization of fibres in the tectum a model very much like that of Cook & Horder (1977) will again have to be seriously considered. Selective interactions between fibres themselves on the basis of their retinal origins will have to be hypothesized and this will require corresponding levels of regional differentiation of the retina.

So far in this review we have assumed that, to the extent that different optic fibres grow in characteristic ways with respect to their contexts, so the regions of retina in which they originate must be differentiated equally. A number of studies have attempted to analyse the pattern and embryonic development of retinal differentiation directly. Examination of the retinotopic restriction of connections formed after ablation of portions of the retina does not necessarily allow us to make inferences about regional differentiation of the retina because, as we have seen in the context of Attardi and Sperry's (1963) experiments, no active fibre selection of terminations may be involved. Particularly if the optic pathway has not been interrupted surgically, as in studies of connection formation in the embryo (De Long & Coulombre, 1965; Crossland et al., 1974), termination patterns may be determined by patterns of fibre access to the tectum. In some species this approach (Székely, 1957; Gaze, 1970; Berman & Hunt, 1975; Feldman & Gaze, 1975 b) may also be complicated by 'regeneration' or abnormal growth of retina (Horder & Spitzer, 1973). However, by transplanting retinal fragments to new positions in the retina, fibres can be given direct access to a variety of paths in the optic nerve so that selective fibre growth can be sought and underlying retinal differentiation then deduced.

For transplantation of retinal grafts we are limited to using embryonic eyes. One such approach, and the simplest, has been the study of compound eyes. These are our principal source of evidence on retinal differentiation and will be discussed in detail to illustrate the complexity of the issues that arise in all studies of this kind and to demonstrate how the question of retinal differentiation in general remains extremely obscure. Compound eyes are formed at early embryonic stages in *Xenopus* by the replacement of one ablated half of the retina by its opposite half, taken from an eye of the other side of the head, so creating three main types; double nasal (NN), double temporal (TT), double ventral (VV) compound eyes. Such eyes form projections which, in many respects, correspond remarkably closely to what would be expected given the way they were constructed surgically: the non-transplanted half projects retinotopically with normal orientation to the tectum while in the transplanted half, after allowing for the rotation involved in transplantation from a contralateral eye, fibres from equivalent retinal locations project to identical positions. The two projections are superimposed. Their line of separation in the retina is sharp and lies appropriately along the vertical or horizontal meridian. To this extent therefore it could be argued that the retinal fragments used to make up these eyes had already acquired, and subsequently retained, regionally distinct differentiation corresponding to that which would be appropriate in a normal, complete eye.

However, in one aspect compound eye projections make it difficult to substantiate this interpretation. The projections occupy the entire extent of the tectum instead of being confined to the expected, appropriate half of the tectum. The only exceptions are VV eyes in which a complete projection occupies the medial half of the tectum (Straznicky, Gaze & Keating, 1974), which appears to receive the entire complement of fibres arriving through the tract (Cook & Horder, 1977), so that, although the inaccessible ventrolateral half has not been mapped, this projection cannot be evenly spread out across the entire medio-ventrolateral extent of the tectum. Double-dorsal compound projections (Hunt & Berman, 1975) do extend into inappropriate tectal regions. This feature of compound eyes other than VV eyes could, of course, be readily accounted for as the result of a process of expansion just as we have described in adult animals with halved retinae. This has proved difficult to establish because of other conceivable variables which may be relevant in the developing embryo. As Sperry (1965) pointed out, if fibres from a compound eye were initially restricted to one part of the tectum, unoccupied tectal areas would undergo atrophy and their place be taken by innervated tectum, which itself might undergo hypertrophy due to the double complement of fibres in the projection: such differential growth is known to be possible during embryonic development (Twitty, 1932).

By two different techniques (by tract transplantation: Gaze, 1970; Straznicky, Gaze & Keating, 1971a, and by using the 'third eye assay'; Jacobson & Hunt, 1973; Hunt & Jacobson, 1974a; Hunt & Berman, 1975) a normal projection has been directed into a tectum hitherto innervated by a TT or NN compound projection and it is found that both projections simultaneously occupy the full rostro-caudal extent of the tectum. This result proves that differential growth cannot be the only variable operating because then a half representation of the normal projection should have been seen. But this does not constitute proof that differential growth does not occur because the result could be explained as due to compression of the normal projection, which is known to occur in amphibians (Straznicky, Gaze & Keating, 1971 *b*; Udin, 1977), if not as predictably as in goldfish (Straznicky *et al.*, 1971 *b*; Meyer & Sperry, 1973; Straznicky, 1973; Ingle & Dudek, 1977). The fact that ventrolateral tectum, though presumably poorly innervated, remains unchanged in its anatomical extent in animals with VV eyes (Horder, unpublished observations) does suggest that differential growth is not a sufficiently powerful phenomenon to explain the extent of compound projections. But it does not follow that differential growth does not play some part. Nasal field is sometimes under-represented in the normal projection to a tectum innervated by an NN eye (Straznicky *et al.*, 1971 *a*; Gaze, 1978). Further, compound projections deflected into previously normally innervated tecta (Straznicky *et al.*, 1971 *a*) show abnormalities, such as non-representation of dorsal field, which suggest that a tectum they have innervated *ab initio* is not equivalent to a normal tectum. Independent evidence suggests the feasibility of the alternative explanation in terms of expansion (Goodman *et al.* 1973; Lund & Lund, 1971; Feldman, Keating & Gaze, 1975) although the evidence for expansion in *Xenopus* embryos (Feldman *et al.*, 1975) does not entirely exclude an effect of differential growth because a partial set of fibres from a normal eye were deflected into an empty tectum shortly after metamorphosis when the tectum is still growing (Kollros, 1953).

The two interpretations considered above both seek to explain the complete tectal coverage by compound projections in a way which is compatible with the two halves of compound eyes being equivalent to corresponding halves of normal eyes. However there are serious reasons to question such an assumption. It has now become evident that compound projections can be caused to develop as a result of a variety of apparently unrelated manipulations of the embryonic eye, including simple transection (Hunt & Jacobson, 1974 *d*), removal of two-thirds of the eyecup (Feldman & Gaze, 1975 *b*; Berman & Hunt, 1975; Hunt & Berman, 1975) and even rotation of the eye (Sharma & Hollyfield, 1974 *a*). The formation of projections identical to those of compound eyes but completely different means suggests that compound eyes may represent an abnormal but stable state of eye development to which the structure is spontaneously inclined to relapse after a number of deflections from the normal course of development.

Thus a vertical line of demarcation of the two halves typical of a surgically compound eye can result after division of the eye at a 45° angle and removal of the 40–60 % fragment centred on the temporodorsal retinal pole (Tosney Hoskins & Hunt, 1978), which indicates that features such as the sharpness and orthogonal alignment of the lines of demarcation do not accurately reflect the origins of fused retinal fragments and do not prove that such regional distinctions exist in a normal retina. The formation of compound eyes from one-third fragments of an eyecup (Feldman & Gaze, 1975 b) proves that the reversal of polarity typical of the transplanted half of a surgically created compound eye is a feature which can be arrived at by processes intrinsic to a damaged eye and not necessrily requiring thea rotation of any part of the eyecup tissue. Hunt & Frank (1975) have shown that different parts of the retina are interdependent ('transrepolarization') at least as late as the time when optic fibres first reach the tectum. There is also evidence in the regenerating retina of the newt that rotated peripheral retina can interact with, and influence the projection from, slower-regenerating non-rotated central retina (Levine & Cronly-Dillon, 1974; Levine, 1975).

In a possible illustration of forces within the eye underlying compound eye formation, Goldberg (1976 a, b) has shown how rotation of the avian eyecup results in a massive but integrated change in the pattern of exit of optic fibres from the eye, a change which might well be expected to lead fibres into new spatial arrangements in the optic pathway and therefore to form new projections. Goldberg (1976 b) reports cases of retinae with duplicated nasal, temporal and ventral halves which may correspond to compound retinae in *Xenopus*. The formation of the double-ventral avian eye can be readily explained because it appears to be linked to the formation of an additional choroid fissure. It seems likely that the integration and alignment of this modification of retinal development could be explained simply in terms of the morphogenetic forces associated with fissure formation which involve shaping of the eyecup as a whole. Since the additional choroid fissure is the result of the changed surroundings of the eye resulting from rotation one can hypothesise that 'polarization' (36) of the retina could be mediated by mechanical forces affecting eye development in such a manner.

Ultimately the greatest problem in the way of our ability to analyse the form and degree of regional differentiation of the retina lies in the fact that the patterns of access of fibres to alternative tectal sites are uncontrolled in experiments of the kind we are discussing. Tay & Straznicky (1977) discuss the evidence that fibres from NN eyes may make anomalous connections in the ipsilateral diencephalon as a result of the positions they occupy in the optic fibre array leaving the eye. It is clear that many features of optic projections can be determined morphogenetically by the developing retina as a result of the way in which fibres are caused to assemble as they enter the optic nerve. Consequently it remains extremely unclear whether any features of compound eye projections, with the possible exception of the control of their orientation, require an explanation in terms of chemospecificity of retinal regions rather than in terms of morphogenetic considerations.

In conclusion, there are strong reasons to think that compound eyes are the result of interactive developmental processes intrinsic to the eye, which are a response to abnormal conditions of development, and which may not directly reflect features of normal retinal differentiation. The uncertainties regarding the rules which govern the distribution of fibres in the embryo mean that the pattern of fibre projection cannot be used as a direct assay of retinal differentiation. In view of these considerations one cannot exclude the possibility that the halves of a compound eye do not correspond in their differentiation to halves of normal eyes. One more explanation (Sperry, 1965) for the simultaneous coverage of the tectum by compound and normal projections (Straznicky et al., 1971 a) is that each half has become developmentally equivalent to a complete retina – the possibility that each half has regenerated its own missing components can be discounted on the grounds that no mitosis occurs at the junction of the halves (Feldman & Gaze, 1972). Against this explanation Straznicky (1976 a) has shown that a complete compound projection can be formed in a half-tectum, presumably by compression. NN and TT eye were shown to be different because neither would transpose into non-corresponding halves of the tectum. One must conclude that none of the available evidence entirely excludes any of the three proposed explanations for complete tectal coverage by a compound projection. It remains possible that all three may play some part.

If these problems arise in the interpretation of experiments with compound eyes which initially appeared to be simple and direct tests of retinal differentiation, they must also apply in attempts to analyse differentiation in more detail on the basis of projections showing smaller patches of anomalously projecting retina (Feldman & Gaze, 1975b; Hunt, 1975a; Hunt & Berman, 1975; Hunt & Frank, 1975; Hunt & Idle, 1977). However, two final considerations make it difficult to believe that high degrees of retinal chemospecificity are required for the formation of a perfectly normal projection. The pigment epithelium of the eye or the non-pigmented iris can give rise to a complete, regenerated retina and this forms an apparently quite normal projection (Cronly-Dillon, 1968; Levine & Cronly-Dillon,

1974). Also against highly elaborate and specific chemical factors is the evidence that eyes can be transplanted between a wide variety of species and, as far as is known, can form normal connections in the tectum of the foreign species (Stone, 1963), as is evidence that intercellular communication involved in polarization of the eye is not species-specific (Hunt & Piatt, 1978), a result which would be unsurprising if polarization is a morphogenetic event alone.

FINAL IMPLICATIONS OF MORPHOGENETICS

The proposals contained in this review are to the effect that the development of the nervous system can be accounted for in terms of the accumulation through time of a succession of morphogenetic events each of which is based on the same, simple principles of neuronal differentiation and axonal outgrowth. We use the term 'morphogenetics' to cover all aspects of the way tissues and cells create and respond to mechanical forces in developing into spatially patterned structures. That enormous complexity of adult structure can be generated out of simple morphogenetic differences, which are changing in time, is dramatically illustrated by the way in which cells from the single source known as neural crest become distributed and diversified through the embryo (Weston, 1970). One can view the whole of embryogenesis as depending on the morphogenetic forces which distribute uncommitted embryonic cells so that they reach new environments which cause their differentiation (Horder, 1976).

The principle of contact guidance offers enormous advantages over other models for the development of nerve connections. It can apply to an axon equally at all points during its outgrowth, as we have attempted to show for the optic nerve fibre, which means that the unity of the cell, and of its growth cone with respect to its cell body, is preserved. Contact guidance applies at the level of the single neurone but equally to large assemblages, as illustrated by fasciculation. Morphogenetic forces can serve as a means of communication between constituent parts in the integration of the formation of patterns within the embryo as a whole. Not only does each neurone respond in its development to mechanical forces in its surroundings but, because it is itself a morphogenetic entity, it contributes to forces applied to neighbouring cells. Mechanical forces integrate pattern formation even on a large scale, as illustrated by the role of intraocular pressure in determining the distribution of ganglion cells in the retina (Coulombre, 1956; Johns, 1977) or by the sequence of events of choroid fissure formation which coordinates invagination of the optic cup with the anatomy of the optic stalk and has profound effects on the pattern of optic fibre growth towards the brain (Goldberg, 1976b; Bunt & Horder, 1977). As mediators of pattern formation during embryogenesis morphogenetic forces have the most significant property that component forces can summate or subtract at any position in the system and through time. The result is the accumulation of complexity during development, each stage of development building on the consequences of all prior morphogenesis. So the tectum can acquire multiple afferent and efferent projections each mapped out, like the retinotopic map, according to its own quite different requirements. This is a feature that a single set of positional cues, as envisaged in the concept of chemospecificity, cannot easily cope with. No overall plan for the laying out of the central nervous system, as would be required by the concept of chemospecificity, has to be explicitly provided in the egg or the genome. The genetic instructions required for brain development are enormously reduced and simplified since all neurones can be virtually identical as differentiated cells and can, for example, use the same mechanisms for axonal outgrowth, myelination and synapse formation. Horder (in preparation) shows that the projection of spinal motoneurones into the limb can be explained using morphogenetic principles similar to those used to account for the development of the retinotectal projection.

It is important to stress that it may not yet be immediately obvious how the principle of contact guidance can be used to account for some of the data concerning the retinotectal projection. But this does not invalidate the principle. Indeed the evidence that optic fibres are subject to purely mechanical forces appears to us to be so overwhelming that it is difficult to see how other influences, such as chemospecific ones, could even play a contributory role. It is difficult to see how mechanical forces, which all fibres must always be experiencing, could be overridden by other considerations. However, our last note must be one of caution. As Harrison (1924) says, with reference to the inconstancy of accepted scientific truths and not, one may be sure, to the paths of growing nerve fibres; 'One who follows historically the subject of the histogenesis of the nervous system cannot but be impressed by the deviousness of the path of progress'.

KEY TO REFERENCE GROUPINGS

Group
number   Subject areas covered

1 General reviews directly concerned with the retinotectal system.
2 Particular theoretical models.
3 Anatomical evidence for the maintenance of retinotopic order throughout the primary optic pathway.
4 Photomicrographs or drawings of the scar region of the optic pathway after regeneration.
5 Topographically ordered retinotectal projections after regeneration of the optic nerve or tract or retina alone.
6 Rotation of the eye, after 'polarization': its effects on the projection formed and on visuomotor behaviour.
7 Surgical deflection of optic fibre bundles into inappropriate regions of the tectum.
8 Contralateral eye transplantation, with or without rotation.
9 Aberrant direct ipsilateral retinal projections regardless of circumstances responsible.
10 Surgical deflection at the chiasma of fibres of the optic nerve into the optic tract of the same side.
11 In-situ rotation of part or whole of the tectum.
12 Translocation of tectal grafts to non-corresponding sites in either tectum.
13 Projections formed after partial retinal ablations.
14 Compound projections such as those formed by 'compound eyes' and other eyes formed by fusion of retinal fragments.
15 Projections formed after partial or complete tectal ablations.
16 Compression of a coherent retinal projection into a smaller than normal tectal area after regeneration or during embryonic development.
17 Compression into an incomplete tectum with an initially intact projection.
18 Abnormal projections, other than compressed projections obtained after partial tectal ablations, including the formation of duplicate projections.
19 Transposition of projections onto foreign tectal regions other than as part of compression or expansion.
20 Expansion of the retinal projection to cover a greater than normal area of tectum.
21 Initially untransposed projections formed after partial tectal or retinal ablation and optic nerve crush or transection.

22 Studies of the time course of changes in the retinotectal projection after various experimental manipulations including the time course regeneration of optic fibres alone.

23 Simultaneous projection of equivalent retinal regions of two, non-compound eyes to the same tectal site.

24 Simultaneous projection of one retinal region, or equivalent regions of both eyes, to different regions of an intact tectum or fragment, excluding developmentally modified and compound eyes.

25 Simultaneous projection of different regions of one eye, or non-equivalent regions of both eyes, to the same region of an intact tectum or fragment, excluding developmentally modified and compound eyes.

26 Unrotated or untransposed projections formed in rotated or transposed tectal tissue, or rotated projections formed by developmentally normal eyes of normal orientation in intact tectum or fragments.

27 Direct ipsilateral retinal projections formed following removal of retina from one eye during regeneration or embryonic development.

28 Direct ipsilateral retinal projections formed, without surgical deflection, at the optic chiasma or other diencephalic commissures.

29 Growth of fibres from one tectum to the other, no surgical deflection having been performed.

30 Studies on the accuracy with which growing regenerating fibres segregate into their normal arrangement within pathways leading to the tectum.

31 Aberrant retinal projections to non-optic nuclei and other sites.

32 Photomicrographs or drawings of fibres which have grown across sites of transection in the spinal cord, brain stem and peripheral nerves, other than optic.

33 In-vitro culture of retinal ganglion cells.

34 Grossly disordered projections formed during embryonic development or regeneration.

35 Effect of lighting conditions on the formation or alteration of the direct contralateral retinotectal projection.

36 'Polarization' of the retina during embryonic development.

REFERENCES

AGRANOFF, B. W., FIELD, P. & GAZE, R. M. (1976). Neurite outgrowth from explanted *Xenopus* retina: an effect of prior nerve section. *Brain Research*, **113**, 225–34.

AITKEN, J. T., SHARMAN, M. & YOUNG, J. Z. (1947). Maturation of regenerating nerve fibres with various peripheral connexions. *Journal of Anatomy*, **81**, 1–22. 33

ANDERS, J. J. & HIBBARD, E. (1974). The optic system of the teleost *Cichlasoma biocellatum. Journal of Comparative Neurology*, **158**, 145–54. 32

ARMSTRONG, J. A. (1951). An experimental study of the visual pathways in a snake (*Natrix natrix*). *Journal of Anatomy*, **85**, 275–88. 3

ARORA, H. L. (1963). Effect of forcing a regenerative optic nerve bundle toward a foreign region of the optic tectum. *Anatomical Record*, **145**, 202. 3

ARORA, H. L. (1966). Regeneration and selective reconnection of optic nerve fibres following contralateral cross in the tectum of the goldfish. *Anatomical Record*, **154**, 311. 7

ARORA, H. L. (1973). Fate of regenerating optic fibers following brain lesions in goldfish. *Anatomical Record*, **175**, 266. 7, 9

ARORA, H. L. & GRINNELL, A. D. (1976). Recovery of visual function in the 'compressed' hemitectum of goldfish. *Anatomical Record*, **184**, 574–5. 15

ARORA, H. L. & SPERRY, R. W. (1962). Optic nerve regeneration after surgical cross-union of medial and lateral optic tracts. *American Zoologist*, **2**, 389. 15, 16, 17, 22, 25

ATTARDI, D. G. & SPERRY, R. W. (1960). Central routes taken by regenerating optic fibers. *Physiologist*, **3**, No. 3, 12. 7

ATTARDI, D. G. & SPERRY, R. W. (1963). Preferential selection of central pathways by regenerating optic fibers. *Experimental Neurology*, **7**, 46–64. 13, 30

BAISINGER, J., LUND, R. D. & MILLER, B. (1977). Aberrant retinothalamic projections resulting from unilateral tectal lesions made in fetal and neonatal rats. *Experimental Neurology*, **54**, 369–82. 13, 22, 30

BALSAMO, J., McDONOUGH, J. & LILIEN, J. (1976). Retinal–tectal connections in the embryonic chick: evidence for regionally specific cell surface components which mimic the pattern of innervation. *Development Biology*, **49**, 338–46. 9, 15, 29, 31

BARBERA, A. J. (1975). Adhesive recognition between developing retinal cells and the optic tecta of the chick embryo. *Developmental Biology*, **46**, 167–91.

BARBERA, A. J., MARCHASE, R. B. & ROTH, S. (1973). Adhesive recognition and retinotectal specificity. *Proceedings of the National Academy of Sciences, U.S.A.*, **70**, 2482–6. 1

BARONDES, S. H. (1970). Brain glycomacromolecules and interneuronal recognition. In *The Neurosciences: Second Study Program*, ed. F. O. Schmitt, pp. 747–60. Rockefeller University Press, New York. 2

BEAZLEY, L. D. (1975a). Factors determining decussation at the optic chiasma by developing retinotectal fibres in *Xenopus*. *Experimental Brain Research*, **23**, 491–504. 5, 8, 9, 23, 25, 28, 34, 36

BEAZLEY, L. D. (1975b). Development of intertectal neuronal connections in *Xenopus*: the effects of contralateral transposition of the eye and of eye removal. *Experimental Brain Research*, **23**, 505–18. 8, 34

BEAZLEY, L. D. (1977). Abnormalities in the visual system of *Xenopus* after larval optic nerve section. *Experimental Brain Research*, **30**, 369–85.   **5, 6, 9, 34**

BERMAN, N. & HUNT, R. K. (1975). Visual projections to the optic tecta in *Xenopus* after partial extirpation of the embryonic eye. *Journal of Comparative Neurology*, **162**, 23–42.   **5, 13, 14, 34**

BERNSTEIN, J. J. (1964). Relation of spinal cord regeneration to age in adult goldfish. *Experimental Neurology*, **9**, 161–74.   **32**

BERTHOUD, E. (1943). Développement des fibres nerveuses dans le pédicule optique chez l'embryon de Poulet. *Revue Suisse de Zoologie*, **50**, 473–84.

BRAY, D. & BUNGE, M. B. (1973). The growth cone in neurite extension. In *Locomotion of Tissue Cells, Ciba Symposium*, **14** (New series), pp. 195–209. Elsevier, Amsterdam.

BROUWER, B. & ZEEMAN, W. P. C. (1926). The projection of the retina in the primary optic neuron in monkeys. *Brain*, **49**, 1–35.   **3**

BUNT, S. M. & HORDER, T. J. (1977). A proposal regarding the significance of simple mechanical events, such as the development of the choroid fissure, in the organization of central visual projections. *Journal of Physiology*, **272**, 10–12P.   **2, 3**

BUNT, S. M., HORDER, T. J. & MARTIN, K. A. C. (1978). Evidence that optic fibres regenerating across the goldfish tectum may be assigned termination sites on a 'first come, first served' basis. *Journal of Physiology*, **276**, 45–6P.   **7, 9, 13, 19, 20, 22 24, 26**

BURGEN, A. S. V. & GRAFSTEIN, B. (1962). Retinotectal connexions after retinal regeneration. *Nature, London* **196**, 898–9.   **5, 6**

CASAGRANDE, V. A., HALL, W. C. & DIAMOND, I. T. (1972). Formation of anomalous projections from the retina to the pulvinar following removal of the superior colliculus in neonatal tree shrews. Society for Neuroscience, Abstracts 2nd Annual Meeting, p. 231.   **31**

CHOW, K. L., MATHERS, L. H. & SPEAR, P. D. (1973). Spreading of uncrossed retinal projection in superior colliculus of neonatally enucleated rabbits. *Journal of Comparative Neurology*, **151**, 307–22.   **20**

CHUNG, S. H. (1974). In search of the rules for nerve connections. *Cell*, **3**, 201–5.   **1**

CHUNG, S. H. & COOKE, J. (1975). Polarity of structure and of ordered nerve connections in the developing amphibian brain. *Nature, London*, **258**, 126–32.   **11, 19, 24–26**

CHUNG, S. H. & FELDMAN, J. D. (1973). Neurospecificity and retinotectal connections: thirty years later. In *Biological Diagnosis of Brain Disorders. The future of the brain sciences*, ed. S. Bogoch, pp. 193–209. Spectrum, New York.   **1**

CHUNG, S. H., GAZE, R. M. & STIRLING, R. V. (1973). Abnormal visual function in *Xenopus* following stroboscopic illumination. *Nature, New Biology, London*, **246**, 186–9.   **5, 35**

CHUNG, S. H., BLISS, T. V. P. & KEATING, M. J. (1974). The synaptic organization of optic afferents in the amphibian tectum. *Proceedings of the Royal Society*, B **187**, 421–47.

CHUNG, S. H., STIRLING, R. V. & GAZE, R. M. (1975). The structural and functional development of the retina in larval *Xenopus*. *Journal of Embryology and Experimental Morphology*, **33**, 915–40.

CLARK, W. E. LE GROS. (1943). The problem of neuronal regeneration in the central nervous system. II. The insertion of peripheral nerve stumps into the brain. *Journal of Anatomy*, **77**, 251–9.   **32**

CLARKE, P. G. H. & WHITTERIDGE, D. (1976). The projection of the retina, including the 'red area' on to the optic tectum of the pigeon. *Quarterly Journal of Experimental Physiology*, **61**, 351–8.

CLEMENTE, C. D. & WINDLE, W. F. (1954). Regeneration of severed nerve fibers in the spinal cord of the adult cat. *Journal of Comparative Neurology*, **101**, 691–731.   32

CONSTANTINE-PATON, M. & CAPRANICA, R. (1975). Central projection of optic tract from translocated eyes in the leopard frog (*Rana pipiens*). *Science*, **189**, 480–2.   31

CONSTANTINE-PATON, M. & CAPRANICA, R. R. (1976). Axonal guidance of developing optic nerves in the frog. 1. Anatomy of the projection from transplanted eye primordia. *Journal of Comparative Neurology*, **170**, 17–32.   31

COOK, J. E. (1978). Interactions between optic fibres controlling the locations of their terminals in the goldfish optic tectum. *Brain Research* (in press).   15–17, 19, 21–22

COOK, J. E. & HORDER, T. J. (1974). Interactions between optic fibres in their regeneration to specific sites in the goldfish tectum. *Journal of Physiology*, **241**, 89–90P.   15, 16, 21, 22

COOK, J. E. & HORDER, T. J. (1977). The multiple factors determining retinotopic order in the growth of optic fibres into the optic tectum. *Philosophical Transactions of the Royal Society* B **278**, 261–76.   1, 2, 15, 16, 30

COOKE, J. (1977). Organizing principles for anatomical patterns and for selective nerve connections in the developing amphibian brain. In *Cell Interactions in Differentiation*, eds. M. Karkinen-Jääskeläinen, L. Saxén & L. Weiss, pp. 111–24. Academic Press, London.   11, 19, 24–26

COULOMBRE, A. J. (1956). The role of intraocular pressure in the development of the chick eye. *Journal of Experimental Zoology*, **133**, 211–25.

COWAN, W. M. (1971). Studies on the development of the avian visual system. In *Cellular Aspects of Neural Growth and Differentiation*, ed. D. C. Pease, pp. 177–218. University of California Press, Berkeley.   11, 13

CRAIN, B. J. & OWENS, M. A. (1976). A determination of the interval between early lesions of the superior colliculus and the appearance of an anomalous retinal projection in the Syrian Hamster. Society for Neuroscience, Abstracts 6th Annual Meeting, **2**, p. 820.   15, 31

CRELIN, E. S. (1952). Excision and rotation of the developing *Amblystoma* optic tectum and subsequent visual behavior. *Journal of Experimental Zoology*, **120**, 547–77.   11

CRONLY-DILLON, J. R. (1968). Pattern of retinotectal connections after retinal regeneration. *Journal of Neurophysiology*, **31**, 410–18.   5, 22,

CRONLY-DILLON, J. R. & GLAIZNER, B. (1974). Specificity of regenerating optic fibres for left and right optic tecta in goldfish. *Nature, London* **251**, 505–7.   7, 9, 13, 15, 19, 24

CROSSLAND, W. J., COWAN, W. M. & ROGERS, L. A. (1975). Studies on the development of the chick optic tectum. IV. An autoradiographic study of the development of retino-tectal connections. *Brain Research*, **91**, 1–23.

CROSSLAND, W. J., COWAN, W. M., ROGERS, L. A. & KELLY, J. P. (1974). The specification of the retino-tectal projection in the chick. *Journal of Comparative Neurology*, **155**, 127–64.   13, 36

CUNNINGHAM, T. J. (1972). Sprouting of the optic projection after cortical lesions. *Anatomical Record*, **172**, 298.

CUNNINGHAM, T. J. (1974). Retinotopic organization of misdirected
retinotectal projections. *Anatomical Record*, **178**, 337.   **9, 26, 27**

CUNNINGHAM, T. J. & SPEAS, G. (1975). Inversion of anomalous uncrossed
projections along the mediolateral axis of the superior colliculus:
implications for retinocollicular specificity. *Brain Research*, **88**,
73–9.   **9, 25–27**

CURRIE, J. & COWAN, W. M. (1974). Evidence for the late development
of the uncrossed retinothalamic projections in the frog, *Rana pipiens*.
Brain Research, **71**, 133–9.

DEAN, G. & USHER, C. H. (1896). Experimental research on the course of
the optic fibres. *Transactions of the Ophthalmological Society*, **16**, 248–
76.   **3**

DE LONG, G. R. & COULOMBRE, A. J. (1965). Development of the retinotectal topographic projection in the chick embryo. *Experimental
Neurology*, **13**, 351–63.   **13, 36**

DE LONG, G. R. & COULOMBRE, A. J. (1967). The specificity of retinotectal connections studied by retinal grafts onto the optic tectum in
chick embryos. *Developmental Biology*, **16**, 513–31.   **7**

DUNN, E. H. (1917). Primary and secondary findings in a series of
attempts to transplant cerebral cortex in the albino rat. *Journal of
Comparative Neurology*, **27**, 565–82.   **32**

EASTER, S. S. & SCHMIDT, J. T. (1977). Reversed visuomotor behavior
mediated by induced ipsilateral retinal projections in goldfish.
Journal of Neurophysiology, **40**, 1245–54.   **9, 10, 23**

FELDMAN, J. D. & GAZE, R. M. (1972). The growth of the retina in
Xenopus laevis: an autoradiographic study. II. Retinal growth in
compound eyes. *Journal of Embryology and Experimental Morphology*,
27, 381–7.

FELDMAN, J. D. & GAZE, R. M. (1975a). The development of the retinotectal projection in *Xenopus* with one compound eye. *Journal of
Embryology and Experimental Morphology*, **33**, 775–87.   **14**

FELDMAN, J. D. & GAZE, R. M. (Appendix by Neil MacDonald) (1975b).
The development of half-eyes in *Xenopus* tadpoles. *Journal of Com-
parative Neurology*, **162**, 13–22.   **13, 14, 34**

FELDMAN, J. D., GAZE, R. M. & KEATING, M. J. (1971a). Delayed innervation of the optic tectum during development in *Xenopus laevis*.
Experimental Brain Research, **14**, 16–23.   **9, 10**

FELDMAN, J. D., KEATING, M. J. & GAZE, R. M. (1975). Retinotectal
mismatch: a serendipitous experimental result. *Nature, London*, **253**,
445–6.   **9, 10, 20**

FERREIRA-BERRUTTI, P. (1951). Experimental deflection of the course of
optic nerve in the chick embryo. *Proceedings of the Society for Experimental Biology and Medicine*, **76**, 302–3.   **31**

FINGER, T. E., ROGERS, L. A. & COWAN, W. M. (1976). Regulation in the
retinotectal system of early chick embryos. Society for Neuroscience,
Abstracts, 6th Annual Meeting, **2**, p. 212.   **15**

FINLAY, B. L. & SO, K-F. (1977). Altered retinotectal topography in
hamsters with neonatal tectal slits. *Investigative Ophthalmology and
Visual Science*, **16**, suppl., 139.

FINLAY, B. L., SCHNEPS, S. E. & SCHNEIDER, G. E. (1976). Compression
of the retinotectal projection after neonatal removal half-tectum
ablation in the hamster. Association for Research in Vision
and Opthalmology, Abstracts of Spring meeting, 1976, p.
91.   **16**

FREEMAN, L. W. (1952). Return of function after complete transection of the spinal cord of the rat, cat and dog. *Annals of Surgery*, **136**, 193–205.

32

FREEMAN, J. A. (1977). Possible regulatory function of acetylcholine receptor in maintenance of retinotectal synapses. *Nature, London*, **269**, 218–22.

5, 16, 17, 22

FROST, D. O. & SCHNEIDER, G. E. (1976). Normal and abnormal uncrossed retinal projections in Syrian hamsters as demonstrated by Fink–Heimer and autoradiographic techniques. Society for Neuroscience, Abstracts, 6th Annual Meeting, **2**, p. 812.

9, 13, 23, 27

FUKUDA, Y. & STONE, J. (1974). Retinal distribution and central projections of Y-, X-, and W-cells of the cat's retina. *Journal of Neurophysiology*, **37**, 749–72.

GAZE, R. M. (1959a). Regeneration of the optic nerve in *Xenopus laevis*. *Quarterly Journal of Experimental Physiology*, **44**, 290–308.

5–7

GAZE, R. M. (1959b). Regeneration of the optic nerve in *Xenopus laevis*. *Journal of Physiology*, **146**, 40–1P.

5, 6

GAZE, R. M. (1960). Regeneration of the optic nerve in amphibia. *International Review of Neurobiology*, **2**, 1–40.

I

GAZE, R. M. (1967). Growth and differentiation. *Annual Review of Physiology*, **29**, 59–86.

I

GAZE, R. M. (1970). *The Formation of Nerve Connections*. Academic Press, New York.

I, 4

GAZE, R. M. (1974). Neuronal specificity. *British Medical Bulletin*, **30**, 116–21.

I

GAZE, R. M. (1978). The problem of specificity in the formation of nerve connections. In *Specificity of Embryological Interactions*, ed. D. Garrod. Chapman & Hall, London (in press).

I

GAZE, R. M. & GRANT, P. (1978). The diencephalic course of regenerating retinotectal fibres in *Xenopus laevis*. *Journal of Embryology and Experimental Morphology*, **44**, 201–16.

30

GAZE, R. M. & HOPE, R. A. (1976). The formation of continuously ordered mappings. *Progress in Brain Research*, **45**, 327–55.

I

GAZE, R. M. & JACOBSON, M. (1963). A study of the retinotectal projection during regeneration of the optic nerve in the frog. *Proceedings of the Royal Society*, B **157**, 420–48.

4, 5, 9, 22, 34

GAZE, R. M. & KEATING, M. J. (1970). Further studies on the restoration of the contralateral retinotectal projection following regeneration of the optic nerve in the frog. *Brain Research*, **21**, 183–95.

5, 9, 22, 34

GAZE, R. M. & KEATING, M. J. (1972). The visual system and 'Neuronal Specificity'. *Nature, London*, **237**, 375–8.

I, 2

GAZE, R. M. & SHARMA, S. C. (1968). Axial difference in the reinnervation of the optic tectum by regenerating optic nerve fibres. *Journal of Physiology*, **198**, 117P.

15, 16

GAZE, R. M. & SHARMA, S. C. (1970). Axial differences in the reinnervation of the goldfish optic tectum by regenerating optic nerve fibres. *Experimental Brain Research*, **10**, 171–81.

5, 15–18, 21, 22, 25, 34,

GAZE, R. M., JACOBSON, M. & SZÉKELY, G. (1963). The retino-tectal projection in *Xenopus* with compound eyes. *Journal of Physiology*, **165**, 484–99.

2, 14

GAZE, R. M., JACOBSON, M. & SZÉKELY, G. (1965). On the formation of connexions by compound eyes in *Xenopus*. *Journal of Physiology*, **176**, 409–17.

14

GAZE, R. M., JACOBSON, M. & SHARMA, S. C. (1966). Visual responses from the goldfish brain following excision and reimplantation of the optic tectum. *Journal of Physiology*, **183**, 38–9P.

GAZE, R. M., KEATING, M. J., SZÉKELY, G. & BEAZLEY, L. (1970). Binocular interaction in the formation of specific intertectal neuronal connexions. *Proceedings of the Royal Society*, B, **175**, 107–47.

GAZE, R. M., CHUNG, S. H. & KEATING, M. J. (1972). Development of the retinotectal projection in *Xenopus*. *Nature, New Biology, London*, **236**, 133–5.

GAZE, R. M., KEATING, M. J. & CHUNG, S. H. (1974). The evolution of the retinotectal map during development in *Xenopus*. *Proceedings of the Royal Society*, B, **185**, 301–30.

GEORGE, S. A. & MARKS, W. B. (1974). Optic nerve terminal arborizations in the frog: shape and orientation inferred from electrophysiological measurements. *Experimental Neurology*, **42**, 467–82.

GLASTONBURY, J. & STRAZNICKY, K. (1977). Delayed functional recovery of regenerated optic fibres in the tectum in *Xenopus*. *Proceedings of the Australian Physiological and Pharmacological Society*, **8**, 22P.   5, 9, 27

GLÜCKSMANN, A. (1940). Development and differentiation of the tadpole eye. *British Journal of Ophthalmology*, **24**, 153–78.

GOLDBERG, S. (1974). Studies on the mechanics of development of the visual pathways in the chick embryo. *Developmental Biology*, **36**, 24–43.   7, 13

GOLDBERG, S. (1976a). The invisible ocular map. *Bulletin of the New York Academy of Medicine*, **52**, 201–11.   1, 36

GOLDBERG, S. (1976b). Polarization of the avian retina. Ocular transplantation studies. *Journal of Comparative Neurology*, **68**, 379–92.   36

GOLDBERG, S. (1976c). Progressive fixation of morphological polarity in the developing retina. *Developmental Biology*, **53**, 126–7.   36

GOLDBERG, S. (1976d). Central nervous system regeneration and ophthalmology. *Survey of Ophthalmology*, **20**, 261–72.   1

GOLDBERG, S. (1977a). Unidirectional, bidirectional and random growth of embryonic optic axons. *Experimental Eye Research*, **25**, 399–404.   31

GOLDBERG, S. (1977b). Realignment of ocular polarity following transplantation of the eye to the optic tectum. *Anatomical Record*, **187**, 589.   36

GOLDBERG, S. & COULOMBRE, A. J. (1972). Topographical development of the ganglion cell fiber layer in chick retina. A whole mount study. *Journal of Comparative Neurology*, **146**, 507–18.

GOODMAN, D. C. & HOREL, J. A. (1966). Sprouting of optic tract projections in the brain stem of the rat. *Journal of Comparative Neurology*, **127**, 71–88.

GOODMAN, D. C., BOGDASARIAN, R. S. & HOREL, J. A. (1973). Axonal sprouting of ipsilateral optic tract following opposite eye removal. *Brain, Behavior and Evolution*, **8**, 27–50.   20

GOTTLIEB, D. I., ROCK, K. & GLASER, L. (1976). A gradient of adhesive specificity in developing avian retina. *Proceedings of the National Academy of Sciences, U.S.A.*, **73**, 410–14.

GRAFSTEIN, B. (1964). Functional organisation in regeneration of amphibian visual pathways. *Boletín del Instituto de Estudias Medicos y Biológicas Universidad Nacional Mexico*, **22**, 217–30.   5, 6, 24

GRAFSTEIN, B. (1971). Role of slow axonal transport in nerve regeneration. *Acta Neuropathologica*, Suppl. V, 144–52.

GRAFSTEIN, B. & BURGEN, A. S. V. (1964). Pattern of optic nerve connections following retinal regeneration. In *Topics in Basic Neurology*, eds. W. Bargmann & J. P. Schade, Elsevier, Amsterdam. *Progress in Brain Research*, 6, 126–38.

GRAINGER, F., JAMES, D. W. & TRESMAN, R. L. (1968). An electron-microscopic study of the early outgrowth from chick spinal cord *in vitro. Zeitschrift für Zellforschung und mikroskopische Anatomie*, 90, 53–67. 5, 6

GRINNELL, A. D. (1977). Specificity of neurons and their interconnections. In *Handbook of Physiology*, section 1, *The Nervous System*, vol. 1, part 2, eds. J. M. Brookhart & V. B. Mountcastle, pp. 803–53. American Physiological Society, Bethesda, Md. 1

HARRISON, R. G. (1901). Ueber die Histogenese des peripheren Nervensystems bei *Salmo salar. Archiv für mikroskopische Anatomie*, 57, 354–444.

HARRISON, R. G. (1910). The outgrowth of the nerve fiber as a mode of protoplasmic movement. *Journal of Experimental Zoology*, 9, 787–848.

HARRISON, R. G. (1924). Neuroblast versus sheath cell in the development of peripheral nerves. *Journal of Comparative Neurology*, 37, 123–205.

HARRISON, R. G. (1935). On the origin and development of the nervous system studied by the methods of experimental embryology. *Proceedings of the Royal Society*, B 118, 155–96.

HEACOCK, A. M. & AGRANOFF, B. W. (1977). Clockwise growth of neurites from retinal explants. Society for Neuroscience, Abstracts, 7th Annual Meeting, 3, p. 524. 33

HEACOCK, A. M. & AGRANOFF, B. W. (1977). Clockwise growth of neurites from retinal explants. *Science*, 198, 64–6. 33

HERRICK, C. J. (1941*a*). Development of the optic nerves of *Amblystoma. Journal of Comparative Neurology*, 74, 473–534.

HERRICK, C. J. (1941*b*). The eyes and optic paths of the catfish, *Ameiurus. Journal of Comparative Neurology*, 75, 255–86.

HERRICK, C. J. (1941*c*). Optic and postoptic systems of fibers in the brain of *Necturus. Journal of Comparative Neurology*, 75, 487–544.

HERRICK, C. J. (1942). Optic and postoptic systems in the brain of *Amblystoma tigrinum. Journal of Comparative Neurology*, 77, 191–353.

HIBBARD, E. (1959). Central integration of developing nerve tracts from supernumerary grafted eyes and brain in the frog. *Journal of Experimental Zoology*, 141, 323–51.

HIBBARD, E. (1963). Regeneration in the severed spinal cord of chordate larvae of *Petromyzon marinus. Experimental Neurology*, 7, 175–85. 32

HIBBARD, E. (1965). Orientation and directed growth of Mauthner's cell axons from duplicated vestibular nerve roots. *Experimental Neurology*, 13, 289–301.

HIBBARD, E. (1967). Visual recovery following regeneration of the optic nerve through the oculomotor nerve root in *Xenopus. Experimental Neurology*, 19, 350–6. 9, 31

HIBBARD, E. & ORNBERG, R. L. (1976). Restoration of vision in genetically eyeless axolotls (*Ambystoma mexicanum*). *Experimental Neurology*, 50, 113–23. 9

HINDS, J. W. & HINDS, P. L. (1974). Early ganglion cell differentiation in the mouse retina: an electron microscopic analysis utilizing serial sections. *Developmental Biology*, 37, 381–416.

HIRSCH, H. V. B. & JACOBSON, M. (1973). Development and maintenance of connectivity in the visual system of the frog. II. The effects of eye removal. *Brain Research*, **49**, 67–74. 13

HIRSCH, H. V. B. & JACOBSON, M. (1975). The perfectible brain: Principles of neuronal development. In *Handbook of Psychobiology*, eds. M. S. Gazzaniga & C. Blakemore, pp. 107–37. Academic Press, New York. 1

HOLMES, W. & YOUNG, J. Z. (1942). Nerve regeneration after immediate and delayed suture. *Journal of Anatomy*, **77**, 63–96. 32

HOPE, R. A., HAMMOND, B. J. & GAZE, R. M. (1976). The arrow model: retinotectal specificity and map formation in the goldfish visual system. *Proceedings of the Royal Society*, B **194**, 447–66. 2, 12, 19, 26

HORDER, T. J. (1971*a*). Retention, by fish optic nerve fibres regenerating to new terminal sites in the tectum, of 'chemospecific' affinity for their original sites. *Journal of Physiology*, **216**, 53–5P. 13, 15, 19, 20

HORDER, T. J. (1971*b*). The course of recovery of the retinotectal projection during regeneration of the fish optic nerve. *Journal of Physiology*, **217**, 53–4P. 5, 22

HORDER, T. J. (1971*c*). 'Path selection in the regeneration of the teleost optic nerve'. Ph.D.Thesis, University of Edinburgh. 3, 5, 13, 15, 19, 20, 22, 30

HORDER, T. J. (1974*a*). Electron microscopic evidence in goldfish that different optic nerve fibres regenerate selectively through specific routes into the tectum. *Journal of Physiology*, **241**, 84–5P. 3, 30

HORDER, T. J. (1974*b*). Changes of fibre pathways in the goldfish optic tract following regeneration. *Brain Research*, **72**, 41–52. 5, 13, 30

HORDER, T. J. (1976). Pattern formation in animal embryos. In *The Developmental Biology of Plants and Animals*, eds. C. F. Graham & P. F. Wareing, pp. 169–97. Blackwells, Oxford.

HORDER, T. J. & MARTIN, K. A. C. (1977*a*). Translocation of optic fibres in the tectum may be determined by their stability relative to surrounding fibre terminals. *Journal of Physiology*, **271**, 23–4P. 13, 15–19, 21, 22, 25, 26, 35

HORDER, T. J. & MARTIN, K. A. C. (1977*b*). Variability among laboratories in the occurrence of functional modification in the intertectal visual projection of *Xenopus laevis*. *Journal of Physiology*, **272**, 90–1P.

HORDER, T. J. & SPITZER, J. L. (1973). Absence of cell mobility across the retina in *Xenopus laevis* embryos. *Journal of Physiology*, **233**, 33–4P.

HOYT, W. F. (1962). Anatomic considerations of arcuate scotomas associated with lesions of the optic nerve and chiasm. A Nauta axon degeneration study in the monkey. *Bulletin of the Johns Hopkins Hospital*, **111**, 57–71. 3

HUGHES, W. F. & LA VELLE, A. (1975). The effects of early tectal lesions on development in the retinal ganglion cell layer of chick embryos. *Journal of Comparative Neurology*, **163**, 265–84.

HUNT, R. K. (1975*a*). Developmental programming for retinotectal patterns. In *Cell Patterning*, eds. R. Porter & J. Rivers, pp. 131–50. Ciba Foundation Symposium, **29** (New Series). Associated Science, Amsterdam. 14, 33, 34, 36

HUNT, R. K. (1975*b*). The cell cycle, cell lineage, and neuronal specificity. In *Cell Cycle and Cell Differentiation*, eds. J. Reinart & H. Holtzer, pp. 43–62. Springer, Berlin and New York. 1

HUNT, R. K. (1976*a*). Informational properties of growing optic axons. *Biophysical Journal*, **16**, 214a. 34, 36

HUNT, R. K. (1976b). Cell lineage and cell interactions in neuronal differentiation. In *Progress in Differentiation Research*, ed. N. Müller-Bérat, pp. 11–24. North Holland, Amsterdam.   **1, 14, 36**

HUNT, R. K. (1976c). Position-dependent differentiation of neurones. In *Developmental Biology: Pattern Formation: Gene Regulation*, eds. D. MacMahon & C. F. Fox, pp. 227–56. INC-UCLA Symposium on Molecular and Cellular Biology. W. A. Benjamin, New York   **1, 9–11, 34**

HUNT, R. K. (1977a). Competitive retinotectal mapping in *Xenopus*. *Biophysical Journal*, **17**, 128a.   **16, 20**

HUNT, R. K. (1977b). Positional signalling and nerve cell specificity. In *Cell Interactions in Differentiation*, eds. M. Karkinen-Jääskeläinen, L. Saxén & L. Weiss, pp. 97–110. Academic Press, London.   **6, 34, 36**

HUNT, R. K. (1978a). Combinatorial specifiers on retinal ganglion cells for retinotectal map assembly in *Xenopus*. *Nature, London* (in press).

HUNT, R. K. (1978b). Target properties in the optic tectum for retinotectal map assembly in *Xenopus*. *Nature, London* (in press).

HUNT, R. K. (1978c). Genetic control of neural development. In *Cell, Tissue and Organ Culture in Neurobiology*, eds. S. Fedoroff & L. Hertz, pp. 369–92. Academic Press, New York.

HUNT, R. K. & BERMAN, N. (1975). Patterning of neuronal locus specificities in retinal ganglion cells after partial extirpation of the embryonic eye. *Journal of Comparative Neurology*, **162**, 43–70. **5, 6, 8, 14, 33, 34, 36**

HUNT, R. K. & FRANK, E. (1975). Neuronal locus specificity: transrepolarization of *Xenopus* embryonic retina after the time of axial specification. *Science*, **189**, 563–5.   **6, 14, 34, 36**

HUNT, R. K. & IDE, C. F. (1977). Radial propagation of positional signals for retinotectal patterns in *Xenopus*. *Biological Bulletin*, **153**, 430–43.

HUNT, R. K. & JACOBSON, M. (1972a). Development and stability of positional information in *Xenopus* retinal ganglion cells. *Proceedings of the National Academy of Sciences, U.S.A.*, **69**, 780–3.   **6, 13, 36**

HUNT, R. K. & JACOBSON, M. (1972b). Specification of positional information in retinal ganglion cells of *Xenopus*: stability of the specified state. *Proceedings of the National Academy of Sciences, U.S.A.*, **69**, 2860–4.   **33, 36**

HUNT, R. K. & JACOBSON, M. (1973a). Specification of positional information in retinal ganglion cells of *Xenopus*: assays for analysis of the unspecified state. *Proceedings of the National Academy of Sciences, U.S.A.*, **70**, 507–11.   **33, 36**

HUNT, R. K. & JACOBSON, M. (1973b). Neuronal locus specificity: altered pattern of spatial deployment in fused fragments of embryonic *Xenopus* eyes. *Science*, **180**, 509–11.   **14, 36**

HUNT, R. K. & JACOBSON, M. (1974a). Neuronal specificity revisited. *Current Topics in Developmental Biology*, **8**, 203–59.   **1**

HUNT, R. K. & JACOBSON, M. (1974b). Rapid reversal of retinal axes in embryonic *Xenopus* eyes. *Journal of Physiology*, **241**, 90–1P.   **33, 34, 36**

HUNT, R. K. & JACOBSON, M. (1974c). Specification of positional information in retinal ganglion cells of *Xenopus laevis*: intra-ocular control of the time of specification. *Proceedings of the National Academy of Sciences, U.S.A.*, **71**, 3616–20.   **6, 36**

HUNT, R. K. & JACOBSON, M. (1974d). Development of neuronal locus specificity in *Xenopus* retinal ganglion cells after surgical eye transection or after fusion of whole eyes. *Developmental Biology*, **40**, 1–15. **6, 14, 34, 36**

HUNT, R. K. & PIATT, J. (1974). Axial specification in salamander embryonic retina. *Anatomical Record*, **178**, 515–6.   34, 36

HUNT, R. K. & PIATT, J. (1978). Cross-species axial signalling, with realignment of retinal axes, in embryonic *Xenopus* eyes. *Developmental Biology*, **62**, 44–51.   6, 34, 36

HUNT, R. K., BERGEY, G. K. & HOLTZER, H. (1978). Identification of a developmental program within *Xenopus* optic cup. *Developmental Biology* (in press).

HUNT, R. K., BODENSTEIN, L. & KOSOFSKY, B. E. (1978). Positional signalling and amphibian neurogenesis. *American Zoologist*, symposium chapter (in press).

INGLE, D. (1973). Two visual systems in the frog. *Science*, **181**, 1053–5. 9, 23, 29

INGLE, D. (1976). Functional consequences of binocular innervation of a single optic tectum in frogs. Society for Neuroscience, Abstracts, 6th Annual Meeting, **2**, p. 1118.

INGLE, D. & DUDEK, A. (1977). Aberrant retinotectal projections in the frog. *Experimental Neurology*, **55**, 567–82.   15, 17, 18, 20, 25, 34

IRWIN, L. N., CHEN, H. & BARRACO, R. A. (1976). Ganglioside, protein, hexose and sialic acid changes in the bisected optic tectum of the chick embryo. *Developmental Biology*, **49**, 29–39.

JACOBSON, M. (1961*a*). The recovery of electrical activity in the optic tectum of the frog during early regeneration of the optic nerve. *Journal of Physiology*, **157**, 27–9P.   5, 22, 30, 34

JACOBSON, M. (1961*b*). Recovery of electrical activity in the optic tectum of the frog during early regeneration of the optic nerve. *Proceedings of the Royal Physical Society of Edinburgh*, **28**, 131–7.   5, 30, 34

JACOBSON, M. (1966). Starting points for research in the ontogeny of behavior. In *Major Problems in Developmental Biology*, ed. M. Locke, pp. 339–83. Academic Press, New York.   1, 30

JACOBSON, M. (1967). Retinal ganglion cells: specification of central connections in larval *Xenopus laevis*. *Science*, **155**, 1106–8.   6, 36

JACOBSON, M. (1968*a*). Specification of neuronal connections during development. In *Physiological and Biochemical Aspects of Nervous Integration*, ed. F. D. Carlson, pp. 195–214. Prentice-Hall, New Jersey.   1

JACOBSON, M. (1968*b*). Development of neuronal specificity in retinal ganglion cells of *Xenopus*. *Developmental Biology*, **17**, 202–18.   3, 36

JACOBSON, M. (1969). Development of specific neuronal connections. *Science*, **163**, 543–7.   6, 14, 36

JACOBSON, M. (1970*a*). Development, specification and diversification of neuronal connections. In *The Neurosciences: Second Study Program*, ed. F. O. Schmitt, pp. 116–29. Rockefeller University Press, New York.   1

JACOBSON, M. (1970*b*). *Developmental Neurobiology*. Holt, Rinehart and Winston, New York.   1

JACOBSON, M. (1971). Formation of Neuronal Connections in Sensory Systems. In *Handbook of Sensory Physiology*, ed. W. R. Loewenstein, vol. 1, pp. 166–190. Springer, Berlin, Heidelberg, New York.

JACOBSON, M. (1973). Genesis of Neuronal Locus Specificity. In *Development and Aging in the Nervous System*, ed. M. Rockstein, pp. 105–19. Academic Press, New York.   1, 36

JACOBSON, M. (1974*a*). Nervous System (Vertebrate). In *McGraw-Hill Yearbook of Science and Technology 1974*, pp. 304–6. McGraw Hill, New York.   1

JACOBSON, M. (1974b). Neuronal Plasticity: Concepts in Pursuit of Cellular Mechanisms. In *Plasticity and Recovery of Function in the Central Nervous System*, eds. D. G. Stein, J. J. Rosen and N. Butters, pp. 31–43. Academic Press, New York.   1

JACOBSON, M. (1974c). Through the jungle of the brain: neuronal specificity and typology revisited. *Annals of the New York Academy of Sciences*, 228, 63–7.   1

JACOBSON, M. (1974d). Differentiation and Growth of Nerve Cells. In *Differentiation and Growth of Cells in Vertebrate Tissues*, ed. G. Goldspink, chapter 2, pp. 53–67. Chapman and Hall, London.   1

JACOBSON, M. (1976a). Histogenesis of retina in the clawed frog with implications for the pattern of development of retinotectal connections. *Brain Research*, 103, 541–5.

JACOBSON, M. (1976b). Premature specification of the retina in embryonic *Xenopus* eyes treated with ionophore X537A. *Science*, 191, 288–90.
                                                                            36
JACOBSON, M. (1976c). Neuronal recognition in the retinotectal system. In *Neuronal Recognition*, ed. S. H. Barondes, pp. 3–23. Chapman & Hall, London.
                                                                             1
JACOBSON, M. (1977). Mapping the developing retinotectal projection in frog tadpoles by a double label autoradiographic technique. *Brain Research*, 127, 55–67.

JACOBSON, M. & GAZE, R. M. (1964). Types of visual response from single units in the optic tectum and optic nerve of the goldfish. *Quarterly Journal of Experimental Physiology*, 49, 199–209.

JACOBSON, M. & GAZE, R. M. (1965). Selection of appropriate tectal connections by regenerating optic nerve fibers in adult goldfish. *Experimental Neurology*, 13, 418–30.   15–21

JACOBSON, M. & HIRSCH, H. V. B. (1973). Development and maintenance of connectivity in the visual system of the frog. 1. The effects of eye rotation and visual deprivation. *Brain Research*, 49, 47–65.   6, 9

JACOBSON, M. & HUNT, R. K. (1973). The origins of nerve-cell specificity. *Scientific American*, 228, February, 26–35.   1, 36

JACOBSON, M. & LEVINE, R. L. (1975a). Plasticity in the adult frog brain: filling the visual scotoma after excision or translocation of parts of the optic tectum. *Brain Research*, 88, 339–45.   12, 15, 17, 19, 26

JACOBSON, M. & LEVINE, R. L. (1975b). Stability of implanted duplicate tectal positional markers serving as targets for optic axons in adult frogs. *Brain Research*, 92, 468–71.   12, 19, 22, 26

JHAVERI, S. R. & SCHNEIDER, G. E. (1974). Retinal projections in Syrian hamsters: normal topography, and alterations after partial tectum lesions at birth. *Anatomical Record*, 178, 383.   15, 16, 19

JOHNS, P. R. (1977). Growth of the adult goldfish eye. III. Source of the new retinal cells. *Journal of Comparative Neurology*, 176, 343–58.

JOHNS, P. R., HEACOCK, A. M. & AGRANOFF, B. W. (1977). Regeneration of retinal ganglion cell axons *in vitro*. Society for Neuroscience, Abstracts, 7th Annual Meeting, 3, p. 524.
                                                                            33
JOHNS, P. R., YOON, M. G. & AGRANOFF, B. W. (1978). Directed growth of optic fibres regenerating *in vitro*. *Nature, London*, 271, 360–2.
                                                                            33
KALIL, R. E. & SCHNEIDER, G. E. (1975). Abnormal synaptic connections of the optic tract in the thalamus after midbrain lesions in newborn hamsters. *Brain Research*, 100, 690–8.
                                                                         15, 31

KARTEN, H. J., FITE, K. V. & BRECHA, N. (1977). Specific projection of displaced retinal ganglion cells upon the accessory optic system in the pigeon (*Columbia livia*). *Proceedings of the National Academy of Sciences, U.S.A.*, **74**, 1753–6.

KATZ, M. J. & LASEK, R. J. (1977). Primitive sensory pathways organize sensory long tracts of hindbrain and spinal cord in *Xenopus* tadpoles. Society for Neuroscience, Abstracts, 7th Annual Meeting, **3**, p. 109.   31

KEATING, M. J. (1976). The formation of visual neuronal connections: an appraisal of the present status of the theory of 'Neuronal Specificity'. In *Neural and Behavioral Specificity. Studies on the development of behavior and the nervous system*, vol. 3, ed. G. Gottlieb, pp. 59–110. Academic Press, New York.   1

KEATING, M. J. (1977). Evidence for plasticity of intertectal connections in adult *Xenopus*. *Philosophical Transactions of the Royal Society*, B **278**, 277–94.

KEATING, M. J. & GAZE, R. M. (1970). Rigidity and plasticity in the amphibian visual system. *Brain, Behavior and Evolution*, **3**, 102–20.   1

KEATING, M. J. & KENNARD, C. (1976). The amphibian visual system as a model for developmental neurobiology. In *The Amphibian Visual System*, ed. K. V. Fite, pp. 267–315. Academic Press, New York.   1, 2

KELLY, J. P. & GILBERT, C. D. (1975). The projections of different morphological types of ganglion cells in the cat retina. *Journal of Comparative Neurology*, **163**, 65–80.

KICLITER, E. (1974). Retinal projections in frogs with regenerated optic nerves. *Anatomical Record*, **178**, 390.   9, 15, 27

KICLITER, E., MISANTONE, L. J. & STELZNER, D. J. (1974). Neuronal specificity and plasticity in frog visual system: anatomical correlates. *Brain Research*, **82**, 293–7.   9, 15, 28

KIRK, E. G. & LEWIS, D. D. (1917). Regeneration in peripheral nerves. An experimental study. *Bulletin of the Johns Hopkins Hospital*, **28**, 71–80.   32

KIRSCHE, W. (1950). Die regenerativen Vorgänge am Rückenmark erwachsener Teleostier nach operativer Kontinuitätstrennung. *Zeitschrift für mikroskopische-anatomische Forschung*, **56**, 190–205.   32

KIRSCHE, W. & KIRSCHE, K. (1961). Experimentelle Untersuchungen zur Frage der Regeneration und Funktion des Tectum opticum von *Carassius carassius* L. *Zeitschrift für mikroskopisch-anatomische Forschung*, **67**, 140–82.

KNOWLTON, V. Y. (1964). Abnormal differentiation of embryonic avian brain centers associated with unilateral anopthalmia. *Acta Anatomica*, **58**, 222–51.

KOLLROS, J. J. (1953). The development of the optic lobes in the frog I. The effects of unilateral enucleation in embryonic stages. *Journal of Experimental Zoology*, **123**, 153–87.

LAND, P. W., POLLEY, E. H. & KERNIS, M. M. (1976). Patterns of retinal projections to the lateral geniculate nucleus and superior colliculus of rats with induced unilateral congenital eye defects. *Brain Research*, **103**, 394–9.   9, 20

LANDRETH, G. E. & AGRANOFF, B. W. (1976). Explant culture of adult goldfish retina: effect of prior optic nerve crush. *Brain Research*, **118**, 299–303.   33

LASHLEY, K. S. (1934). The mechanism of vision. VII. The projection of the retina upon the primary optic centers in the rat. *Journal of Comparative Neurology*, **59**, 341–73.

LÁZÁR, G. (1971). The projection of the retinal quadrants on the optic centres in the frog. A terminal degeneration study. *Acta Morphologica Academiae Scientiarum Hungaricae*, **19**, 325–34.

LÁZÁR, G. (1973). The development of the optic tectum in *Xenopus laevis*: a Golgi study. *Journal of Anatomy*, **116**, 347–55.

LEVICK, W. K. (1975). Formation and function of cat retinal ganglion cells. *Nature, London*, **254**, 659–62.

LEVINE, R. (1975). Regeneration of the retina in the adult newt, *Triturus cristatus*, following surgical division of the eye by a limbal incision. *Journal of Experimental Zoology*, **192**, 363–80.

LEVINE, R. & CRONLY-DILLON, J. R. (1974). Specification of regenerating retinal ganglion cells in the adult newt, *Triturus cristatus*. *Brain Research*, **68**, 319–29.   6

LEVINE, R. & JACOBSON, M. (1974). Deployment of optic nerve fibers is determined by positional markers in the frog's tectum. *Experimental Neurology*, **43**, 527–38.   11, 26

LEVINE, R. & JACOBSON, M. (1975). Discontinuous mapping of retina onto tectum innervated by both eyes. *Brain Research*, **98**, 172–6.
                                                                10, 15, 23, 28, 31

LLINÁS, R. & PRECHT, W. (1976). *Frog Neurobiology. A Handbook*. Springer, Berlin.

LOPRESTI, V., MACAGNO, E. R. & LEVINTHAL, C. (1973). Structure and development of neuronal connections in isogenic organisms: cellular interactions in the development of the optic lamina of *Daphnia*. *Proceedings of the National Academy of Sciences, U.S.A.*, **70**, 433–7.

LUBIŃSKA, L. (1964). Axoplasmic streaming in regenerating and in normal nerve fibres. *Progress in Brain Research*, **13**, 1–66.

LUBSEN, J. (1921). Over de projectie het netvlies op het tectum opticum bij een beenvisch. *Nederland Tijdsch. Geneesk.* **67**, 1258–60.   3

LUND, R. D. (1972). Anatomic studies on the superior colliculus. *Investigative Ophthalmology*, **11**, 434–40.   9, 27

LUND, R. D. (1975). Variations in the laterality of the central projections of retinal ganglion cells. *Experimental Eye Research*, **21**, 193–203.   1

LUND, R. D., CUNNINGHAM, T. J. & LUND, J. S. (1973). Modified optic projections after unilateral eye removal in young rats. *Brain, Behavior and Evolution*, **8**, 51–72.   9, 27

LUND, R. D. & LUND, J. S. (1971). Synaptic adjustment after deafferentation of the superior colliculus of the rat. *Science*, **171**, 804–7.   20, 27

LUND, R. D. & LUND, J. S. (1973). Reorganization of the retinotectal pathway in rats after neonatal retinal lesions. *Experimental Neurology*, **40**, 377–90.   9, 13, 28

LUND, R. D. & LUND, J. S. (1975). Spatially inappropriate retinotectal connections in rats. *Anatomical Record*, **181**, 416.   9, 13, 16, 24

LUND, R. D. & LUND, J. S. (1976). Plasticity in the developing visual system: the effects of retinal lesions made in young rats. *Journal of Comparative Neurology*, **169**, 133–54.   9, 13, 16, 24, 27–29

LUND, R. D. & HAUSCHKA, S. D. (1976). Transplanted neural tissue develops connections with host rat brain. *Science*, **193**, 582–4.   12

LUND, R. D. & MILLER, B. F. (1975). Secondary effects of fetal eye damage in rats on intact central optic projections. *Brain Research*, **92**, 279–289.   9, 13

McGILL, J. I., POWELL, T. P. S. & COWAN, W. M. (1966). The retinal representation upon the optic tectum and isthmo-optic nucleus in the pigeon. *Journal of Anatomy*, **100**, 5–33.

23

McLACHLAN, E. M. (1974). The formation of synapses in mammalian sympathetic ganglia reinnervated with preganglionic or somatic nerves. *Journal of Physiology*, **237**, 217–42.

McQUARRIE, I. G., GRAFSTEIN, B. & GERSHON, M. D. (1977). Axonal regeneration in the rat sciatic nerve: effect of a conditioning lesion and of dbcAMP. *Brain Research*, **132**, 443–53.

MARCHASE, R. B., BARBERA, A. J. & ROTH, S. (1975). A molecular approach to retinotectal specificity. In *Cell Patterning*, eds. R. Porter & J. Rivers, pp. 315–27. Ciba Foundation Symposium, No. 29 (New Series). Associated Science, Amsterdam. 2

MARLSBURG, C. V.D. & WILLSHAW, D. J. (1976). A mechanism for producing continuous neural mappings, ocularity dominance stripes and ordered retino-tectal projections. *Experimental Brain Research*, Suppl. 1, 463–9. 2

MARLSBURG, C. V.D. & WILLSHAW, D. J. (1977). How to label nerve cells so that they can interconnect in an orderly fashion. *Proceedings of the National Academy of Sciences, U.S.A.*, **74**, 5176–8. 2

MAROTTE, L. R., WYE-DVORAK, J. & MARK, R. F. (1977). Ultrastructure of reorganising visual projections in half tecta of carp kept in constant light. *Neuroscience*, **2**, 767–80. 15, 17, 25, 35

MARTIN, K. A. C. (1978). Combination of fibre–fibre competition and regional tectal differences accounting for the results of tectal graft experiments in goldfish. *Journal of Physiology*, **276**, 44–5P. 12, 15, 19, 26, 34

MATTHEY, R. (1926). La Greffe de l'oeil. 1. Étude histologique sur la greffe de l'oeil chez la larve de Salamandre (*Salamandra maculosa*). *Revue Suisse de Zoologie*, **33**, 317–34. 4

MATURANA, H. R. (1958). Efferent fibres in the optic nerve of the toad (*Bufo bufo*). *Journal of Anatomy*, **92**, 21–7.

MATURANA, H. R., LETTVIN, J. Y., McCULLOCH, W. S. & PITTS, W. H. (1959). Evidence that cut optic nerve fibers in a frog regenerate to their proper places in the tectum. *Science*, **130**, 1709–10. 5

MATURANA, H. R., LETTVIN, J. Y., McCULLOCH, W. S. & PITTS, W. H. (1960). Anatomy and physiology of vision in the frog (*Rana pipiens*). *Journal of General Physiology*, **43**, suppl. 2, 129–75.

MAY, R. M. & DETWILER, S. R. (1925). The relation of transplanted eyes to developing nerve centers. *Journal of Experimental Zoology*, **43**, 83–103. 31

MEYER, R. L. (1974). 'Factors affecting regeneration of the retinotectal projection'. Doctoral dissertation, California Institute of Technology. *Dissertation Abstracts*, **35**, 1510B (Ann Arbor, Michigan University Microfilms No. 74-21603). 7, 9

MEYER, R. L. (1975). Tests for regulation in the goldfish retinotectal system. *Anatomical Record*, **181**, 427. 13, 15, 19

MEYER, R. L. (1976a). Tests for field regulation in the retinotectal system of goldfish. In *Developmental Biology: Pattern Formation: Gene Regulation*, eds. D. McMahon & C. F. Fox, pp. 257–75. Benjamin, New York. 7, 9, 13, 15, 17, 19, 20, 24, 27, 34

MEYER, R. L. (1976b). Electrophysiological analysis of the retinotectal projection in goldfish following uncrossing of selected optic radiation fibres. Society for Neuroscience, Abstracts, 6th Annual Meeting, 2, p. 1127. 7, 9, 16, 19, 24

MEYER, R. L. (1977). Eye-in-water electrophysiological mapping of goldfish with and without tectal lesions. *Experimental Neurology*, **56**, 23–41. 15–18, 21, 22, 25, 35

MEYER, R. L. & SCOTT, M. Y. (1977). Failure of continuous light to inhibit compression of retinotectal projection in goldfish. *Brain Research*, **128**, 153–7.   **15, 17, 25, 35**

MEYER, R. L. & SPERRY, R. W. (1973). Tests for neuroplasticity in the anuran retinotectal system. *Experimental Neurology*, **40**, 525–39. **5, 15, 21, 22**

MEYER, R. L. & SPERRY, R. W. (1974). Explanatory models for neuroplasticity in retinotectal connections. In *Plasticity and Recovery of Function in the Central Nervous System*, eds. D. G. Stein, J. J. Rosen & N. Butters, pp. 45–63. Academic Press, New York.   **1, 2**

MEYER, R. L. & SPERRY, R. W. (1976). Retinotectal specificity: chemoaffinity theory. In *Neural and Behavioral Specificity. Studies on the development of behavior and the nervous system*, vol. 3, ed. G. Gottlieb, pp. 111–49. Academic Press, New York.   **1, 2**

MILLER, B. F. (1975). Effects of superior collicular lesions in fetal rats on the development of the retinotectal projection. *Anatomical Record*, **181**, 428.   **9, 19, 23, 24, 29, 31**

MILLER, B. F. & LUND, R. D. (1975). The pattern of retinotectal connections in albino rats can be modified by fetal surgery. *Brain Research*, **91**, 119–25.   **9, 19, 24, 29, 31**

MISANTONE, L. J. & STELZNER, D. J. (1974*a*). Competition of retinal endings for sites in doubly innervated frog optic tectum. *Anatomical Record*, **178**, 419.   **9, 15, 27, 29**

MISANTONE, L. J. & STELZNER, D. J. (1974*b*). Behavioral manifestations of competition of retinal endings for sites in doubly innervated frog optic tectum. *Experimental Neurology*, **45**, 364–76.   **9, 15, 27, 29, 31**

MOSCONA, A. A. (1974). Surface specification of embryonic cells: lectin receptors, cell recognition, and specific cell ligands. In *The Cell Surface in Development*, ed. A. A. Moscona, pp. 67–99. Wiley, New York.

MURRAY, M. (1976). Regeneration of retinal axons into the goldfish optic tectum. *Journal of Comparative Neurology*, **168**, 175–96.

MURRAY, M. (1977). Delayed regeneration of the retino-diencephalic projections in the goldfish, *Carassius auratus*. *Brain Research*, **125**, 149–53.

MURRAY, M. & SHARMA, S. C. (1977). Reinnervation of partially ablated tectum in adult goldfish. *Society for Neuroscience, Abstracts*, 7th Annual Meeting, **3**, p. 429.   **15, 17**

NAKAI, J. (1960). Studies on the mechanism determining the course of nerve fibers in tissue culture. II. The mechanism of fasciculation. *Zeitschrift für Zellforschung und mikroskopische Anatomie*, **52**, 427–49.

OPPENHEIMER, J. M. (1950). Anomalous optic chiasma in *Fundulus* embryos. *Anatomical Record*, **108**, 477–83.   **28**

ORTIN, L. & ARCAUTE, L. R. (1913). Procesos regenerativos del nervio óptico y retina con ocasión de ingertos nerviosos. *Trabajos del Laboratorio de Investigaciones Biologicas, Madrid*, **11**, 239–54.   **4**

PARSONS, J. H. (1902). Degenerations following lesions of the retina in monkeys. *Brain*, **25**, 257–69.   **3**

PATON, M. C. & CAPRANICA, R. R. (1974). Evidence for directed growth of the optic tract in foreign nervous tissue. *Society for Neuroscience, Abstracts*, 4th Annual Meeting, p. 369.   **31**

PIATT, J. (1949). A study of the development of fiber tracts in the brain of *Amblystoma* after excision or inversion of the embryonic dimesencephalic region. *Journal of Comparative Neurology*, **90**, 47–94.   **11**

PIATT, J. (1955). Regeneration of the spinal cord in the salamander. *Journal of Experimental Zoology*, **129**, 177–207.   **32**

PICK, A. (1896). Untersuchungen über die topographischen Beziehungen zwischen Retina, Opticus und gekreuztem Tractus opticus beim Kaninchen. *Nova Acta Academiae Caesareae Leopoldino-Carolinae Germanicae Naturae Curiosorum*, **66**, 1–24. 3

POLYAK, S. (1957). *The Vertebrate Visual System.* University of Chicago Press, Chicago. 3

POMERANZ, B. (1972). Metamorphosis of frog vision: changes in ganglion cell physiology and anatomy. *Experimental Neurology*, **34**, 187–99.

POTTER, H. D. (1976). Axonal and synaptic lamination in the optic tectum. *Experimental Brain Research*, suppl. 1, 506–11.

PRESTIGE, M. C. & WILLSHAW, D. J. (1975). On a role for competition in the formation of patterned neural connexions. *Proceedings of the Royal Society*, B **190**, 77–98. 2

PURO, D. G., DE MELLO, F. G. & NIRENBERG, M. (1977). Synapse turnover: the formation and termination of transcient synapses. *Proceedings of the National Academy of Sciences, U.S.A.*, **74**, 4977–81.

RAFFIN, J.-P. (1972). Quelques effets de l'ablation uni- ou bilaterale de l'ebauche optique sur le tectum superficiel du poulet (*Gallus domesticus*). *Acta embryologiae et morphologiae experimentalis*, 45–63. 9, 28

RAFFIN, J.-P. (1975). Ipsilateral optic pathways in uniopthalm chicken (*Gallus domesticus* L.). *Experimental Brain Research*, **23** (suppl.) 168. 9, 27, 28

RAFFIN, J.-P. & REPÉRANT, J. (1975). Etude expérimentale de la spécificité des projections visuelles d'embryons et de poussin de *Gallus domesticus* L. micropthalmes et monopthalmes. *Archives d'anatomie microscopique et de morphologie expérimentale*, **64**, 93–111. 9, 27, 28

RAGER, G. (1976). Basic developmental processes in the retino-tectal system of the chicken. *Experimental Brain Research*, suppl. 1, 526–32.

RAGER, G. & RAGER, U. (1976). Generation and degeneration of retinal ganglion cells in the chicken. *Experimental Brain Research*, **25**, 551–3.

RAMÓN Y CAJAL, S. (1928). *Degeneration and Regeneration of the Nervous System.* Translated by R. M. May. Hafner, New York (1959).

RAMÓN Y CAJAL, S. (1960). *Studies in Vertebrate Neurogenesis.* Translated by L. Guth. Charles C. Thomas, Springfield, Ill. pp. 432.

RANSON, S. W. (1912). Degeneration and regeneration of nerve fibres. *Journal of Comparative Neurology*, **22**, 487–546.

REIER, P. J. & WEBSTER, H. DE F. (1974). Regeneration and myelination of *Xenopus* tadpole optic nerve fibres following transection or crush. *Journal of Neurocytology*, **3**, 591–618. 4

REIER, P. J. (1978). An ultrastructural analysis of early axonal outgrowth during optic nerve regeneration in *Xenopus laevis* tadpoles. *Anatomical Record*, **190**, 519.

ROGERS, K. T. (1952). Optic nerve pattern evidence for fusion of eye primordia in cyclopia in *Fundulus heteroclitus. Journal of Experimental Zoology*, **120**, 287–309.

ROGERS, L. A., FINGER, T. E. & COWAN, W. M. (1977). Plasticity in the visual system of the developing chick. Society for Neuroscience, Abstracts, 7th Annual Meeting, **3**, p. 430. 9, 15, 28

ROMESKIE, M. & SHARMA, S. C. (1977). Effects of prior long-term tectal denervation on the retinal projection in goldfish. Society for Neuroscience, Abstracts, 7th Annual Meeting, **3**, p. 431. 9, 11, 15, 16, 29

ROTH, R. L. (1967). Patterns of tectal reinnervation by regenerating optic nerve fibers in the goldfish. *Anatomical Record*, **157**, 312. 9, 13, 15, 19, 29, 31

ROTH, R. L. (1972). 'Normal and regenerated retino-tectal projections in the goldfish'. D. Phil. Thesis, Case Western Reserve University, Cleveland, Ohio. *Dissertation Abstracts*, **33**, 4085–6B (Ann Arbor, Michigan, University Microfilms no. 73-6335). **5, 9, 13, 15, 19, 20, 29, 30**

ROTH, R. L. (1974). Retinotopic organization of goldfish optic nerve and tract. *Anatomical Record*, **178**, 453. **3**

ROTH, S. & MARCHASE, R. B. (1976). An *in vitro* assay for retinotectal specificity. In *Neuronal Recognition*, ed. S. H. Barondes, pp. 227–48. Chapman & Hall, London.

RUTISHAUSER, U., THIERY, J.-P., BRACKENBURY, R., SELA, B.-A. & EDELMAN, G. M. (1976). Mechanisms of adhesion among cells from neural tissues of the chick embryo. *Proceedings of the National Academy of Sciences, U.S.A.*, **73**, 577–81.

SANDERSON, K. J., GUILLERY, R. W. & SHACKELFORD, R. M. (1974). Congenitally abnormal visual pathways in mink (*Mustela vison*) with reduced retinal pigment. *Journal of Comparative Neurology*, **154**, 225–48.

SCALIA, F. & FITE, K. (1974). A retinotopic analysis of the central connections of the optic nerve in the frog. *Journal of Comparative Neurology*, **158**, 455–78.

SCHMATOLLA, E. (1974). Retino-tectal course of optic nerves in cyclopic and synopthalmic Zebrafish embryos. *Anatomical Record*, **180**, 377–84.

SCHMIDT, J. T. (1975). Retinal afferents relabel the tectum in goldfish. Society for Neuroscience, Abstracts, 5th Annual Meeting, **1**, p. 766.

SCHMIDT, J. T. (1978). Retinal fibers alter tectal positional markers during the expansion of the half retinal projection in goldfish. *Journal of Comparative Neurology*, **177**, 279–300. **2, 9, 10, 13, 15, 20–5**

SCHMIDT, J. T., CICERONE, C. M. & EASTER, S. S. (1974). Reorganisation of the retinotectal projection in goldfish. Society for Neuroscience, Abstracts, 4th Annual Meeting, p. 413. **13, 15, 17, 20**

SCHMIDT, J. T., CICERONE, C. M. & EASTER, S. S. (1978). Expansion of the half-retinal projection to the tectum in goldfish: an electrophysiological and anatomical study. *Journal of Comparative Neurology*, **177**, 257–78. **13, 20–2**

SCHMITT, F. O., ADELMAN, G. & WORDEN, F. G. (eds.) (1973). *Neurosciences Research Symposia Summaries*, **7**, 253–367. MIT Press, Cambridge, Mass.

SCHNEIDER, G. E. (1970). Mechanisms of functional recovery following lesions of visual cortex or superior colliculus in neonate and adult hamsters. *Brain, Behavior and Evolution*, **3**, 295–323. **15, 31**

SCHNEIDER, G. E. (1971). Competition for terminal space in formation of abnormal retinotectal connections, and a functional consequence. *Anatomical Record*, **169**, 420. **9, 15, 27, 29**

SCHNEIDER, G. E. (1973). Early lesions of superior colliculus: factors affecting the formation of abnormal retinal projections. *Brain, Behavior and Evolution*, **8**, 73–109. **9, 16, 24, 29, 31**

SCHNEIDER, G. E. (1974). Anomalous axonal connections implicated in sparing and alteration of function after early lesions. In *Functional recovery after lesions of the nervous system*, eds. E. Eidelberg & D. G. Stein, *Neurosciences Research Program Bulletin*, **12**, 222–7. **9, 15, 29, 31**

SCHNEIDER, G. E. & JHAVERI, S. R. (1974). Neuroanatomical correlates of spared or altered function after brain lesion in the newborn hamster. In *Plasticity and Recovery of Function in the Central Nervous System*, eds. D. G. Stein, J. J. Rosen & N. Butters, pp. 65–109. Academic Press, New York. **9, 15, 16, 24, 27, 29, 31, 34**

SCHNEIDER, G. E. & NAUTA, W. J. H. (1969). Formation of anomalous retinal projections after removal of the optic tectum in the neonate hamster. *Anatomical Record*, **163**, 258.   **9, 15, 29, 31**

SCHNEIDER, G. E., SINGER, D. A., FINLAY, B. L. & WILSON, K. G. (1975). Abnormal retinotectal projections in hamsters with unilateral neonatal tectum lesions: topography and correlated behavior. *Anatomical Record*, **181**, 472.   **15, 27, 34**

SCOTT, M. Y. (1975). Functional capacity of compressed retinotectal projection in goldfish. *Anatomical Record*, **181**, 474.   **15**

SCOTT, M. Y. (1977). Behavioral tests of compression of retinotectal projection after partial tectal ablation in goldfish. *Experimental Neurology*, **54**, 579–90.   **17, 22**

SCOTT, M. Y. & MEYER, R. L. (1976). Effect of continuous light exposure on retinotectal compression after excision of caudal tectum in gold-fish. Society for Neuroscience, Abstracts, 6th Annual Meeting, **2**, p. 836. **15–17, 35**

SCOTT, T. M. (1974). The development of the retino-tectal projection in *Xenopus laevis*: an autoradiographic and degeneration study. *Journal of Embryology and Experimental Morphology*, **31**, 409–14.

SCOTT, T. M. & LÁZÁR, G. (1976). An investigation into the hypothesis of shifting neuronal relationships during development. *Journal of Anatomy*, **121**, 485–96.

SHARMA, S. C. (1969). Restoration of the visual projection from the retina to the rotated tectal reimplants in adult goldfish. *Physiologist*, **12**, 354.   **11**

SHARMA, S. C. (1970). Plasticity in the restoration of retino-tectal projection in adult goldfish: a gradient hypothesis. *Physiologist*, **13**, 307.   **15, 16**

SHARMA, S. C. (1972 a). Retinotectal connexions of a heterotopic eye. *Nature New Biology, London*, **238**, 286–7.   **23**

SHARMA, S. C. (1972 b). Redistribution of visual projections in altered optic tecta of adult goldfish. *Proceedings of the National Academy of Sciences, U.S.A.*, **69**, 2637–9.   **15–17, 21, 22**

SHARMA, S. C. (1972 c). Reformation of retinotectal projections after various tectal ablations in adult goldfish. *Experimental Neurology*, **34**, 171–82.   **15–18, 21, 22, 25**

SHARMA, S. C. (1972 d). Restoration of the visual projection following tectal lesions in goldfish. *Experimental Neurology*, **35**, 358–65.   **22**

SHARMA, S. C. (1973 a). Anomalous retinal projections after tectal removal in adult goldfish. *Physiologist*, **16**, 449.   **9, 15, 23**

SHARMA, S. C. (1973 b). Anomalous retinal projection after removal of contralateral optic tectum in adult goldfish. *Experimental Neurology*, **41**, 661–9.   **9, 15, 23, 27–9**

SHARMA, S. C. (1975 a). Visual projection in surgically created 'compound' tectum in adult goldfish. *Brain Research*, **93**, 497–501.   **12, 15, 16,19, 21 22, 26, 34**

SHARMA, S. C. (1975 b). Retinotectal specificity in adult goldfish. In *Vision in Fishes – New Approaches in Research*, ed. M. A. Ali, pp. 145–50. Plenum Press, New York.   **1**

SHARMA, S. C. & GAZE, R. M. (1971). The retinotopic organization of visual responses from tectal reimplants in adult goldfish. *Archives Italiennes de Biologie*, **109**, 357–66.   **11, 34**

SHARMA, S. C. & HOLLYFIELD, J. G. (1974 a). Specification of retinal central connections in *Rana pipiens* before the appearance of the first post-mitotic ganglion cells. *Journal of Comparative Neurology*, **155**, 395–408.   **6, 14, 36**

SHARMA, S. C. & HOLLYFIELD, J. (1974 b). The retinotectal projection in

Xenopus laevis following left–right exchanges of the eye rudiment. Society for Neuroscience, Abstracts, 4th Annual Meeting, p. 421.   6, 8

SHARMA, S. C. & HOLLYFIELD, J. G. (1974 c). Specification of the retinal central connections in *Xenopus laevis*: A re-evaluation. *Proceedings of the International Union of Physiological Sciences*, 11, 221.   36

SHARMA, S. C. & ROMESKIE, M. (1977 a). Immediate compression in a tectum devoid of degenerating myelin. *International Union of Physiology, Paris*, 13, 689.   15

SHARMA, S. C. & ROMESKIE, M. (1977 b). Immediate 'compression' of the goldfish retinal projection to a tectum devoid of degenerating debris. *Brain Research*, 134, 1–4.   9, 15, 16, 21, 22, 29, 34

SHARMA, S. C. & TUNG, Y. L. (1978). Competition between nasal and temporal heteronymous hemiretinal fibers in adult goldfish tectum. *Journal of Physiology* (in press).

SHAWE, G. D. H. (1955). On the number of branches formed by regenerating nerve-fibres. *British Journal of Surgery*, 42, 474–88.

SITTHI-AMORN, C. (1973). Ontogenetic development of retinotopic organisation of the superior colliculus of rabbits. Society for Neuroscience, Abstracts, 3rd Annual Meeting, No. 303.

SO, K.-F., SCHNEIDER, G. E. & FROST, D. O. (1977). Normal development of the retinofugal projections in Syrian hamsters. *Anatomical Record*, 187, 719.

SPEIDEL, C. C. (1932). Studies of living nerves. I. The movements of individual sheath cells and nerve sprouts correlated with the process of myelin sheath formation in amphibian larvae. *Journal of Experimental Zoology*, 61, 279–331.

SPEIDEL, C. C. (1933). Studies of living nerves. II. Activities of ameboid growth cones, sheath cells and myelin segments, as revealed by prolonged observation of individual nerve fibers in frog tadpoles. *American Journal of Anatomy*, 52, 1–79.

SPERRY, R. W. (1942). Reestablishment of visuomotor coordinations by optic nerve regeneration. *Anatomical Record*, 84, 470.   6

SPERRY, R. W. (1943 a). Visuomotor coordination in the newt (*Triturus viridescens*) after regeneration of the optic nerve. *Journal of Comparative Neurology*, 79, 33–55.   4, 6

SPERRY, R. W. (1943 b). Effect of 180 degree rotation of the retinal field on visuomotor coordination. *Journal of Experimental Zoology*, 92, 263–79.   6

SPERRY, R. W. (1944). Optic nerve regeneration with return of vision in anurans. *Journal of Neurophysiology*, 7, 57–69.   6

SPERRY, R. W. (1945 a). The problem of central nervous reorganization after nerve regeneration and muscle transposition. *Quarterly Review of Biology*, 20, 311–69.

SPERRY, R. W. (1945 b). Restoration of vision after crossing of optic nerves and after contralateral transplantation of eye. *Journal of Neurophysiology*, 8, 15–28.   8, 9, 10

SPERRY, R. W. (1948 a). Patterning of central synapses in regeneration of the optic nerve in teleosts. *Physiological Zoology*, 21, 351–61.   4, 6

SPERRY, R. W. (1948 b). Orderly patterning of synaptic associations in regeneration of intracentral fiber tracts mediating visuomotor coordination. *Anatomical Record*, 102, 63–75.   32

SPERRY, R. W. (1949). Reimplantation of eyes in fishes (*Bathygobius soporator*) with recovery of vision. *Proceedings of the Society for Experimental Biology and Medicine*, 71, 80–1.   6

SPERRY, R. W. (1950). Neuronal specificity. In *Genetic Neurology*, ed. P. Weiss, pp. 232–9. University of Chicago Press, Chicago.   1

SPERRY, R. W. (1951a). Mechanisms of neural maturation. In *Handbook of Experimental Psychology*, ed. S. S. Stevens, pp. 236–80. Wiley, New York.   1, 4

SPERRY, R. W. (1951b). Regulative factors in the orderly growth of neural circuits. *Growth*, **15**, supplement, 10th Symposium, 63–87.   1

SPERRY, R. W. (1951c). Developmental patterning of neural circuits. *Chicago Medical School Quarterly*, **12**, 66–73.   1

SPERRY, R. W. (1955). Functional regeneration in the optic system. In *Regeneration in the Central Nervous System*, ed. W. F. Windle, pp. 66–76. Charles C. Thomas, Springfield, Ill.   1, 4, 30

SPERRY, R. W. (1956). The eye and the brain. *Scientific American*, **194**, 48–52.   1, 4

SPERRY, R. W. (1963a). Chemoaffinity in the orderly growth of nerve fiber patterns and connections. *Proceedings of the National Academy of Sciences, U.S.A.*, **50**, 703–10.   1

SPERRY, R. W. (1963b). Evidence behind chemoaffinity theory of synaptic patterning. *Anatomical Record*, **145**, 288.   1

SPERRY, R. W. (1965). Embryogenesis of behavioral nerve nets. In *Organogenesis*, eds. R. L. DeHaan & H. Ursprung, pp. 161–86. Holt, Rinehart and Winston, New York.   1

SPERRY, R. W. (1966). A selective communication in nerve nets: Impulse specificity versus connection specificity. *Neurosciences Research Symposia Summaries*, eds. F. O. Schmitt & T. E. Melnechek, **1**, 213–19. MIT Press, Cambridge, Mass.   1

SPERRY, R. W. (1971). How a developing brain gets itself properly wired for adaptive function. In *The Biopsychology of Development*, eds. E. Tobach, L. A. Aronson & E. Shaw, pp. 27–42. Academic Press, New York.

SPERRY, R. W. (1975). Models, new and old, for growth of retino-tectal connections. In *From Theoretical Physics to Biology*, ed. M. Marois, pp. 191–8. North Holland, Amsterdam.   1

SPERRY, R. W. & HIBBARD, E. (1968). Regulative factors in the orderly growth of retino-tectal connexions. In *Growth of the Nervous System*, eds. G. E. W. Wolstenholme & M. O'Connor, pp. 41–52. Ciba Foundation Symposium, Churchill, London.   7

SPERRY, R. W. & MINER, N. (1949). Formation within sensory nucleus V of synaptic associations mediating cutaneous localization. *Journal of Comparative Neurology*, **90**, 403–23.   32

SPRINGER, A. D. & AGRANOFF, B. W. (1977). Effect of temperature on rate of goldfish optic nerve regeneration: a radioautographic and behavioral study. *Brain Research*, **128**, 405–15.

SPRINGER, A. D. & LANDRETH, G. E. (1977). Direct ipsilateral retinal projections in goldfish (*Carassius auratus*). *Brain Research*, **124**, 533–7.

SPRINGER, A. D., HEACOCK, A. M. & AGRANOFF, B. W. (1976). Radioautographic, electrophysiological and behavioral analyses of optic nerve regeneration in goldfish. Society for Neuroscience, Abstracts, 6th Annual Meeting, **2**, p. 836.   9, 27, 28

SPRINGER, A. D., HEACOCK, A. M., SCHMIDT, J. T. & AGRANOFF, B. W. (1977). Bilateral tectal innervation by regenerating optic nerve fibers in goldfish: a radioautographic, electrophysiological and behavioral study. *Brain Research*, **128**, 417–27.   9, 27, 28

STELZNER, D. J. (1976). An autoradiographic analysis of ipsilateral retinal projections of the intact or regenerated optic nerve of *Rana pipiens* and the effect of enucleations of the opposite eye. Society for Neuroscience, Abstracts, 6th Annual Meeting, **2**, p. 817.   **9, 27, 28**

STENSAAS, L. J. & FERINGA, E. F. (1976). Axon regeneration in the newt optic nerve. Society for Neuroscience, Abstracts, 6th Annual Meeting, **2**, p. 1137.

STEVENSON, J. A. (1977). Cell proliferation in the 'compressing' optic tectum of goldfish. Society for Neuroscience, Abstracts, 7th Annual Meeting, **3**, p. 432.

STONE, L. S. (1944). Functional polarization in retinal development and its reestablishment in regenerated retina of rotated grafted eyes. *Proceedings of the Society for Experimental Biology and Medicine*, **57**, 13–14.   **6, 8, 36**

STONE, L. S. (1948). Functional polarization in developing and regenerating retinae of transplanted eyes. *Annals of the New York Academy of Science*, **49**, 856–65.   **1, 6, 8, 36**

STONE, L. S. (1960). Polarization of the retina and development of vision. *Journal of Experimental Zoology*, **145**, 85–95.   **6, 36**

STONE, L. S. (1963). Vision in eyes of several species of adult newts transplanted to adult *Triturus v. viridescens*. *Journal of Experimental Zoology*, **153**, 57–67.

STRAZNICKY, K. (1973). The formation of the optic fibre projection after partial tectal removal in *Xenopus*. *Journal of Embryology and Experimental Morphology*, **29**, 397–409.   **15, 34**

STRAZNICKY, K. (1976 a). Reorganization of retinotectal projection of compound eyes after various tectal lesions in *Xenopus*. *Journal of Embryology and Experimental Morphology*, **35**, 41–57.   **14, 15**

STRAZNICKY, K. (1976 b). Retinotopically organised direct ipsilateral retinotectal projection after partial tectal ablation in *Xenopus*. *Proceedings of the Australian Physiological and Pharmacological Society*, **7**, 154P.   **9, 15, 29**

STRAZNICKY, K. (1978). The acquisition of tectal positional specification in *Xenopus*. *Proceedings of the Australian Physiological and Pharmacological Society* (in press).

STRAZNICKY, K. & GAZE, R. M. (1971). The growth of the retina in *Xenopus laevis*: an autoradiographic study. *Journal of Embryology and Experimental Morphology*, **26**, 67–79.

STRAZNICKY, K. & GAZE, R. M. (1972). The development of the tectum in *Xenopus laevis*: an autoradiographic study. *Journal of Embryology and Experimental Morphology*, **28**, 87–115.

STRAZNICKY, K. & TAY, D. (1977). Retinal growth in normal double dorsal and double ventral eyes in *Xenopus*. *Journal of Embryology and Experimental Morphology*, **40**, 175–85.

STRAZNICKY, K., GAZE, R. M. & KEATING, M. J. (1971 a). The retinotectal projections after uncrossing the optic chiasma in *Xenopus* with one compound eye. *Journal of Embryology and Experimental Morphology*, **26**, 523–42.   **9, 10, 14**

STRAZNICKY, K., GAZE, R. M. & KEATING, M. J. (1971 b). The establishment of retinotectal projections after embryonic removal of rostral or caudal half of the optic tectum in *Xenopus laevis* toad. *Proceedings of the International Union of Physiological Sciences*, **9**, 540.   **9, 10, 15, 16**

STRAZNICKY, K., GAZE, R. M. & KEATING, M. J. (1974). The retinotectal projection from a double-ventral compound eye in *Xenopus laevis*. *Journal of Embryology and Experimental Morphology*, **31**, 123–37.   **14**

STRAZNICKY, K., TAY, D. & LUNAM, Christine. (1977). Reorganisation of

the retinotectal connections following partial retinal ablation in adult
Xenopus. *Proceedings of the Australian Physiological and Pharmacological Society*, 8, 125P.   13, 20

STRÖER, W. F. H. (1939). Über den Faserverlauf in den optischen Bah-
nen bei Amphibien. *Proc. Kon. Nederl. Wetensch.* 42, 649–56.   3

STRÖER, W. F. H. (1940a). Zur vergleichenden Anatomie des primären
optischen Systems bei Wirbeltieren. *Zeitschrift für Anatomie und
Entwicklungsgeschichte*, 110, 301–21.   3

STRÖER, W. F. H. (1940b). Das optische System beim Wassermolch
(*Triturus taeniatus*). *Acta Neerlandica Morphologiae Normalis et
Pathalogicae*, 3, 178–95.   3

SUGAR, O. & GERARD, R. W. (1940). Spinal cord regeneration in the rat.
Journal of Neurophysiology, 3, 1–19.   32

SWISHER, J. E. & HIBBARD, E. (1967). The course of Mauthner axons in
Janus-headed *Xenopus* embryos. *Journal of Experimental Zoology*,
165, 433–39.

SZÉKELY, G. (1954a). Untersuchung der Entwicklung optischer Reflex-
mechanismen an Amphibienlarven. *Acta Physiologica Academiae scientiarum Hungaricae*, 6, suppl., p. 18.   14

SZÉKELY, G. (1954b). Zur Ausbildung der lokalen funktionellen Spezifität
der Retina. *Acta Biologica Academiae scientiarum Hungaricae*, 5, 157–
67.   6, 8, 36

SZÉKELY, G. (1957). Regulationstendenzer in der Ausbildung der 'Funktionellen Spezifitat' der Retinoanlage bei *Triturus vulgaris*. *Wilhelm
Roux' Archiv für Entwicklungsmechanik der Organismen*, 150, 48–60. 8, 13, 36

SZÉKELY, G. (1966). Embryonic determination of neural connections.
Advances in Morphogenesis, 5, 181–219.   1

SZENTÁGOTHAI, J. & SZÉKELY, G. (1956). Zum Problem des Kreuzung
von Nervenbahnen. *Acta Biologica Academiae scientiarum Hungaricae*,
6, 215–29.

TAY, D. & STRAZNICKY, K. (1977). The formation of direct ipsilateral
retino-diencephalic connections from compound eyes in *Xenopus*.
Brain Research, 125, 345–50.   9, 14

TELLO, J. F. (1907). La régénération dans les voies optiques. *Trabajos del
Laboratorio de Investigaciones Biologicas de la Universidad de Madrid*,
5, 237–48.

TELLO, J. F. (1923). Les différenciations neuronales dans l'embryon du
poulet, pendant les premiers jours de l'incubation. *Trabajos del
Laboratorio de Investigaciones Biologicas de la Universidad de Madrid*,
21, 1–93.

TOSNEY, K., HOSKINS, S. & HUNT, R. K. (1978). Testing a radial model
for positional information in the development of frog retinotectal
specificity. *Biophysical Journal*, 21, 110a.   13, 14

TUNG, Y. L. & SHARMA, S. C. (1975a). Kinetics of neuronal plasticity in
retinotectal connectivity in lower vertebrates. *Federation Proceedings*,
34 (3), 440.   2

TUNG, Y. L. & SHARMA, S. C. (1975b). Competition between nasal and
temporal heteronymous hemiretinal fibers in adult goldfish tectum.
Society for Neuroscience, Abstracts, 5th Annual Meeting, 1, p.
101.   9, 13, 15, 20, 22, 29, 34

TURNER, J. E. & GLAZE. K. A. (1977). Regenerative repair in the severed
optic nerve of the newt (*Triturus viridescens*): effect of nerve growth
factor. *Experimental Neurology*, 57, 687–97.

TWITTY, V. C. (1932). Influence of the eye on the growth of its associated structures, studied by means of heteroplastic transplantation. *Journal of Experimental Zoology*, **61**, 333–74.

UDIN, S. B. (1976). Progressive alterations in optic tract and retinotectal topography during optic nerve regeneration in *Rana pipiens*. Society for Neuroscience, Abstracts, 6th Annual Meeting, **2**, p. 838.

UDIN, S. B. (1977). Rearrangements of the retinotectal projection in *Rana pipiens* after unilateral caudal half-tectum ablation. *Journal of Comparative Neurology*, **173**, 561–82.   **9, 15, 16, 18, 20, 25, 34**

UDIN, S. B. (1978). Permanent disorganization of the regenerating optic tract in the frog. *Experimental Neurology*, **58**, 455–70.

VAN BUREN, J. M. (1963). *The Retinal Ganglion Cell Layer*. Thomas, Springfield, Ill.

WEDDELL, G. (1942). Axonal regeneration in cutaneous nerve plexuses. *Journal of Anatomy*, **77**, 49–62.

WEISS, P. A. (1934). In vitro experiments on the factors determining the course of the outgrowing nerve fiber. *Journal of Experimental Zoology*, **68**, 393–448.

WEISS, P. A. (1945). Experiments on cell and axon orientation *in vitro*: the role of colloidal exudates in tissue organization. *Journal of Experimental Zoology*, **100**, 353–86.

WEISS, P. A. (1955). Nervous system (Neurogenesis). In *Analysis of Development*, eds. B. H. Willier, P. A. Weiss & V. Hamburger, pp. 346–401. Saunders, Philadelphia.

WESTERMAN, R. A. (1965). Specificity in regeneration of optic and olfactory pathways in teleost fish. In *Studies in Physiology*, eds. D. R. Curtis & A. K. McIntyre, pp. 263–9. Springer, Berlin.   **13, 22**

WESTON, J. A. (1970). The migration and differentiation of neural crest cells. *Advances in Morphogenesis*, **8**, 41–114.

WIEMER, F.-K. (1955). Mittelhirnfunktion bei Urodelen nach Regeneration und Transplantation. *Wilhelm Roux' Archiv für Entwicklungsmechanik der Organismen*, **147**, 560–633.   **11**

WILLSHAW, D. J. & V.D. MALSBURG, C. (1976). How patterned neural connections can be set up by self-organisation. *Proceedings of the Royal Society*, B **194**, 431–45.   **2**

WILSON, M. A. (1971). Optic nerve fibre counts and retinal ganglion cell counts during development of *Xenopus laevis* (Daudin). *Quarterly Journal of Experimental Physiology*, **56**, 83–91.

WINDLE, W. F. & CHAMBERS, W. W. (1950). Regeneration in the spinal cord of the cat and dog. *Journal of Comparative Neurology*, **93**, 241–57.   **32**

WINDLE, W. F., CLEMENTE, C. D. & CHAMBERS, W. W. (1952). Inhibition of formation of a glial barrier as a means of permitting a peripheral nerve to grow into the brain. *Journal of Comparative Neurology*, **96**, 359–69.   **32**

YAGER, D. & SHARMA, S. C. (1975a). Anomalous ipsilateral visuotectal projection is functional. Society for Neuroscience, Abstracts, 5th Annual Meeting, **1.**   **9**

YAGER, D. & SHARMA, S. C. (1975b). Evidence for visual function mediated by anomalous projection in goldfish. *Nature, London*, **256**, 490–1. **9, 15, 29**

YAGER, D. & SHARMA, S. C. (1976). Retinofugal pathways in tectumless goldfish. Society for Neuroscience, Abstracts, 6th Annual Meeting, **2**, 841.   **15, 31**

YAGER, D., SHARMA, S. C. & GROVER, B. G. (1977). Visual function in goldfish with unilateral and bilateral tectal ablation. *Brain Research* **137**, 267–75.   **9, 15**

Yoon, M. G. (1971*a*). Reorganization of retinotectal projection following surgical operations on the optic tectum in goldfish. *Experimental Neurology*, **33**, 395–411.   5, 15–17

Yoon, M. G. (1971*b*). Biaxial plasticity of the optic tectum in goldfish. *Biophysical Journal*, **11**, 40*a*.   15, 17

Yoon, M. G. (1972*a*). Development of retinotectal reorganisation in goldfish. *Biophysical Journal*, **12**, 273*a*.   17

Yoon, M. G. (1972*b*). Specificity and plasticity of the visual projection from the eye to the brain in goldfish. *Journal of the Optical Society of America*, **62**, 714.   17

Yoon, M. G. (1972*c*). Synaptic plasticities of the retina and of the optic tectum in goldfish. *American Zoologist*, **12**, xxii, Abstract 106.   13, 15, 19

Yoon, M. G. (1972*d*). Retinal projection onto the 180° rotated tectal reimplant in goldfish. *American Zoologist*, **12**, 696, Abstract 312.   11

Yoon, M. G. (1972*e*). Reversibility of the reorganization of retinotectal projection in goldfish. *Experimental Neurology*, **35**, 565–77.   17, 22

Yoon, M. G. (1972*f*). Transposition of the visual projection from the nasal hemi-retina onto the foreign rostral zone of the optic tectum in goldfish. *Experimental Neurology*, **37**, 451–62.   5, 13, 15, 19, 26

Yoon, M. G. (1973*a*). Formation of retinotectal projections during light or dark deprivations in goldfish. Society for Neuroscience, Abstracts, 3rd Annual Meeting, p. 258.   5, 15, 35

Yoon, M. G. (1973*b*). Retention of the original topographic polarity by the 180° rotated tectal reimplant in young adult goldfish. *Journal of Physiology*, **233**, 575–88.   11, 17

Yoon, M. G. (1974*a*). Morphological changes in the tectal layers followed by reorganisation of retinotectal projection in goldfish. Society for Neuroscience, Abstracts, 4th Annual Meeting, p. 491.   15, 17

Yoon, M. G. (1974*b*). Continual input of visual stimuli delays reorganization of retinotectal projection in goldfish. *American Zoologist*, **14**, 1253, Abstract 54.   15–17, 22, 35

Yoon, M. G. (1975*a*). Time course of the topographic regulation of a halved optic tectum in adult goldfish. Society for Neuroscience, Abstracts, 5th Annual Meeting, **1**, No. 115.   15

Yoon, M. G. (1975*b*). Topographic polarity of the optic tectum studied by reimplantation of the tectal tissue in adult goldfish. In *Cold Spring Harbor Symposia on Quantitative Biology*, **40**, The Synapse, pp. 503–19.   1, 11, 15

Yoon, M. G. (1975*c*). Effects of post-operative visual environments on reorganization of retinotectal projection in goldfish. *Journal of Physiology*, **246**, 673–94.   15–17, 22, 35

Yoon, M. G. (1975*d*). Readjustment of retinotectal projection following reimplantation of a rotated or inverted tectal tissue in adult goldfish. *Journal of Physiology*, **252**, 137–58.   11, 22

Yoon, M. G. (1975*e*). Reorganization of the topographic pattern of visual projection in goldfish. *Experimental Brain Research*, **23**, (suppl.), 244.   15, 16, 21, 22

Yoon, M. G. (1976*a*). Induction of the field compression within a rotated tectal reimplant in goldfish. Society for Neuroscience, Abstracts, 6th Annual Meeting, **2**, p. 818.   11, 15, 17

Yoon, M. G. (1976*b*). Progress of topographic regulation of the visual projection in the halved optic tectum of adult goldfish. *Journal of Physiology*, **257**, 621–43.   15, 16, 21, 22, 35

Yoon, M. G. (1977*a*). Visual projection following transplantation of cerebellar tissue into the optic tectum of goldfish. *Investigative Ophthalmology and Visual Science*, **16**, suppl., 137.   31

YOON, M. G. (1977*b*). Induction of compression in the re-established visual projections on to a rotated tectal reimplant that retains its original topographic polarity within the halved optic tectum of adult goldfish. *Journal of Physiology*, **264**, 379–410.   11, 15–17

YOON, M. G. (1977*c*). Specificity and plasticity in the topographic patterns of visual projections in goldfish. In *From Molecular Recognition to Perception, 5th International Conference; From Theoretical Physics to Biology*, Versailles, France (in press).

YOON, M. G. (1977*d*). Retinotectal projection after translocation of two tectal grafts within the same optic tectum in goldfish. Society for Neuroscience, Abstracts, 7th Annual Meeting, **3**, p. 385.   12

SUPPLEMENTARY REFERENCES
(*added in proof*)

CAMPBELL, J. B., BASSETT, C. A. L., HUSBY, J. & NOBACK, C. R. (1957). Axonal regeneration in the transected adult feline spinal cord. *Surgical Forum*, **8**, 528–32.   32

DEVOR, M. & SCHNEIDER, G. E. (1975). In *Aspects of Neural Plasticity*, eds. F. Vital-Durand & M. Jeannerod, pp. 191–200, I.N.S.E.R.M., Paris.   31

FELDMEN, J. D., GAZE, R. M. & KEATING, M. J. (1971). The post-metamorphic innervation by optic fibres of a virgin tectum in *Xenopus laevis*. *Journal of Physiology*, **213**, 34P.   9, 10

FRASER, S. & HUNT, R. K. (1978). Neuroplasticity in *Xenopus*. *Biophysical Journal*, **21**, 110*a*.   9, 27

FREEMAN, L. W. (1962). Experimental observations upon axonal regeneration in the transected spinal cord of mammals. *Clinical Neurosurgery*, **8**, 294–316.   32

GAZE, R. M. & JACOBSON, M. (1959). The response of the frog's optic lobe to stimulation of the eye by light, after section and regeneration of the optic nerve. *Journal of Physiology*, **148**, 45–6P.   5

GAZE, R. M. & JACOBSON, M. (1962). Anomalous contralateral retino-tectal projection in frogs with regenerated optic nerves. *Journal of Physiology*, **163**, 39–41P.   5

GAZE, R. M. & KEATING, M. J. (1969). The depth distribution of visual units in the tectum of the frog following regeneration of the optic nerve. *Journal of Physiology*, **200**, 128–9P.   5

GAZE, R. M., KEATING, M. J. & STRAZNICKY, K. (1970). The re-establishment of retinotectal projections after uncrossing the optic chiasma in *Xenopus laevis* with one compound eye. *Journal of Physiology*, **207**, 51–2P.   9, 10, 14

GAZE, R. M., KEATING, M. J. & STRAZNICKY, K. (1971). The retinotectal projection from a double-ventral compound eye in *Xenopus*. *Journal of Physiology*, **214**, 37–8P.   14

GLASTONBURY, J. & STRAZNICKY, K. (1978). Aberrant ipsilateral retino-tectal projection following optic nerve section in *Xenopus*. *Neuroscience Letters*, **7**, 67–72.   9, 28

IDE, C. F. & HUNT, R. K. (1978). Positional signalling in chimeric *Xenopus* retinae. *Biophysical Journal*, **21**, 110*a*.   36

KAO, C. C. (1974). Comparison of healing process in transected spinal

cords grafted with autogenous brain tissue, sciatic nerve, and
nodose ganglion. *Experimental Neurology*, **44**, 424–39.   32

KLINE, D. G., HAYES, G. J. & MORSE, A. S. (1964). A comparative study
of response of species to peripheral nerve injury. *Journal of Neurosurgery*, **21**, 968–79 & 980–8.   32

KOSOFSKY, B. E., IDE, C. F. & HUNT, R. K. (1978). Fractionation of
signals for retinotectal patterning in *Xenopus*. *Biophysical Journal*,
21, 110*a*.   14, 36

MACDONALD, N. (1977). A polar co-ordinate system for positional information in the vertebrate neural retina. *Journal of Theoretical Biology*,
69, 153–65.   2

MCLOON, S. C. & HUGHES, W. F. (1978). Effects of lesions on the
development of visual pathways in the chick embryo. *Anatomical
Record*, **190**, 476–7   35

MURRAY, G., UGRAY, E. & GRAVES, A. (1965). Regeneration in injured
spinal cord. *American Journal of Surgery*, **109**, 406–9.   32

RHO, J. H. (1978). Cell interactions in retinotectal mapping. *Biophysical
Journal*, 21, 137*a*.   11, 12, 26

SCHMIDT, J. T. & EASTER, S. S. (1978). Independent biaxial reorganization
of the retinotectal projections: a reassessment. *Experimental Brain
Research*, **31**, 155–62.   13, 15, 17, 20

SCHNEIDER, G. E. (1977). In *Neurosurgical Treatment in Psychiatry, Pain
and Epilepsy*, eds. W. H. Sweet, Obrador, S. & J. G. Martin-
Rodrigues, pp. 5–26. University Park Press, Baltimore.   1

SCHWARTZ, E. L. (1977). The development of specific visual connections
in the monkey and the goldfish: outline of a geometric theory of
receptotopic structure. *Journal of Theoretical Biology*, **69**, 655–83.   2

So, K.-F. (1978). Development of abnormal recrossing retinotectal
projections after superior colliculus lesions in newborn Syrian
hamsters. *Anatomical Record*, **190**, 546–7.   9, 15, 29

So, K.-F. & SCHNEIDER, G. E. (1978). Abnormal recrossing retinotectal
projections after early lesions in Syrian hamsters: age related
effects. *Brain Research*, **142**, 277–95.   7

STENSAAS, L. J. & FERINGA, E. F. (1977). Axon regeneration across the
site of injury in the optic nerve of the newt, *Triturus pyrrhogaster*.
Cell and Tissue Research, **178**, 501–16.   5

TARLOV, I. M. & BOERNSTEIN, W. (1948). Nerve regeneration: a comparative experimental study following suture by clot and thread.
Journal of Neurosurgery, **5**, 62–83.   32

TAY, D. & STRANZICKY, K. (1976). The formation of direct ipsilateral
retinodiencephalic projections in *Xenopus*. *Proceedings of the Australian Physiological and Pharmacological Society*, **7**, 155P.   14

TURNER, J. E. & SINGER, M. (1974). The ultrastructure of regeneration in
the severed newt optic nerve. *Journal of Experimental Zoology*, **190**,
249–68.   5

WEISS, P. (1943). Nerve reunion with sleeves of frozen-dried artery in
rabbits, cats and monkeys. *Proceedings of the Society for Experimental Biology and Medicine*, **54**, 274–7.   32

WEISS, P. A. (1943). Functional nerve regeneration through frozen-dried
nerve grafts in cats and monkeys. *Proceedings of the Society for
Experimental Biology and Medicine*, **54**, 277–9.   32

WEISS, P. A. & TAYLOR, A. C. (1943). Histochemical analysis of nerve
reunion in the rat after tubular splicing. *Archives of Surgery*, **47**,
419–47.

POSSIBLE CLUES TO THE MECHANISM UNDERLYING THE SELECTIVE MIGRATION OF LYMPHOCYTES FROM THE BLOOD*

By W. L. FORD, M. E. SMITH and P. ANDREWS

Department of Pathology, University Medical School,
Manchester, U.K.

INTRODUCTION

In mammals and probably in other vertebrates (Bell & Lafferty, 1972; Ellis & de Sousa, 1974) small lymphocytes leave the bloodstream in large numbers to enter the extravascular compartment of many organs and then return to the blood within a period of hours by an anatomical route which is usually particular to each organ. In mice and in rats there is a good deal of quantitative information on the fluxes of recirculating lymphocytes through different anatomical compartments and the transit times from blood to tissue to blood (Sprent, 1977; Rannie & Ford, 1978). However this contribution is not primarily concerned with a description of the kinetics of lymphocyte recirculation but rather with the mechanisms which determine the unique migratory behaviour of lymphocytes.

One function of lymphocyte recirculation is to redistribute continuously antigen-sensitive cells between the anatomically dispersed components of the peripheral lymphoid system including the spleen, lymph nodes and gut-associated lymphoid tissues. In primary immune responses this redistribution facilitates the recruitment into a small focus of antigen deposition the tiny minority of lymphocytes in the recirculating pool which are specifically reactive to the particular antigen. Later in the primary response immunological memory, defined as the capacity to mount an augmented secondary response, is disseminated throughout the body by lymphocyte recirculation. Another function of lymphocyte migration is to allow access of sensitized or effector lymphocytes to depots of antigen in non-lymphoid tissue such as the skin (Ford, 1975).

We conceive that the physiological recirculation of small lymphocytes between the blood and the tissues can be analysed into three distinct processes. *First* lymphocytes leave the bloodstream by adhering to the surface

---

* Abbreviations used in this paper: DAB, Dulbecco's solution with calcium and magnesium; FCS, Foetal calf serum; TDL, Thoracic duct lymphocytes; HEV, High-endothelial venules; Hy-ase, Ovine testicular hyaluronidase; LPS, Bacterial lipopolysaccharide.

of vascular endothelium and subsequently crossing the wall of small vessels. This probably occurs in all tissues but at a much greater rate in some tissues than others; for example the rate at which lymphocytes leave the blood within the spleen is at least 10^4 times greater than within the brain (on a per gram basis). The lymph nodes, gut-associated lymphoid tissue and the bone marrow are the other major sites of lymphocyte re-circulation and most non-lymphoid tissues such as skin, muscle and endo-crine tissue are sites of lymphocyte recirculation on a much smaller scale. *Second* within lymphoid tissue thymus-derived (T) lymphocytes segregate from thymus-independent (B) lymphocytes after each has crossed the same vessel wall to reach the extravascular compartment. *Third* each cell type returns to the blood by migrating along pre-existing channels within the tissues. Lymphocytes are actively motile cells and the simplest con-ception of this third stage is that they return to the blood by taking the path of least resistance through labyrinthine but ultimately open passages. The rate at which lymphocytes are able to negotiate these passages may be partly governed by the activity of the macrophages which line the cortical and medullary sinuses (Spry, Lane & Vyakarnam, 1977) but otherwise there is no reason to believe that the movement of lymphocytes through extravascular channels is more complicated than this; by contrast each of the first two processes hinge upon poorly understood examples of cell–cell recognition.

In the first process the cells within the blood perfusing certain micro-vascular beds, for example those in the deep zone of the lymph-node cortex or the marginal zone of the spleen, are subject to a rapid selection process such that a high proportion of small lymphocytes are removed from the blood but the many other cell types are ignored. Two sorts of factors may contribute to the mechanism of cell selection – haemodynamic and a physico-chemical interaction between blood cells and the vascular endo-thelium. Although the microcirculation of the spleen (Gall & Maegraith, 1950; Dubreuil, Herman, Tilney & Mellins, 1975) and lymph nodes (Anderson & Anderson, 1975; Herman, Lyonnet, Fingerhut & Tuttle, 1976) has been carefully studied, little or nothing can be said about the possibility that haemodynamic peculiarities may influence the selective migration of lymphocytes. Of course there can be no doubt that blood flow is one factor controlling the numbers of lymphocytes entering an organ. For example following an antigenic stimulus the blood flow through the sheep popliteal lymph node increased over a few days by a factor of three; the entry of labelled lymphocytes from the blood increased to the same extent and followed the same temporal pattern (Hobbs & Hay, 1977). However rheological factors could only sort out different leukocytes on the basis of gross physical characteristics such as size and density and there is considerable overlap between different cells in these respects especially in the case of lymphocytes and monocytes. Because of this general consideration we believe that the selective migration of lymphocytes is governed by a specific interaction between lymphocytes and specialized vascular endothelium. This functional specialization is apparently associated with a distinctive cell morphology in the deep cortex of lymph nodes, in the thymus-dependent areas of gut-associated lymphoid tissue and as an acquired change in chronic inflammatory lesions (Åström, Webster & Arnason, 1968; Smith, McIntosh & Morris, 1970, Graham & Shannon, 1972). In these sites the endothelial cells of post-capillary venules are differentiated from other endothelial cells in that they are plump or cuboidal, have pale nuclei with a distinct rim of chromatin and include many mitochondria, ribosomes and a prominent Golgi complex in their cytoplasm. Lymphocytes enter these tissues by adhering to and migrating between these specialized endothelial cells in the high-endothelial venules (HEVs). These endothelial cells are not seen in the spleen or bone-marrow in which case there is other evidence that the mechanism of lymphocyte selection may be somewhat different (Rannie, Smith & Ford, 1977). This paper focusses on the selection of small lymphocytes within the HEVs of the deep cortex of lymph nodes as an example of cell–cell recognition between two different cell types – the small lymphocyte and the specialized endothelial cell.

An important factor in considering the nature of this recognition is that the few seconds taken for blood cells to traverse a microvascular bed demands that selection and adhesion occur rather quickly and moreover the adhesion must be comparatively strong because the marginated lymphocytes project into the bloodstream and must withstand considerable hydrodynamic shearing forces until they sink into the endothelium (Åström et al., 1968; Anderson & Anderson, 1976). By contrast the second process in lymphocyte migration – the segregation of B and T cells into their own distinctive areas – takes place at a much more leisurely pace. It does not begin until B and T cells have entered by the same route (Gutman & Weissman, 1973; Nieuwenhuis & Ford, 1976) and the two cell types are not perceptibly in different locations until at least an hour after they have left the blood. The segregation is not complete for a few hours more. Thus this process takes place over hours rather than the seconds required for the selection of lymphocytes within the blood vessels and there is no indication that strong forces are necessary. A plausible mechanism for the remarkably efficient segregation of B and T cells was suggested by Curtis & de Sousa (1973) and is elaborated by Curtis in this volume (pp. 51–82). It remains

24

to be resolved what part (if any) elements other than the B and T cells themselves, such as the reticulin meshwork and macrophages, play in the segregation process.

The notion that the selective migration of lymphocytes into lymph nodes begins with, and ultimately depends on, selective adhesion to the endothelial cells of HEVs arises from histological observation of lymph-node sections in which lymphocytes can be seen to be concentrated within the lumens of HEVs to such an extent that they may outnumber erythro-cytes although the latter are about 1000 times more frequent in arterial blood. Of the lymphocytes in the lumen many are in contact with the endothelium, others are in contact only with attached lymphocytes and a third category appear not to be in contact with other nucleated cells. Pos-sibly the last have become detached from the endothelium during the fixation of the tissue. The idea that these free lymphocytes can be account-ed for by a reverse stream of lymphocytes, that is from the lymph node into the vascular lumen (Sainte-Marie, Yin & Chan, 1967) was not sup-ported by a study on the isolated mesenteric lymph node of the rat in which the traffic of lymphocytes across the HEVs was found to be strictly one-way (Sedgley & Ford, 1976). After adhesion the lymphocytes cross the endothelium as first described by Gowans & Knight (1964). Contrary to suggestions that most migrating lymphocytes traverse the cytoplasm of endothelial cells (Marchesi & Gowans, 1964; Åström et al., 1968) several recent studies have indicated that most if not all lymphocytes pass be-tween the endothelial cells (e.g. Schoefl, 1972).

The experiments to be described are an example of an approach to selective lymphocyte migration which was pioneered by Gesner & Gins-burg (1964). This consists of modifying lymphocytes or specialized endo-thelial cells with a number of agents about which there exists some under-standing of their mode of action, and then measuring the consequent altera-tions in migratory behaviour. Many contributors to this symposium have applied the same principle and the same tools to diverse examples of cell–cell recognition. In general lymphocyte migration studies are complicated by the number of anatomical compartments involved and in particular the problem that an increased localization of lymphocytes in organ A may be either the cause or the result of decreased localization in organ B. For this reason study of the effect of any agent on lymphocyte migration must be rather comprehensive in terms of both time and anatomical compartments. In recent years Gesner's approach has been extended by Freitas & de Sousa (1975, 1976 a, b, c, 1977) using mice and by ourselves using rats (Ford, Sedgley, Sparshott & Smith, 1976; Smith, Sparshott & Ford, 1977; Ford, Andrews & Smith, 1978). These results are reviewed in the discussion along with the data presented here, which are concerned with the treatment of either lymphocytes or of recipients with dextran sulphate or with hyaluronidase and an attempt to pinpoint the site of action of dextran sulphate by determining the localization of labelled material.

Many sulphated polysaccharides, including heparin and dextran sulphate, are known to induce an excess of lymphocytes in the blood by impeding the physiological migration of lymphocytes from the blood into lymph nodes (Sasaki, 1967; Bradfield and Born, 1974). Other workers have found that the specialized endothelium of the HEVs in the lymph-node paracortex stains metachromatically with toluidine blue (Smith & Henon, 1959) and positively with alcian blue (Anderson, Anderson & Wyllie, 1976) suggesting the presence of acidic (possibly sulphated) proteoglycan. Each of these observations would seem to be worthwhile clues in tracking down the mechanism by which lymphocytes selectively adhere to specialized endothelium and they might conceivably be related if it were possible to show that dextran sulphate inhibited lymphocyte/endothelial interaction by competing with a sulphated proteoglycan displayed by the endothelial cell for a recognition structure on the lymphocyte surface. Although no evidence was found to support this idea some further clues have been suggested by the results.

MATERIALS AND METHODS

Principle

Thoracic duct lymphocytes (TDL) were radioactively labelled *in vitro* before injection into a series of syngeneic recipients which were killed at strategic intervals up to 24 h in order to assess the distribution of lymphocytes between various organs including the blood. The effect of two agents, dextran sulphate and hyaluronidase (Hy-ase) on lymphocyte migration was studied either by exposing lymphocytes *in vitro* to one of the agents or by injecting the agent into a recipient.

Rats

Three highly inbred strains were used – AO, HO and HO.B2. Each experiment involved only one strain. Lymphocyte donors were males of 180–270 g and recipients were females of 150–190 g.

Reagents

The sodium salt of dextran sulphate ($\sim$ 500000 daltons; S content $17 \pm 1 \%$) was obtained from Pharmacia (Lot 2909). Preliminary tests confirmed that it was about as effective in inducing lymphocytosis as was described by Bradfield & Born (1974). TDL were treated *in vitro* by adding either 100 μg ml^{-1} or 500 μg ml^{-1} to a suspension at 5×10^7 ml^{-1} in medium RPMI with 10 % FCS (both Gibco–Biocult) for 1 h at 37 °C. Alternatively labelled (^{3}H/^{14}C) control lymphocytes were incubated in the same medium without dextran sulphate. Both populations were centrifuged, resuspended in Dulbecco's solution (DAB) + 2 % FCS and centrifuged again before combining them for injection.

Recipients which were treated with dextran sulphate received an i.v. injection at a dose level of 17 mg kg^{-1} 5–10 min before the labelled lymphocytes.

Hyaluronidase (Hy-ase) prepared from ovine testes was obtained from Boehringer (Mannheim). The activity was approximately 1000 units mg^{-1}. TDL were treated by incubating washed cells in PBS to which was added either 75 units or 250 units ml^{-1}. The cells were incubated at 5×10^7 ml^{-1} for 15 min at 37 °C and then centrifuged, combined with alternatively labelled control cells which had been incubated under the same conditions except for hyaluronidase. The combined populations were centrifuged together and then resuspended for injection.

The recipients which were pretreated with hyaluronidase were given either 10000 units or 20000 units i.v. at 10 min or 30 min respectively before the injection of lymphocytes which had been labelled with [^{3}H]uridine or ^{51}Cr respectively.

Measurement of the distribution of labelled lymphocytes

Each recipient was injected with 100–200 $\times 10^6$ labelled TDL in a volume of 2 ml. The intervals at which the recipients were killed – usually 0.5 h or 1 h, 2.5 h and 24 h – were chosen because at the earliest interval the blood concentration is still falling while the tissue activities are already significant; at 2.5 h the blood level is at a nadir and the splenic localization is near its peak; while at 24 h the injected cells are approaching equilibrium with the recirculating pool (Ford *et al.*, 1976).

Before each recipient was killed it was anaesthetized with ether, a 2 ml sample of blood was taken for separation of the leukocytes by the ficoll/hypaque method (Boyum, 1968) and the rat was bled out by incision of the aorta. The spleen and the group of six superficial cervical lymph nodes were removed and samples of liver, lung and small intestine from which Peyer's patches had been excised and discarded. The samples were weighed and either processed for liquid scintillation counting (Ford & Hunt, 1973) or in the case of ^{51}Cr-containing samples put into a tube for gamma counting.

In experiments in which alternatively labelled (^{3}H/^{14}C) treated and untreated cells were injected together simultaneous experiments were performed with reversal of the isotopes and compensation was made for the differential loss of ^{3}H and ^{14}C as described by Atkins & Ford (1975).

Radioactive labelling

Radionuclides were obtained from the Radiochemical Centre, Amersham. The methods of labelling lymphocytes with [5-^{13}H]uridine (TRA.178) at 7 μCi ml^{-1}; [^{14}C]uridine (U) (CFB.51) at 1 μCi ml^{-1} and sodium [^{51}Cr]-chromate (CJS.1) at 10 μCi ml^{-1} have all been fully described (Ford & Hunt, 1973).

Tritiated dextran sulphate was specially prepared by the Radiochemical Centre. Approximately 2 g of the material already described was exposed to tritium gas. The radioactivity incorporated was about 20 μCi g^{-1}. [^{3}H]dextran sulphate was injected either i.v. or s.c. into the footpad. In other experiments sodium [^{35}S]sulphate (SJS.2P) was injected by the same routes at a dose of 1 μCi g^{-1} body weight.

Other methods

Thoracic duct cannulation, the collection of TDL, autoradiography and scintillation counting of radioactivity were performed as described by Ford & Hunt (1973). Autoradiography of specimens prepared conventionally for electron microscopy was performed by a loop-applied liquid emulsion technique.

Presentation of results

The results of cell distribution studies can be expressed as the fraction of the injected (cell-associated) radioactivity present in different organs at different times as in Fig. 1. However most of the results are expressed as the ratio of treated cell localization to untreated cell localization so that a ratio above unity indicates a surplus and below unity a deficit (Tables 1–6). The approximate fractional distribution of treated cells can be calculated by multiplying the ratio with the fractional distribution of untreated cells as recently published (Ford *et al.*, 1976).

RESULTS

(1) Dextran sulphate treatment of lymphocytes

Thoracic duct lymphocytes (TDL) were not perceptibly damaged when inspected by phase-contrast microscopy after incubation with dextran sulphate for 1 h. Their distribution in recipients at 24 h after i.v. injection was almost the same as that of untreated cells (Table 1, Fig. 1) confirming

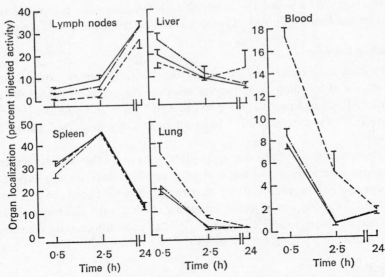

Fig. 1. Effect of dextran sulphate on lymphocyte localization in five compartments. ——, Untreated lymphocytes in untreated recipients; — · —, dextran sulphate treated lymphocytes in untreated recipients; – – –, untreated lymphocytes in dextran sulphate treated recipients. For clarity the plot of treated lymphocytes in treated recipients is omitted because it almost coincides with the previous group. Mean ± s.e. For numbers see Tables 1 and 3.

Table 1. *Treatment of lymphocytes* in vitro *with dextran sulphate before injection into normal recipients*

$$\frac{\text{Ratio of radioactivity associated with treated cells}}{\text{radioactivity associated with untreated cells*}}$$

Mean and s.d. ($n = 6$ pairs)

Time after injection (h)	Lymph nodes	Spleen	Liver	Blood	Lung	Small intestine −pp†
0.5	0.56	0.87	1.23‡	1.10	1.12	0.97‡
	0.07	0.04	0.05	0.05	0.08	0.09
2.5	0.62	1.01	1.17‡	0.91	0.91	0.80‡
	0.06	0.03	0.14	0.12	0.09	0.21
24	0.98	1.02	1.06‡	1.02	1.05‡	0.90‡
	0.01	0.03	0.07	0.02	0.14	0.15

* Relative to injection sample = 1.00.
† Small intestine with Peyer's patches removed.
‡ $n = 4$ pairs.

that their ability to survive had not been compromised. At 0.5 h after injection there was a deficit of treated cells in the lymph nodes amounting to 44 % of the radioactive label carried by control cells into the nodes and at the same time there was a marginal but consistent surplus (10 %) in the blood. This suggested that the treated population had been partially impaired in its ability to migrate into lymph nodes and this effect could not be attributed to sequestration in the liver or lungs, first of all because the localization of treated cells was only marginally increased in these organs and secondly because such a sequence of events would have produced a deficit of cells in the blood as was the case with neuraminidase treatment (Ford *et al.*, 1976). There was no significant difference in the partial inhibition of lymph-node localization following dextran sulphate treatment at 100 μg ml^{-1} (2 experiments) or 500 μg ml^{-1} (4 experiments) and the results were therefore pooled, as shown in Fig. 1 and Table 1.

(2) *Dextran sulphate treatment of recipients*

Rats were injected i.v. with dextran sulphate (17 mg kg^{-1}) followed after 5–10 min by an i.v. injection of radiolabelled syngeneic TDL. The distribution of radioactivity was compared to that found in untreated recipients of another sample of the same labelled population. At half an hour after injection of the TDL there was a deficit of radioactivity in the lymph node corresponding to 89 % of the activity in the control recipients; at the same time the radioactivity associated with blood lymphocytes was more than double the control value and this rose to almost sevenfold after 2.5 h (Table 2, Fig. 1). The localization of TDL in the spleens of untreated recipients at 0.5 and 2.5 h was almost precisely the same as in control recipients. However, the numbers in the spleen should be compared with

Table 2. *Treatment of recipients with dextran sulphate before injection of labelled lymphocytes*

Time after injection (h)	Ratio of radioactivity recovered in treated animals / radioactivity recovered in untreated animals*					
	Lymph nodes	Spleen	Liver	Blood	Lung	Small intestine − ppt†
0.5	0.11	0.98	0.82	2.22	1.87	0.48
2.5	0.18	0.99	0.97	6.66	2.58	0.35
24	0.81	0.93	2.85	1.14	1.08	0.64

* Ratio of mean (d.p.m. g^{-1}/d.p.m. injected) for injected rats ($n = 6$) compared to the same function in controls ($n = 12$).
† Small intestine with Peyer's patches removed.

the level in the blood since it has been shown that the influx of lymphocytes into the spleen is directly proportional to the concentration in the
blood (Ford, 1969). If this principle is accepted then the conclusion is that
the migration of lymphocytes into the spleen is impaired by dextran sulphate treatment but to a slightly lesser extent than is migration into lymph
nodes. This was firmly supported by results described later (Section 7).

By 24 h after injection the distribution of TDL in treated recipients
approached that in control recipients except for a surplus of cells in the
liver which was not present earlier. The results as expressed in Fig. 1
show that a small minority of cells are sufficient to account for this surplus
which is somewhat exaggerated in Table 2 because the denominator of the
ratio – the hepatic localization in untreated recipients – is small at this time
(Ford *et al.*, 1976).

In a second series of experiments dextran sulphate (17 mg kg^{-1}) or PBS
was injected into four recipients which had been subjected to thoracic duct
cannulation on the previous day. The radioactivity associated with the
thoracic duct lymphocytes of the recipients was measured in 3–6 h collections over the next 30 h. There was a period of about 12 h following injection during which very few labelled cells were recovered from the treated recipients (Fig. 2) as had been expected from the deficient lymph-node
localization. After this period the recovery of labelled lymphocytes in-

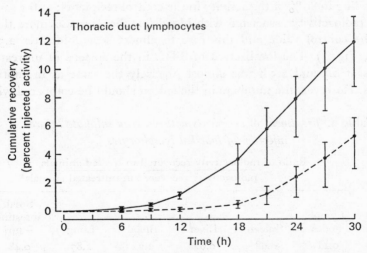

Fig. 2. Effect of dextran sulphate injection on the recirculation of labelled lymphocytes from blood to lymph. ———, Cumulative recovery of cell-associated radioactivity from thoracic duct lymph of normal recipients after the i.v. injection of
labelled lymphocytes; – – – – –, cumulative recovery from dextran sulphate treated
recipients. The mean and range of three recipients in each group are plotted.

creased with the same slope as they followed in untreated recipients; the delay imposed by dextran sulphate treatment was about 10 h.

The total output of lymphocytes, of which the great majority were the recipient's own unlabelled lymphocytes, was not clearly affected by dextran sulphate but small effects could have been obscured by the variation between rats in their output of TDLs and its short-term fluctuation. In the non-cannulated recipients dextran sulphate increased the level of lymphocytes in the blood to an average of 26.5×10^6 mm^{-3} compared to the control level of 9.0×10^6 mm^{-3} as found by others (Sasaki, 1967; Bradfield & Born, 1974). These observations confirmed that the main effect of dextran sulphate is to inhibit lymphocyte migration from the blood rather than to mobilize directly cells from lymphoid tissues.

(3) *Dextran sulphate treatment of lymphocytes and of recipients combined*

Dextran sulphate treated recipients were injected with alternatively labelled (^{3}H/^{14}C) control lymphocytes or lymphocytes which had been treated with dextran sulphate *in vitro* as described in section (1) and thus the distribution of treated and untreated cells in dextran sulphate recipients was compared (Table 3). At half an hour after injection the distribution of treated and untreated cells was almost identical; in particular there was no deficit of the treated cells in the recipients' lymph nodes. There was a

Table 3. *Treatment of lymphocytes* in vitro *with dextran sulphate before injection into dextran sulphate treated recipients*

Time after injection (h)	Ratio of radioactivity associated with treated lymphocytes / radioactivity associated with control lymphocytes* Mean and s.d. ($n = 3$ pairs)					
	Lymph nodes	Spleen	Liver	Blood	Lung	Small intestine −ppt†
0.5	1.02	0.87	1.00‡	0.92	1.08	0.96‡
	0.20	0.03	0.14	0.03	0.03	0.19
2.5	0.85	1.01	1.03‡	0.83	0.91	1.14‡
	0.14	0.04	0.06	0.08	0.04	0.25
24	0.95	1.03	0.80‡	1.00	0.90‡	1.09‡
	0.03	0.02	0.32	0.02	0.11	0.16

* Relative to injection sample = 1.00.
† Small intestine with Peyer's patches removed.
‡ $n = 2$ pairs.

marginal deficit in the lymph nodes at 2.5 h but this reflected an equally small deficit in the blood. The effects of treating cells and of treating recipients were clearly not synergistic or even additive.

(4) *Hyaluronidase treatment of lymphocytes*

In five experiments TDL were exposed *in vitro* to ovine testicular hyaluronidase (three at 75 units ml^{-1}, two at 250 units ml^{-1}). The cells were not visibly damaged by the treatment as was again confirmed by their distribution relative to that of control cells at 24 h after i.v. injection (Table 4). (The finding that the mean ratios in the three major compartments – spleen, lymph nodes and blood – were 1.00 or 1.01 provided reassurance of the precision of this method.) At the earlier intervals the localization of the treated cells was nearly identical to that of the control cells. The same results were found after each concentration. It was concluded that hyaluronidase treatment of lymphocytes had no effect on their migration.

Table 4. *Treatment of lymphocytes* in vitro *with hyaluronidase before injection into normal recipients*

Ratio of radioactivity associated with treated cells

radioactivity associated with untreated cells*

Mean and s.d. ($n = 5$ pairs)

Time after injection (h)	Lymph nodes	Spleen	Liver	Blood	Lung	Small intestine −pp†
0.5	1.04	0.99	1.01	1.00	1.02	0.98
	0.07	0.01	0.04	0.02	0.02	0.10
2.5	1.07	1.00	0.99	1.00	1.03	0.94
	0.08	0.01	0.09	0.01	0.04	0.08
24	1.01	1.00	0.97	1.00	1.00	0.95
	0.03	0.01	0.05	0.01	0.02	0.04

* Relative to injection sample = 1.
† Small intestine with Peyer's patches removed.

(5) *Hyaluronidase treatment of recipients*

Three series of experiments were performed in which recipients of radioactively labelled TDL were pretreated with an i.v. injection of hyaluronidase. In the first version 10000 units of hyaluronidase was injected i.v. 10 min before cells which had been labelled *in vitro* with [³H]uridine, and in the second version 20000 units of hyaluronidase was injected i.v. 30 min before cells which had been labelled with sodium [⁵¹Cr]chromate.

After both doses oedema of the feet, snout and ears became apparent within half an hour and gradually waned over the next few hours. Differential polymorph counts on blood samples taken at various intervals after injection showed that the levels reached 3–5 times that of control rats by 3 h after injection. Blood lymphocyte levels were only slightly higher in the treated rats. Despite the obvious effects of hyaluronidase administration the distribution of labelled lymphocytes was only perturbed to a limited extent. At 0.5 h or 1 h after injection the radioactivity in lymph nodes was approximately 70% of control values (Table 5). There was a slight excess of labelled cells in the blood of treated recipients and a very slight excess in the lungs and liver. The localization in the spleen was slightly reduced. By 2.5 h after injection the distribution of ^{3}H-labelled cells had almost returned to normal (Table 5) but there was still a deficit of ^{51}Cr-labelled cells in lymph nodes probably because double the dose of enzyme had been given in these experiments. (Because this was the only difference between the two series, detailed results of the ^{51}Cr experiments are not presented in tabular form.)

Table 5. *Treatment of recipients with hyaluronidase before injection of labelled lymphocytes*

Ratio of $\dfrac{\text{radioactivity recovered in treated recipients}}{\text{radioactivity recovered in untreated recipients}}$*

Time after injection (h)	Lymph nodes	Spleen	Liver	Blood	Lung	Small intestine −pp†
0.5‡	0.68	0.93	1.11	1.32	1.40	0.95
1§	0.68	0.92	0.89	0.91	0.74	0.83
2.5§	1.01	1.16	1.02	1.20	1.30	0.93
24‡	1.13	1.15	1.05	1.03	1.05	1.07

* Ratio of mean (d.p.m. g^{-1}/d.p.m. injected) for injected rats, compared to same function in controls.
† Small intestine with Peyer's patches removed.
‡ $n = 6$ treated recipients, 3 controls.
§ $n = 4$ treated recipients, 2 controls.

The third version of the experiment entailed injecting labelled TDL into recipients which had been subjected to thoracic duct cannulation on the previous day. The recipients were injected i.v. with 10000 units of hyaluronidase just before the cells and 500 units every 10 min over the next 200 min giving a total dose of 20000 units. The treatment slightly reduced the numbers of cells recovered from the thoracic duct as had been expected from the lower activities in lymph nodes observed at early intervals after injection. There was no substantial delay of lymphocytes within lymph nodes (Fig. 3).

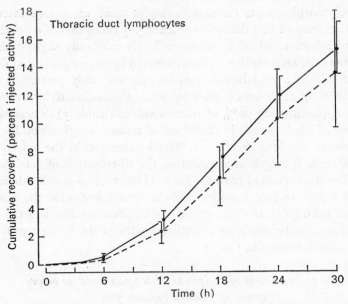

Fig. 3. Effect of hyaluronidase injection on the recirculation of labelled lympho-cytes from blood to lymph. ———, Cumulative recovery of cell-associated radio-activity from thoracic duct lymph of normal recipients after the i.v. injection of labelled lymphocytes; – – –, cumulative recovery from hyaluronidase-treated recipients. The mean and the range of four recipients in each group are plotted.

(6) *The combined effect of hyaluronidase treatment of recipients and dextran sulphate treatment of lymphocytes*

Labelled lymphocytes were exposed *in vitro* to a high concentration of dextran sulphate (500 μg ml⁻¹) and injected into recipients after combining with alternatively labelled ($^3H/^{14}C$) control lymphocytes from the same collection. The labelled cells were injected i.v. either into recipients which had been given i.v. injection of hyaluronidase or into untreated recipients. Both the dextran sulphate treatment of lymphocytes and the hyaluronidase treatment of recipients had a modest inhibitory effect on the early lymph-node localization of the labelled cells which in each case was very slightly more pronounced than in previous experiments (Tables 1, 5 and 6). In combination these agents were precisely additive (Tables 6a and 6b). The inhibition produced by dextran sulphate treatment was identical in hy-aluronidase-treated recipients and in control recipients (6a) and the inhibi-tion induced by hyaluronidase treatment was equal with respect to dextran sulphate treated and control lymphocytes (6b).

Table 6. *Combined treatment of lymphocytes with dextran sulphate and recipients with hyaluronidase*

(a) *Relative localization of treated and control cells*

Time after cell injection (h)	Recipient treatment at −10 min	Ratio of radioactivity associated with treated cells† / radioactivity associated with untreated cells* Mean of two experiments					
		Lymph nodes	Spleen	Liver	Blood	Lung	Small intestine‡ −pp
0.5	⎫ 1 ml	⎧ 0.41	0.84	1.30	1.16	1.17	0.83
2.5	⎬ PBS	⎨ 0.45	1.02	1.12	0.73	0.98	1.00
24	⎭ i.v.	⎩ 0.98	1.05	1.03	1.01	0.96	1.03
0.5	⎧ 10000 units	⎧ 0.42	0.82	1.27	0.95	1.13	0.87
2.5	⎨ hyaluroni-	⎨ 0.43	0.99	1.06	0.71	1.00	0.96
24	⎩ dase i.v. in 1 ml	⎩ 0.98	1.03	0.99	0.99	1.02	1.11

* Relative to injection sample = 1.00.

† Treated cells = 1 h incubation with 500 μg ml^{-1} dextran sulphate in RPMI +10 % FCS.

‡ Small intestine with Peyer's patches removed.

(b) *Relative localization in treated and control recipients*

Time after injection (h)	Cell treatment after labelling	Ratio* of radioactivity recovered in treated animals† / radioactivity recovered in control animals, Mean of two experiments					
		Lymph nodes	Spleen	Liver	Blood	Lung	Small intestine −pp‡
0.5	⎫ 1 h in	⎧ 0.57	0.90	0.81	1.27	1.20	0.92
2.5	⎬ control	⎨ 0.72	0.97	0.93	1.07	1.17	0.90
24	⎭ medium	⎩ 0.94	0.98	0.83	0.83	0.87	0.77
0.5	⎧ 1 h in	⎧ 0.57	0.88	0.78	1.15	1.18	1.16
2.5	⎨ 500 μg ml^{-1}	⎨ 0.71	0.95	0.91	1.04	1.16	0.86
24	⎩ dextran sulphate	⎩ 1.08	0.97	0.79	0.83	0.91	0.78

* Each ratio standardized with respect to injection sample.

† Recipient treatment: 10000 units hyaluronidase 10 min before cells.

‡ Small intestine with Peyer's patches removed.

(7) Does dextran sulphate treatment of recipients impede a 'lymph-node-seeking' but spare a 'spleen-seeking' population?

The proportion of lymphocytes recovered from the spleens of dextran sulphate treated recipients was, within the limits of measurement, the same as in untreated recipients at 0.5 and 2.5 h after injection although the values in the lymph nodes were much lower and in the blood were much higher than in untreated recipients (Table 2). A possible interpretation is that the migration from the blood of a 'lymph-node-seeking' population is impaired but a 'spleen-seeking' population is unaffected. This notion was tested by performing secondary transfer of labelled lymphocytes from the blood of three groups of primary recipients of TDLs which had been labelled *in vitro* with [³H] or [¹⁴C]uridine. Approximately 10 ml of blood was obtained when the primary recipients were killed at 2.5 h after injection, the lymphocytes were separated from the blood by the ficoll–hypaque method (Boyum, 1968) and the secondary recipients were also killed at 2.5 h for measurements of radioactivity in the spleen and lymph nodes. It would be predicted on the assumption of separate subpopulations that the cells transferred from the dextran sulphate treated recipients would be predominantly 'lymph-node-seeking' so that the 'spleen-seeking' cells would be diluted down; conversely the lymphocytes transferred from the splenectomized primary recipients would be predominantly 'spleen-seeking'.

The results decisively rejected these predictions (Table 7). The splenic localization of lymphocytes transferred from the blood of all three groups was identical. The localization in lymph nodes of lymphocytes from the dextran sulphate treated rats was reduced to 38 % rather than increased. This is no doubt because dextran sulphate injected into the primary recipient exerted the same effect as does dextran sulphate applied *in vitro*. Two other significant points are that ficoll–hypaque separation does not affect the localization in the spleen or lymph nodes either by selection or otherwise and lymphocytes transferred from splenectomized recipients have a much lower localization in lymph nodes for reasons which are not clear. The main conclusion is that dextran sulphate treatment of recipients inhibits both lymph node and splenic localization causing an accumulation of lymphocytes in the blood but the splenic localization is inhibited to a lesser extent so that the two factors – increased numbers in the blood and diminished capacity to enter the spleen – happen to balance each other exactly.

Table 7. *Distribution of radioactivity in secondary recipients at 2.5 h after transfer from the blood of primary recipients*

Primary recipient	% of injected radioactivity recovered (mean and s.d.)	
	Spleen	Lymph nodes*
Normal ($n = 4$)	42.0	9.75
	3.7	5.10
Dextran sulphate treated ($n = 4$)	43.1	3.70
	5.2	0.18
Splenectomized ($n = 3$)	42.1	4.28
	12.7	2.56
None – cells held *in vitro* ($n = 4$)	36.6	12.19
	1.7	1.72
None – ficoll–hypaque separation† ($n = 3$)	37.3	13.26
	7.6	2.54

* Activity in lymph nodes measured per milligram and expressed as per 700 mg (the approximate weight of the aggregated lymph nodes in these recipients).

† Lymphocytes held *in vitro* as in previous group and then recovered from interface before injection.

(8) *Autoradiographic localization of labelled lymphocytes in recipients treated with dextran sulphate*

Thoracic duct lymphocytes labelled *in vitro* with [³H]uridine were injected i.v. into a series of recipients. These had either been injected i.v. 10 min previously with dextran sulphate at 17 mg kg⁻¹ or with Dulbecco's solution as controls. The spleens and lymph nodes were removed for processing at intervals of 10 min, 30 min and 3.25 h after injection.

In the spleens the localization of the labelled cells was not perceptibly different at 10 min or 3.25 h. Whether or not the recipients had received dextran sulphate most of the labelled cells were present in the marginal zone at the earlier time and in the periarteriolar lymphoid sheath (PALS) at the later time. However there was a clear difference of localization at 30 min when in the controls a substantial majority of the labelled cells had cleared the marginal zone to enter the PALS, whereas in the dextran sulphate treated recipient a bare majority of the labelled cells were still in the marginal zone. This delay in the transit of lymphocytes may be attributed to two factors (*a*) slower entry into the splenic marginal zone from the blood and (*b*) a slower transit of the marginal zone into the PALS. From similar observations on mice Freitas & de Sousa (1977) concluded that the passage of lymphocytes through the marginal zone was decelerated by dextran sulphate treatment.

In the lymph nodes very few labelled lymphocytes were visible at any of the time intervals. There was no obvious abnormality in the distribution of these cells, in particular no accumulation of cells were noted in the base of the endothelium in HEVs as was described in mice which had been injected with pentosan sulphate 90 min before labelled cell transfer (Freitas & de Sousa, 1977). Some swollen, damaged cells with nuclear and cytoplasmic vacuolation were observed in the HEVs of dextran sulphate treated recipients (Plate 4b) but this was variable and its significance is uncertain.

(9) *Localization of [³⁵S]sulphate in lymph nodes*

The histochemical evidence that HEVs contain acidic proteoglycan has been disputed (Röpke, Jørgensen & Cläesson, 1972). To seek evidence of the synthesis of sulphated compounds a series of rats were injected with sodium [³⁵S]sulphate at a dose of 1μCi g^{-1} body weight either into the footpad or intravenously. By 15 min after footpad injection localization of radioactivity was marked over the HEVs in the draining popliteal lymph nodes (Plates 1b and 1c). This was less prominent at 30 min after injection and continued to wane while the incorporation of ³⁵S into mast cells became obvious later. Autoradiography of electron microscope preparations showed that the radioactivity at 15 min was concentrated in the cytoplasm between the nucleus and the luminal surface of the cell and was clearly associated with the Golgi complex (Plates 2a and b, 3a and b). No localization of radioactivity was perceptible after i.v. injection of [³⁵S]sulphate at this dosage and applying a solution of [³⁵S]sulphate to frozen sections of lymph nodes under various conditions failed to reproduce the selective localization. No other cell-type in the lymph node incorporated sulphate as early as did the HEV cells.

(10) *Localization of [³H]dextran sulphate*

The inhibitory effect of dextran sulphate on lymphocyte transport across HEVs and the selective localization of sodium sulphate in these cells might be associated if the endothelial cells had a receptor for a sulphated compound which was competitively inhibited by dextran sulphate. In this case dextran sulphate would be expected to bind to the specialized endothelial cells.

A solution of [³H]dextran sulphate was injected i.v. into recipients at a dose of 40μCi (2 mg); other recipients were given footpad injections of 4μCi (0.2 mg) and in other experiments a solution of [³H]dextran sulphate in Dulbecco's solution was perfused through the isolated mesenteric lymph node from artery to vein. Inspection of autoradiographs of lymph nodes taken at time intervals ranging from 3 min to 24 h after injection

PLATE I

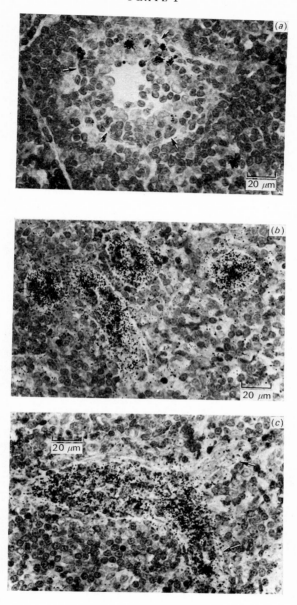

For explanation see p. 391

PLATE 2

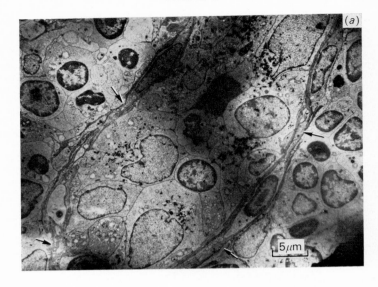

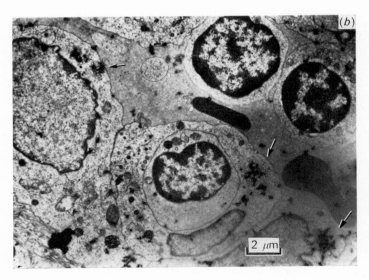

For explanation see p. 391

PLATE 3

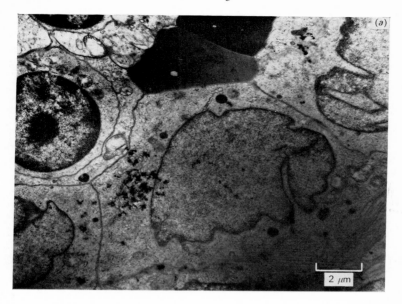

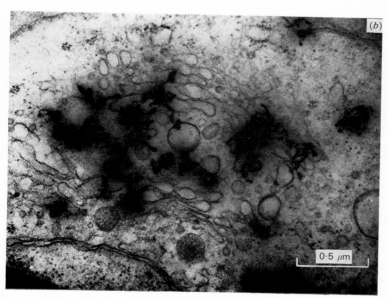

For explanation see p. 392

PLATE 4

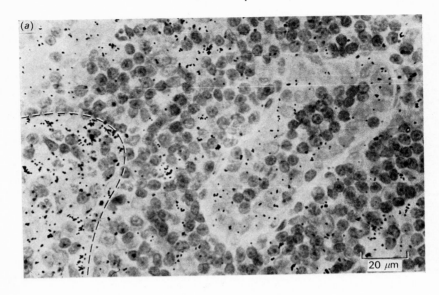

(a)

20 μm

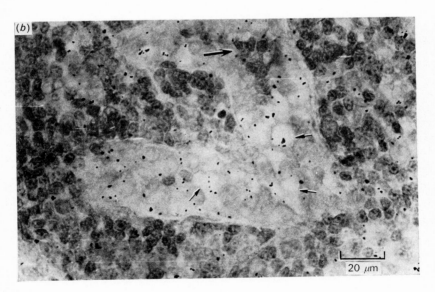

(b)

20 μm

For explanation see p. 392

failed to reveal any binding to or concentration by the specialized endo-thelial cells (Plate 4a and b). After footpad injection radioactivity was con-centrated in the subcapsular sinus, the cortical sinuses and in the medullary sinuses but little or none had infiltrated into vascular endothelium as has been described for other macromolecules (Anderson & Anderson, 1976). After i.v. injection radioactivity was present, as expected, in the lumens of all blood vessels and was also concentrated in the medullary sinuses (Plate 4a). Scintillation counting of whole organs did not reveal an exceptionally high uptake by lymph nodes compared to the kidney and liver. The potency of the tritiated dextran sulphate in inducing a lymphocytosis after i.v. injection was the same as that of the unlabelled preparation. Thus no evidence was obtained that dextran sulphate acts by binding to a receptor on the specialized endothelial cells.

DISCUSSION OF PRESENT RESULTS

The most striking finding from the cell kinetic experiments is that the administration of dextran sulphate to recipients reduced the localization of lymphocytes in lymph nodes to 10–15 % of control values at the same time as the concentration in the blood was more than double the control values (Table 2). This confirmed the work of Bradfield & Born (1974) showing that one of the mechanisms by which sulphated polysaccharides provoke a lymphocytosis is to impede the transport of lymphocytes from the blood into the tissues. Another polyanion which produces lymphocy-tosis when administered to animals is polymethacrylic acid (Ormai & de Clerq, 1969; Ross, Martens & van Bekkum, 1975). However this substance promotes a huge increase in thoracic duct output, contrary to what was observed in the present experiments. Clearly these two polyanions do not have identical actions on lymphocyte recirculation. In the case of dextran the degree of sulphation appears to be critical for the inhibitory effect on lymphocyte migration (Bradfield & Born, 1974).

The migration of lymphocytes into lymph nodes was also inhibited when they were exposed to dextran sulphate *in vitro* but the effect was much less than when the receipients were treated. Moreover in treated recipients the treated cells were no more inhibited from entering lymph nodes than were normal cells. These results suggested that the main site of action of dextran sulphate was on the endothelial cell rather than the lymphocyte, although another explanation is that to be fully effective this reagent has to be continuously present in the vicinity of the lymphocyte–endothelium interaction. The slight inhibition of lymph-node entry after exposure of lymphocytes *in vitro* can be speculatively attributed to general

25

effects such as an increased negative charge or disruption of the structure of the glycocalyx.

Hyaluronidase treatment of lymphocytes had no effect and hyaluronidase treatment of recipients had only a moderate effect on the lymph-node localization of labelled cells. The two treatments which produced relatively minor perturbations of lymphocyte migration (dextran sulphate on cells; hyaluronidase on recipients) were precisely additive when they were applied together suggesting that they were independent and acted on different mechanisms. Possibly the effect of hyaluronidase is a consequence of reduced blood flow to lymph nodes since the oedema in the recipients must have been associated with a diminished blood volume.

These experiments were intended to test the idea that sulphated proteoglycans may play a part in the selective migration of lymphocytes from the blood. A specific proteoglycan might be present on the surface of the lymphocyte, of the specialized endothelial cell or on both. One cell type might recognize the other by a complementary receptor for the specific proteoglycan and such a receptor might be competitively inhibited by dextran sulphate. The present results fail to lend support to any of these possibilities but none of them is firmly excluded. Hyaluronidase cleaves the bond between D-glucuronate residues and 2-acetamido-2-desoxy-D-glucose in the following common proteoglycans – hyaluronic acid, chondroitin, chondroitin 4S and 6S and dermatan sulphate (Meyer, 1971). The reasons that proteoglycans are not excluded are that a hyaluronidase-resistant substance might be involved and inhibitors of hyaluronidase are present in serum (Mathews & Dorfman, 1955).

The possibility remains that HEV endothelial cells express a receptor for a sulphated molecule on lymphocytes and that this receptor is competitively inhibited by dextran sulphate. If this were so some binding of [^{3}H]dextran sulphate to the surface of HEVs would have been expected but no trace of such a concentration was seen autoradiographically after heavily labelled material was administered by various routes. It must be accepted that the mechanism of action of dextran sulphate at the molecular level remains a mystery.

It remains to comment on the significance of the concentration of sodium [^{35}S]sulphate in the HEV cells after footpad injection. The observation that sulphate had no affinity for HEV cells in frozen sections, whether gluteraldehyde-fixed or not, suggested that active metabolism was necessary for the sulphate concentration rather than the alternative of binding to a preformed receptor. The electron microscopic localization of the sulphate concentration to the Golgi complex suggested that the secretion of a sulphated molecule is one of the properties of the HEV cells. The observation that sulphate was not concentrated either by other endothelial cells or in the Golgi complex of plasma cells prompts the speculation that this phenomenon is associated with the transport of lymphocytes, but so far there is no direct evidence on this point and the material into which sulphate is incorporated has still to be established.

REVIEW OF OTHER RECENT EVIDENCE

(1) *Treatment of lymphocytes or of recipients with enzymes, lectins and other reagents*

Within the past few years the eleven agents listed in Table 8 have been thoroughly studied by a number of workers for their possible effect on lymphocyte migration. The table places the agents into four categories – (*a*) marked inhibition of entry into lymph nodes; (*b*) slight inhibition of entry into lymph nodes; (*c*) prolonged retention in one or more organs and (*d*) no effect.

Table 8. *Studies of the migration of lymphocytes after exposure in vitro to various reagents*

Reagent	Inhibition of migration into:* Lymph nodes	Spleen	Retention in specified organ	Selected references
Trypsin	+ + +	–	–	Woodruff & Gesner, 1968; Ford et al., 1976; Freitas & de Sousa, 1976a
Papain	+ + +	–	–	Authors' unpublished data
Sodium azide	+ +	–	–	Ford et al., 1978
Phospholipase A	+	–	Lungs	Freitas & de Sousa, 1976c
Phospholipase C	+	–	–	Freitas & de Sousa, 1976c
Dextran sulphate	+	±	–	Bradfield & Born, 1974; Freitas & de Sousa, 1977; this paper
Neuraminidase	–	–	Liver	Ford et al., 1976; Freitas & de Sousa, 1976a
Concanavalin A	–	–	All organs	Freitas & de Sousa, 1975; Smith et al., 1977
Bacterial lipopolysaccharide	–	–	Spleen	Freitas & de Sousa, 1976b
Hyaluronidase	–	–	–	This paper
Thrombin	±	–	–	Authors' unpublished data

* + + +, Inhibition to ≤ 10 %; + +, inhibition to 10–50 %; +, inhibition to ⩾ 50 %; ±, marginal effect; –, no effect.

Trypsin treatment of lymphocytes *in vitro* profoundly inhibits their capacity to migrate into lymph nodes (Woodruff & Gesner, 1968). An essential point is that trypsin does not compromise the survival of the cells because their distribution in different compartments of injected recipients returns very nearly to that of untreated cells by 24 h after injection. The failure of the trypsinized cells to enter lymph nodes is not a consequence of sequestration in another compartment for three reasons; (1) for several hours after injection they are in surplus in the blood compared to control cells (Ford *et al.*, 1976; Freitas & de Sousa, 1976a); (2) they are not found in excessive numbers in any of the likely sites of sequestration such as the liver, lung, spleen or bone marrow and (3) the deficient localization of trypsinized lymphocytes was detectable when they were perfused through the isolated mesenteric lymph-node chain (Ford *et al.*, 1976). The inhibitory effect of trypsin on migration into lymph nodes is thus firmly established as being direct, consistent and reversible. However it can no longer be taken for granted that trypsin acts by cleaving a surface receptor for the endothelial cell because mild trypsinization is now known to produce internal changes in lymphocytes including an increase in cyclic AMP (Shneyour, Pott & Trainin, 1976).

The selective effect of trypsin treatment inhibiting entry into lymph nodes but sparing entry into the spleen (Woodruff & Gesner, 1968) has been fully confirmed and extended by studying the migration of trypsinized lymphocytes into other sites (Rannie *et al.*, 1977). Migration into Peyer's patches and into an adjuvant granuloma is also inhibited by trypsin but migration into other tissues including the bone marrow is unaffected. The three examples of trypsin-sensitive migration from the blood are characterized by morphologically distinct high-endothelial venules and in each case lymphocytes return to the blood by long lymphatic channels; by contrast HEVs are not found in trypsin-resistant examples of lymphocyte migration such as the spleen and bone marrow. In each of these examples entry is across a flat, sinusoidal endothelium (Hudson & Yoffey, 1966; Goldschneider & McGregor, 1968) and lymphocytes return to the venous blood within the organ (Ford, 1969).

The second agent which has a clear-cut inhibitory effect on migration into lymph nodes is sodium azide. At 30 min after the injection of treated lymphocytes a reduction in the lymph-node localization by a factor of 3 was apparent in the face of an increased concentration in the blood (Ford *et al.*, 1978). There was a small decrease in the 24 h survival of azide treated cells but, at about 20 %, the deficit due to cell death was much less than the deficit in lymph nodes early after injection. Incubation with sodium azide arrests much of the metabolic activity of cells by inhibiting oxidative phosphorylation. In general it does not stop the physical binding of ligands to receptors on the cell surface although it does prevent those consequences of binding which require an active response by the cell including the redistribution of receptors into a polar cap (Unanue, Karnovsky & Engers, 1973). Thus if it is accepted that azide inhibits the adhesion of lymphocytes to HEVs it can be concluded that a more complicated mechanism than the interlocking of complementary receptors is involved even at the initial stage of lymphocyte–endothelium interaction.

Nevertheless the possibility remains that the adhesion of lymphocytes to HEVs might be unaffected by azide which might solely interfere with the translation of lymphocytes across the endothelium. It could be envisaged that if the treated cells adhered but failed to migrate they would sooner or later be swept on into the venous bloodstream. However this notion was not supported by autoradiographic analysis of the distribution of azide-treated lymphocytes in lymph nodes at 10, 30 and 150 min after injection. Counts were made of the proportion of labelled cells in the lumen of the HEVs, in the walls and outside (see Plate 1a). These confirmed the expectation that azide-treated cells entering lymph nodes were rather slower at crossing the vessel walls but there was only a moderate excess in the lumen of the HEVs, whereas if azide had uncoupled adhesion and translation by inhibiting *only* the latter then the 30 % of azide-treated cells in the lymph nodes early after injection would have been predominantly intravascular and there would have been very few outside the vessel walls (authors' unpublished observations).

Another proteolytic enzyme, papain, was found to have a similar and equal effect to that of trypsin in modifying lymphocytes so that they failed to localize in the lymph nodes in the face of an excess in the blood (authors' unpublished data). However lymphocytes were killed when the concentration of papain was raised from 2 units ml^{-1} to 50 units ml^{-1} with 30 min exposure at 37 °C in presence of 5 mM cysteine. By contrast a 25-fold range of trypsin concentrations effectively inhibited migration without compromising the survival of the cells (Rannie *et al.*, 1977). Papain is apparently a blunter instrument than is trypsin, in that its specificity for peptide bonds is less and the exposure is more critical.

The treatment of lymphocytes *in vitro* with four reagents (hyaluronidase, thrombin, neuraminidase and concanavalin A) had no effect on their capacity to migrate into the spleen or lymph nodes according to our interpretation of data on the localization of treated cells at sequential time intervals in the rat and the mouse. The interpretation of experiments with hyaluronidase and thrombin is straightforward because neither had any measurable effect on lymphocyte migration *in vivo*. On the other hand both neuraminidase and concanavalin A treatment of lymphocytes *in vitro* profoundly altered their distribution *in vivo*. The primary effect of neuraminidase treatment is to promote the sequestration of lymphocytes in the liver as a consequence of which the numbers in the blood and other normal sites of localization are reduced. When the treated cells were given the opportunity to reach the specialized endothelium, as in the isolated, perfused lymph node, they migrated across HEVs at least as well as did control cells (Ford *et al.*, 1976).

The treatment of lymphocytes *in vitro* with concanavalin A perturbed their migration *in vivo* in a more complicated way. The earliest studies were interpreted as showing that con A treatment increased migration to the spleen at the expense of migration into lymph nodes (Gillette, McKenzie & Swanson, 1973; Schlesinger & Israel, 1974), but two recent studies have suggested that the previous interpretations are wrong (Freitas & de Sousa, 1975; Smith *et al.*, 1977). Our results were consistent with the propositions that (*a*) con A treated lymphocytes migrated from the blood into the same tissues as did untreated lymphocytes and (*b*) the treated lymphocytes were retained for a considerably longer period in normal sites of migration, as was clearly shown to be the case for the lymph nodes and spleen. It must be admitted that the capacity of con A treated cells to cross HEVs as efficiently as do control cells is not so firmly established as is the case with neuraminidase-treated cells. However if our interpretation of neuraminidase and con A treatment is accepted then the widespread assumption that glycosides or glycoproteins on the lymphocyte surface play an essential role in their migration *in vivo* is not supported by any substantial evidence; nevertheless it remains highly plausible.

The series of experiments by Freitas & de Sousa (1975, 1976*a*, *b*, *c*, 1977) on lymphocyte migration in mice have already been cited because with respect to trypsin, concanavalin A and dextran sulphate they reached broadly similar conclusions to our own based on rat experiments. They have also originated studies on the effects of several other agents on lymphocyte migration. Phospholipase A was shown to alter the distribution of lymphocytes presumably by interfering with the structure of the lipid bilayer. The altered localization of the treated cells was attributed to two effects (1) increased localization in the lungs and, to a lesser extent, in the liver; (2) interference with migration into lymph nodes across HEVs, as occurs with trypsin treatment (Freitas & de Sousa, 1976*a*). The observation that the concentration of treated cells in the blood was lower than was the concentration of control cells (Freitas & de Sousa, 1976*c*) suggested that a major effect was the sequestration of the enzyme-treated cells in the lungs. By 72 h after injection the distribution of phospholipase A treated cells had returned almost to that of normal cells showing that they had not been irreversibly damaged. Phospholipase C also had a small but significant inhibitory effect on lymphocyte migration into lymph nodes. By contrast with phospholipase A there was no increased localization of lymphocytes in the lung and associated with this there was a definite surplus in the blood (Freitas & de Sousa, 1976c).

The second substance studied by Freitas & de Sousa (1976b) was bacterial lipopolysaccharide (LPS). This had little or no effect on the early distribution of lymphocytes but at 24 h after injection there was a moderate surplus in the spleen and a moderate deficit in lymph nodes. Except for the normal values for lung and liver localization this is a similar situation to that found after con A treatment. The explanation could be prolonged retention in the spleen preventing the physiological redistribution of lymphocytes to lymph nodes. This interpretation was confirmed by an experiment in which LPS-treated cells were injected into splenectomized recipients. There was no difference in their distribution at any time compared to control cells.

As far as it is possible to draw any broad conclusion from all these experiments on modification of the lymphocyte *in vitro* it is that impairment of migration from the blood into lymph nodes is *not* a general consequence of any modification of the cell surface. Con A, LPS and neuraminidase all have dramatic effects exerted at the cell surface when measured in other ways such as mitogenesis or electrophoretic mobility but none apparently interferes with the ability of lymphocytes to localize in lymph nodes. Of all the agents which have been screened so far our current prejudice is that the most significant clues lie in the inhibitions induced by trypsin, sodium azide and dextran sulphate. It is notable that as far as localization in the spleen is concerned the only inhibitory agent known is dextran sulphate administered to the recipient and even then the effect is much less than on lymph node localization. None of the agents applied to lymphocytes *in vitro* have any substantial effect on the highly selective migration into the spleen.

(2) *Techniques for studying lymphocyte–endothelium interaction* in vitro

Lymphocyte populations can be isolated, exposed to a variety of agents and functionally tested but the specialized endothelial cells in HEVs are much less accessible to experimental manipulation. Attempts have been made to modify endothelium in intact animals for example by the i.v. injection of neuraminidase (Gasic & Gasic, 1962), or of dextran sulphate or hyaluronidase as reported here but, especially when a positive effect is obtained, it is difficult to be sure that it is a consequence of an alteration of the endothelial cell and not attributable to another factor such as altered blood flow. The isolated perfused lymph-node preparation might be thought to allow the possibility of modifying the endothelial cells but in practice this has not been successful because perfusion with serum-free medium, as is necessary to investigate the effect of most agents such as trypsin, led to apparently aberrant lymphocyte localization (Ford et al., 1978).

Up to the present a method for isolating or culturing specialized endothelial cells from HEVs has not been reported. One of the drawbacks is the lack of a specific cell marker and in this respect the uptake of [^{35}S]sulphate described here may find a useful application.

A recent method of promise is the fruit of long perseverance by Woodruff and her colleagues (Stamper & Woodruff, 1976). Lymphocyte adhesion to specialized endothelial cells is measured by layering a cell suspension over a frozen, fixed section of lymph node. After washing off the non-adherent cells those remaining in contact with HEVs can be counted. In the hands of Stamper & Woodruff (1977) selective adhesion to HEVs is only found if the frozen sections have been fixed in glutaraldehyde, optimally 3.0%, but another group have achieved selective adhesion without fixation of frozen sections (E. Butcher & I. L. Weissman, personal communication). The optimum temperature for the adhesion is 7 °C (Stamper & Woodruff, 1977) but despite these surprising features of the system it can be accepted as reflecting the first stage of the physiological interaction between lymphocytes and endothelial cells because (1) lymphocyte adhesion is selective for the endothelial cell – per unit area about 300 times as many lymphocytes adhere to HEVs as to other tissue and (2) adhesion is specific for recirculating lymphocytes, since thymocytes or trypsinized lymph-node cells adhere less well. The failure of neuraminidase treatment to inhibit lymphocyte adhesion to HEVs was also confirmed in this system although the treated cells became more adherent to tissues other than HEVs (Woodruff, Katz, Lucas & Stamper, 1978).

The use of frozen sections has suggested two important clues to the role of the specialized endothelial cell. Firstly selective adhesion may occur even when that cell is thoroughly fixed, unlike the situation with the lymphocyte, since glutaraldehyde-treated lymphocytes do not stick. Secondly the lymphocytes adhere not just to the luminal surface of the endothelial cell but, according to our tentative interpretation of published photomicrographs, across the entire cut surface of the cell. Each of these points is telling evidence against the idea of complementary receptors confined to the surface of each cell type.

A number of agents have already been tested for their effect on the capacity of lymphocytes to adhere to HEVs in frozen sections (Table 9).

Not only did trypsin and neuraminidase have the effects predicted from *in-vivo* experiments but so too did sodium azide (Woodruff *et al.*, 1978). Moreover another metabolic inhibitor, sodium iodoacetate, also profoundly inhibited lymphocyte adhesion. Perhaps most interesting of all the microfilament inhibitor, cytochalasin B, had a profound though reversible inhibitory effect whereas colchicine, which interferes with the function of microtubules, was without effect. Taken along with our own observations on the failure of membrane vesicles to adhere to HEVs in the isolated lymph node (Sedgley & Ford, 1976) and the inhibition of lymph-node localization caused by sodium azide treatment of lymphocytes, there is a strong presumption that an active response on behalf of the lymphocyte is necessary even for the initial adhesion stage.

Table 9. *Effect of reagents on lymphocyte adhesion to HEVs in vitro (Stamper & Woodruff, 1977; Woodruff et al., 1978)*

Reagent	Effect on adhesion
Trypsin	Inhibition
Neuraminidase	None*
EDTA	Inhibition
EGTA	Inhibition
Ionophore A-23187	Inhibition†
Glutaraldehyde	Inhibition
Cold (1 °C)	Inhibition
Sodium azide	Inhibition
Sodium iodoacetate	Inhibition
Cytochalasin B	Inhibition
Colchicine	None

* Slight increase in lymphocytes adhering to HEVs but adhesion to other tissues also increased.
† Less profound inhibition than with other agents.

(3) Transmission and scanning electron microscope observations

Another clue that the lymphocyte may respond actively to a signal generated by the endothelial cell has come from observations made on lymph nodes by the scanning electron microscope. Van Ewijk, Brons & Rozing (1975) noted that the majority of lymphocytes, whether T or B, display an abundance of microvilli while in the blood but extravascular lymphocytes in lymphoid tissues, including the lymph nodes, spleen and thymus, are almost all lacking in microvilli. They produced startling images of lymphocytes adhering to and sinking into the walls of HEVs. The adherent cells in the lumen had numerous microvilli and it appeared (not surprisingly) that the earliest contact might be made by the tips of the microvilli. Inspection of lymphocytes which had begun to sink into the endothelium suggested that withdrawal of microvilli might proceed simultaneously with migration out of the blood vessel. This observation was essentially confirmed by Anderson & Anderson (1976) who noted that lymphocytes lost their microvilli when they had penetrated deep into the endothelium. Lymphocytes which are adherent to the endothelium are difficult to detach by perfusion (Van Ewijk et al., 1975). Thus it seems a strong possibility that modulation of lymphocyte membrane activity following contact with HEVs results first in increased adhesiveness and second in loss of microvilli.

The scanning electron microscope appearance of the HEVs from the luminal aspect has produced no surprises. It presents a cobblestone appearance because the plump endothelial cells project into the lumen. Lymphocytes generally adhere in the clefts between the cells (Anderson & Anderson, 1976). At the stage of pentration the endothelial cell appears to play an active role by sending out processes to embrace the lymphocyte (Marchesi & Gowans, 1964; Anderson & Anderson, 1976).

Transmission electron microscope studies of lymphocyte migration across HEVs have been preoccupied with the question of whether lymphocytes pass between the endothelial cells or through their cytoplasm. Although the question seems to have been settled in favour of the former route (Cläesson, Jørgensen & Röpke, 1971; Schoefl, 1972; Anderson & Anderson, 1976) the different interpretations provide an object lesson of how an apparently simple exercise in electron microscopy can be beset with technical difficulties. In particular the appearance of the HEVs is very dependent on fixation (Nieuwenhuis & Ford, 1976).

The study of Anderson, Anderson & Wyllie (1976) provided a detailed description of the complex basement membrane enveloping the HEVs. This consists of overlapping reticular cell plates to which collagen bundles are attached externally. Lymphocytes require to perform contortions so as to insinuate themselves through the pockets formed by these plates whereas between the endothelial cells the lymphocytes are usually nearly circular in section (Plates 2a and b). The plausible suggestion has been made that this elaborate sheath around the HEVs functions as a barrier to certain macromolecules which are able to pass between the endothelial cells with the lymphocytes (Anderson & Anderson, 1976).

Mitochondria are abundant in the cytoplasm of the endothelial cell and there is a prominent Golgi complex (Cläesson et al., 1971) which is often situated on the luminal aspect of the cell near an intercellular junction rather than at the apex of the cell. Ribosomes are plentiful in the cytoplasm but endoplasmic reticulum and lysosomes are relatively scanty. These ultrastructural features suggest a metabolically active cell secreting a sulphated molecule (see *Results*) but not synthesizing large quantities of protein for export.

The areas of contact between adherent lymphocytes and endothelial cells do not appear at first sight to afford any insight as to the mechanism of adhesion. There is a remarkably regular electron–light gap between the membranes of the lipid bilayers of about 16 nm or about 3 times the breadth of the lipid bilayer itself. The dimensions and regularity of this gap are maintained when the lymphocyte is fully within the endothelium (Plates 2*a* and *b*, 3*a*). The most plausible idea is that the regular gap is attributable to the compressed glycocalyces of each cell type. Fibrillar material with affinity for alcian blue was observed within the gap by Anderson & Anderson (1976) but the difficulty of interpreting such a reaction has already been stressed (Röpke *et al.*, 1972).

In conclusion the unique migratory pathway of lymphocytes through lymph nodes and eventually into efferent lymph is governed by a strong and rapidly developing adhesion of lymphocytes to the luminal surface of HEVs. Although the nature of this reaction remains uncertain the evidence does not favour either complementary receptors nor identical receptors bound together by a bivalent ligand. A more plausible mechanism is that the lymphocyte responds to a specific signal generated by the specialized endothelial cell. The response may require the activity of microfilaments leading to increased adhesiveness, retraction of microvilli and migration into the endothelium. The glycocalyx of one or both cell types seems to be modulated in some way to achieve the adhesion because the two lipid bilayers remain widely separate and the cell membranes seldom interdigitate. The nature of the endothelial signal is unclear but it is likely to be a chemical substance which is present in the endothelial cell cytoplasm and is resistant to glutaraldehyde fixation. It could act enzymatically on the lymphocyte surface or by ion transfer. The Golgi complex of the endothelial cell contains a sulphated molecule which we speculatively associate with the specific signal.,

It will be of great interest to determine whether the same mechanism that operates in lymph nodes is responsible for lymphocyte migration into sites of cell-mediated immunity. The observation that trypsin treatment of lymphocytes *in vitro* inhibits their augmented localization in sites of cell-mediated immunity as well as in lymph nodes, but does not affect migration to other sites, suggests that this may be so (Rannie *et al.*, 1977).

We wish to thank Drs Judith Woodruff, E. Butcher and I. Weissman for permission to refer to their unpublished results. We acknowledge the excellent technical assistance of Ms T. Aslan, S. Sparshott, S. Walker and Mr G. Humberstone and the secretarial help of Ms G. Brocklehurst.

REFERENCES

ANDERSON, A. O. & ANDERSON, N. D. (1975). Studies on the structure and permeability of the microvasculature in normal rat lymph nodes. *American Journal of Pathology*, 80, 387–412.

ANDERSON, A. O. & ANDERSON, N. D. (1976). Lymphocyte emigration from high endothelial venules in rat lymph nodes. *Immunology*, 31, 731–48.

ANDERSON, N. D., ANDERSON, A. O. & WYLLIE, R. G. (1976). Specialized structure and metabolic activities of high endothelial venules in rat lymphatic tissues. *Immunology*, 31, 455–73.

ÅSTRÖM, K. E., WEBSTER, H. F. & ARNASON, B. G. (1968). The initial lesion in experimental allergic neuritis. A phase and electron microscopic study. *Journal of Experimental Medicine*, 128, 469–96.

ATKINS, R. C. & FORD, W. L. (1975). Early cellular events in a systemic graft vs. host reaction. I. The migration of responding and non-responding donor lymphocytes. *Journal of Experimental Medicine*, 141, 664–80.

BELL, R. G. & LAFFERTY, K. J. (1972). The flow and cellular composition of cervical lymph from unanaesthetized ducks. *Australian Journal of Experimental Biology and Medical Science*, 50, 611–23.

BOYUM, A. (1968). Separation of leucocytes from blood and bone marrow. *Scandinavian Journal of Clinical and Laboratory Investigation*, 21, Supplement 97, 77–89.

BRADFIELD, J. W. B. & BORN, G. V. R. (1974). Lymphocytosis produced by heparin and other sulphated polysaccharides in mice and rats. *Cellular Immunology*, 14, 22–32.

CLÄESSON, M. H., JØRGENSEN, O. & RÖPKE, C. (1971). Light and electron microscopic studies of the paracortical post-capillary high-endothelial venules. *Zeitschrift für Zellforschung*, 119, 195–207.

CURTIS, A. S. G. & DE SOUSA, M. A. B. (1973). Factors influencing adhesion of lymphoid cells. *Nature, New Biology, London*, 244, 45–7.

DUBREUIL, A. E., HERMAN, P. G., TILNEY, N. L. & MELLINS, H. Z. (1975). Microangiography of the white pulp of the spleen. *American Journal of Roentgenology*, 123, 427–32.

ELLIS, A. E. & DE SOUSA, M. A. B. (1974). Phylogeny of the lymphoid system. I. A study of the fate of circulating lymphocytes in plaice. *Immunology*, 4, 338–43.

FORD, W. L. (1969). The kinetics of lymphocyte recirculation within the rat spleen. *Cell and Tissue Kinetics*, 2, 171–91.

FORD, W. L. (1975). Lymphocyte migration and immune responses. *Progress in Allergy*, 19, 1–59.

FORD, W. L. & HUNT, S. V. (1973). In *Handbook of Experimental Immunology*, ed. D. M. Weir, chapter 23, pp. 1–27. Blackwell, Oxford and Edinburgh.

FORD, W. L., SEDGLEY, M., SPARSHOTT, S. M. & SMITH, M. E. (1976). The migration of lymphocytes across specialized vascular endothelium. II. The contrasting consequences of treating lymphocytes with trypsin or neuraminidase. *Cell and Tissue Kinetics*, 9, 351–61.

FORD, W. L., ANDREWS, P. & SMITH, M. E. (1978). Lymphocyte migration in relation to the pathophysiology of inflammatory rheumatic disease. In *Research into Rheumatoid Arthritis and Allied Diseases*, eds. D. C. Dumonde & R. N. Maini. M.T.P. Lancaster (in press).

FREITAS, A. A. & DE SOUSA, M. A. B. (1975). Control mechanisms of lymphocyte traffic. Modification of the traffic of ^{51}Cr-labelled mouse lymph-node cells by treatment with plant lectins in intact and splenectomized mice. *European Journal of Immunology*, 5, 831–8.

FREITAS, A. A. & DE SOUSA, M. A. B. (1976a). The role of cell interactions in the control of lymphocyte traffic. *Cellular Immunology*, 22, 345–50.

FREITAS, A. A. & DE SOUSA, M. A. B. (1976b). Control mechanism of lymphocyte traffic. Altered distribution of ^{51}Cr-labelled mouse lymph-node cells pre-treated *in vitro* with lipopolysaccharide. *European Journal of Immunology*, 6, 269–73.

FREITAS, A. A. & DE SOUSA, M. A. B. (1976c). Control mechanism of lymphocyte traffic. Altered migration of ^{51}Cr-labelled mouse lymph-node cells pretreated *in vitro* with phospholipases. *European Journal of Immunology*, 6, 703–711.

FREITAS, A. A. & DE SOUSA, M. A. B. (1977). Control mechanism of lymphocyte traffic. A study of the action of two sulphated polysaccharides on the distribution of ^{51}Cr and (^{3}H)-adenosine-labelled mouse lymph-node cells. *Cellular Immunology*, 31, 62–77.

GALL, D. & MAEGRAITH, B. G. (1950). The splenic circulation. Latex cast studies in the rat. *Annals of Tropical Medicine and Parasitology*, 44, 331–8.

GASIC, G. & GASIC, T. (1962). Removal of sialic acid from the cell coat in tumor cells and vascular endothelium and its effect on metastasis. *Proceedings of the National Academy of Sciences, U.S.A.*, 48, 1172–7.

GESNER, B. M. & GINSBURG, V. (1964). Effect of glycosidases on the fate of transfused lymphocytes. *Proceedings of the National Academy of Sciences, U.S.A.*, 52, 750–5.

GILLETTE, R. V., McKENZIE, G. O. & SWANSON, K. H. (1973). Effect of concanavalin A on the homing of labelled T lymphocytes. *Journal of Immunology*, 111, 1902–5.

GOLDSCHNEIDER, I. & McGREGOR, D. D. (1968a). Migration of lymphocytes and thymocytes in the rat. I. The route of migration from blood to spleen and lymph-nodes. *Journal of Experimental Medicine*, 127, 155–168.

GOWANS, J. L. & KNIGHT, E. J. (1964). The route of recirculation of lymphocytes in the rat. *Proceedings of the Royal Society, London*, B 159, 257–82.

GRAHAM, R. C. & SHANNON, S. L. (1972). Peroxidase arthritis. II. Lymphoid cell–endothelial interactions during a developing immunologic inflammatory response. *American Journal of Pathology*, 69, 7–24.

GUTMAN, G. A. & WEISSMAN, I. L. (1973). Homing properties of thymus-independent follicular lymphocytes. *Transplantation*, 16, 621–9.

HERMAN, P. G., LYONNET, D., FINGERHUT, R. & TUTTLE, R. N. (1976). Regional blood flow to the lymph-node during the immune response. *Lymphology*, 9, 101–4.

HOBBS, B. B. & HAY, J. B. (1977). The flow of blood to lymph-nodes and its relation to lymphocyte traffic and the immune response. *Journal of Experimental Medicine*, 145, 31–44.

HUDSON, G. & YOFFEY, J. M. (1966). The passage of lymphocytes through the sinusoidal endothelium of guinea-pig bone marrow. *Proceedings of the Royal Society, London*, B 165, 486–96.

MARCHESI, V. T. & GOWANS, J. L. (1964). The migration of lymphocytes through the endothelium of venules in lymph-nodes: an electron microscopic study. *Proceedings of the Royal Society, London*, B, 159, 283–90.

MATHEWS, M. B. & DORFMAN, A. (1955). Inhibition of hyaluronidase. *Physiological Reviews*, 35, 381–402.

MEYER, K. (1971). Hyaluronidases. In *The Enzymes*, vol. V, ed. P. D. Boyer, 3rd edition, chap. 11, pp. 307–20. Academic Press, New York.

NIEUWENHUIS, P. & FORD, W. L. (1976). Comparative migration of B- and T-lymphocytes in the rat spleen and lymph-nodes. *Cellular Immunology*, **23**, 254–67.

ORMAI, S. & DE CLERQ, E. (1969). Polymethacrylic acid: Effects on lymphocyte output of the thoracic duct in rats. *Science*, **158**, 471–2.

RANNIE, G. H. & FORD, W. L. (1978). Physiology of lymphocyte recirculation in animal models. In *Proceedings of the Ninth Course on Transplantation, and Clinical Immunology, Lyon 1977*, ed. J. L. Touraine, pp. 165–72 Excerpta Medica, Amsterdam.

RANNIE, G. H., SMITH, M. E. & FORD, W. L. (1977). Lymphocyte migration into cell-mediated immune lesions is inhibited by trypsin. *Nature, London* **267**, 520–2.

RÖPKE, C., JØRGENSEN, O. & CLÄESSON, M. H. (1972). Histochemical studies of high-endothelial venules of lymph nodes and Peyer's patches in the mouse. *Zeitschrift für Zellforschung*, **131**, 287–97.

ROSS, W. M., MARTENS, A. C. & VAN BEKKUM, D. W. (1975). Polymethacrylic acid: Induction of lymphocytosis and tissue distribution. *Cell and Tissue Kinetics*, **8**, 467–77.

SAINTE-MARIE, G., YIN, Y. M. & CHAN, C. (1967). The diapedesis of lymphocytes through post-capillary venules of rat lymph nodes. *Revue Canadienne de Biologie*, **26**, 141–57.

SASAKI, S. (1967). Production of lymphocytosis by polysaccharide polysulphates (heparinoids). *Nature, London*, **214**, 1041–2.

SCHLESINGER, M. & ISRAEL, E. (1974). The effect of lectins on the migration of lymphocytes in vivo. *Cellular Immunology*, **14**, 66–79.

SCHOEFL, G. I. (1972). The migration of lymphocytes across the vascular endothelium in lymphoid tissue. A re-examination. *Journal of Experimental Medicine*, **136**, 568–88.

SEDGLEY, M. & FORD, W. L. (1976). The migration of lymphocytes across specialized vascular endothelium. I. The entry of lymphocytes into the isolated mesenteric lymph-node of the rat. *Cell and Tissue Kinetics*, **9**, 231–43.

SHNEYOUR, A., PATT, Y. & TRAININ, N. (1976). Trypsin-induced increase in intracellular cyclic AMP of lymphocytes. *Journal of Immunology*, **117**, 2143–9.

SMITH, C. & HENON, B. K. (1959). Histological and histochemical study of high endothelium of post-capillary veins of the lymph node. *Anatomical Record*, **135**, 207–13.

SMITH, J. B., MCINTOSH, G. H. & MORRIS, B. (1970). The migration of cells through chronically inflamed tissues. *Journal of Pathology*, **100**, 21–9.

SMITH, M. E., SPARSHOTT, S. M. & FORD, W. L. (1977). The migration of lymphocytes across specialized vascular endothelium. III. Concanavalin A delays lymphocytes in normal traffic areas. *Experimental Cell Biology*, **45**, 9–23.

SPRENT, J. (1977). Recirculating lymphocytes. In *The Lymphocyte: Structure and Function*, ed. J. J. Marchalonis, pp. 43–111. Marcel Dekker, New York.

SPRY, C. J. F., LANE, J. T. & VYAKARNAM, A. (1977). The effects of complement activation by Cobra venom factor on the migration of T and B lymphocytes into rat thoracic duct lymph. *Immunology*, **32**, 947–54.

STAMPER, H. B. & WOODRUFF, J. J. (1976). Lymphocyte homing into lymph-nodes: in vitro demonstration of the selective affinity of recirculating lymphocytes for high-endothelial venules. *Journal of Experimental Medicine*, **144**, 828–33.

STAMPER, H. B. & WOODRUFF, J. J. (1977). An in vitro model of lymphocyte homing. I. Characterization of the interaction between thoracic duct lymphocytes and specialized high-endothelial venules of lymph-nodes. *Journal of Immunology*, **119**, 772–80.

UNANUE, E. R., KARNOVSKY, M. J. & ENGERS, H. D. (1973). Ligand-induced movement of lymphocyte membrane macromolecules. III. Relationship between the formation and fate of anti-Ig surface complexes and cell metabolism. *Journal of Experimental Medicine*, **137**, 675.

VAN EWIJK, W., BRONS, H. H. C. & ROZING, J. (1975). Scanning electron microscopy of homing and recirculating lymphocyte populations. *Cellular Immunology*, **19**, 245–61.

WOODRUFF, J. & GESNER, B. M. (1968). Lymphocytes: circulation altered by trypsin. *Science*, **161**, 176–8.

WOODRUFF, J. J., KATZ, I. M., LUCAS, L. E. & STAMPER, H. B. (1978). An in vitro model of lymphocyte homing. Membrane and cytoplasmic events involved in lymphocyte adherence to specialized high-endothelial venules of lymphnodes. *Journal of Immunology*, (in press).

EXPLANATION OF PLATES

PLATE I

(a) Autoradiograph of high endothelial venules in the lymph node of a rat which had received an i.v. injection of ^{3}H-labelled lymphocytes 10 min previously. The gap defined by the four arrows corresponds to the basement membrane which is not stained. Most of the labelled cells are within the vessel wall but one is in the lumen and one (lower right) is outside. Note that most of the lymphocytes in the wall (labelled and unlabelled) are close to the basement membrane. Methyl green–pyronin with yellow filter.

(b) Autoradiograph of part of the paracortex of a popliteal lymph-node removed 15 min after footpad injection of sodium [^{35}S]sulphate. Radioactivity is concentrated over HEVs of which four sections are included in this field. Methyl green–pyronin with yellow filter, 4 weeks' exposure.

(c) Autoradiograph of junction of HEVs in lymph node removed 15 min after injection of sodium [^{35}S]sulphate. Two tributary vessels cut in oblique section are arrowed. Radioactivity is concentrated over the main vessel (left) and the lower tributary. There is little activity over the upper tributary possibly because only the basal part of the endothelial cells which do not contain the Golgi complexes are included in the section (see Plates 2a and b, 3a and b). Methyl green–pyronin with yellow filter, 4 weeks' exposure.

PLATE 2

(a) Electron micrograph-autoradiograph of HEV in lymph node removed 15 min after injection of sodium [^{35}S]sulphate. The outer limits of the obliquely cut vessel are defined by the prominent basement membrane (arrowed). Radioactivity is concentrated in discrete areas of the cytoplasm of the plump endothelial cells. These areas have vesicular complexes and four out of the five concentrations of radioactivity in this field are situated on the luminal side of the nucleus. There is very little activity over lymphocytes. Uneven illumination is due to differences in emulsion thickness. 3 months' exposure.

(b) Electron micrograph–autoradiograph of part of HEV from a lymph node 15 min after s.c. injection of [^{35}S]sulphate. The lumen of the vessel crosses the field from the top left to the right. Two lymphocytes within the lumen are present in the upper right quadrant. On the lower side of the lumen parts of three plump endothelial cells are visible with a lymphocyte in close relation to the middle cell. The three arrows indicate concentrations of radioactivity in the cytoplasm of each endothelial cell. In each case the concentration is over a vesicular complex on the luminal side of the nucleus. 3 months' exposure.

PLATE 3

(a) Electron micrograph–autoradiograph of part of HEV from a lymph node 15 min after injection of [³⁵S]sulphate. The centre of the field is dominated by a plump endothelial cell. The lumen of the vessel is visible (top centre) and basement membrane on lower right. A lymphocyte which is clearly migrating between endothelial cells is seen (top left). Radioactivity is almost confined to one small area of endothelial cytoplasm. Uneven illumination is due to uneven A.R. emulsion thickness. 3 months' exposure.

(b) Higher power view from Plate 3a of concentration of radioactivity. Radioactivity is localized to the Golgi complex. 3 months' exposure.

PLATE 4

(a) Autoradiography of cortico-medullary junctional area in a lymph node from a recipient of i.v. [³H]dextran sulphate 2.5 h previously. The area below and to the left of the broken line is a small part of a medullary sinus. Radioactivity was concentrated in both the medullary and cortical sinuses. To the right is a large HEV in which there is no marked concentration of radioactivity. There are many lymphocytes within the lumen of the vessel but very few within the vessel wall (cf. Plate 1a). Methyl green–pyronin with yellow filter, 4 weeks' exposure.

(b) Autoradiograph of junction of HEVs in a lymph node of a recipient of i.v. [³H]dextran sulphate 10 min previously. Several lymphocytes are clustered within the lumen of the upper tributary (broad arrow). There are few lymphocytes within the vessel wall. The endothelial cells are swollen and several large ill-defined vesicles are present in their cytoplasm (three indicated by thin arrows). There was considerable radioactivity in the blood within all vessels at this stage and no tendency of radioactivity to accumulate on the luminal surface of the HEVs was evident. Methyl green–pyronin with yellow filter, 4 weeks' exposure.

LYMPHOID CELL POSITIONING: A NEW PROPOSAL FOR THE MECHANISM OF CONTROL OF LYMPHOID CELL MIGRATION

By MARIA DE SOUSA

Laboratory of Cell Ecology, Sloan Kettering Institute for
Cancer Research, New York, N.Y. 10021, U.S.A.

INTRODUCTION

'A train really isn't a train if it can't go anywhere'*

R. M. Pirsig (1974)

'T' and B lymphocytes circulate continuously between blood and lymph through the peripheral lymphoid organs, utilizing common routes of entry and exit in and out of the spleen, lymph node and Peyer's patches. Within these organs, however, the two major classes of lymphocytes migrate and arrange themselves in separate territories, a phenomenon which has been named ecotaxis (de Sousa, 1971, 1973, 1976, 1977). This capacity of lymphocytes places them among the natural models of study of cell sorting out in higher vertebrates. In contrast with other cell sorting-out systems, however, few authors have approached the question of the control mechanism of specific lymphoid cell positioning (Gesner, 1966; de Sousa, 1971, 1973; Curtis, 1976).

In the present paper, the current ideas on the mechanisms of control of lymphoid cell circulation and positioning are reviewed. Additional data enlarging the scope of one of the ideas are presented, leading to the formulation of a new proposal of control of lymphoid cell migration. For the purpose of clarity, the paper has been divided in three sections: in the first, the current ideas on control of lymphocyte traffic are briefly reviewed; in the second section preliminary data on the migration of the Ly subsets in the mouse and on the modulation of lymphocyte migration by iron salts is presented; in the third section a new proposal for the control of lymphoid cell migration is put forward.

* This paper is dedicated to my friend M.G. who introduced me to R. M. Pirsig, and thus across the high country to the bottom of the ocean with Phaedrus.

CURRENT IDEAS ON CONTROL OF LYMPHOCYTE
TRAFFIC

Principal role of the lymphocyte–post-capillary venule (PCV)
interaction

In mammals, the majority of circulating lymphocytes leave the blood and
enter the lymph through a well-defined section of the post-capillary venules
in the mid cortex of the lymph nodes (Gowans & Knight, 1964; Herman,
Yamamoto & Mellins, 1972). In normal animals, this section of the venule
is characterized by the presence of high, cuboidal endothelial cells. In
mice and rats depleted of circulating T lymphocyte, however, the venule's
endothelium is ill-defined, consisting of flat endothelial cells (Parrott, de
Sousa & East, 1966; Goldschneider & McGregor, 1968a; de Sousa &
Pritchard, 1974). In depleted animals reconstituted with lymphocyte suspensions (Goldschneider & McGregor, 1968b) or with thymus grafts (de
Sousa & Pritchard, 1974) the PCV (post-capillary venule) endothelium
rapidly assumes its typical high cuboidal morphology. This morphology
was first elegantly described by Gowans & Knight (1964) who at the same
time demonstrated that lymphocytes enter the lymph through these venules.
These observations led to the present dominant proposal for the control
of lymphocyte circulation. This proposal is concerned strictly with lymphocyte entry into lymph thus focusing on the interaction of circulating
lymphocytes with the endothelium of the PCV found in the mid cortical
area of the lymph node (Gesner & Ginsburg, 1964; Ford, this volume, pp.
359–92), despite the fact that lymph nodes are first seen in mammals
and that lymphocytes have since been shown to recirculate between blood
and lymph in fish (Ellis & de Sousa, 1971) and birds (Bell & Lafferty, 1972).
    It was first formulated by Gesner & Ginsburg (1964) as follows:

> However, it is also possible that the integrity of the sugars on lympho-
> cytes is necessary for these cells to traverse their unique route through
> the body by acting as sites recognized by complementary structures on
> the surface of endothelial cells in the post-capillary venules of lymphoid
> tissue. This interaction could be a critical event which controls the
> selective emigration of lymphocytes from the blood into lymphoid
> tissue.

Gesner & Ginsburg's proposal has inspired a substantial amount of
work, which is reviewed in detail by Ford in this volume (pp. 359–92).
Its greatest limitation as originally conceived is its narrowness, not only
from the phylogenetic point of view mentioned already, but also in trying
to explain exclusively the lymphocytes' entry into the lymph node, it

ignores the importance of lymphocyte interactions with other vascular endothelia and other cells in other organs, and it makes no attempt to explain the migratory patterns of the cells *within* the lymphoid organs.

The role of other cell interactions in the control of lymphocyte traffic

The possibility that other cell–cell interactions played a role in the control of lymphoid cell positioning and tempo of circulation, was first raised in a study of T and B lymphocyte adhesive interactions *in vitro* (Curtis & de Sousa, 1973, 1975).

In this study it was found that T and B lymphocytes had different adhesiveness, and that when the two populations were mixed, the adhesion of the total population was constantly less than the one expected from the addition of the adhesion values of the two separate populations. Supernatants from either cell population, when added to the unlike population, effectively diminish its adhesion. From these observations, it was suggested that adhesive interactions between the two major lymphocyte populations might play a role in the control of lymphocyte positioning in lymphoid organs and that the presence of one cell population might influence the tempo of circulation of the unlike population.

One of the attractions of this work, was the fact that it took lymphocytes away from the 'immunological shrine' where immunologists had kept them over the last twenty years, and placed them in the wider world of cell biology where other cells display similar kinds of interactions (Curtis, 1976). The papers by Curtis & de Sousa (1973, 1975) omitted, however, the alternative interactions of the circulating lymphocytes with other cells in tissues, namely, endothelial, stromal and phagocytic cells. This was partly corrected in the later work of Freitas & de Sousa (1975, 1976a, b, c, 1977) on modulation of lymphocyte traffic in splenectomized and intact recipients by enzymes and plant lectins (see review by de Sousa, 1976).

Pretreatment of lymphocytes *in vitro* with neuraminidase, trypsin phospholipases or plant lectins always results in decreased lymph node entry. With the exception of trypsin, this is not due to a failure of lymphocytes to enter the lymph node but to their delay elsewhere in the recirculatory pathway (de Sousa, 1976). Thus, from these observations it was concluded (Freitas & de Sousa, 1976a): 'the present results though not excluding the importance of the interaction between circulating lymphocytes and the PCV, strongly support the view that traffic of lymphocytes is controlled by cell interactions within lymphoid and non-lymphoid organs.'

In a more recent autoradiographic study of the distribution of [3H]-adenosine-labelled lymph node cells pre-treated *in vitro* with dextran sulphate (Freitas & de Sousa, 1977a), it was clearly demonstrated that failure of the labelled cells to enter the lymph node concurred with their slower migration through the lungs and spleen, thus confirming the view that failure of lymphocytes to enter the lymph node is not the reflection of abnormal interaction with the PCV but the reflection of modified interactions with alveolar macrophages in the lung and with macrophages and other lymphocytes in the spleen.

The work on the *in-vitro* treatment of ^{51}Cr-labelled lymphocytes, though clearly helping to shift the emphasis from the PCV–lymphocyte interaction and opening up the way to interest in other cell interactions, still leaves us uncertain about the mechanism of ecotaxis of normal lymphocyte subpopulations.

The ecotaxis hypothesis

The ecotaxis hypothesis was first formulated apropos the specific migration of thymus-derived lymphocytes (de Sousa, 1973) as follows: 'It is thus possible that as the progeny of stem cells differentiate within the thymus, they are not only induced to synthetize a certain set of surface antigens but at the same time acquire the chemical and physical surface make-up that will determine their unique migration behavior'.

According to this hypothesis, specific cell positioning was viewed as pre-programmed in cells expressing different phenotypes; its plausibility, however, could only be tested by the study of the migratory patterns of lymphoid cells expressing clearly distinct phenotypes. This has recently become possible with the separation of T cell subpopulations *in vitro* with the use of specific anti-*Ly* antisera (Cantor & Boyse, 1975). A summary of preliminary results of experiments tracing the fate of ^{51}Cr-labelled [^{3}H]adenosine-labelled *TLy* 1 and *TLy* 2.3 cells in syngeneic recipients is presented in the following section.

As will be seen, the data, though supporting the ecotaxis hypothesis, still give no clues as to the molecular basis of specific lymphoid cell migration.

As in so many instances in biomedical research we sought the clues in a pathological situation. In this case, a situation characterized by lymphocyte maldistribution, Hodgkin's disease (de Sousa *et al.*, 1977). The clues from the study of Hodgkin's disease (de Sousa, Smithyman & Tan, 1978), led us in turn to consider the possible role of iron and iron-binding proteins in the control of lymphoid cell migration.

The preliminary results of experiments on the effect of iron salts in lymphocyte migration in the mouse are also presented in the following section.

PRELIMINARY DATA ON THE MIGRATION OF THE Ly SUBSETS IN THE MOUSE AND ON MODULATION OF LYMPHOCYTE TRAFFIC BY IRON SALTS

The present section is a preliminary account of results of experiments to be published in detail later (de Sousa, Huber, Freitas, Shen, Cantor & Boyse, in preparation; de Sousa & Bognaki, in preparation).

Materials and methods

Cell suspensions. Preparation of pure suspensions of Ly 1 and Ly 2.3 T cells was carried out according to Cantor & Boyse (1975). In the experiments with metal salts, C_3HAn females aged between 8 and 16 weeks were used. Single-cell suspensions were prepared from pooled axillary, inguinal and mesenteric lymph nodes in cold Eagle's medium, adjusted to a concentration of 10^8 cells ml^{-1}.

Radioisotope labelling. The labelling procedures with ^{51}Cr and with [^{3}H]-adenosine used routinely in lymphocyte traffic experiments are published in detail elsewhere (Freitas & de Sousa, 1977).

In-vitro treatment with metal salts. In the experiments with metal salts, after labelling, the cells were washed twice in cold medium and the cell concentration adjusted to 2×10^6 cells ml^{-1} in Hank's medium supplemented with 10 % Fcs (foetal calf serum). Ferric chloride, ferric citrate or sodium citrate was added to the medium at concentrations of 0.3×10^{-5} M or 1.0×10^{-5} M (ferric citrate and sodium citrate) and 3×10^{-7} M (ferric chloride). The cells were incubated for 30 min at 37 °C. A third control group was incubated in medium without added ferric citrate or sodium citrate, for the same length of time. The pH was adjusted to 7.4. After incubation, the cells were washed three or four times to remove any excess metal and checked for viability by the trypan blue (0.1 %) exclusion test. The labelled treated or control untreated cells were injected into the tail vein of syngeneic recipients (10^7 cells in a volume of 0.2 ml).

Distribution of labelled cells. The distribution of the ^{51}Cr-labelled cells, and the position occupied by the [^{3}H]adenosine-labelled cells in tissues as detected by autoradiography, were determined according to procedures described previously (Freitas & de Sousa, 1977).

Distribution patterns of ^{51}Cr and [^3H]adenosine-labelled Ly 1 and Ly 2.3 mouse T lymphocytes

A detailed account of the study of the fate of ^{51}Cr and [^{3}H]adenosine labelled Ly 1 and Ly 2.3 cells is presently in preparation. Here only the points of relevance to the ecotaxis hypothesis are summarized. They are as follows:

    (a) ^{51}Cr-labelled *Ly* 1 and *Ly* 2.3 cells differ in their overall patterns of distribution in syngeneic hosts.

    (b) At 24 h after intravenous injection, the autoradiographic study of the distribution of [^{3}H]adenosine-labelled cells revealed that the *Ly* 1 cells have a distribution pattern indistinguishable from that of T cells, whereas a proportion of *Ly* 2.3 cells is found within B areas (Fig. 1).

These observations confirm the proposal that heterogeneity of lymphoid cell positioning is pre-programmed in cells expressing different phenotypes. A similar heterogeneity has been demonstrated in immunological function: *Ly* 1 cells exert helper effects whereas *Ly* 2.3 cells act as suppressor and cytotoxic cells (Cantor & Boyse, 1975).

Clues leading to, and results from, the study of action of iron
salts on lymphocyte traffic

As mentioned earlier neither the work on lymphocyte migration nor the work on immunological function has disclosed any decisive clues about the molecular basis of the migratory behavior of lymphoid cell populations. Some clues were found, however during the study of spleen cell population, in Hodgkin's disease. Details of the clinical and pathological aspects of the study are published elsewhere (de Sousa *et al.*, 1978).

    Recent studies of the spleen, removed at staging laparotomy from patients with Hodgkin's disease, have revealed the presence of abnormally high numbers of T lymphocytes at all stages of the disease, i.e. at a time when no involvement of the spleen by the disease can be detected by conventional pathology (Kaur, Spiers, Catovsky & Dalton, 1974; Payne, Jones, Haegert, Smith & Wright, 1976; de Sousa *et al.*, 1977). In *in-vitro* tests of immunological function, the spleen T cells behave normally in contrast with the abnormal function of peripheral blood lymphocytes in the same patients (Matchett, Huang & Kremer, 1973; de Sousa *et al.*, 1977). We interpreted the findings of high numbers of T cells in the spleen of these patients as the result of sequestration of normal circulating T cells.

    If normal circulating T cells were sequestered in these spleens, the next question was to find a reason for the sequestration, in the spleen environment. Earlier studies of the HD spleen in search of tumor-associated antigens had unravelled the presence of unusually high amounts of ferritin in the spleen and in other tissues involved by the disease (Bieber & Bieber,

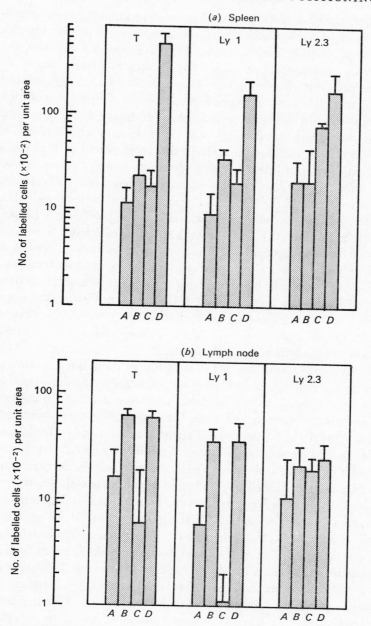

Fig. 1. (a) Distribution of [³H]adenosine-labelled T, Lyl and Ly23 in the four major areas of the spleen: (A) red pulp, (B) marginal zone, (C) outer layer of the Malpighian nodule ('B' area), (D) periarteriolar, thymus-dependent area. (b) In the lymph node: (A) medulla, (B) cortico-medullary junction, (C) 'B' area, (D) thymus-dependent area.

1973; Eshhar, Order & Katz, 1974). More recently the findings of excessive amounts of stainable ferric iron has been reported in involved lymph nodes (Dumont, Ford & Becker, 1976). Early in our own analysis of unstained frozen spleen sections under the fluorescence microscope we detected the presence of very high numbers of cells which autofluoresced in the porphyrin region and confirmed quantitatively the presence of high amounts of porphyrin in HD-involved tissue (de Sousa et al., 1978). All these somewhat unrelated findings and other clinical observations of abnormalities of iron metabolism early in the development of Hodgkin's disease (Beamish et al., 1972) led us to a systematic search of the presence of ferric and ferrous iron, and of iron-binding proteins, namely transferrin, lactoferrin and ferritin in the spleen and lymph nodes of patients with HD. This search confirmed and extended the earlier findings (Bieber & Bieber, 1973, Eshhar et al., 1974). Unusually high amounts of ferritin were seen in association with areas of disease involvement, in addition, deposits of ferric iron were also present in these areas. But the increase was not confined exclusively to ferritin; large numbers of transferrin- and lactoferrin-containing cells were also found in these spleens. It is not within the scope of this paper to relate these findings to the pathogenesis of the disease. The relevant question here is whether these findings can be related to the sequestration of circulating T lymphocytes.

The simplest way to visualize the mechanism of the sequestration was to predict that lymphocytes had receptors either for the iron-binding proteins or for the metal–protein complex, and were thus able to migrate to and lodge in metalloprotein-rich sites. If this was correct, pre-treatment in vitro with iron salts in the presence of serum would result in blindfolding of the receptor and consequent failure of the cells to migrate normally.

The results of experiments using ferric chloride and ferric citrate are shown in Fig. 2. As predicted, ^{51}Cr-labelled lymph node cells pre-treated in vitro with the iron salts failed to enter the lymph nodes and spleen, and were recovered from the blood and first capillary network crossed after an i.v. injection, namely, the pulmonary capillary network. No labelled pre-treated cells were found in abnormal numbers in the liver, indicating that the cells had not been killed by the in-vitro treatment. Treatment in vitro with sodium citrate had no effect on lymphocyte distribution; furthermore, experiments comparing the inhibitory effect of ferric citrate on the migration of ^{51}Cr-labelled enriched mouse T or B lymphocytes indicate that migration of both populations is affected by exposure to Fe^{3+} (de Sousa & Bognaki, unpublished observations).

At the time we developed the working models of interaction between lymphocytes and iron-binding proteins, we were unaware of recent work

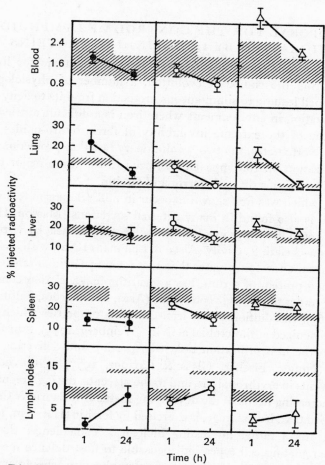

Fig. 2. Distribution of ^{51}Cr-labelled mouse lymph node cells, treated *in vitro* with ferric chloride (●) and two doses of ferric citrate, 0.3×10^{-5} M (○), and 1×10^{-5} M (△) at 1 and 24 h after intravenous injection in syngeneic recipients. Shaded areas represent the control range (mean ± 1 S.D.), i.e. the distribution of ^{51}Cr-labelled untreated cells. Each point represents the mean ($\pm$ S.D.) of the results obtained with groups of 5–10 mice.

demonstrating that normal human peripheral blood lymphocytes and a mouse lymphoma cell line (Phillips, 1976; Yang Hu, Gardner, Aisen & Skoultchi, 1977) have receptors for transferrin, and that mouse lymphocytes have receptors for lactoferrin (van Snick & Masson, 1976).

The knowledge that such receptors do exist, made the task of developing a new proposal for mechanism that might direct lymphoid cell migration easier.

NEW PROPOSAL FOR THE CONTROL OF LYMPHOID
CELL MIGRATION BY IRON-BINDING PROTEINS

Iron, like most metals, is simultaneously essential and toxic for living organisms. During the course of evolution, a number of physiological systems developed leading to simultaneous protection from its toxicity and successful utilization in environments where iron is a difficult nutrient to obtain by virtue of the extreme insolubility of ferric hydro-oxides and salts. In higher vertebrates, the two 'evolutionary tasks' of protection and successful utilization of iron appear epitomized in the three major iron-binding proteins: lactoferrin, transferrin and ferritin.

Lactoferrin, which was first shown to occur in milk (Sørenson & Sørenson, 1939) and is also found in many external secretions (Masson, Heremans & Dive, 1966), is clearly involved in the protection task, by virtue of its powerful iron-chelating activity (Aisen & Leibman, 1972), with an iron-association constant 300-fold greater than that for transferrin. Transferrin, the iron-binding protein of serum, has a logarithmic association constant as high as 10^{36} under physiologic conditions (Aasa, Malmström, Saltman & Vänngård, 1963). In addition to the iron-chelating properties which play a role in its recognized antimicrobial activity (Weinberg, 1975), transferrin participates in the iron utilization task, transporting and delivering it to reticulocytes (Nuñez, Fischer, Glass & Lavidor, 1977) in the normal sequence of events in erythropoiesis, and to the placenta, during pregnancy, to meet the increasing needs of the developing foetus (McLaurin & Cotter, 1967). The third task, iron storage, is exercised by ferritin (Harrison, 1977).

It is, therefore, a rather interesting biological 'coincidence' that the three major circulating cell types indispensable to host defence mechanisms, namely leukocytes, macrophages and lymphocytes, should participate in the synthesis and recognition of the three major iron-binding proteins. Thus, neutrophilic leukocytes synthesize lactoferrin (Masson *et al.*, 1969), macrophages synthesize transferrin (Phillips & Thorbecke, 1966), and ferritin (Okhuma, Noguchi, Amano, Mizuno & Yasuda, 1976). Moreover, receptors for lactoferrin have been demonstrated on the surface of macrophages and lymphocytes (van Snick & Masson, 1976), receptors for transferrin have been demonstrated on the surface of human peripheral blood lymphocytes (Phillips, 1976) and in one mouse lymphoma cell line (Yang Hu *et al.*, 1977) and ferritin has been found on the surface of a sub-population of T lymphocytes in human peripheral blood (Moroz, Giler, Kupfer & Urca, 1977).

Such a 'coincidence' is most likely the expression of some evolutionary adaptation of survival value. It may seem intriguing for instance that degranulation of the specific granules of leukocytes containing lactoferrin (Baggiolini, De Duve, Masson & Heremans, 1970) precedes de-granulation of the azurophil granules (Bainton, 1973) during the early stages of the inflammation process. But not so, if lactoferrin is seen playing a role in directing the subsequent migration of cells with receptors for lactoferrin to the inflammatory site, i.e., macrophages and lymphocytes, thus serving the function of 'amplifying and localizing an inflammatory response' as originally speculated by Wright & Malawista (1972). The further migration of lymphocytes usually seen at the later stages of inflammation is also easily understood if local synthesis of transferrin and ferritin by macrophages is directing the migration of lymphocytes with receptors for those proteins.

Thus, by suggesting that the iron-binding proteins play a role in directing the migration of cells with appropriate receptors, the cell 'cavalcade' normally seen at a site of inflammation (de Sousa & Parrott, 1969; Spector, 1969) becomes easily explainable (Fig. 3).

We are left with the question of normal lymphocyte circulation and positioning. If we look at the inflammatory process, and at the Hodgkin's disease results referred to earlier, as naturally occurring experimental models of what is happening normally, the prediction is that lymphocytes, possibly T rather than B lymphocytes, normally circulate in iron–iron-binding protein gradients.

The continuous lymph–blood–lymph cycle can be visualized as follows: cells emerging from the lymph compartment with transferrin on the surface enter the blood and migrate in the direction of the splenic marginal zone and red pulp areas rich in iron-containing macrophages. Further, indirect evidence for the presence of transferrin on the lymphocyte surface has been provided by the recent study of Rannie, Thakur & Ford (1977) comparing the cell-labelling and elution patterns of ^{111}In, ^{51}Cr and ^{99m}Tc using rat thoracic duct lymphocytes. Rannie *et al.* (1977) found ^{111}In to be the best of the three radiolabels tested, with the lowest rate of elution *in vitro*. Indium has been shown to bind specifically to the iron-binding site of transferrin (Beamish & Brown, 1974).

However, cells possibly attracted to splenic iron-rich sites, as a result of their surface transferrin, do not remain there but move on to separate sites of the white pulp. T lymphocytes cross the periphery of the Malpighian body to lodge in the thymus-dependent area around the central arteriole, and B lymphocytes migrate to the periphery of the Malpighian body.

The mechanism of the specific lodging of T and B lymphocytes within the white pulp is not fully understood, but it is probable that interaction modulation factors (IMF) (Curtis, this volume, pp. 51–82) play a role

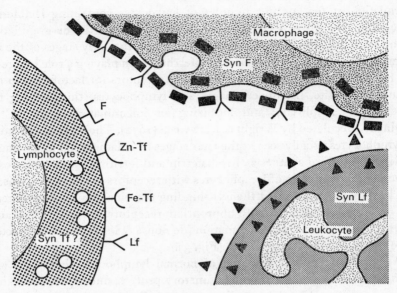

Key: Lf,▲,lactoferrin;Y, receptor. F,▰▰ , ferritin; ⊤ , receptor. Tf,○,
transferrin;Y, receptor.

Fig. 3. Diagram summarizing data on the association of iron-binding proteins and cells present at sites of inflammation. Polymorphonuclear leukocytes (Leuk), which are the cells first found at an inflammatory site, synthesize lactoferrin (LF: ▲). Macrophages and lymphocytes follow the polymorphs. Both macrophages and lymphocytes have recently been shown to have receptors for LF (▲). Thus, their accumulation at the site may be related to the release of lactoferrin by the polymorphs. Furthermore, macrophages synthesize ferritin (F: ▰) and in some species, transferrin is also synthesized by macrophages (not indicated). Lymphocytes, by virtue of their receptors for transferrin and ferritin may thus be attracted or induced to remain in sites rich in transferrin (○) and ferritin (▰). Human lymphocytes have also been shown to synthesize transferrin (Syn Tf?).

in this form of *in-vivo* cell sorting out. How IMF relate to surface transferrin is not clear at this stage. It is conceivable, however, that in an iron-rich area the lymphocyte surface is in some way altered, and that as the result of this alteration the cells move on to areas of 'receptor reconstruction'. Evidence for the existence of different metabolic patterns of cells migrating within lymphoid organs and cells in non-lymphoid organs derives from earlier findings in the rat (Goldschneider & McGregor, 1968b) and plaice (Ellis & de Sousa, 1974), indicating that [³H]uridine-labelled cells recovered from lymphoid organs have a more active pattern of RNA synthesis than that of similarly labelled cells recovered from non-lymphoid organs (see also de Sousa, 1973) It is of interest to consider the possibility that these differences in RNA metabolism relate to a step of 'receptor reconstruction' in lymphocyte circulation. Once this step occurs, cells move on to compartments rich in unsaturated protein, namely serum and lymph (Vaerman, Andre, Bazin & Heremans, 1973; see also de Sousa, 1976). In higher vertebrates, this brings us to the point of entry into the lymph node – the PCV (see Ford, this volume, pp. 359–97). We have no direct clues whether metals are present in this endothelium. However, in a recent study of an experimental model of leprosy infection in mice, we found that *Mycobacteria lepremurium* show a marked preference for pro-liferation in the endothelium of the post-capillary venules in the mid cor-tex (Nayak, de Sousa & Good, unpublished observations, see Plate 1). I would like to speculate that the preference of mycobacteria for this site reflects its hitherto unsuspected metal content. Lymphocytes with metal-binding proteins on their surface may be attracted to this specific section of the post-capillary venule by the presence of metals therein.

Recently, Woodruff and her co-workers (Woodruff, Katz, Lucas & Stamper, 1977) showed that the binding of rat thoracic duct lymphocytes to the endothelium of the post-capillary venule *in vitro* is inhibited by exposure of the lymphocytes to EDTA and EGTA. This finding was interpreted as indicating that the binding step is calcium-dependent. Calcium, however, did not appear to play a role in maintaining the physical linkage between lymphocytes and endothelium, since exposure of endo-thelium-bound lymphocytes to 10 mM EGTA did not diminish the adhe-sion. Unfortunately the effect of EDTA on the lymphocytes bound to the endothelium was not studied. EDTA is also an iron chelator (Bates, Billups & Saltman, 1967), a fact which is frequently overlooked.

The latest work from Woodruff's laboratory is of considerable historical significance in the field of lymphocyte circulation for other reasons. In it, Woodruff, who for so long attributed paramount importance to the role of sialic acid in the lymphocyte–endothelium interaction (Woodruff & Gesner, 1969; Woodruff, 1974; Woodruff & Woodruff, 1974), has con-firmed independently the observations of Freitas & de Sousa (1975) in the mouse, and of Ford, Sedgley, Sparshott & Smith (1976) in the rat, that neuraminidase-treated lymphocytes interact normally with PCV endo-thelium.

It is pertinent to recall in the context of the present proposal that the carbohydrate moiety of transferrin consists of N-acetyl-neuraminic acid, N-acetylglucosamine, galactose and mannose. However, removal of sialic acid or of galactose from transferrin, though changing its electrophoretic mobility does not affect its iron binding capacity (Jamieson, 1966; Morgan, Marsaglia, Giblett & Finch, 1967; Kornfeld, 1968).

I shall conclude by suggesting that experiments designed to test the effect of highly specific iron chelators on lymphocytes interactions with endothelia will help to clarify the practical validity of the ideas put forward in the present paper. Preliminary analysis of electron micrographs of human peripheral blood lymphocytes treated *in vitro* with ferric citrate has indicated that the metal binds to the cell surface in discrete patches and does not cause damage to the cells.

Work by the author was done with financial support from the following sources: NIH, Zelda R. Weintraub Foundation and Special Projects Committee of Memorial Sloan–Kettering Cancer Center.

I wish to acknowledge J. Bognaki, A. M. Smithyman, K. Nishiya, G. Munn and E. Newcomb for much of the discussion that constitutes the basis of the present paper. I acknowledge all those that expressed their protective concern over my being right or wrong. But above all, I wish to thank J. Hirsch from whose comments I derived the strength to submit the manuscript in its final form. For it was he who reassured me that the value of a hypothesis is still not to be right or wrong but to show the way to new experiments.

REFERENCES

AASA, R., MALMSTRÖM, B. G., SALTMAN, P. & VÄNNGÂRD, T. (1963). The specific binding of iron (III) and copper (II) to transferrin and conalbumin. *Biochimica et Biophysica Acta*, **75**, 203–22.

AISEN, P. & LEIBMAN, A. (1968). Citrate-mediated exchange of ^{54}Fe among transferrin molecules. *Biochemical and Biophysical Research Communications*, **32**, 220–6.

BAGGIOLINI, M., DE DUVE, C., MASSON, P. L. & HEREMANS, J. F. (1970). Association of lactoferrin with specific granules in rabbit heterophil leukocytes. *Journal of Experimental Medicine*, **131**, 559–70.

BAINTON, D. F. (1973). Sequential degranulation of the two types of polymorphonuclear leukocyte granules during phagocytosis of microorganisms. *Journal of Cell Biology*, **58**, 249–64.

BATES, G. W., BILLUPS, C. & SALTMAN, P. (1967). The kinetics and mechanism of iron (III) exchange between chelates and transferrin. II. The presentation and removal with ethylenediamine tetraacetate. *Journal of Biological Chemistry*, **242**, 2816–21.

BEAMISH, M. R. & BROWN, E. B. (1974). A comparison of the behavior of ^{111}In and ^{59}Fe-labelled transferrin on incubation in the human and rat reticulocytes. *Blood*, **43**, 703–11.

BEAMISH, M. R., ASHLEY JONES, P., TREVETT, D., HOWELL EVANS, I. & JACOBS, A. (1972). Iron metabolism in Hodgkin's disease. *British Journal of Cancer*, **26**, 444–52.

BELL, E. R. & LAFFERTY, K. J. (1972). The flow and cellular characteristics of cervical lymph from unanaesthetized ducks. *Australian Journal of Experimental Biology and Medicine*, **50**, 611–23.

BIEBER, C. P. & BIEBER, M. M. (1973). Detection of ferritin as a circulating tumor associated antigen in Hodgkin's disease. *National Cancer Institute Monographs*, **36**, 147–53.

CANTOR, H. & BOYSE, E. A. (1975). Functional subclasses of T lymphocytes bearing different Ly antigens. 1. The generation of functionally distinct T cell subclasses is a differentiative process independent of antigen. *Journal of Experimental Medicine*, **141**, 1376–89.

CURTIS, A. S. G. (1976). Le positionnement cellulaire et la morphogenèse. *Bulletin de la Société Zoologique de France*, **101**, 9–21.

CURTIS, A. S. G. & DE SOUSA, M. (1973). Factors influencing the adhesion of lymphoid cells. *Nature, New Biology, London*, **244**, 45–7.

CURTIS, A. S. G. & DE SOUSA, M. A. B. (1975). Lymphocyte interactions and positioning 1. Adhesive interactions. *Cellular Immunology*, **19**, 282–97.

DE SOUSA, M. (1971). Kinetics of the distribution of thymus and bone marrow cells in the peripheral organs of the mouse: ecotaxis. *Clinical and Experimental Immunology*, **9**, 371–80.

DE SOUSA, M. (1973). The ecology of thymus-dependency. In *Contemporary Topics in Immunobiology*, eds. A. J. S. Davies & R. L. Carter, vol. 2, pp. 119–36. Plenum, New York.

DE SOUSA, M. (1976). Cell traffic. In *Receptors and Recognition*, eds. P. Cuatrecasas & M. Greaves, Series A, vol. 2, pp. 106–63. Chapman & Hall, London.

DE SOUSA, M. (1977). Ecotaxis, ecotaxopathy and lymphoid malignancy. In *Immunopathology of Lymph Neoplasms*, eds. R. A. Good & J. Twomey. Plenum Press, New York (in press).

DE SOUSA, M. & PARROTT, D. M. V. (1969). Induction and recall in contact sensitivity. Changes in skin and draining lymph nodes of intact and thymectomized mice. *Journal of Experimental Medicine*, **130**, 671–90.

DE SOUSA, M. & PRITCHARD, H. (1974). The Cellular Basis of Immunological recovery in nude mice after thymus grafting. *Immunology*, **26**, 769–76.

DE SOUSA, M., SMITHYMAN, A. & TAN, C. (1978). Suggested models of ecotaxopathy in lymphoreticular malignancy. *American Journal of Pathology* **90**, 497–520.

DE SOUSA, M., YANG, M., LOPEZ-CORRALES, G., TAN, C., HANSEN, J. A., DUPONT, B. & GOOD, R. A. (1977). Ecotaxis: the principles and its application to the study of Hodgkin's disease. *Clinical and Experimental Immunology*, **27**, 143–51.

DUMONT, A. E., FORD, R. J. & BECKER, F. F. (1976). Siderosis of lymph nodes in patients with Hodgkin's disease. *Cancer*, **38**, 1247–52.

ELLIS, A. E. & DE SOUSA, M. (1974). Phylogeny of the lymphoid system: study of the fate of circulating lymphocytes in plaice. *European Journal of Immunology*, **4**, 338–43.

ESHHAR, Z., ORDER, S. E. & KATZ, D. H. (1974). Ferritin: a Hodgkin's disease associated antigen. *Proceedings of the National Academy of Sciences, U.S.A.*, **71**, 3956–60.

FORD, W. L., SEDGLEY, M., SPARSHOTT, S. M. & SMITH, M. E. (1976). The migration of lymphocytes across specialized vascular endothelium. II. The contrasting consequences of treating lymphocytes with trypsin or neuraminidase. *Cell and Tissue Kinetics*, **9**, 351–61.

FREITAS, A. A. & DE SOUSA, M. (1975). Control mechanisms of lymphocyte traffic. Modification of the traffic of ^{51}Cr-labeled mouse lymph node cells by treatment with plant lectins in intact and splenectomised hosts. *European Journal of Immunology*, **5**, 831–8.

FREITAS, A. A. & DE SOUSA, M. (1976a). The role of cell interactions in the control of lymphocyte traffic. *Cellular Immunology*, **22**, 345–50.

FREITAS, A. A. & DE SOUSA, M. (1976b). Control mechanism of lymphocyte traffic. Altered migration of ^{51}Cr labeled mouse lymph node cells pretreated in vitro with lipopolysaccharide. *European Journal of Immunology*, **6**, 269–73.

FREITAS, A. A. & DE SOUSA, M. (1976c). Control mechanism of lymphocyte traffic. Altered migration of ^{51}Cr-labeled mouse lymph node cells pretreated in vitro with phospholipases. *European Journal of Immunology*, **6**, 703–11.

FREITAS, A. A. & DE SOUSA, M. (1977). Control mechanism of lymphocyte traffic. A study of the action of two sulphated polysaccharides on the distribution of ^{51}Cr and [^{3}H]adenosine labeled mouse lymph node cells. *Cellular Immunology*, **31**, 62–76.

GESNER, B. M. (1966). Cell surface sugars as sites of cellular reaction. Possible role in physiological processes. *Annals of the New York Academy of Science*, **129**, 758–66.

GESNER, B. M. & GINSBURG, V. (1964). Effect of glycosidases on the fate of transfused lymphocytes. *Proceedings of the National Academy of Sciences, U.S.A.*, **52**, 750–5.

GOLDSCHNEIDER, I. & McGREGOR, D. D. (1968a). Migration of lymphocytes and thymocytes in the rat. I. The route of migration from blood to spleen and lymph nodes. *Journal of Experimental Medicine*, **127**, 155–68.

GOLDSCHNEIDER, I. & McGREGOR, D. D. (1968b). Migration of lymphocytes and thymocytes in the rat. II. Circulation of lymphocytes and thymocytes from blood to lymph. *Laboratory Investigation*, **18**, 397–406.

GOWANS, J. L. & KNIGHT, G. J. (1964). The route of recirculation of lymphocytes in the rat. *Proceedings of the Royal Society*, B, **159**, 257–82.

HARRISON, P. M. (1977). Ferritin: an iron storage molecule. *Seminars in Haematology*, **14**, 55–70.

HERMAN, R. G., YAMAMOTO, I. & MELLINS, H. Z. (1972). Blood microcirculation in the lymph node during the primary immune response. *Journal of Experimental Medicine*, **136**, 697–714.

JAMIESON, G. A. (1966). Reaction of diptheria toxin neuraminidase with human transferrin. *Biochimica Biophysica Acta*, **121**, 326–37.

KAUR, J., SPIERS, A. S. D., CATOVSKY, D. & DALTON, D. A. G. (1974). Increase of T lymphocytes in the spleen in Hodgkin's disease, *Lancet*, ii, 800–802.

KORNFELD, S. (1968). The effects of structural modifications on the biologic activity of human transferrin. *Biochemistry*, **7**, 945–54.

McLAURIN, L. P., JR. & COTTER, JR. (1967). Placental transfer of iron. *American Journal of Obstetrics and Gynecology*, **98**, 931–7.

MASSON, P. L., HEREMANS, J. F. & DIVE, C. (1966). An iron-binding protein common to many external secretions. *Clinica Chemica Acta*, **14**, 735–9.

MASSON, P. L., HEREMANS, J. F. & SCHONNE, E. (1969). Lactoferrin, an iron-binding protein in neutrophilic leukocytes. *Journal of Experimental Medicine*, **130**, 643–58.

MATCHETT, K. M., HUANG, A. T. & KREMER, W. B. (1973). Impaired lymphocyte transformation in Hodgkin's disease. Evidence for depletion of circulation T lymphocytes. *Journal of Clinical Investigation*, **52**, 1908–17.

MORGAN, E. H., MARSAGLIA, G., GIBLETT, E. R. & FINCH, C. A. (1967). A method of investigating iron exchange utilizing two types of transferrin. *Journal of Laboratory and Clinical Medicine*, **69**, 370–81.

MOROZ, C., GILER, S., KUPFER, B. & URCA, I. (1977). Lymphocytes bearing surface ferritin in patients with Hodgkin's disease and breast cancer. *New England Journal of Medicine*, **297**, 1172–3.

NUÑEZ, M. T., FISCHER, S., GLASS, J. & LAVIDOR, L. (1977). The crosslinking of ^{125}I-labeled transferrin to rabbit reticulocytes. *Biochimica et Biophysica Acta*, **490**, 87–93.

OKHUMA, S., NOGUCHI, H., AMANO, F., MIZUNO, D. & YASUDA, T. (1976). Synthesis of apoferritin in mouse peritoneal macrophages. Characterization of 20S particles. *Journal of Biochemistry*, **79**, 1365–76.

PARROTT, D. M. V., DE SOUSA, M. A. B. & EAST, J. (1966). Thymus-dependent areas in the lymphoid organs of neonatally thymectomized mice. *Journal of Experimental Medicine*, **123**, 191–204.

PAYNE, S. V., JONES, D. B., HAEGERT, D. G., SMITH, J. L. & WRIGHT, D. H. (1976). T and B lymphocytes and Reed-Sternberg cells in Hodgkin's disease lymph nodes and spleens. *Clinical and Experimental Immunology*, **24**, 280–6.

PHILLIPS, J. L. (1976). Specific binding of zinc transferrin to human lymphocytes. *Biochemical and Biophysical Research Communications*, **72**, 634–9.

PHILLIPS, M. E. & THORBECKE, G. J. (1966). Studies on the serum proteins of chimeras. *International Archives of Allergy and Applied Immunology*, **29**, 553.

PIRSIG, R. M. (1974). *Zen and the Art of Motorcycle Maintenance*. The Bodley Head, London.

RANNIE, G. H., THAKUR, M. L. & FORD, W. L. (1977). An experimental comparison of radioactive labels with potential application to lymphocyte migration studies in patients. *Clinical and Experimental Immunology*, **29**, 509–14.

SØRENSEN, M. & SØRENSEN, S. P. L. (1939). The proteins in whey. *Comptes rendus des travaux du Laboratoire de Carlsberg*, **23**, 55; cited in Masson, Heremans & Dive, 1966.

SPECTOR, W. G. (1969). The granulomatous inflammatory exudate. *International Review of Experimental Pathology*, **8**, 1–55.

VAERMAN, J. P., ANDRE, C., BAZIN, H. & HEREMANS, J. F. (1973). Mesenteric lymph as a major source of IgA in guinea pigs and rats. *European Journal of Immunology*, **3**, 580–4.

VAN SNICK, J. L. & MASSON, P. L. (1976). The binding of human lactoferrin to mouse peritoneal cells. *Journal of Experimental Medicine*, **144**, 1568–80.

WEINBERG, E. B. (1974). Iron and susceptibility to infectious disease. *Science*, **184**, 952–84.

WOODRUFF, J. J. (1974). Role of lymphocyte surface determinants in lymph node homing. *Cellular Immunology*, **13**, 378–84.

WOODRUFF, J. J. & GESNER, B. M. (1969). The effect of neuraminidase on the fate of transfused lymphocytes. *Journal of Experimental Medicine*, **129**, 551–67.

WOODRUFF, J. J. & WOODRUFF, J. F. (1974). Virus-induced alterations of lymphoid tissues. IV. The effect of Newcastle disease virus on the fate of radiolabeled thoracic duct lymphocytes. *Cellular Immunology*, **10**, 78–85.

WOODRUFF, J. J., KATZ, I. M., LUCAS, L. E. & STAMPER, H. B. (1977). An *in vitro* model of lymphocyte homing. II. Membrane and cytoplasmic events involved in lymphocyte adherence to specialized high-endothelial venules of lymph nodes. *Journal of Immunology*, **119**, 1603–10.

WRIGHT, D. G. & MALAWISTA, S. E. (1972). The mobilization and extracellular release of granular enzymes from human leukocyte, during phagocytosis. *Journal of Cell Biology*, **53**, 788–97.

YANG HU, H.-Y., GARDNER, J., AISEN, P. & SKOULTCHI, A. I. (1977). Inducibility of transferrin receptors via friend erythroleukemic cells. *Science*, **197**, 559–61.

EXPLANATION OF PLATE

Section of lymph node removed from a C3H An mouse infected with *M. lepremurium* two months previously. Ziel–Nielsen stain: note that few mycobacteria (lightly stained red) are present in the section of the post-capillary venule crossing the B cell area, in marked contrast with the presence of numerous microorganisms in the section of the venule in the thymus-dependent area (TDA). The clear-cut presence of microorganisms in this area is indicated by the two arrows. (Unpublished data of R. Nyak, M. de Sousa, R. A. Good).

PLATE I

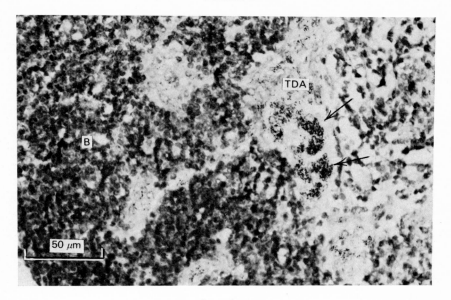

For explanation see p. 409

SELF-RECOGNITION AS A MEANS OF CELL COMMUNICATION IN THE IMMUNE SYSTEM

By DAVID H. KATZ

The Department of Cellular and Developmental Immunology,
Scripps Clinic and Research Foundation, La Jolla,
California 92037, U.S.A.

INTRODUCTION

During the past decade, there have been several revolutionary conceptual developments in our understanding of cellular and molecular mechanisms involved in the functions of the immune system. This era was ushered in by the observations of Claman, Chaperon & Triplett (1966) that first demonstrated that the development of antibody responses in certain instances required cooperative interactions between two distinct classes of lymphocytes, those derived from the thymus (known as T cells) and those derived from bone marrow (known as B cells). These findings were subsequently confirmed and extended by others (Davies, Leuchars, Wallis, Marchant & Elliot, 1967; Mitchell & Miller, 1968) who furthermore documented that B cells were the actual precursors of antibody-forming cells while T cells provided an important, although undefined, auxiliary function necessary for B cell precursors to fully develop into mature antibody-secreting cells. Such observations therefore initiated evolution of our thinking away from the previously widely-held *unicellular* concept of immune recognition and function in which the prevailing thought centered on the capacity of a single antigen-specific cell to recognize antigen, and upon so doing, differentiate into a functionally mature effector cell mediating either humoral (antibody) immunity in the case of the bone marrow system or cell-mediated (delayed hypersensitivity, transplantation reactions) immunity in the case of the thymus system.

Very shortly after these observations were made, a series of observations indicated that among the advantages of such multicellular involvement in a given immune response was the establishment of complex regulatory control mechanisms capable of guiding each response in appropriate directions at any given time. By the early 1970s, it was clear that T lymphocytes exerted sophisticated regulatory effects on other lymphocytes of both classes and in this way dictated the quality and magnitude of most immune

[411]   27-2

responses (reviewed in Katz & Benacerraf, 1972). This concept represented a departure from earlier thinking which ascribed many regulatory effects to antibody molecules alone. That is not to say that antibody molecules do not, indeed, exert regulatory feed-back effects on various immune responses, but rather that T-cell regulation provides an additional (and perhaps primary) basic control mechanism necessary for the integrity of the system. It should be stated that T-cell regulation of the immune system spans the entire spectrum from enhancement to suppression and involves complex cellular interactions between T cells and macrophages, T cells and T cells, and T cells and B cells; moreover, T cells participating in different regulatory functions or exerting different effector roles in cell-mediated immunity are now known to belong to distinct subclasses identifiable by their phenotypic expressions of distinct cell-surface antigenic markers (for comprehensive discussions see Gershon, 1973; Cantor & Weissman, 1976; Cantor & Boyse, 1977; Katz, 1977b). In terms of antibody-mediated feedback control, a novel concept has been fostered by Jerne (1975) in which feedback control of immune responses is considered to involve a series of anti-idiotypic responses directed against receptor binding sites (idiotypes) of responding immunocompetent cells. This network of idiotype–anti-idiotype recognition is perceived to serve as an intricate control system to contain immune responses and to avoid, under normal circumstances, the inadvertent loss of self-tolerance.

Another revolutionary series of concepts that have evolved during the past decade have been those concerned with the role of the major histocompatibility gene complex (MHC) in the immune system. For many years, gene products of the MHC (histocompatibility antigens) were studied in terms of their widely recognized function as major determinants of compatibility and incompatibility in transplantation reactions. Extensive analyses have been made of the MHC in several species, and particularly the *HLA* system of humans (reviewed by Bach & van Rood, 1976) and the *H-2* system of mice (reviewed in Klein, 1976). In the mid 1960s fundamental discoveries were made by Benacerraf and McDevitt and their respective coworkers that demonstrated for the first time a function of MHC genes in the control of immune responses to antigens other than those related in any obvious way to transplantation antigens and/or reactions (see McDevitt & Benacerraf, 1969; Benacerraf & McDevitt, 1972; Benacerraf & Katz, 1975 for review). The basic observations of these investigators were that individual strains of guinea pigs and mice differed in their capacities to develop immune responses to defined synthetic antigens. Responder strains and non-responder strains were distinguished by their ability or inability to manifest both cell-mediated (delayed hypersensitivity) and humoral (antibody production) immune responses to the relevant antigens.

Studies utilizing congenic-resistant strains of mice which differed from one another only at the *H-2* gene complex established the linkage of responder or non-responder status to genes in the MHC, and the MHC genes controlling such responses were termed immune response or *Ir* genes (McDevitt & Benacerraf, 1969; Benacerraf & McDevitt, 1972). Such genes were shown to be inherited in Mendelian fashion as autosomal dominant traits in most instances, although more recently the responder phenotype has been shown in selected cases to result from gene complementation, thus implicating the involvement of at least two *Ir* genes in the control of such responses (reviewed in Benacerraf & Katz, 1975). Analyses designed to locate the position of *Ir* genes in the MHC were conducted with recombinant strains of mice differing from one another by defined intra-*H-2* cross-over events. These studies established the fact that *Ir* genes were not linked to the terminal *K* (left) or *D* (right) regions of the complex which had previously been shown to contain the genes encoding the major transplantation antigens detected most readily by serological and tissue grafting techniques. Thus, *Ir* genes were found to reside in a new region, adjacent and to the right of the *K* region, which was appropriately denoted as the *I* region (McDevitt *et al.*, 1972).

In view of the facts that (1) all responses controlled by *Ir* genes involved the participation of T lymphocytes; (2) the specificity of such control appeared to be of a high order; and (3) the nature of the molecules functioning as receptors on T cells was uncertain and appeared to be different from the immunoglobulin molecules serving as receptors on B lymphocytes, Benacerraf & McDevitt proposed several years ago that *Ir* genes encoded the molecular products functioning as T-cell receptors (Benacerraf & McDevitt, 1972). Indeed, this notion prompted considerable speculation and experimentation during the past five years and is still regarded by a few immunologists as a valid possibility (see discussion in Katz & Benacerraf, 1976); however, the bulk of both direct and indirect evidence presently available contradicts the validity of this possibility (see Katz, 1977*b*). Thus, it has been established by studies performed initially by Ramseier & Lindenmann (1972), and more recently by Binz & Wigzell (1977) and Rajewsky *et al.* (1976) that T-cell receptors possess identical idiotypic determinants as found on immunoglobulin receptors on B cells. Since idiotypic determinants are characteristic of the region comprising the specific antigen-combining site, these findings strongly indicate that the variable (*V*) region genes encoding specific binding sites for antigen in immunoglobulins are also responsible for dictating specificity of T-cell

receptors. Additional information about T-cell antigen receptors possessing such idiotypic determinants indicates that the molecules do not possess
conventional immunoglobulin markers and are not linked in any obvious
way to the MHC. Of considerable importance is the recent finding that
inheritance of the relevant idiotype on T-cell receptors is clearly linked to
immunoglobulin heavy chain genes.

INVOLVEMENT OF MAJOR HISTOCOMPATIBILITY COMPLEX GENES IN CELL–CELL INTERACTIONS

Another major conceptual advance during recent years has been that concerned with the role of major histocompatibility complex genes in governing cell–cell interactions and communication in the development of immune responses. These ideas arose during the course of studies designed
to ascertain the nature of the mechanism(s) by which different populations
and subpopulations of lymphocytes interact with one another as well as
with macrophages. As stated above, it is now well documented that such
cellular interactions underlie much of the regulatory control over the
immune system (Katz & Benacerraf, 1972). T lymphocytes, B lymphocytes, and macrophages are capable of intercommunicating with one
another by various means. At the moment, it appears that T lymphocytes
exert the most wide-ranging regulatory effects on the system in general,
since it is now clear that subpopulations within this class communicate
with one another as well as with macrophages and B lymphocytes (Katz,
1977b). The precise mechanism of cell–cell communication, namely
whether this involves direct cell–cell contact, activity of secreted (or
released) molecules, or a combination of cell contact and mediator release
has not yet been definitively established.

The basic concept that histocompatibility gene products are integrally
involved in the mechanism of regulatory cell interactions initially arose
from certain unexpected observations made in experiments conducted
six to seven years ago. These observations stemmed from the finding that
transfer of histoincompatible T cells to previously immunized recipient
animals resulted in circumvention of the normal requirement for antigen-
specific helper T cells in secondary antibody responses. This phenomenon
was termed the 'allogeneic effect' and was shown to reflect the development of an active graft-versus-host reaction in recipient lymphoid organs
(reviewed in Katz, 1972). Extensive analyses of this phenomenon revealed
that the allogeneic effect displayed remarkable similarities to normal
regulatory T–B cell interactions in isogeneic or syngeneic combinations,
and hence was postulated to be an analogous model for regulatory cell
interactions in the immune response (Katz, 1972).

The fact that the final pathway in the allogeneic effect involved inter-actions at histocompatibility molecules on the cell surface of the target T or B cells employed in the system suggested that perhaps precisely the same pathway was involved in syngeneic interactions, occurring perhaps by similar molecular mechanisms. In order to ascertain the validity of this possibility, experiments were conducted to investigate whether physio-logic T–B cell interactions in the mouse involved genetic restrictions associated with MHC genes. Indeed, utilizing a technique in which primed T and B lymphocytes were adoptively transferred to irradiated recipients under conditions especially designed to circumvent detectable non-specific allogeneic effects (see Katz, Hamaoka & Benacerraf, 1973 a for full description of the system), the basic observation was that antigen-specific T cells were capable of providing specific helper function for primed B cells of histocompatible, but *not* of histoincompatible, donor origin in secondary antibody responses of the IgG class (Katz *et al.*, 1973 a; Katz, Hamaoka, Dorf & Benacerraf, 1973 b). In other words, such studies established the fact that T cells from a mouse of strain A were capable of interacting with B cells also obtained from a mouse of strain A, but not with B cells derived from a donor mouse of strain B genotype. It was further established that cooperative interactions could be obtained between reci-procal mixtures of F_1 hybrid and parental T and B cells (Katz *et al.*, 1973 a) and, moreover, that the presence of primed histoincompatible T cells did not appreciably alter cooperative interactions between histocompatible T and B cells (Katz, Hamaoka, Dorf & Benacerraf, 1974; Katz, Chiorazzi, McDonald & Katz, 1976) thus arguing against the possibility of non-specific or specific suppressive effects contributing to the observed genetic restrictions in the systems.

Studies utilizing congenic-resistant mouse strains established that this genetic restriction in cooperative T–B cell interactions is linked to the *H-2* gene complex (Katz *et al.*, 1973 b), and very importantly, the critical genetic locus or loci involved in controlling such interactions map in the *I* region of the histocompatibility complex (Katz, Graves, Dorf, Dimuzio & Bena-cerraf, 1975) which, as mentioned above, is precisely where the *Ir* genes are located. The original interpretation of the aforementioned observations was that these genetic restrictions reflected the existence of cell surface molecules, distinct from antigen-specific receptors, that play a crucial role in mediating effective cell–cell interactions between T and B lymphocytes (Katz *et al.*, 1973 a, b). These membrane components were assumed to be gene products of the *H-2* complex, and it was postulated that genetic identity between the T cell and the B cell was necessary for the relevant T cell surface molecules to bind to the corresponding B cell molecules, which

were termed 'acceptor' sites, for effective interactions between primed T
and B lymphocytes in development of antibody responses. Subsequently,
the respective molecules were defined as cell interaction (CI) molecules
and the *I* region genes coding them as *CI* genes (Katz & Benacerraf, 1975).
The fact that both *CI* and *Ir* genes map in the same genetic region of the
H-2 complex may be merely coincidental, or it may be that these respective
genes are one and the same – this has yet to be sorted out.

Another example of involvement of histocompatibility gene products in
cell–cell interactions of an entirely different type concerns the ability of
cytotoxic T lymphocytes (CTL) to effectively lyse virus-infected or chemi-
cally-modified target cells. In such circumstances it has been found that
CTL are most efficient in lysing target cells derived from a similar histocompatibility genotype (reviewed by Shearer, Schmitt-Verhulst & Rehn,
1976; Zinkernagel & Doherty, 1976). The critical genetic locus or loci
involved in controlling interactions between CTL and target cells map in
the *K* and *D* regions of the histocompatibility complex, thus differing from
those involved in T–B cell interactions which are located in the *I* region.

Although it was initially believed that the mechanisms underlying
genetic control of T–B cell interactions and interactions between CTL and
target cells may be different, recent studies have clearly indicated that the
two systems reflect the same phenomenon (Katz, 1977*a*, *b*; Zinkernagel,
1976). Hence, it is attractive to consider that cell interaction structures are
present on the surface membranes of lymphoid cells and macrophages and
it is via some type of molecular interaction at these sites that the communication between the relevant cells takes place. This, in turn, either
results in a signal for induction of differentiation in the case of helper T
cell interactions with B cells or with other T cells, or a signal to initiate
cytolysis in the case of CTL interactions with target cells.

A major focus of interest now centres on elucidating the molecular
mechanisms by which such cell interaction structures interact on the cell
surface membrane between communicating cells. Essentially two alternative mechanisms are presently considered (Katz, 1977*a*). In one, some
form of homologous or like–like interaction between identical molecules
may be envisaged to occur in one or more of a variety of ways. In the second, molecular interaction has been envisaged to occur between complementary structures in a manner analogous to receptor–ligand interaction.
It is clearly not possible at the present time to make a valid conclusion as to
whether interactions at the CI sites under normal circumstances occur by
homology or complementarity, and distinguishing between these two
possibilities will surely require molecular approaches in order to be answered in the future. It is nevertheless germane to point out that the in-

ferences of the two alternative mechanisms for CI molecular interactions are substantial in that homologous interactions imply that the CI molecules on the respective cells are products of *identical* genes, whereas complementarity implies the reaction of a receptor entity on the T cell with a CI molecule on the corresponding target B cell (or T cell or macrophage), the respective molecules (i.e. receptor and CI site) being products of *distinct* genes.

SELF-RECOGNITION AND CELL–CELL COMMUNICATION

We can now turn to the major point of this paper, namely that self-recognition is the critical mechanism by which cell–cell communication takes place, certainly in the immune system, and perhaps as well in other control processes of differentiation and organogenesis. Since the basis for this concept has largely derived from experience with cell–cell interactions in the immune system, we will limit our discussion to the relevant cells of this system. The rationale for directing our thinking along such lines of considering self-recognition as an integral mechanism for cell communication in immune responses, stems largely from the collective implications arising from extensive studies of the allogeneic effect, described above, together with the genetic restrictions observed in cell–cell interactions occurring between isogeneic cells of the same individual. In essence, the point to be made when one considers these two seemingly disparate circumstances in the same context, is that the critical common feature of these two types of interactions is the molecular interaction occurring at a relevant histocompatibility (CI) site on the target cell, whether this be a B cell, T cell or macrophage, resulting in a requisite stimulus for further differentiation of the target cell. In the case of the allogeneic effect, this has been firmly established in view of the fact that the phenomenon depends on an active interaction of histoincompatible T cells with the appropriate cell surface alloantigen(s) on the target responding cell in a manner known to be analogous to what occurs in a typical in-vitro mixed lymphocyte reaction.

It is pertinent to point out that in recent years firm evidence has been obtained documenting that the strongest alloantigen target sites in such reactions are themselves gene products of the *I* region of the MHC (reviewed in Katz, 1977*b*). Since genetic mapping studies of the loci restricting physiologic interactions between syngeneic cells have located the relevant genes also in the *I* region of the *H-2* complex of the mouse, it is difficult to imagine that the relevant target sites in the allogeneic effect and those involved in syngeneic interactions are products of genes of the same region by mere chance alone. Rather, it seems more plausible to consider that the respective phenomena occur by identical or very similar mechanisms. This, in turn, strongly implies that normal cell–cell communication within a given individual occurs by a process analogous to, or identical with, the molecular interactions occurring in a typical transplantation-type reaction, such as the mixed lymphocyte reaction.

As stated above, the respective CI molecules on the T cell and the target cell with which it communicates may interact by complementarity in a manner analogous to receptor–ligand interactions, in which case the CI molecule with specific binding capabilities would be a receptor-like molecule; it may, in fact, be the same receptor-like molecule mediating conventional alloaggressive transplantation reactions which occur between histoincompatible effector and target cells. Alternatively, the T cell CI molecule may be identical in structure to the corresponding CI molecule on the interacting target cell, in which case a like–like molecular interaction would be involved. Nevertheless, this structure which may interact by homology with CI molecules on histocompatible target cells could also be the same structure capable of interacting by complementarity with I region-coded alloantigens on histoincompatible target cells. Discrimination between these possibilities requires further study.

A concept of self-recognition as a means for normal cell communication in the immune systems of higher vertebrates, which has also been postulated in a somewhat different form by Cohen (1976), seems at first glance to contradict the classical dogma that the sophisticated immune system of higher animals evolved in a manner to allow an individual to distinguish self from non-self. In this classical concept, the immunological apparatus of a given individual has been perceived to evolve in order to *allow* recognition of foreign (i.e. non-self) antigens while at the same time being *incapable* of recognizing self-antigenic determinants; in other words, inherent in the development of immunocompetence is a process by which self-tolerance is achieved to purposefully exclude the capacity of cells within the individual from recognizing one another. If one considers, however, that self recognition is, in fact, an integral mechanism for purposeful cell–cell communication in the immune system of a given individual, then one would suspect that a suitable adaptation has occurred during ontogeny of an individual's immune system to allow such self recognition to occur in a constructive way while prohibiting deleterious (i.e. autoaggressive) reactions against oneself under normal circumstances. On the other hand, the ability to recognize aberrant cells within the individual would lead to an advantageous reaction thereby eliminating them from the system.

It should not be entirely surprising that a mechanism of self recognition would form the basis of cell–cell communication and interaction in the immune system of higher vertebrates. Indeed, similar mechanisms of self-recognition have been well established to exist in lower colonial marine forms as well as in flowering plants for purposes of governing fertilization and/or colony formation. A conceptual association was developed several years ago by Burnet (1971) between primitive mechanisms of self recognition in colonial marine forms and flowering plants and more sophisticated (?) and diversified methods of immunological recognition in higher animals. Although such associations were drawn as a form of speculation for the evolution of recognition of non-self in immunological systems of vertebrates, one could easily interpret such associations in precisely the reverse manner, namely that of evolution of self-recognition for cell communication in the immune system of such higher animals. Burnet's notions stemmed in part from the extensive studies of Oka (1970) on the colonial tunicate, *Botryllus*, and those of Theodor (1970) on the colonial anthozoan, *Eunicella*. *Botryllus* is a colonial form in which each colony consists of genetically identical units derived from a single larva. Each colony clearly appears to be endowed with the capability of recognizing and distinguishing individuals derived from the same colony or from a different colony within the same species. Thus if a given colony I is divided into two and the parts allowed to grow separately and then brought into contact at a later time, the two daughter colonies will fuse together to reconstitute a single colony. In contrast, an unrelated colony II of the same species will evoke a positive rejection-type reaction if brought into contact with colony I. Since *Botryllus* is hermaphrodite, the mechanisms of sexual reproduction involve control of self fertilization. Oka's studies (1970) have shown that fusion or rejection between colonies depends upon a single locus with many alleles which have been referred to as recognition genes. Similarly, prevention of self fertilization appears to be controlled by these same genes. Since self fertilization is prevented, all natural colonies of *Botryllus* are heterozygous for such recognition genes. Successful fusion of colonies requires the existence of at least one common allele; lack of a common gene between two colonies will lead to rejection if placed in contact with one another. In contrast, self-fertilization is prevented when sperm and ova have a common gene. A very similar situation exists in *Eunicella* as described by Theodor (1970). Although the precise basis of the recognition phenomena in these colonial marine forms has not been ascertained, it has been speculated that a system involving cell surface complementary structures, products of the respective recognition genes, permit the respective cells to distinguish between identity or non-identity in these individuals.

A very analogous situation exists in the mechanisms preventing self-fertilization in flowering plants. Studies of this type have been reviewed and detailed by Lewis (1963) and will be briefly summarized here in the following context. In essence, the situation is very similar to that described above for *Botryllus* and *Eunicella* in that associated with the mechanism of pollination of flowering plants are genetic loci which permit recognition of self or non-self. Thus pollen from an individual plant is incapable of setting seed and flowers of that plant or in any other plants of the same incompatibility group, that is, with a common self-incompatibility allele. However, pollen from one self-incompatible group is fully active in fertilizing any other plant of the species not possessing the same allele. When pollen from one self-incompatibility group interacts with the style of a flower of the same self-incompatibility group, a positive recognition results in a type of rejection reaction preventing the pollen tube from growing. From the genetic point of view, it appears that these reactions reflect the existence of two gene products, one each in the pollen and style, respectively. Lewis has hypothesized about the nature of this process in two ways, both of which consider that one allele codes for proteins in two different sites. In the first hypothesis, it is considered that the allele codes for two distinct proteins, one for the pollen tube and one for the style. The respective proteins have a complementary configuration analogous to antigen–antibody or enzyme–substrate. In the second hypothesis, and the one Lewis favors, the allele is considered to be a complex which codes for a specific protein pattern common to both sites; in addition, he postulates the existence of an activator for protein production in the pollen tube and a separate activator for protein production in the style. In this hypothesis, Lewis postulates that the primary gene product takes the form of a dimer; when this makes contact with an identical dimer on the interacting surface, union occurs (perhaps involving an allosteric molecule) and the tetramer functions as a growth inhibitor.

It is quite interesting that these respective hypotheses of Lewis were proposed some years ago to explain possible mechanisms of self-recognition in flowering plants, and to date it has not been possible to definitively decide between the two possibilities proposed: namely interaction of two identical proteins (or polypeptide determinants) or of sterically complementary patterns. This is precisely the same dilemma that we face today in speculating about mechanisms of cell–cell interactions in the immune system where, as discussed above, similar speculations exist concerning whether or not this involves complementary interactions or like–like interactions. Interestingly, a very recent report by Clarke *et al.* (1977) provides some evidence in favor of the hypothesis proposing that interactions occur between identical proteins in the case of flowering plants, since isolation of the major structures present on the surface of pollen and style, respectively, have indicated that they are antigenically indistinguishable from one another; this, in turn, strongly implies that the respective products are derived from identical genes.

Nevertheless, irrespective of the nature of the molecular interactions occurring in such circumstances, the main point to be emphasized here is that self-recognition appears to be a fundamental biological process concerned with control of many types of developmental and differentiation events. It seems highly unlikely, therefore, that as evolution proceeded, a mechanism so important for primitive life forms would have been eliminated and, in fact, exactly reversed in the development of cognitive immunological function in higher vertebrates. Rather, as discussed above, it is easier to envisage the evolution of self-recognition as an integral part of cell–cell interaction in the immune system and the concomitant evolution of a damping mechanism to limit the autoaggressive nature of such self-recognition reactions. Loss of this normal damping process could well account for the development of various states of autoaggression and autoimmunity that could be incited by a variety of different exogenous stimuli, such as viruses, drugs, etc.

THE CONCEPT OF ADAPTIVE DIFFERENTIATION IN THE MECHANISM OF SELF–RECOGNITION IN CELL–CELL INTERACTIONS IN THE IMMUNE RESPONSE

Sometime after the initial description of genetic restrictions in T–B cell interactions in the development of antibody responses in mice, there appeared several reports of observations which were initially interpreted to argue against the CI molecule concept as summarized in preceding paragraphs. In all instances of such reports, good cooperative T–B cell interactions were obtained across major histocompatibility differences. The first such studies were those of Bechtol, Freed, Herzenberg & McDevitt (1974a) and Bechtol et al. (1974b) in which tetraparental mice, produced by fusion of embryos of strains which are responders and non-responders, respectively, to the random synthetic polypeptide antigen poly-L-(tyrosine, glutamic acid)-poly-D,L-alanine–poly-L-lysine, or (T,G)-A–L, produced anti-(T,G)-A–L antibodies of both responder and non-responder IgG allotype. The production of non-responder allotype antibodies was interpreted to indicate cooperation across the I-region differences between responder T cells and non-responder B cells. Other examples of successful interaction between histoincompatible T and B cells were those studies performed with cells from allogeneic bone marrow chimeras prepared by transferring equal numbers of anti-θ serum-treated bone marrow cells from two parent strains into lethally irradiated F_1 recipients (von Boehmer, Hudson & Sprent, 1975 b; Waldmann, Pope & Munro, 1975; von Boehmer, 1976). Such chimeras possess approximately 50–50 mixtures of parental lymphoid cells and a state of stable mutual tolerance exists among the chimera T cells (von Boehmer, Sprent & Nabholz, 1975 a). Utilizing in-vivo adoptive secondary or in-vitro secondary antibody responses it has been shown that primed chimeric T cells bearing H-2 determinants of only one parental strain, provided comparable activity for both syngeneic B cells and allogeneic B cells of the same H-2 haplotype to which the T cells had been tolerized in this chimeric environment. The original interpretation of these observations was that the capacity of histoincompatible T and B cells to cooperatively interact is determined by unknown determinants which under normal circumstances prevent such interactions, whereas in conditions of mutual tolerance this interference is circumvented. This interpretation is, of course, quite different from the concept proposed in the cell interaction hypothesis; the latter concept envisages the ability to interact successfully under normal circumstances as reflecting a mechanism of self-recognition whereas, conversely, the inability of histoincompatible T and B cells to effectively interact reflects the *absence* of a necessary self-recognition capability.

One is left then with explaining the differences observed when primed lymphoid cells are obtained from normal mouse donors, in which cases histoincompatible T and B cells fail to interact, versus when they are obtained from tetraparental or bone marrow chimera mice. The glaring difference in these respective circumstances is that in the former case T cells and B cells have differentiated in distinct genetic environments (i.e. haplotype A and B) whereas in the latter case cells of the independent haplotypes have differentiated under conditions of long-term cohabitation. These differences in results could be explained by a hypothesis which considers that the process of differentiation of stem cells and their progeny may be critically regulated by histocompatibility molecules on cell surfaces, and that such differentiation may be 'adaptive' to the environment in which it takes place (Katz, 1976; Katz, 1977 a). In other words, 'adaptive differentiation' results in preferential interactions among cells which have undergone their differentiative process in the same (or similar) genotypic environment (i.e. they have learned to successfully communicate with one another).

This concept predicts that interactions between lymphoid cells very early in ontogeny will promote adaptive changes in one or both cell types to permit successful communication and interactions between them at later stages of development. Such adaptation would fully explain the observations in tetraparental or bone marrow chimeric mice. The genetic restrictions in studies performed with cells primed in independent donors may be considered to occur between mixtures of T and B lymphocytes which have been previously primed by antigen and also perhaps irretrievably 'locked' into set phenotypic expression of the *functional* cell surface CI molecules required for successful cell interactions. Thus, it is conceivable that interactions between genetically histoincompatible T and B lymphocytes (or other cells for that matter) during very early stages of differentiation result in either some type of selective and/or inductive events that determine the ultimate haplotype preference expressed by such cells in subsequent communications and interactions. The nature of such selective and/or inductive events, the role of crucial microenvironmental influences of primary and/or secondary lymphoid organs and the expression of their consequences at the molecular level are questions that need to be explored in the future; it is clearly premature to attempt to define this concept in molecular terms.

CONCLUDING REMARKS

In preceding sections of this paper, attention has been drawn to several areas in which major conceptual changes have taken place in our understanding of the immune system during the past decade. Particular emphasis was given to the striking parallels which exist between the phenomenon known as the allogeneic effect and physiologic cell–cell interactions that occur in normal immune responses, and the view has been expressed that physiologic regulation of the immune system takes place in large part by a sequence of anti-'self' reactions in a manner analogous to that observed in the allogeneic effect. Prevention of deleterious autoreactivity appears to merely require the selection of a suitable 'damping' mechanism, which could be easily performed by either suppressor T lymphocytes, anti-idiotypic antibodies or comparable mechanisms, to maintain homeostatic control. This would provide some meaningful explanation for the many observations indicating the development of auto-reactivity under suitable conditions both *in vivo* and *in vitro*, the high frequency of alloreactive cells and the selective pressure to maintain, at the same time: (1) homology among *H-2K* and *H-2D* gene products among different mouse haplotypes as well as in man, and (2) extensive polymorphism within the MHC of various species. Adaptive differentiation may conceivably involve modification of the 'damping' mechanism to permit effective, but not deleterious, anti-self reactivity to permit survival of the individual (such as occurs in tetraparental chimeras).

Whether this concept of communication by self-recognition in the immune system is valid or not and, irrespective of this, whether effective cell–cell interactions occur by molecular mechanisms involving complementarity or homology are questions that appear to be relevant not only to students of lymphocyte biology but also to those interested in developmental processes in other areas of biology. It remains to be seen whether or not we are in a suitable position to distinguish between these possibilities in terms of lymphocyte interactions more so than has been the case in other systems, such as flowering plants and lower colonial marine forms. Nevertheless, as stated above, self-recognition appears to be a fundamental biological process concerned with control of many types of developmental and differentiation events. Elucidation of the genetic and molecular mechanisms underlying such recognition processes in lymphocytes could very well serve to enlighten our understanding of similar phenomena in other biological systems.

In summarizing, the reader should crystallize a single theme from the foregoing discussion. *The immune system appears to have evolved with inherent mechanisms FOR self-recognition, rather than having developed mechanisms to prevent such recognition, and in this sense has capitalized on the most basic form of cell–cell communication known in biology; the relevant molecules involved in such self-recognition events in the immune system are gene products of the MHC.* If one could predict the next conceptual revolution in immunological thought, it would most probably concern the role of integrated viral genomes in determining the extent of MHC polymorphisms and, hence, the extent and complexity of cell–cell interactions in such systems.

Studies performed in our laboratory have been supported by Grant AI-13874 from the National Institutes of Health. This is publication no. 43 from the Department of Cellular and Developmental Immunology and publication no. 1387 from the Immunology Departments, the Research Institute of Scripps Clinic, La Jolla, Calif. We are grateful to Ms Judy Henneke for excellent secretarial assistance in preparation of the manuscript.

REFERENCES

BACH, F. H. & VAN ROOD, J. J. (1976). The major histocompatibility complex. *New England Journal of Medicine*, 295, 806–12, 872–8, 927–34.
BECHTOL, K. B., FREED, J. H., HERZENBERG, L. A. & McDEVITT, H. O. (1974a). Genetic control of the antibody response to poly-L-Tyr, Glu-poly-D,L-Ala-poly-L-Lys in C3H↔CWB tetraparental mice. *Journal of Experimental Medicine*, 140, 1660–73.
BECHTOL, K. B., WEGMANN, T. G., FREED, J. H., GRUMET, F. C., CHESEBRO, B. W., HERZENBERG, L. A. & McDEVITT, H. O. (1974b). Genetic control of the immune response to (T,G)-A-L in C3H↔C57 tetraparental mice. *Cellular Immunology*, 13, 264–75.

BENACERRAF, B. & KATZ, D. H. (1975). The nature and function of histocompatibility-linked immune response genes. In *Immunogenetics and Immunodeficiency*, ed. B. Benacerraf, pp. 117–77. Medical and Technical Publishing, London.

BENACERRAF, B. & McDEVITT, H. O. (1972). Histocompatibility-linked immune response genes. *Science*, **175**, 273–85.

BINZ, H. & WIGZELL, H. (1977). Antigen binding, idiotypic receptors from T lymphocytes. An analysis as to their biochemistry, genetics and use as immunogens to produce specific immune tolerance. In *Origins of Lymphocyte Diversity – Cold Spring Harbor Symposium on Quantitative Biology XLI*, p. 275.

BURNET, F. M. (1971). 'Self-recognition' in colonial marine forms and flowering plants in relation to the evolution of immunity. *Nature, London*, **232**, 230–38.

CANTOR, H. & BOYSE, E. A. (1977). Regulation of cellular and humoral immune responses by T cell subclasses. In *Origins of Lymphocyte Diversity – Cold Spring Harbor Symposium on Quantitative Biology, XLI*, p. 23.

CANTOR, H. & WEISSMAN, I. (1976). Development and function of subpopulations of thymocytes and T lymphocytes. *Progress in Allergy*, **20**, 1–56.

CLAMAN, H. N., CHAPERON, E. A. & TRIPLETT, R. F. (1966). Thymus–marrow cell combinations. Synergism in antibody production. *Proceedings of the Society for Experimental Biology and Medicine*, **122**, 1167–71.

CLARKE, A. E., HARRISON, S., KNOX, R. B., RAFF, J., SMITH, P. & MARCHALONIS, J. J. (1977). Common antigens and male–female recognition in plants. *Nature, London*, **265**, 161–3.

COHEN, I. R. (1976). Autimmunity, self-recognition, and blocking factors. In *Autoimmunity*, ed. N. Talal. Academic Press, New York.,

DAVIES, A. J. S., LEUCHARS, E., WALLIS, V., MARCHANT, R. & ELLIOTT, E. V. (1967). The failure of thymus–derived cells to produce antibody. *Transplantation*, **5**, 222–31.

GERSHON, R. K. (1973). T cell control of antibody production. *Contemporary Topics in Immunology*, **3**, 1–42.

JERNE, N. K. (1975). The immune system: A web of V-domains. *The Harvey Lectures*, Series 70, 1974–1975. Academic Press, New York.

KATZ, D. H. (1972). The allogeneic effect on immune responses. Model for regulatory influences of T lymphocytes on the immune system. *Transplantation Reviews*, **12**, 141–72.

KATZ, D. H. (1976). The role of the histocompatibility gene complex in lymphocyte differentiation. In Proceedings of the First International Symposium on the Immunobiology of Bone Marrow Transplantation. *Transplantation Proceedings*, **8**, 405–12.

KATZ, D. H. (1977*a*). The role of the histocompatibility gene complex in lymphocyte diversity. In *Origins of Lymphocyte Diversity – Cold Spring Harbor Symposium on Quantitative Biology XLI*, p. 611.

KATZ, D. H., HAMAOKA, T. & BENACERRAF, B. (1973*a*). Cell interactions between histoincompatible T and B lymphocytes. II. Failure of physiologic cooperative interactions between T and B lymphocytes from allogeneic donor strains in humoral response to hapten-protein conjugates. *Journal of Experimental Medicine*, **137**, 1405–20.

KATZ, D. H., HAMAOKA, T., DORF, M. E. & BENACERRAF, B. (1973*b*). Cell interactions between histoincompatible T and B lymphocytes. III. Demonstration that the *H-2* gene complex determines successful physiologic lymphocyte interactions. *Proceedings of the National Academy of Sciences, U.S.A.*, **70**, 2626–33.

28

KATZ, D. H., HAMAOKA, T., DORF, M. E. & BENACERRAF, B. (1974). Cell inter-actions between histoincompatible T and B lymphocytes. V. Failure of histoincompatible T cells to interfere with physiologic cooperation between T and B lymphocytes. *Journal of Immunology*, **112**, 855–62.

KATZ, D. H., GRAVES, M., DORF, M. E., DIMUZIO, H. & BENACERRAF, B. (1975). Cell interactions between histoincompatible T and B lymphocytes. VII. Cooperative responses between lymphocytes are controlled by genes in the *I* region of the *H-2* complex. *Journal of Experimental Medicine*, **141**, 263–8.

KATZ, D. H., CHIORAZZI, N., MCDONALD, J. & KATZ, L. R. (1976). Cell interactions between histoincompatible T and B lymphocytes. IX. The failure of histoincompatible cells is not due to suppression and cannot be circumvented by carrier-priming T cells with allogeneic macrophages. *Journal of Immunology*, **117**, 1853–68.

KLEIN, J. (1976). An attempt at an interpretation of the mouse *H-2* complex. In *Contemporary Topics in Immunobiology*, ed. W. O. Weigle, vol. 5, pp. 297–340. Plenum, New York.

KATZ, D. H. (1977b). *Lymphocyte Differentiation, Recognition and Regulation*. Academic Press, New York.

KATZ, D. H. & BENACERRAF, B. (1972). The regulatory influence of activated T cells on B cell responses to antigen. *Advances in Immunology*, **15**, 1–93.

KATZ, D. H. & BENACERRAF, B. (1975). The function and inter-relationships of T cell receptors, Ir genes and other histocompatibility gene products. *Transplantation Reviews*, **22**, 175–95.

KATZ, D. H. & BENACERRAF, B. (eds.) (1976). Interrelationship between products of the major histocompatibility complex and their relevance to disease. In *The Role of the Histocompatibility Gene Complex in Immune Responses*. Proceedings of an International Conference at Brook Lodge, Michigan, November, 1976, pp. 715–80. Academic Press, New York.

LEWIS, D. (1963). A protein dimer hypothesis on incompatibility. In *Genetics Today*, ed. S. J. Geerts, vol. 3, pp. 657–75. International Congress of Genetics.

MCDEVITT, H. O. & BENACERRAF, B. (1969). Genetic control of specific immune responses. *Advances in Immunology*, **11**, 31–74.

MCDEVITT, H. O., DEAK, B. D., SHREFFLER, D. C., KLEIN, J., STIMPFLING, J. H. & SNELL, G. D. (1972). Genetic control of the immune response. Mapping of the *Ir-1* locus. *Journal of Experimental Medicine*, **135**, 1259–72.

MITCHELL, G. F. & MILLER, J. F. A. P. (1968). Cell to cell interaction in the immune response. II. The source of hemolysin-forming cells in irradiated mice given bone marrow and thymus or thoracic duct lymphocytes. *Journal of Experimental Medicine*, **128**, 821–34.

OKA, H. (1970). Colony specificity in compound ascidians. In *Profiles of Japanese Science and Scientists*, ed. H. Yukawa, pp. 196–206. Kodansha, Tokyo.

RAJEWSKY, K., HÄMMERLING, G. J., BLACK, S. J., BEREK, C. & EICHMANN, K. (1976). T lymphocyte receptor analysis by anti-idiotypic stimulation. In *The Role of Products of the Histocompatibility Gene Complex in Immune Responses*, eds. D. H. Katz & B. Benacerraf, pp. 445–53. Academic Press, New York.

RAMSEIER, H. & LINDENMANN, J. (1972). Aliotypic antibodies. *Transplantation Review*, **10**, 57–75.

SHEARER, G. M., SCHMITT-VERHULST, A.-M. & REHN, T. G. (1976). Bifunctional histocompatibility-linked regulation of cell mediated lympholysis to modified autologous cell surface components. In *The Role of the Products of the Histo-compatibility Gene Complex in Immune Responses*, eds. D. H. Katz & B. Benacerraf, pp. 133–41. Academic Press, New York.

THEODOR, J. L. (1970). Distinction between 'self' and 'not-self' in lower inverte-brates. *Nature, London*, **227**, 690–702.

VON BOEHMER, H. (1976). A possible function for products of the major histo-compatibility complex in humoral immunity. In *The Role of Products of the Histocompatibility Gene Complex in Immune Responses*, eds. D. H. Katz & B. Benacerraf, pp. 307–15. Academic Press, New York.

VON BOEHMER, H., SPRENT, J. & NABHOLZ, M. (1975 a). Tolerance to histocompati-bility determinants in tetraparental bone marrow chimeras. *Journal of Experi-mental Medicine*, **141**, 322–30.

VON BOEHMER, H., HUDSON, L. & SPRENT, J. (1975 b). Collaboration of histo-incompatible T and B lymphocytes using cells from tetraparental bone mar-row chimeras. *Journal of Experimental Medicine*, **142**, 989–98.

WALDMANN, H., POPE, H. & MUNRO, A. J. (1975). Cooperation across the histo-compatibility barrier. *Nature, London*, **258**, 728–31.

ZINKERNAGEL, R. M. (1976). *H-2* restriction of virus-specific cytotoxicity across the *H-2* barrier. Separate effector T-cell specificities are associated with self-*H-2* and with tolerated allogeneic *H-2* in chimeras. *Journal of Experimental Medicine*, **144**, 933–43.

ZINKERNAGEL, R. M. & DOHERTY, P. C. (1976). Does the apparent *H-2* compati-bility requirement for virus-specific T cell-mediated cytolysis reflect T cell specificity to 'altered-self' or physiological interaction mechanisms? In *The Role of the Histocompatibility Gene Complex in Immune Responses*, eds. D. H. Katz & B. Benacerraf, pp. 203–11. Academic Press, New York.

CELL SURFACE STRUCTURES, DIFFERENTIATION AND MALIGNANCY IN THE HAEMOPOIETIC SYSTEM

By MELVYN GREAVES

Membrane Immunology Laboratory, Imperial Cancer
Research Fund, Lincoln's Inn Fields, London, U.K.

The mammalian haemopoietic system consists of 5 or 6 distinctive cell lineages which are all derived from a common precursor – the so-called (in mice) pluripotential stem cell (Abramson, Miller & Phillips, 1977). An incomplete but none the less impressive picture has been built up of cellular interactions and soluble regulators (e.g. erythropoietin, 'granulopoietin' and various thymic 'hormone' preparations) and feedback control involved in lymphoid, erythroid and myeloid cell maturation. However, little is known or seriously conjectured concerning early differentiation events in haemopoiesis.

INSTRUCTIVE OR SELECTIVE CELL INTERACTIONS?

It has been proposed that the commitment of haemopoietic stem cell progeny to different cell lineages occurs, and is induced, within specific tissue microenvironments (Wolf & Trentin, 1968), presumably via some form of complementary interaction of juxtapositioned cell surfaces or short-range soluble factors. This recalls other more well known (if not well understood) 'inductive' embryological systems in which cells *appear* to be imposed upon to differentiate according to their position and their near neighbours (Grobstein, 1968; Saxen *et al.*, 1977).

This traditional embryological paradigm is perhaps rather Lamarckian in terms of the central tenets of molecular biology and it is quite possible that cell surface receptors and cell interactions may have no qualitative control over the critical decision-making processes in differentiation. Clearly where a cell finds itself is of considerable consequence, and 'specific' microenvironments are essential for developmental processes. This point is well illustrated in the haemopoietic system by animal models as well as by clinical disorders in man (Table 1). However, many or perhaps all of these obligatory environmental conditions may be permissive and selective (i.e. Darwinian) rather than specifically instructive. There is a useful analogy (and lesson) to be gleaned in this context from an historical per-

Table 1. *Developmental lesions in the haemopoietic system*

Intrinsic cellular defects
    CFUs block W/W^r (1)

Micro-environmental defects

(A) T lineage maturation block	Congenital thymic aplasia*† (2), (3)	⎫ Lack of essential extracellular maturation inducing/ regulating factors
Myeloid block	$Sl/Sl^{d\ +}$ (4)	⎭

(B) Erythroid stem cell block	Diamond–Blackfan Syndrome* (5)	⎫
Myeloid stem cell block	Aplastic anaemia* (6)	Suppressor T cell activity
B lymphocyte	Acquired agammaglobuli-naemia* (2), (7), (8)	⎭

* Man. † Mouse.
(1), (3) and (4) McCullough, Mak, Price & Till (1974).
(2) Cooper & Seligmann (1977).
(5) Hoffman *et al.* (1976).
(6) Ascensáo *et al.* (1976).
(7) Waldmann, Durm & Broder (1974).
(8) Waldmann & Broder (1977).

spective of views on the role (now seen as purely 'selective') of antigen in influencing the diversification of B lymphocyte clones.

As emphasised previously in the context of the haemopoietic system (Lajtha & Schofield, 1974) it is essential to discriminate between differentiation, in the sense of the genetic commitment of a cell to a given lineage, and the active, controlled expression of this option (i.e. maturation). In terms of cell interactions one can then consider a model in which, as Holtzer proposes, choices of lineage commitment are severely restricted (perhaps to two) and the adoption of one particular option is made on an essentially random basis which is qualitatively independent of the cells' local environment (Holtzer, 1968; Dienstman & Holtzer, 1975). Commitment would require 'selection' of a specific gene set(s) and initially this might be expressed solely by synthesis of those cell surface receptors complementary to the extrinsic signals which are available (and selective for) the regulation of maturation in that particular lineage. For example in the haemopoietic system the first lineage-specific genes expressed by a T cell might be those coding for thymic 'hormone' receptors, and in the case of erythroid cells for erythropoietin receptors. Direct cell interactions and soluble ligands (including hormones) certainly play an essential role in these developmental responses, but this regulatory power may be exercised primarily through rate control rather than quality control of gene transcription.

MONOCLONAL MALIGNANCIES AS MODEL SYSTEMS

Despite the considerable cellular complexity of haemopoietic tissue it does offer some attractions as a system in which to study the role of cell-surface events in differentiation and maturation. Functional clonal assays are becoming available for most of the cell lineages and highly selective (or 'luxury') gene transcription, and the protein products of this activity (e.g. haemoglobulin, immunoglobulin) may be detectable at an early stage of maturation. One limiting factor is the relative scarcity of stem cells in haemopoietic tissues and the lack of in-vitro assays for pluripotential stem cells in particular.

One approach which avoids the complications of cell population heterogeneity has been to use malignant haemopoietic cells (e.g. leukaemias). These have the very substantial advantage that they are usually monoclonal and in many instances can now be maintained as established cell lines *in vitro*. At least some appear to be blocked or 'frozen' in an early stage of development yet can be 'induced' to mature *in vitro* by the addition of appropriate ligands or chemicals (Lotem & Sachs, 1974; Gallagher *et al.*, 1975; Ichikawa, Maeda & Horiuchi, 1976). If such monoclonal lines could be shown to be capable of giving rise to progenitors of two (or more) distinctive cell lineages then this system would be ideal for analysing the significance of cell surface structures and cell interactions in haemopoiesis.

Friend-virus-induced erythroleukaemia has, for example, provided a most useful system for the analysis of transcriptional and translational control of haemoglobin synthesis (Paul, 1976). However it is doubtful if such a system can be employed to successfully resolve a (and probably *the*) central problem of differentiation since the cells under study have already made a genetic commitment to the erythroid lineage and have thus made haemoglobin genes susceptible to 'inducible' or enhanced transcription (and have no other options!).

MEMBRANE PHENOTYPES OF HUMAN ACUTE
LEUKAEMIC CELLS

Acute lymphoid, myeloid and so called 'undifferentiated' leukaemias in man offer some special attractions for this type of study. These monoclonal populations (Fialkow, 1976) are readily available in quantity, can be (in a few cases) maintained as established cultures with stable gene expression (Nilsson & Ponten, 1975; Minowada, Tsubota, Greaves & Walters, 1977; Minowda *et al.*, 1978) and in two instances provide a distinctive chromosome correlate of malignancy – the t(22q⁻, 9q⁺) in Chronic Myeloid Leukaemia (Rowley, 1973) and t(14q⁺, 8q⁻) in B cell malignancies (Zech, Haglund, Nilsson & Klein, 1976; Kaiser-McCaw, Epstein, Kaplin & Hecht, 1977). More significantly these cells may represent stem-cell derivatives blocked at key 'bifurcation points' in differentiation (Greaves *et al.*, 1976; Janossy, Roberts & Greaves, 1976).

Over the past four years our group, and others, have carried out an extensive phenotypic analysis of leukaemias in terms of their cell-surface antigens and receptors and of some constitutive enzymes (reviewed in Thierfelder, Rodt & Thiel, 1976). Part of the incentive here has been to provide more rational and precise clinical diagnostic tests (Greaves, 1977). From screening several hundred acute leukaemias we can now provide a reasonably accurate profile of the cellular heterogeneity of these diseases (Fig. 1). In the context of this discussion the most relevant observation is that the majority of acute non-myeloid leukaemias which appear (morphologically, cytochemically) either as lymphoid or 'undifferentiated' do not appear to express on their surface specific gene products of T *and* B lymphocytes (i.e. T cell antigens and immunoglobulin respectively.) Therefore we conclude either that Acute Lymphoblastic Leukaemia cells (ALL cells) have lost such expression (i.e. 'de-differentiated') or they are frozen at an earlier pre-T, pre-B stage of development. The latter alternative is by the far the most likely to be correct (see below).

Although only a few established cell lines are available which are bona fide leukaemia derivatives (as determined by unique karyotypes) we have been able to study several lines which have a stable (for more than one year at least) immunological and enzymatic phenotype which is precisely the same as that of the common ALL and the Thy–ALL groups (Fig. 1) (Minowada *et al.*, 1977; Rosenfeld *et al.*, 1977).

Ph¹-POSITIVE LEUKAEMIA: A MODEL FOR HAEMOPOIESIS?

Study of chronic myeloid leukaemia (CML) has been particularly revealing in terms of possible relationships with normal haemopoiesis (Janossy *et al.*, 1976). This disease is monoclonal (Fialkow, 1976) and characterised by a unique chromosomal marker, the Philadelphia chromosome: Ph¹ t(22q⁺, 9q⁺) (Lawler, 1976). Patients usually first present at hospital with a chronic disease manifested by a great overproduction of relatively mature granu-

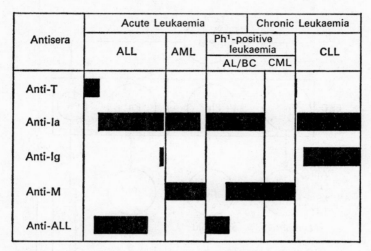

Fig. 1. Major membrane phenotypes of human leukaemic cells.

Antisera. Anti-T: rabbit anti-human T lymphocyte (membrane antigen/s). Anti-Ia: rabbit anti-human B cell gp 28, 33 'Ia-like' by analogy with murine B cell antigens. Anti-Ig: sheep anti-human immunoglobulin. Anti-M: rabbit anti-myelo-monocytic leukaemia. Defines a granulocyte and monocytic lineage restricted membrane antigen. Anti-ALL: rabbit antisera to Acute Lymphoblastic Leukaemia of the non-T, non-B variety.

Leukaemias. ALL: acute lymphoblastic leukaemia. AML: acute myeloblastic leukaemia. AL/BC: acute leukaemia/blast crisis. CML: chronic myeloid leukaemia. CLL: chronic lymphocytic leukaemia.

Horizontal bars represent relative numbers of positive cases (between 150 and 1000 individual blood and/or bone marrow samples from untreated or relapse patients tested with the various markers).

locytic elements. Karyotypic analysis reveals, however, that monocyte, erythroid and platelet precursors also carry the Ph¹ translocation indicating that an early stem cell must have provided the target for the disease (reviewed in Lawler, 1976). There is also suggestive though at present inconclusive evidence for Ph¹ in T and B cells which, if substantiated, would 'force' the target cell into the pluripotential stem cell category. The chronic disease is usually transitory and frequently superseded by an acute phase blast crisis characterised by less mature cells usually with chromosomal abnormalities in addition to Ph¹. The cell populations observed in blast crisis can be predominantly 'myeloid', 'lymphoid' or mixed by morphological and histo-, cyto-chemical criteria (Beard *et al.*, 1976). When we examined these cells immunologically we found that some cases were indistinguishable from common (Ph¹ negative) acute *lymphoblastic* leukaemia and others were similar or identical to (Ph¹ negative) acute myeloid leukaemia (Janossy *et al.*, 1976; Janossy *et al.* (1977*b*). We believe

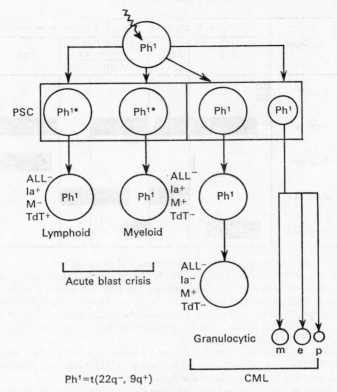

Fig. 2. Stem cell progeny and their phenotypes in chronic myeloid leukaemia with the *Philadelphia chromosome*. A schematic representation of the origin of different lymphoid or myeloid subclones from the initial single target cell. Either the CML phase or acute blast crisis (seen as Ph^1-positive acute leukaemia) can be present first. Mixtures of the different subclones are common and treatment (chemotherapy) influences the observed cellular patterns. The Ph^1 clone as a whole may be either totally or partially dominant over all other normal Ph^1-negative clones.

Ph^1: with the Philadelphia chromosome translocation. CML: chronic myeloid leukaemia. * additional chromosomal changes. PSC: pluripotential stem cells. m: monocytes. e: erythrocytes. p: platelets.

Phenotype markers (+ *present*, — *absent*). ALL: Acute Lymphoblastic Leukaemia associated membrane antigen (Greaves *et al.*, 1976). Ia: HLA associated membrane molecule (pp. 28, 33) (Schlossman, Chess, Humphreys & Strominger, 1976; Janossy *et al.*, 1977). M: Myeloid/monocyte maturation linked membrane antigen (Roberts & Greaves, 1978); all three antigens defined by rabbit antisera. TdT: Terminal deoxynucleotidyl transferase enzyme (McCaffrey, Harrison, Parkman & Baltimore, 1975; Greenwood *et al.*, 1977; Hoffbrand *et al.*, 1977).

these observations are very revealing in terms of the cellular origins of leukaemia (Fig. 2). In brief, we have suggested that: (1) CML can frequently involve transformation of a pluripotential stem cell. (2) That since different leukaemias (Ph^1-positive and Ph^1-negative) can have similar or identical phenotypes then the components of the phenotype (e.g. membrane antigens) are reflections of *normal* gene activity, i.e. they are precursor or *stem cell* markers. (3) That CML can be used as a monoclonal system to map normal cellular relationships and developmental sequences in haemopoiesis.

LEUKAEMIC MEMBRANE ANTIGENS ARE STEM CELL ANTIGENS

It is well established that malignant cells can continue to express cell-lineage-specific or differentiation-linked membrane antigens (Boyse & Old, 1969; Goldenberg, Pant & Dahlman, 1976; Hager & Tompkins 1976; Akeson, 1977) and there are several clear examples of this in the acute leukaemias and in the more 'mature' chronic leukaemias (e.g. T cell and myeloid antigens, B cell immunoglobulin). From the phenotypic profiles available in acute leukaemia (Fig. 1) two membrane antigens stand out as being of special interest: the ALL-associated antigen and the 'Ia-like' antigen. The former has been considered as a candidate leukaemia-specific antigen (Brown, Capellaro & Greaves, 1975*a*; Brown, Hogg & Greaves, 1975*b*) while the latter is found also on mature B lymphocytes, some T cells, macrophages (and several other non-haemopoietic cell types) (Möller, 1976) where it is thought to play a crucial role in cell interactions (see Katz, this volume, pp. 411–27).

Table 2. *Leukaemia-associated membrane antigens**

(1) 'ALL' antigen: single glycosylated polypeptide MW $\sim$ 105000 daltons
(2) Ia-antigen: 2 non-covalently linked polypeptides MW $\sim$ 33000 daltons (HLA/Chr. 6) $\sim$ 28000 daltons†

* Sutherland, Smart & Greaves (*Leukaemia Research*, in press).
† These molecular characteristics are shared with B cell Ia structures (Springer *et al.*, 1977).

The 'ALL' antigen and the 'Ia-like' antigen have recently been isolated from leukaemic cell lines and shown to be distinct glycosylated polypeptides (Table 2). We now believe that both are indeed normal gene products of the 'target' stem cells for acute leukaemia transformation and that they may play an important role in early haemopoiesis. The detailed evidence for this statement is being published elsewhere (Greaves & Janossy, 1978) but can be summarised as follows: although cells with the ALL$^+$ Ia$^+$ (SmIg$^-$ T antigen$^-$) phenotype cannot be identified in normal adult haemopoietic tissue they are demonstrable in foetal tissues in newborn and in the bone marrow of children (e.g. post-chemotherapy or post-irradiation and transplantation). Now these are precisely the circumstances under which one can detect the expression of 'onco-foetal' or 'carcino-embryonic' antigens (e.g. CEA, AFP; Möller, 1973). The conventional interpretation of such observations is that 'de-repression' of foetal genes is occurring.

An alternative interpretation which we favour (Greaves & Janossy, 1978) is that malignancy and regeneration both involve selective proliferation (monoclonal or polyclonal respectively) of stem cells which are more common in foetuses (and children) than in adults and that the antigenic properties of these cells are effectively normal, i.e. 'ALL' and 'Ia-like' antigens are normal stem cell antigens, comparable to the F9 antigen of normal germ cells, early embryo cells and teratocarcinomas (Jacob, 1977). A parallel situation exists with haemoglobin gene expression. Foetal haemoglobin is produced in increased amounts in regenerating marrow (Alter et al., 1976) and leukaemia, particularly juvenile CML and some erythro-leukaemias (Sheridan et al., 1976) and is probably the product of the selective proliferation of infrequent stem cells that are committed to foetal type haemoglobin and demonstrable in normal (adult) individuals (Papayannopoulou, Brice & Stamatoyannopoulos, 1977).

ALL$^+$ Ia$^+$ non-T, non-B cells (i.e. the ALL phenotype) can be demonstrated in normal 'young', active haemopoietic tissue (Table 3) as can another population of larger ALL$^-$ Ia$^+$ non-T, non-B cells (i.e. AML phenotype) (Greaves, Janossy, Francis & Minowada, 1978). We suspect that the ALL$^+$ Ia$^+$ cell is a lymphoid stem cell but as yet have no direct evidence for this view. Since we now have several established monoclonal leukaemia cell lines available with the non-T, non-B (and presumed pre-T, pre-B) phenotype (Minowada et al., 1978; Rosenfeld et al., 1977) we possibly have access to a critical cell type in lymphoid development. If the differentiation block in these cells is reversible in vitro then we may be able to investigate some key questions relating to the differentiation and immunological diversity of T and B lymphocytes.

The possible presence of Ia$^+$ on functional haemopoietic precursors has been investigated recently using complement mediated cytotoxicity and fluorescence-activated cell sorting.

We found that the pluripotential CFU(s) in mice were Ia negative (Basch, Janossy & Greaves, 1977). Most of these cells are however in G_0 at the time of cell transfer and we do not exclude that Ia is expressed on

Table 3. *Presence of 'ALL' membrane phenotype in non-leukaemic tissues*

(1) Normal Haemopoietic tissue:*   Foetal
                                    Newborn, children   } 1–20 %
                                    Adult: extremely rare or absent (< 1 in 10^4)

(2) Regenerating haemopoietic tissue: Post chemotherapy†
                                         Post XR + trans-
                                         plantation‡   } 5–50 %

(3) Proliferative disorders: Neonatal 'leukaemoid' reactions§

  * Bone marrow (and liver in foetus).
  † Leukaemia or non-haematological malignancies (e.g. neuroblastoma).
  ‡ Patients with aplastic anaemia.
  § Characterised by excessive numbers of small lymphocytes. Usually remits spontaneously.
  Taken from Greaves, Janossy, Francis & Minowada (1978).

these cells when they are in cell cycle. When human bone marrow cells are fractionated into Ia+ and Ia− populations using the FACS then we were able to demonstrate CFU-c (myeloid colony formation) activity in the Ia+ ALL− and Ia− ALL− but not in the ALL+ Ia+ cells (Janossy & Francis unpublished). This suggests as anticipated that the ALL+ cells are not myeloid progenitors and that Ia is expressed on at least some stem cells committed to the granulocytic-macrophage pathway. It is relevant in this context to note that in CML, Ia antigen expression correlates inversely with myeloid maturation within a single clone indicating that the antigen is lost as cells mature (Janossy *et al.*, 1976). These results establish that Ia structures which are thought to be HL-A (probably D locus) coded (or controlled) glycosylated polypeptide molecules (Möller, 1976; Springer, Kaufman, Terhorst & Strominger, 1977; Snary *et al.*, 1977) are present on acute leukaemias and on normal stem cells from which these malignancies almost certainly originate.

In the mouse Ia antigens are coded (or otherwise controlled) by the I (immune response gene) region of the *H-2* locus and are present on the surface of B cells, macrophages and some T cells (suppressor?) as well as being components of T cell soluble factors which regulate immunological effector function (see Katz, this volume, pp. 411–27). Could they serve a similar regulatory role in early haemopoiesis? This can now be investigated using both CFU-c assays and the leukaemic cell lines. In the meantime there is clinical evidence which suggests that T cells can regulate haemopoietic development (of granulocytic and erythroid lineages) (see Table 4). We can therefore consider the possibility that regulatory cell interactions

438   MELVYN GREAVES

Table 4. *Evidence for regulatory interactions between T cells and haemopoietic stem cells*

(1) Thymectomy (mice): deficient CFU(s) migration from marrow to spleen*

(2) Amplification of CFU(s) activity by:
Syngeneic T cells, allogeneic (T + B)
Lymphocytes and culture supernatants from activated T cells†‡§

(3) Inhibition of myeloid and erythroid stem cells by T cells:
Aplastic anaemia‖
Diamond–Blackfan Syndrome¶

* Petrov, Khaitov, Aleinikova & Gulak (1977).
† Basford & Goodman (1974).
‡ Hara *et al.* (1971; 1974).
§ Barr, Whang-Peng & Perry (1977).
‖ Ascensáo *et al.* (1976).
¶ Hoffman *et al.* (1976).

controlled by the major histocompatibility gene locus have a central significance in haemopoiesis and are not restricted to mature immunocompetent cells. A further clue supporting this possibility is provided by the linkage of congenital neutropenia (an autosomal recessive disease characterised by maturation arrest at the promyelocyte stage of granulocyte maturation) to HLA genes (Hansen, Dupont, Esperance & Good, 1977). These speculations should come as no surprise since *H-2*-linked genes in the mouse (T,t loci), coding for cell surface structures, are strongly implicated in early embryogenesis (Bennett, 1975; Klein & Hämmerberg, 1977).

REFERENCES

ABRAMSON, S., MILLER, R. G. & PHILLIPS, R. A. (1977). The identification in adult bone marrow of pluripotent and restricted stem cell of the myeloid and lymphoid systems. *Journal of Experimental Medicine*, **145**, 1567–79.
AKESON, R. (1977). Human lung organ – specific antigens on normal lung, lung tumors, and a lung tumor cell line. *Journal of the National Cancer Institute*, **58**, 863–8.
ALTER, B. P., RAPPEPORT, J. M., HUISMAN, T. H. J., SCHROEDER, W. A. & NATHAN, D. G. (1976). Fetal erythropoiesis following bone marrow transplantation. *Blood*, **48**, 843–53.
ASCENSÁO, J., PAHWA, R., KAGAN, W., HANSEN, J., MOORE, M. & GOOD, R. A. (1976). Aplastic anaemia: evidence for an immunological mechanism. *Lancet*, i, 669–71.
BARR, R. D., WHANG-PENG, J. & PERRY, S. (1977). Regulation of human hemopoietic stem cell proliferation by syngeneic thymus derived lymphocytes. *Acta haematologica*, **58**, 74–8.
BASCH, R. S., JANOSSY, G., GREAVES, M. F. (1977). Murine pluripotential stem cells (CFUS) lack Ia antigen. *Nature, London*, **270**, 520.

BASFORD, N. L. & GOODMAN, J. W. (1974). Effect of lymphocytes from the thymus and lymph nodes on differentiation of hemopoietic spleen colonies in irradiated mice. *Journal of Cellular and Comparative Physiology*, **84**, 37–48.

BEARD, M. E. J., DURRANT, J., CATOVSKY, D., WILTSHAW, E. I., AMESS, J. L., BREARLEY, R. L., KIRK, B., WRIGLEY, P. F. M., JANOSSY, G., GREAVES, M. F. & GALTON, D. A. G. (1976). Blast crisis of chronic myeloid leukaemia. *British Journal of Haematology*, **34**, 167–78.

BENNETT, D. (1975). The T-locus of the mouse. *Cell*, **6**, 441–54.

BOYSE, E. A. & OLD, L. J. (1969). Some aspects of normal and abnormal cell surface genetics. *Annual Reviews in Genetics*, **3**, 269–90.

BROWN, G., CAPELLARO, D. & GREAVES, M. F. (1975a). Leukaemia associated antigens in man. *Journal of the National Cancer Institute*, **55**, 1281–9.

BROWN, G., HOGG, N. & GREAVES, M. F. (1975b). A candidate human leukaemia antigen. *Nature, London*, **258**, 454–6.

COOPER, M. D. & SELIGMANN, M. (1977). B and T lymphocytes in immunodeficiency and lymphoproliferative diseases. In *B and T Cells in Immune Recognition*, eds. F. Loor & G. E. Roelants, pp. 377–400. Wiley, New York.

DIENSTMAN, S. R. & HOLTZER, H. (1975). Myogenesis: A cell lineage interpretation. In *Cell Cycle and Differentiation*, eds. J. Reinert & H. Holtzer, pp. 1–26, Springer-Verlag, Berlin and New York.

FIALKOW, P. J. (1976). Clonal origin of human tumors. *Biochimica et Biophysica Acta*, **458**, 283–321.

GALLAGHER, R. E., SALAHUDDIN, S. Z., HALL, W. T., McCREDIE, K. B. & GALLO, R. C. (1975). Growth and differentiation in culture of leukaemia and reidentification of type C virus. *Proceedings of the National Academy of Sciences, U.S.A.*, **72**, 4137–41.

GOLDENBERG, D. M., PANT, K. D. & DAHLMAN, H. L. (1976). Antigens associated with normal and malignant gastro-intestinal tissues. *Cancer Research*, **36**, 3455–63.

GREAVES, M. F. (1977). Recent progress in the immunological characterisation of leukaemic cells. *Blut*, **34**, 349–56.

GREAVES, M. F. & JANOSSY, G. (1978). Patterns of gene expression and the cellular origins of human leukaemias. *Biochimica Biophysica Acta, Reviews in Cancer* (in press).

GREAVES, M. F., JANOSSY, G., ROBERTS, M., RAPSON, N. T., ELLIS, R. B., CHESSELLS, J., LISTER, T. A. & CATOVSKY, D. (1976). Membrane phenotyping: diagnosis, monitoring and classification of acute 'lymphoid' leukaemias. In *Immunological Diagnosis of Leukaemias and Lymphomas*, eds. S. Thierfelder, H. Rodt & E. Thiel, p. 61. Springer, Berlin and New York.

GREAVES, M. F., JANOSSY, G., FRANCIS, G. & MINOWADA, J. (1978). Membrane phenotypes of human leukaemic cells and leukaemic cell lines. In *Cell Proliferation*, vol. 5, eds. B. Clarkson, J. Till & P. Marks. Cold Spring Harbor Press (in press).

GREENWOOD, M. F., COLEMAN, M. S., HUTTON, J. J., LAMPKIN, B., KRILL, C., BOLLUM, F. J. & HOLLAND, P. (1977). Terminal deoxynucleotidyl transferase distribution in neoplastic and hematopoietic cells. *Journal of Clinical Investigation*, **59**, 889–99.

GROBSTEIN, C. (1968). The problem of the chemical nature of embryonic inducers. In *Cell Differentiation*, eds. A. V. S. De Reuck & J. Knight, pp. 131–47. J. & A. Churchill, London.

HAGER, J. C. & TOMPKINS, W. A. F. (1976). Antibodies in normal rabbit serum that react with tissue-specific antigens on the plasma membranes of human adenocarcinoma cells. *Journal of the National Cancer Institute*, **56**, 339–42.

440   MELVYN GREAVES

HANSEN, J. A., DUPONT, B., ESPERANCE, P. L. & GOOD, R. A. (1977). Congenital Neutropenia: abnormal neutrophil differentiation associated with HLA. *Immunogenetics*, **4**, 327–32.

HARA, H., KITAMURA, Y., KAWATA, T., KANAMURA, A. & NAGAI, K. (1971). Colony-stimulating factor and allogeneic lymphocytes. *Acta Haematologica, Japan*, **34**, 517–18.

HARA, H., KITAMURA, Y., KAWATA, T., KANAMURA, A. & NAGAI, K. (1974). Synergism between lymph node and bone marrow cells for production of granulocytes. II. Enhanced colony-stimulating activity of sera. *Experimental Haematology*, **2**, 43–9.

HOFFBRAND, V., GANESHAGURU, K., JANOSSY, G., GREAVES, M. F., CATOVSKY, D. & WOODRUFF, R. K. (1977). Diagnostic value of terminal transferase levels and membrane phenotypes in acute leukaemia. *Lancet*, ii, 520–3.

HOFFMAN, R., ZANJANI, E. D., VILA, J., ZALUSKI, R., LUTTON, J. D. & WASSERMAN, L. R. (1976). Diamond-Blackfan syndrome: lymphocyte mediated suppression of erythropoiesis. *Science*, **193**, 899–900.

HOLTZER, H. (1968). Induction of Chondrogenesis: A concept in quest of mechanisms. In *Epithelial–Mesenchymal Interactions*, ed. R. Billingham, pp. 152–64. Williams and Wilkins, Baltimore.

ICHIKAWA, Y., MAEDA, M. & HORIUCHI, M. (1976). *In vitro* differentiation of Rauscher-virus induced myeloid leukaemia cells. *International Journal of Cancer*, **17**, 789–97.

JACOB, F. (1977). Mouse teratocarcinoma and embryonic antigens. *Immunological Reviews*, **33**, 3–32.

JANOSSY, G., ROBERTS, M. & GREAVES, M. F. (1976). Target cell in chronic myeloid leukaemia and its relationship to acute lymphoid leukaemia. *Lancet*, ii, 1058–60.

JANOSSY, G., GOLDSTONE, A. H., CAPELLARO, D., GREAVES, M. F., KULENKAMPFF, PIPPARD, M. & WELSH, K. (1977a). Differentiation-linked expression of p. 28, 33 (Ia-like) structures on human leukaemic cells. *British Journal of Haematology*, **37**, 391.

JANOSSY, G., GREAVES, M. F., SUTHERLAND, R., DURRANT, J. & LEWIS, C. (1977b). Membrane phenotypes of acute lymphoblastic leukaemia and chronic myeloid leukaemia in blast crisis. *Leukaemia Research*, **1**, 289.

KAISER-McCAW, B., EPSTEIN, A. L., KAPLIN, H. S. & HECHT, F. (1977). Chromosome 14 translocation in African and North American Burkitt's lymphoma. *International Journal of Cancer*, **19**, 482–6.

KLEIN, J. & HAMMERBERG, C. (1977). The control of differentiation by the T complex. *Immunological Reviews*, **33**, 70–104.

LAJTHA, L. G. & SCHOFIELD, R. (1974). On the problem of differentiation in haemopoiesis. *Differentiation*, **2**, 313–20.

LAWLER, S. D. (1976). The cytogenetics of chronic granulocytic leukaemia. *Clinics in Haematology*, **6**, 55–75.

LOTEM, J. & SACHS, L. (1974). Different blocks in the differentiation of myeloid leukaemic cells. *Proceedings of the National Academy of Sciences, U.S.A.*, **71**, 3507–11.

McCAFFREY, R., HARRISON, T. A., PARKMAN, R. & BALTIMORE, D. (1975). Terminal deoxynucleotidyl transferase activity in human leukaemic cells and in normal human thymocytes. *New England Journal of Medicine*, **292**, 775–80.

McCULLOUGH, E. A., MAK, T. W., PRICE, G. B. & TILL, J. E. (1974). Organisation and communication in populations of normal and leukaemic hemopoietic cells. *Biochimica et Biophysica Acta*, **355**, 260–99.

MINOWADA, J., TSUBOTA, T., GREAVES, M. F. & WALTERS, T. R. (1977). A non-T, non-B human leukaemia cell line (NALM-1): establishment of the cell line and presence of leukaemia-associated antigens. *Journal of the National Cancer Institute*, **59**, 83–7.

MINOWADA, J., JANOSSY, G., GREAVES, M. F., TSUBOTA, T., SRIVASTAVA, S., MORIKAWA, S. & TATSUMI, E. (1978). The expression of acute lymphoblastic leukaemia antigen in human leukaemia-lymphoma. *Journal of the National Cancer Institute* (in press).

MÖLLER, G. (ed.) (1973). Tumor-associated embryonic antigens. *Transp antation Review*, **20**.

MÖLLER, G. (ed.) (1976). Biochemistry and Biology of Ia antigens. *Transplantation Review*, **30**.

NILSSON, K. & PONTEN, J. (1975). Classification and biological nature of established human hematopoietic cell lines. *International Journal of Cancer*, **15**, 321–41.

PAPAYANNOPOULOU, TH., BRICE, M. & STAMATOYANNOPOULOS (1977). Hemoglobin F sythesis *in vitro*: evidence for control at the level of primitive erythroid stem cells. *Proceedings of the National Academy of Sciences, U.S.A.*, **74**, 2923–7.

PAUL, J. (1976). Molecular mechanisms in erythroid differentiation. *Blut*, **19**, 125–35.

PETERSON, L. C., YOUNG, R. C., BLOOMFIELD, C. D., BURNING, R. D. (1976). Blast crisis as an initial or terminal manifestation of chronic myeloid leukaemia. *American Journal of Medicine*, **60**, 209–20.

PETROV, R. V., KHAITOV, R. M., ALEINIKOVA, N. V. & GULAK, L. V. (1977). Factors controlling stem cell recirculation. III. Effect of the thymus on the migration and differentiation of hemopoietic stem cells. *Blood*, **491**, 865–72.

ROBERTS, M. & GREAVES, M. F. (1978). A myeloid differentiation-linked membrane antigen. *British Journal of Haematology* (in press).

ROSENFELD, C., GOUTNER, A., CHOQUER, C., VENUAT, A. M., KAYIBAUDA, B., PICO, J. L. & GREAVES, M. F. (1977). Phenotypic characterisation of a unique non-T, non-B acute lymphoblastic leukaemia cell line. *Nature, London*, **267**, 841–3.

ROWLEY, J. D. (1973). A new consistent abnormality in chronic myelogenous leukaemia identified by quinacrine fluorescence and Giemsa staining. *Nature, London*, **243**, 290–3.

SAXEN, L., KARKINEN-JÄÄKELÄINEN, M., LEHTONEN, E., NORDLING, S. & WARTIO-VAARA, J. (1977). Inductive tissue interactions. In *The Cell Surface in Animal Embryogenesis and Development*, eds. G. Poste & G. L. Nicholson. North Holland, Amsterdam.

SCHLOSSMAN, S. M., CHESS, L., HUMPHREYS, R. E. & STROMINGER, J. L. (1976). Distribution of Ia-like molecules on the surface of normal and leukaemic human cells. *Proceedings of the National Academy of Sciences, U.S.A.*, **73**, 1288–92.

SHERIDAN, B. L., WEATHERALL, D. J., CLEGG, J. B., PRITCHARD, J., WOOD, W. G., CALLENDER, S. T., DURRANT, I. J., McWHIRTER, W. R., ALI, M., PARTRIDGE, W. & THOMPSON, E. N. (1976). The patterns of fetal haemoglobin production in leukaemia. *British Journal of Haematology*, **32**, 487–506.

SNARY, D., BARNSTABLE, L. J., BODMER, W. F., GOODFELLOW, P. N. & CRUMPTON, M. J. (1977). Cellular distribution, purification and molecular nature of human Ia antigens. *Scandinavian Journal of Immunology*, **6**, 439–52.

SPRINGER, T. A., KAUFMAN, J. F., TERHORST, C. & STROMINGER, J. L. (1977). Purification and structural characterisation of human *HLA*-linked B-cell antigens. *Nature, London*, **268**, 213–18.

THIERFELDER, S., RODT, H. & THIEL, E. (eds.) (1976). *Immunological Diagnosis of Leukaemias and Lymphomas*. Springer, Berlin and New York.

WALDMANN, T. A. & BRODER, S. (1977). Suppressor cells in the regulation of the immune response. *Progress in Clinical Immunology*, 3, 155–84.

WALDMANN, T. A., DURM, M., BRODER, S., BLACKMAN, M., BLAESE R. M. & STROBER, W. (1974). Role of suppressor T cells in pathogenesis of common variable hypogammaglobulinaemia. *Lancet*, ii, 609–13.

WOLF, N. S. & TRENTIN, J. J. (1968). Hemopoietic colony studies. *Journal of Experimental Medicine*, 127, 205–14.

ZECH, L., HAGLUND, U., NILSSON, K. & KLEIN, G. (1976). Characteristic chromosomal abnormalities in biopsies and lymphoid cell lines from patients with Burkitt and non-Burkitt lymphomas. *International Journal of Cancer*, 17, 47–56.

POSTER SESSION ABSTRACTS

The following contributed to the poster session held at the Symposium:

J. E. Beesley, J. D. Pearson and J. L. Gordon
ARC Institute of Animal Physiology, Babraham, Cambridge CB2 4AT
Interactions of blood leukocytes with cultured vascular cells

B. P. Hayes and J. D. Feldman
Institute of Ophthalmology and University College, London
Fine structure of a retinotectal cell adhesion assay for adult frog

G. P. Bolwell, J. A. Callow, M. E. Callow and L. V. Evans
Department of Plant Sciences, University of Leeds, Leeds
Evidence for complementary receptors involved in fertilisation in brown seaweeds

M. D. J. Davies,* W. S. Haston,† C. W. Evans and A. S. G. Curtis*
Departments of *Cell Biology and †Bacteriology and Immunology, University of Glasgow
Recognition and adhesion: two events contributing to cell traffic

Paul A. W. Edwards
Department of Biochemistry, University of Oxford, South Parks Road, Oxford, OX1 3QU
Do cells adhere by non-specific interactions between their surface glycoproteins?

W. S. Haston
Department of Bacteriology and Immunology, University of Glasgow
An inhibitor of lymphocyte DNA synthesis produced by thymocytes in vitro

B. E. Jones
University of Aberdeen
Lectin binding to sexual cells of the fungus Mucor mucedo

G. E. Jones
Queen Elizabeth College, University of London
Adhesive interactions between cells of the embryonic chick neural retina

G. E. Jones and T. Partridge
Queen Elizabeth College and Charing Cross Hospital Medical School,
University of London
Regulation of cell adhesion

Ann M. Lackie
Department of Zoology, University of Glasgow
Cellular recognition of 'not-self' in insects

F. Sless and D. M. V. Parrott
Department of Bacteriology and Immunology, University of Glasgow
The effect of Concanavalin A on the tissue localization of lymphoblasts

R. P. C. Smith
Department of Cell Biology, University of Glasgow
In-vitro adhesive interactions of neutrophil granulocytes (PMNs) and cultured endothelial cells

Author Index

Figures in bold type refer to entries in the lists of references

Schnarrenberger, C., 153, **157**
Schneider, G. E., 298, 299, 305, 308, 309, 311, **336, 337, 343, 349, 350, 357, 358**
Schneps, S. E., **336**
Schoefl, G. I., 362, 386, **390**
Schofield, R., 430, **440**
Schonne, E., 402, **408**
Schroeder, T. E., 26, **48**
Schroeder, W. A., 436, **438**
Schubiger, K., 169, **171**
Schwartz, E. L., **358**
Schwarz, H., 175, **199**
Scott, M. Y., 287, **347, 350**
Scott, T. K., 133, **137**
Scott, T. M., 316, 320, **350**
Seaman, G. V. F., 236, **238**
Sears, E. R., 130, 134, **137**
Sedgley, M., 67, **80**, 362, 364, 365, 367, 368, 379, 380, 382, 385, **388, 390,** 405, **407**
Sela, B.-A., 322, **349**
Seligmann, M., 430, **439**
Shackelford, R. M., 308, 309, **349**
Shaffer, B. M., 180, 193, **201**
Shand, F. L., 72, **79**
Shannon, S. L., 361, **389**
Shapleigh, E., 151, **157**
Sharma, S. C., 278, 283, 284, 285, 287, 294, 297, 298, 299, 300, 308, 310, 311, 312, 317, 318, 321, 323, 326, **337, 338, 347, 348, 350, 351, 354, 355**
Sharman, M., 305, **333**
Sharon, N., 151, 155, **158, 159,** 250, **255**
Shaw, G., 123, **134**
Shawe, G. D. H., 305, 306, **351**
Shearer, G. M., 416, **426**
Sheridan, B. L., 436, **441**
Shields, R., 222, **239**, 246, **257**
Shimoda, C., 96, **102**, 107, 112, **118**
Shivanna, K. R., 126, 128, 130, **136, 137**
Shneyour, A., 380, **390**
Shoemaker, D. W., 87, 88, **103**
Shreffler, D. C., 413, **426**
Shur, B., 270, **273**
Sieja, T. W., 189, 192, **198**
Simmons, S. J., 67, **80**
Simpson, D. L., 155, **159,** 175, 176, **201**
Simpson, T. L., 8, 18, **23**
Sinensky, M., 148, 149, **159**
30

Sing, V., 110, **118**
Singer, M., **358**
Singer, D. A., 299, **350**
Singer, S. J., 249, **255**
Sitthi-Amorn, C., **351**
Skoultchi, A. I., 401, 402, **409**
Smets, L. A., 221, **239**
Smith, C., 363, **390**
Smith, C. G., 241–60, 245, 252, 254, **256**, 260
Smith, J. B., 361, **390**
Smith, J. L., 398, **409**
Smith, M. E., 66, 67, **80**, 359–92, 361, 362, 364, 365, 367, 368, 379, 380, 381, 382, 384, 387, **388, 390,** 405, **407**
Smith, P., 127, 129, **136,** 155, **158,** 420 **425**
Smithyman, A., 396, 398, 400, **407**
Snary, D., 437, **441**
Snell, G. D., 413, **426**
Snell, W. J., 95, **102**
So, K. F., 309, **336, 351, 358**
Sonneborn, D. R., 192, **201**
Sørensen, M., 402, **409**
Sørensen, S. P. L., 402, **409**
Southworth, D., 151, **159**
Sparshott, S. M., 67, **80**, 362, 364, 365, 367, 368, 379, 380, 382, **388, 390,** 405, **407**
Spear, P. D., **334**
Speas, G., 293, **336**
Spector, W. G., 403, **409**
Speidel, C. C., 302, 303, 304, 305, 306, 313, 317, **351**
Speman, H., 174, **201**
Sperry, R. W., 262, 263, **273,** 275, 276, 278, 279, 281, 282, 283, 284, 286, 287, 288, 290, 291, 292, 295, 298, 302, 311, 312, 324, 325, 326, 328, **333, 347, 351, 352**
Spiers, A. S. D., 398, **408**
Spitzer, J. L., 324, **340**
Spooner, B. S., 26, **48, 49**
Sprent, J., 75, **81,** 359, **390,** 422, **427**
Springer, A. D., 291, 308, 309, 310, 317, **352**
Springer, T. A., 435, 437, **441**
Spry, C. J. F., 360, **390**
Sreevalsan, T., 225, **238**
Srivastava, S., 432, 436, **441**
Stamatoyannopoulos, G., 436, **441**
Stamper, H. B., jr., 75, **81,** 384, 385 **390, 391,** 405, **409**

Subject Index

acetylgalactosaminidase: sensitivity to, of *Chlamydomonas* mating-type substance, 89–90

acetylglucosamine UDP–galactosyl transferase (control enzyme), equally distributed in dorsal and ventral neural retinal cells, 270, 271

actomyosin system
  components of, in attachment sites of fibroblasts, 241–2, 243, 245, 246, 250, 252, 253
  in ovarian granulosa cells, effect of concanavalin A on, 250

adaptive differentiation, leading to preferential interaction between cells, 422–3, 424

adhesion, cell to cell
  binding energy of, 45
  cell surface modulations and, *see under* cell surfaces
  chronospecific, 173
  of *Chlamydomonas* flagella in mating, *see under Chlamydomonas*
  differences in, and mixing or sorting out of two kinds of cell, 41, 59
  differences in number of sites for, produce immiscibility and mutual spreading between cell populations, 37–8, 186
  differential, 3, 32–3, 37, 41, 58, 59, 186
  glycoproteins in, *see* glycoproteins
  inhibitors of: produced by slime moulds, 176–80, 181, 197, and by sponges, 10; *see also* interaction modulation factors
  interaction modulation factors and, *see* interaction modulation factors
  interaction between enzymes and substrates in, 221–2

measurement of, 222
microtubules not involved in, 245, 385
of polymorphonuclear leukocytes to endothelium, *see* polymorphonuclear leukocytes
in positioning of cells, 56, 57–8, 59–60
rates of, reflect only initial, not subsequent, contacts, 44–5
of retina neural cells to optic tectum, *see* retina neural cells
reversible work of, related to surface tension, 39–40
sites for, on slime moulds, *see* contact sites
specific, 56, 58, 59, 63; of mating types, 83, 105; of species, 9–10, 173; of tissues, 63, 173
stabilization of, 7
surface charge effects in, 236–7
surface tension differences and, 32–7, 45–6, 180
temporary, in redistribution of embryonic cells, 2, 3, 4, 5, 7
tissue configurations depend on affinities, abundance, and distribution of sites for, 41–4
and viscosity of cell aggregates, 31

adhesion, cell to substrate
  of cultured cells, fatty acid effects on, 226–30, 232–3
  of fibroblasts, *see* fibroblasts
  of slime moulds, effects of inhibitor on, 181
  of sponge (*Ephydatia*), affected by substance from opposite strain type, 61

agammaglobulinaemia, acquired, 430

agglutination factors, *see under Chlamy-*

Ir genes, *see* immune response genes

Iridaceae

proteins of: from pollen, 125; from stigma, 127

removal of stigma surface in, prevents pollen tube penetration, 130

iron, both essential and toxic for living organisms, 402

ferric: deposited in spleen in Hodgkin's disease, 400; lymphocytes treated with salts of, fail to enter lymph nodes and spleen, 400, 401

isoagglutinins, *see Chlamydomonas* mating-type substances

lactoferrin, iron-binding protein, synthesized by neutrophilic leukocytes 402, 403, and PMN, 404

receptors for, on macrophages, 402, 403, and on mouse lymphocytes, 401

in spleen in Hodgkin's disease, 400

lactoperoxidase-catalysed iodination of externally exposed proteins, as marker for cell surface components, 149

lectins (phytohaemagglutinins, carbohydrate-binding proteins), 151

all-*β*, in seeds, 153–4

bound by stigma surface, 127, 129, 155

cross-linking of fibroblast cell-surface proteins by, affecting cytoskeleton, 249, 250

as glycoproteins, 155

possible function of, in plants, 151

possibly involved in *Chlamydomonas* mating-type reaction, 98, and in compatibility of plant grafts, 151–4, 155

on surfaces of slime moulds, and adhesiveness, 4, 175–6, 186, 197

treatment of lymphocytes with, delays their passage through tissues, 395

see also concanavalin A

Leguminosae, lectins in establishment of symbiotic relation between *Rhizobium* and, 151–2

LETS, *see* globulin, cold-insoluble

leukaemias

acute leukaemia/blast crisis, 433, 434

acute lymphoblastic, cells frozen at early stage of development? 432, 433, 434

acute myeloblastic, 433

acute myeloid, 433

cell-surface antigens in different types of, 432, 433, 435–8

chronic lymphatic, abnormally low amounts of IMFs from B and T cells in serum in, 76

chronic lymphocytic, 433

chronic myeloid, with chromosome marker, 433, 434, 435

leukocytes

IMFs produced by, 62

movement of, through tissues, 55, 76

neutrophilic, synthesize lactoferrin, 402, 403

see also polymorphonuclear leukocytes

lichens, positioning of, 54

ligands, 17, 41–3

Lineus (nemertine), second-set grafts in, 215–16

linoleic acid

decreases adhesion of tissue culture cells to substrates, 126–8, and diminishes growth, 228–30

effects of, on galactosyl transferase activity and adhesiveness in embryonic cells, 232–3

incorporation of, into cell phospholipids, 228, 229, 230

linolenic acid, decreases adhesion of tissue culture cells to substrates, 231

lipids, in cell membranes, 225

lipopolysaccharide, bacterial: effect of treating lymphocytes with, on their distribution, 379, 383

liquids

equilibrium shapes of drops of, subjected to centrifugal force, found also in cell aggregates, 30–1, 34–7

flow of cells operates on thermodynamic principles applying to, 27

multicellular assembly as analogue of assembly of, 27–9, 45, 46

reversible works of adhesion and cohesion of, determine most stable configuration of system, 40

stable when binding energy is maximal and interfacial free energy minimal, 28

subunits adhere in, but are mobile with respect to one another, 28

surface tension of, 32–3